Innovation, Technology, and Knowledge Management

Series Editor
Elias G. Carayannis

For further volumes:
http://www.springer.com/series/8124

Marzenna Anna Weresa
Editor

Innovation, Human Capital and Trade Competitiveness

How Are They Connected and Why Do They Matter?

Comparing Countries in Europe, North America, and Asia

Springer

Editor
Marzenna Anna Weresa
Warsaw School of Economics
World Economy Research Institute
Warsaw
Poland

This work is to be published in Polish by the Warsaw School of Economics Publishing Office with the following title: Kapitał ludzki i innowacyjność jako czynniki długookresowych przewag konkurencyjnych w handlu międzynarodowym. Wnioski dla Polski.
This monograph is a result of a research project number 3926/B/H03/2011/40 that has been financed by the National Science Centre (Narodowe Centrum Nauki) and conducted by the research team of the World Economic Research Institute of the Collegium of World Economy at the Warsaw School of Economics.

ISSN 2197-5698 ISSN 2197-5701 (electronic)
ISBN 978-3-319-02071-6 ISBN 978-3-319-02072-3 (eBook)
DOI 10.1007/978-3-319-02072-3
Springer Cham Heidelberg New York Dordrecht London

Library of Congress Control Number: 2013955018

Printed on acid-free paper

Springer is part of Springer Science+Business Media (www.springer.com)

Series Foreword

The Springer book series *Innovation, Technology, and Knowledge Management* was launched in March 2008 as a forum and intellectual, scholarly "podium" for global/local, transdisciplinary, transsectoral, public–private, and leading/"bleeding"-edge ideas, theories, and perspectives on these topics.

The book series is accompanied by the Springer *Journal of the Knowledge Economy,* which was launched in 2009 with the same editorial leadership.

The series showcases provocative views that diverge from the current "conventional wisdom" that are properly grounded in theory and practice, and that consider the concepts of ***robust competitiveness,***[1] ***sustainable entrepreneurship,***[2] and ***democratic capitalism,***[3] central to its philosophy and objectives. More specifically, the aim of this series is to highlight emerging research and practice at the dynamic intersection of these fields, where individuals, organizations, industries, regions, and nations are harnessing creativity and invention to achieve and sustain growth.

---

[1] We define *sustainable entrepreneurship* as the creation of viable, profitable, and scalable firms. Such firms engender the formation of self-replicating and mutually enhancing innovation networks and knowledge clusters (innovation ecosystems), leading toward robust competitiveness (E. G. Carayannis, *International Journal of Innovation* and *Regional Development* 1(3), 235–254, 2009).

[2] We understand *robust competitiveness* to be a state of economic being and becoming that avails systematic and defensible "unfair advantages" to the entities that are part of the economy. Such competitiveness is built on mutually complementary and reinforcing low-, medium-, and high-technology and public and private sector entities (government agencies, private firms, universities, and nongovernmental organizations) (E. G. Carayannis, *International Journal of Innovation and Regional Development* 1(3), 235–254, 2009).

[3] The concepts of *robust competitiveness and sustainable entrepreneurship* are pillars of a regime that we call "*democratic capitalism*" (as opposed to "popular or casino capitalism"), in which real opportunities for education and economic prosperity are available to all, especially—but not only—younger people. These are the direct derivatives of a collection of top-down policies as well as bottom-up initiatives (including strong research and development policies and funding, but going beyond these to include the development of innovation networks and knowledge clusters across regions and sectors) (E. G. Carayannis and A. Kaloudis, *Japan Economic Currents*, p. 6–10 January 2009).

Books that are part of the series explore the impact of innovation at the "macro" (economies, markets), "meso" (industries, firms), and "micro" levels (teams, individuals), drawing from such related disciplines as finance, organizational psychology, research and development, science policy, information systems, and strategy, with the underlying theme that for innovation to be useful it must involve the sharing and application of knowledge.

Some of the key anchoring concepts of the series are outlined in the figure below and the definitions that follow (all definitions are from E. G. Carayannis and D. F. J. Campbell, *International Journal of Technology Management,* 46, 3–4, 2009).

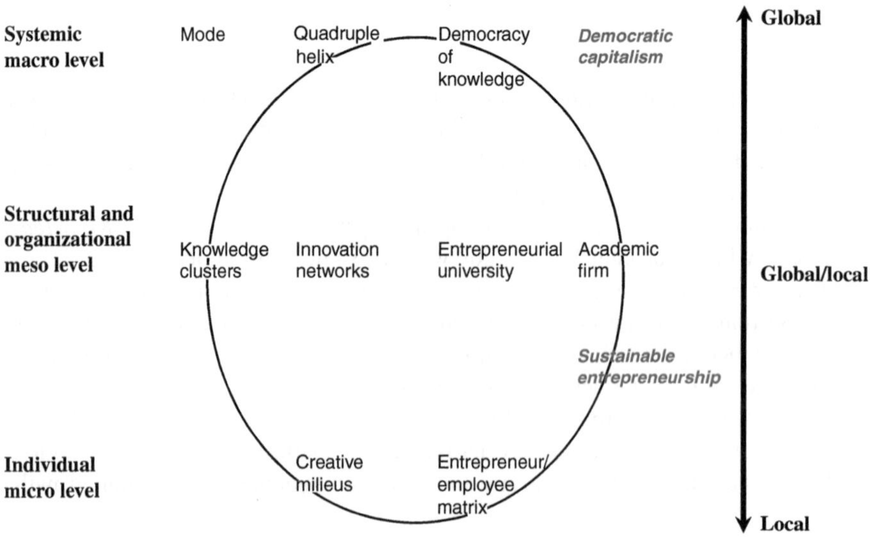

Conceptual profile of the series *Innovation, Technology,* and *Knowledge Management*

- The "Mode 3" Systems Approach for Knowledge Creation, Diffusion, and Use: "Mode 3" is a multilateral, multinodal, multimodal, and multilevel systems approach to the conceptualization, design, and management of real and virtual, "knowledge-stock" and "knowledge-flow," modalities that catalyze, accelerate, and support the creation, diffusion, sharing, absorption, and use of cospecialized knowledge assets. "Mode 3" is based on a system-theoretic perspective of socioeconomic, political, technological, and cultural trends and conditions that shape the coevolution of knowledge with the "knowledge-based and knowledge-driven, global/local economy and society."
- Quadruple Helix: Quadruple helix, in this context, means to add to the triple helix of government, university, and industry a "fourth helix" that we identify as the "media-based and culture-based public." This fourth helix associates with "media," "creative industries," "culture," "values," "life styles," "art," and perhaps also the notion of the "creative class."

- Innovation Networks: Innovation networks are real and virtual infrastructures and infratechnologies that serve to nurture creativity, trigger invention, and catalyze innovation in a public and/or private domain context (for instance, government–university–industry public–private research and technology development coopetitive partnerships).
- Knowledge Clusters: Knowledge clusters are agglomerations of cospecialized, mutually complementary, and reinforcing knowledge assets in the form of "knowledge stocks" and "knowledge flows" that exhibit self-organizing, learning-driven, dynamically adaptive competences, and trends in the context of an open systems perspective.
- Twenty-First Century Innovation Ecosystem: A twenty-first century innovation ecosystem is a multilevel, multimodal, multinodal, and multiagent system of systems. The constituent systems consist of innovation metanetworks (networks of innovation networks and knowledge clusters) and knowledge metaclusters (clusters of innovation networks and knowledge clusters) as building blocks and organized in a self-referential or chaotic fractal knowledge and innovation architecture,[4] which in turn constitute agglomerations of human, social, intellectual, and financial capital stocks and flows as well as cultural and technological artifacts and modalities, continually coevolving, cospecializing, and cooperating. These innovation networks and knowledge clusters also form, reform, and dissolve within diverse institutional, political, technological, and socioeconomic domains, including government, university, industry, and nongovernmental organizations and involving information and communication technologies, biotechnologies, advanced materials, nanotechnologies, and next-Generation energy technologies.

Who is this book series published for ? The book series addresses a diversity of audiences in different settings:

1. *Academic communities*: Academic communities worldwide represent a core group of readers. This follows from the theoretical/conceptual interest of the book series to influence academic discourses in the fields of knowledge, also carried by the claim of a certain saturation of academia with the current concepts and the postulate of a window of opportunity for new or at least additional concepts. Thus, it represents a key challenge for the series to exercise a certain impact on discourses in academia. In principle, all academic communities that are interested in knowledge (knowledge and innovation) could be tackled by the book series. The interdisciplinary (transdisciplinary) nature of the book series underscores that the scope of the book series is not limited a priori to a specific basket of disciplines. From a radical viewpoint, one could create the hypothesis that there is no discipline where knowledge is of no importance.
2. *Decision makers—private/academic entrepreneurs and public* (*governmental, subgovernmental*) actors: Two different groups of decision makers are being addressed simultaneously: (1) private entrepreneurs (firms, commercial firms, academic firms) and academic entrepreneurs (universities), interested in optimizing

[4] E. G. Carayannis, *Strategic Management of Technological Learning,* CRC Press, 2000.

knowledge management and in developing heterogeneously composed knowledge-based research networks; and (2) public (governmental, subgovernmental) actors that are interested in optimizing and further developing their policies and policy strategies that target knowledge and innovation. One purpose of public *knowledge and innovation policy* is to enhance the performance and competitiveness of advanced economies.

3. *Decision makers in general*: Decision makers are systematically being supplied with crucial information, for how to optimize knowledge-referring and knowledge-enhancing decision-making. The nature of this "crucial information" is conceptual as well as empirical (case-study-based). Empirical information highlights practical examples and points toward practical solutions (perhaps remedies), conceptual information offers the advantage of further driving and further-carrying tools of understanding. Different groups of addressed decision makers could be decision makers in private firms and multinational corporations, responsible for the knowledge portfolio of companies; knowledge and knowledge management consultants; globalization experts, focusing on the internationalization of research and development, science and technology, and innovation; experts in university/business research networks; and political scientists, economists, and business professionals.
4. *Interested global readership*: Finally, the Springer book series addresses a whole global readership, composed of members who are generally interested in knowledge and innovation. The global readership could partially coincide with the communities as described above ("academic communities," "decision makers"), but could also refer to other constituencies and groups.

Elias G. Carayannis
Series Editor

Preface

The processes that accompany the shift from industrial economics to a knowledge-based economy are currently the object of interest of both scientists and economic politicians, as well as entrepreneurs. In many countries, the traditional sources of socioeconomic development are waning. These sources used to be as follows: low labor costs, availability of inexpensive raw materials, or favorable geographic location. These economies are searching for new sources of competitive advantage that will allow for maintaining growth, among other things by boosting participation in international trade. Theoretical studies show that competitive factors alter their impact on the economy of a given country in line with the changing level of economic development and the developments in institutional environmental conditions in which business entities operate (cf., e.g., Porter 1990; Bieńkowski et al. 2010; Misala 2011). The theory also confirms that there has been a change in the significance of many competitiveness factors resulting from technological progress, globalization, and economic integration. A number of processes occurring in the global economy and particular countries increased the importance of nontraditional competitiveness factors that once were of lesser importance, i.e., human capital, innovativeness level, capability to create and employ newest technologies, level of infrastructure development and quality, and industry clusters. Moreover, the changing structure of global demand enlarges the significance of higher-quality and cutting-edge technology in international trade. Only those countries that are capable of rapidly accumulating and using knowledge resources can meet this demand. This, in turn, strengthens the role of nontraditional competitive advantages connected with knowledge, innovation, quality of production and management processes, and institutions.

This monograph focuses on the evaluation of the significance of two factors that are currently gaining importance in the international competitiveness of economies. These factors are: human capital and innovation. In this book, the international competitiveness of economies is considered in the narrow sense and understood as maintaining long-term competitive advantages in international trade. In this work, these advantages are analyzed by means of export competitiveness indexes (trade structure, revealed relative advantage, etc.). The process of developing competitiveness on foreign markets has been evaluated in three cross sections by means of quality-based industry taxonomies that refer to such criteria as:

- Technological level of processing industries—this taxonomy was developed for the Organisation for Economic Co-operation and Development (OECD) (Hatzichronoglou 1997) to enable the evaluation of technological innovativeness;
- Utilization level of input factors—this classification is used to perform European Union (EU) competitiveness analyses (Peneder 1999) and allows for portraying not only technological innovation but also organizational and marketing innovations;
- Skills of the labor force that constitutes the basis of the given industry—this classification allows for determining the significance of human capital in foreign trade competitiveness (Peneder 1999).[1]

In the following chapters of this monograph, we have shown how human capital is accumulated and how changes in innovativeness influence the development of competitive advantages in international trade in terms of both theory and practice. The following two hypotheses have been subjected to verification:

- For highly developed countries to maintain a competitive position and the share in international trade, it is necessary to ensure human capital development and a high level of technological and organizational innovativeness. Accumulation and absorption of knowledge and the innovative use of technology are the main factors influencing the development of new competitive advantages in trade in relation to countries and regions with low unit labor costs.
- Relatively highest advantages in international goods and services trading were developed by those states that succeeded in employing human capital accumulation and technological and organizational innovations to create competitive advantages.

The national innovation system (NIS) concept was employed to analyze the interrelations between innovation and human capital. Thanks to adopting such a common framework that encompasses innovativeness-related issues and the characteristic features of human capital, it was possible to determine whether and to what extent the operation of the NIS determines the competitive advantages of particular countries in the domain of trading.

This book is divided into two parts. The first part (Chaps. 1–4) focuses on theoretical issues and the methodology of the study on international competitiveness, human capital, and innovativeness and presents the interrelations between these phenomena.

Chapter 1 presents controversial issues connected with the concept of competitiveness and the means of its measurement by referring to the achievements of the appropriate orthodox and heterodox theory of normative nature. Special attention was dedicated to the methods of evaluating comparative advantages and the poli-

[1] These classifications have been discussed in greater detail in the Annex.

cy of influencing such advantages (so-called competitive advantages) in foreign trade of particular countries and their groups in modern global economy. Chapter 2 focuses on the theoretical discussion of human capital and innovativeness and the methods of measuring these phenomena.

As a result, common analysis frameworks for both issues were established for particular countries—the concept of the NIS.

Its more detailed characteristics and connection with competitiveness were described in Chap. 3 of the monograph. The interrelations between the NIS and competitiveness of economies are analyzed by means of Porter's concept of competitive advantages of countries, expanded by aspects of international dependencies indicated on the basis of the theory of, among others, Dunning. What is new about the discussion presented in this chapter is that the approach to innovativeness adopted herein is systemic and attempts to describe the most important features of the two-way link between the innovation system and competitiveness in foreign trade. On this basis, three dimensions of interrelations between innovation and competitiveness systems are recognized: technological, institutional, and international. The chapter concludes with a taxonomy of NISs in the global economy that is based on the criteria related to innovativeness and development level of human capital in particular countries. Six main types of NISs were discerned. Taking into account their characteristic features, in the next—fourth chapter—we verify the hypothesis concerning the diverse influence of competitiveness' determinants in particular groups of countries (including especially human capital and innovation) on competitive advantages in international trade.

The second part of the monograph (Chaps. 5–8) is of empirical nature. This section contains case studies of selected countries that represent models of various NISs. The objective of this part of the monograph is to illustrate the theoretical assumption that, in a nutshell, accumulation and absorption of knowledge and innovative applied technologies can be a factor that enables effective competition between states and regions with low unit labor costs. In this empirical part of the monograph, 12 countries representing various types of NISs were analyzed. Two types of NIS were selected from the group of developed innovation systems: dynamic and stably functioning. The dynamic NIS and the manner in which its functioning influences competitive advantages in international trading are illustrated by the case studies of three countries: Finland, Ireland, and Switzerland, discussed in Chap. 5. International competitiveness and changes in this area in states with stably functioning NISs were evaluated in Chap. 6, where the experiences of Germany, Austria, Spain, and the USA were presented. Another group of countries was discerned, namely the so-called "catching-up" innovation system states. This type is represented by Poland and Hungary, whose competitive advantages in trading were analyzed in terms of human resources development and innovativeness in Chap. 7. The most varied group of states has been discussed in Chap. 8. What these countries have in common is that they are characterized by an unbalanced NIS. In the case studies drafted for these three states, i.e., Russia, China, and Mexico, the authors attempt to detect the connection between human capital, innovativeness, and competitiveness in foreign trade in these states that until now compete mainly by means of low labor costs.

Both parts of the monograph—the theoretical and the empirical—are summed up by final conclusions and recommendations concerning the economic policy, especially including economic policy tools that support the competitive position of economies in the medium- and long-term. These conclusions refer to countries with varied types of innovation systems, but they are of special and key importance for countries such as Poland. In Poland (similarly, as in other developed countries), possibilities of competing on the basis of traditional factors, such as low labor costs are already waning. Therefore, the recommendations presented in the summary concerning economic policy tools that stimulate human capital accumulation and innovativeness might prove useful when implementing a growth strategy enabling sustainable improvement of international competitiveness in the long term.

The studies discussed in this monograph were conducted in the years 2011–2012 at the World Economy Research Institute of the Warsaw School of Economics as part of research project no. 3926/B/H03/2011/40 titled "Human capital and innovativeness as factors of long-term competitive advantages in international trade. Conclusions for Poland" financed from the funds of the National Science Center. The authors would like to express their gratitude toward the authorities and employees of the Warsaw School of Economics, especially their partners of the Collegium of World Economy for organizational support and the creative and inspiring discussions held at various stages of research task performance. The authors of this monograph would also like to thank their colleagues from foreign academic centers concerned with research on competitiveness, especially the researchers operating within the *Microeconomics of Competitiveness* network of the Harvard Business School for the debates that took place during conferences and seminars, as they contributed greatly to the interpretation of the results of analyses carried out as part of this project.

Marzenna Anna Weresa

Contents

Contributors

Andreas Bielig World Economy Research Institute, Warsaw School of Economics, Warsaw, Poland

Ireneusz Bil World Economy Research Institute, Warsaw School of Economics, Warsaw, Poland

Ziemowit Czajkowski World Economy Research Institute, Warsaw School of Economics, Warsaw, Poland

Krzysztof Falkowski World Economy Research Institute, Warsaw School of Economics, Warsaw, Poland

Marcin Gomułka World Economy Research Institute, Warsaw School of Economics, Warsaw, Poland

Günter Heiduk World Economy Research Institute, Warsaw School of Economics, Warsaw, Poland

Oskar Kowalewski World Economy Research Institute, Warsaw School of Economics, Warsaw, Poland

Arkadiusz Michał Kowalski World Economy Research Institute, Warsaw School of Economics, Warsaw, Poland

Agnieszka McCaleb World Economy Research Institute, Warsaw School of Economics, Warsaw, Poland

Beata Michorowska World Economy Research Institute, Warsaw School of Economics, Warsaw, Poland

Józef Misala World Economy Research Institute, Warsaw School of Economics, Warsaw, Poland

Tomasz Napiórkowski World Economy Research Institute, Warsaw School of Economics, Warsaw, Poland

Józef Olszyński World Economy Research Institute, Warsaw School of Economics, Warsaw, Poland

Mariusz-Jan Radło World Economy Research Institute, Warsaw School of Economics, Warsaw, Poland

Marzenna Anna Weresa World Economy Research Institute, Warsaw School of Economics, Warsaw, Poland

Grażyna Wojtkowska-Łodej Foreign Trade and European Studies Institute, Warsaw School of Economics, Warsaw, Poland

Part I
Human Capital and Innovations vs. Competitive Advantages in Foreign Trade: Theoretical and Model Approach

Chapter 1
Theoretical Grounds of the Development of Long-Term Competitive Advantages in International Trade

Józef Misala

Nowadays, we are witnessing an increasingly rapid process of internationalization of economic life. Such a state of affairs stems from many causes; however, striving for the welfare of populations of individual countries and regions may be considered the prevailing one. This causes increasingly intense mutual competition between various economic entities (households, enterprises, associations thereof, etc.), both on domestic markets and on international markets. In addition, authorities of individual countries and their groups participate in this struggle by way of determining the institutional and instrumental (systemic) frameworks for competition of the said entities. It is accompanied by theoretical research aimed at the development of a kind of subdiscipline of economics, generally defined as the theory and politics of international competitiveness of national economy.

The main goal of this chapter is to discuss the theoretical grounds of development of international competitiveness of the national economy of a specific country, and, more specifically, the development of its long-term competitive advantages in international trade with particular focus on their nature, determinants, and methods of analysis. Apart from that, the aim is also to draw conclusions for international competitiveness analysts and economic policymakers at the central level. It is necessary to present an outline of the available achievements of relevant normative orthodox and heterodox theories and to pay particular attention to the issues of natural (objective) shaping of comparative advantages and the policy of shaping such advantages (the so-called competitive advantages) in international trade of individual countries and groups thereof, and thus, in summary, also in international trade as a part of the contemporary global economy.

In investigations contained in subsequent subchapters, attention is paid to theoretical issues of competition on a macroscale, and so at the level of countries. It does not mean that problems of international competition of other types of economic entities, households, and enterprises of various types, in particular, are virtually

J. Misala (✉)
World Economy Research Institute, Warsaw School of Economics,
Al. Niepodleglosci 162, 02-554 Warsaw, Poland
e-mail: mweres@sgh.waw.pl

M. A. Weresa (ed.), *Innovation, Human Capital and Trade Competitiveness*, Innovation, Technology, and Knowledge Management, DOI 10.1007/978-3-319-02072-3_1,

passed over. It is assumed that such problems are background ones, which is not tantamount to failing to recognize their significance, which is altogether substantial and growing.

1.1 Essence of the Theory of International Competitiveness of a National Economy

1.1.1 The Concept of International Competitiveness

It is quite common to consider that each economic theory should play three basic functions, namely explanatory function, decision-making function, and the so-called determinant optimization function. In the case of the theory of international competitiveness of national economy, the explanatory function consists in facilitating the understanding of the process of shaping this competitiveness with regard to individual national economies in the past; the decision-making function in providing clues related to intenseness and structure of competition of respective countries and/or their groups in the future; and the determinant optimization function in providing hints for decisions taken within the scope of systemic solutions (e.g., ownership rights and their enforcement, formation of monetary, fiscal, structural policy, etc., in a given country or a group of countries), which can and should be conducive to increasing the international competitive ability and current international competitiveness of the national economy of a given country, also called its pure international competitiveness and/or its international current competitive advantage[1].

Nowadays, we are still far from a complete and consistent theory of international competitiveness of national economy of a given country, a theory which would fulfill the three functions mentioned above and which would be understood as a set of relevant assumptions, an unambiguous and commonly accepted definition of this competitiveness and its constituents, a specification of determinants, methods, and measures of analysis, and unquestionable conclusions for economic policy. However, we have substantial achievements at our disposal, making it possible and at the same time facilitating the process of development of this theory, including also its essence and basic principles. The outline of these achievements is presented in Fig. 1.1.

[1] For a more detailed discussion of these issues, see *inter alia* Bossak (1984, 2000, 2006); Bossak and Bieńkowski (2001, 2004); Dołęgowski (2002); Wysokińska (2004); Misala (2007, 2009, 2011a); Borowski (2008); Bieńkowski et al. (ed., 2010).

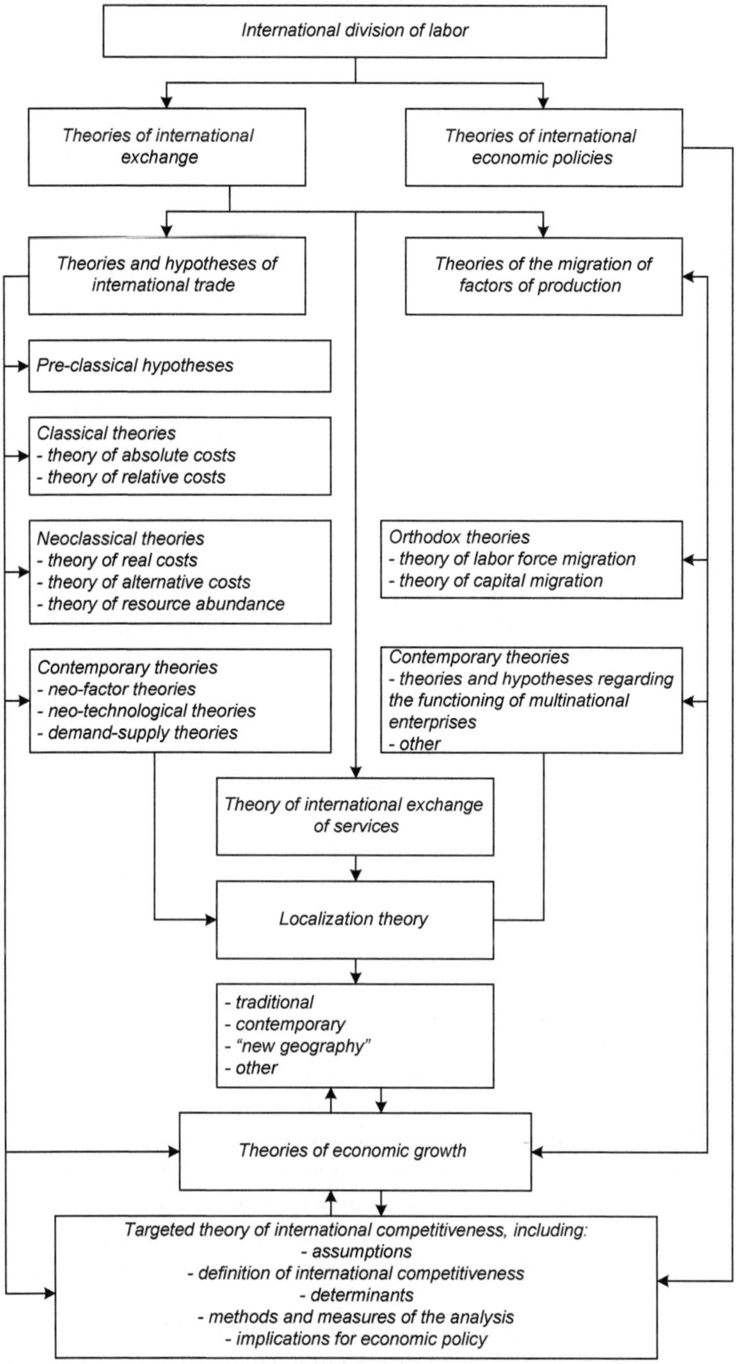

Fig. 1.1 Diagram of connections between international division of labor theory, economic growth theories, and international competitiveness theory. (Source: Author's own study)

The gradually developed theory of international competitiveness of national economy of a given country may be divided into two essential parts: the normative (descriptive) theory of this competitiveness and the theory of policy of an efficient international-scale competition between countries and/or their groups. As far as the normative (descriptive) theory of international competitiveness of national economy is concerned, there are two quite distinct categories in economic literature, namely international competitive ability and pure international competitiveness of national economy.

There are many general definitions of the international competitive ability of individual national economies. From a purely theoretical point of view, the definition by J. W. Bossak (1984, p. 37) is very convincing: "(…) The international competitive ability may be (…) defined simply as the ability to struggle, compete for benefits with the participation of a country in the international division of labor. This ability is of relative nature at least in two ways; firstly, in relation to other countries; and secondly, in relation to the international competition features characteristic of a given stage of development."

The international competitive ability is defined most often as a given country's ability to struggle, compete for benefits stemming from the development of the social division of labor and international economic exchange (Bossak 1984, p. 37). On purely theoretical grounds, it is possible to further distinguish two fundamental, changing components of the international competitive ability of national economy of a given country. It is, on the one hand, the so-called real component, understood as an actual (objectified) sphere of economic management, while on the other, the so-called institutional and instrumental component, understood as a sphere of regulations, i.e., an economic systemthat is also called the economic operating system.

It is obvious that, from the point of view of formation of the international competitive ability of a given country, each of the above-mentioned components, and constituents thereof, is of particular importance, including those that are so frequently underestimated, such as climate, hard infrastructure (e.g., roads, bridges, and ports) and soft infrastructure (e.g., telecommunications connections), ethical and moral standards, or religion. Furthermore, these components and constituents thereof are at any moment mutually interconnected and codependent (Sulmicki 1977; Misala 2004, 2007, 2011a).

From the economic point of view, important issues include skills of and propensity to saving. They are important, because savings create broadly understood capital, which is an extremely vital production factor.

Further, skills of and propensity to inventions and innovations are of importance, as they provide a basis for another, very important production factor. The latter is technical knowledge (technology), which in addition is a relatively rare and, thus, expensive, factor. It makes it easier to manage not only capital but also the environment.

The third significant factor is constituted by human skills of and propensity to self-organizing. They influence the shaping of the political and economic systems. It is societies and nations capable of and prone to self-organizing that are able to introduce and adequately develop such political and economic systems that are conducive to strengthening the ability of and propensity to saving and also inventions and innovations.

From the notion of international competitive ability, one should definitely distinguish the notion of pure international competitiveness. It constitutes only a part of the broadly understood international competitiveness discussed earlier, which, as it is known, international competitiveness *sensu largo*, has not yet been defined in full. Under the notion of pure international competitiveness, one understands the present state and directions of changes of the real and the institutional components of the international competitive ability of a given country in its struggle for benefits of participation in the international division of labor (Bossak 1984, 2002; Bossak and Bieńkowski 2004; Misala 2007, 2009, 2011a).

On purely theoretical grounds, it is possible to speak of the occurrence of specific relations between the formation of international competitive ability and pure international competitiveness of the national economy of a given country. These are feedbacks whose development is also influenced by relevant economic policy.

1.1.2 Signs of Pure International Competitiveness

Scientific deliberations on signs of pure international competitiveness at the macroeconomic level were initiated by representatives of the so-called mercantilism, and, in particular, by classical authors of political economics, especially A. Smith and D. Ricardo[2]. It were their achievements that were referred to by one of the contemporary precursors of the still forged theory of international competitiveness at the national level, namely Balassa (1962). He recognized the broadly understood ability of a given country to sell as the main manifestation of this competitiveness. Later on, other economists, and especially Trabold (1995), Reiljan et al. (2000); Verblane (2006); Misala (2007, 2011) added four other abilities to these signs, namely the ability to attract mobile production factors, ability to invent and to innovate, ability to adjust to various so-called internal and external shocks, and the ability to earn. As the final result, a pyramid of development concepts and signs of formation of the pure international competitiveness of national economy of a given country came into being (Fig. 1.2).

It should be noted, however, that relationships between the abilities indicated in Fig. 1.2—even setting aside the problem of a given country's society's ability to self-organize, i.e., the problem of the so-called systemic competitiveness—can be and usually are shaped in a diverse manner, if their development is examined over time, and, more specifically, in different temporal sections (Table 1.1).

Table 1.1 indicates explicitly that, first, the importance of factors determining the formation of pure international competitiveness of the national economy changes

[2] For a more detailed discussion of these issues see, *inter alia*, Reinert (1994); Kibritcioglu (2002); Misala (2009).

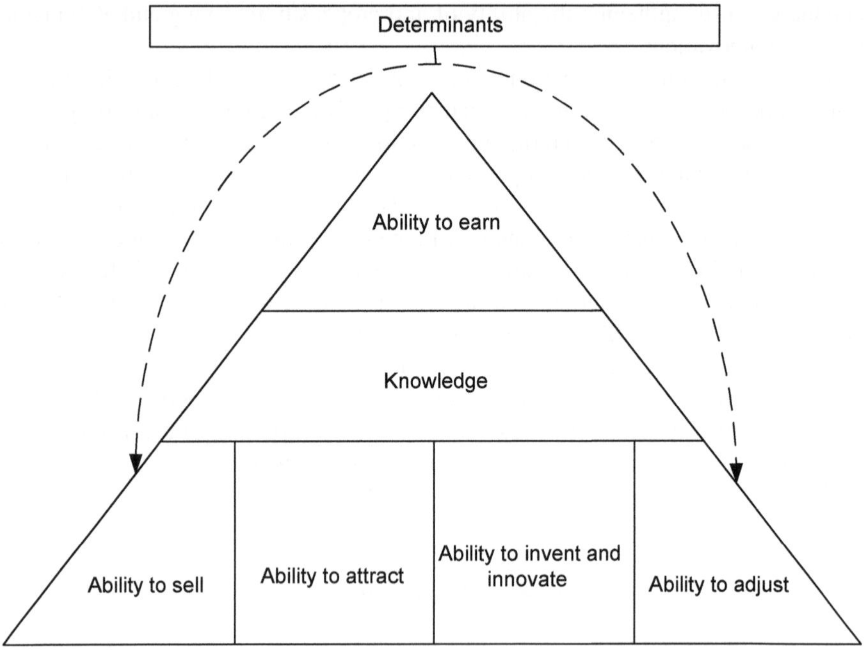

Fig. 1.2 Formation of mutual correlations between the international competitive ability and international competitiveness over time. (Source: Author's own study on the basis of Trabold (1995); Reiljan et al. (2000); Verblane (2006))

over time. Second, even in a relatively short time, the intenseness of impact of respective factors and groups thereof (e.g., ability to sell and ability to adjust) is different in countries with diversified levels of economic development. The above fact is recognized, among others, by Sala-I-Martin and Artadi (2004), who explicitly suggest the need for distinguishing the strength of influence of determinants in various groups of countries[3].

What is meant here is the relative importance of the diverse factors determining the formation of pure international competitiveness (and also the international competitive ability) in forerunner countries and in catching-up countries. On purely theoretical grounds, the essence of the problem comes down—among other things, but not exclusively—to the fact that, at each moment "t," the production function is shaped differently in each of the countries, the same being true with regard to endowment of basic resources. It is possible to conclude on the basis of this observation that the strength of influence of human capital and innovation on competitiveness in foreign trade is closely connected with the level of development achieved by a given country.

[3] This topic is discussed in more detail in Chap. 3.

Timeframe	Ability to earn	Ability to sell	Ability to attract	Ability to innovate	Ability to adjust
Short-term	Level of per capita wealth is dependent on the development of foreign demand The role of investments in the development of global demand influences the development of changes in the stock of national capital	The acquisition of foreign demand positively influences the development of domestic production		Has influence on the development of the non-price (quality) competitiveness	Has influence of the development of investor's and consumer's expectations
Medium-term	Level of per capita wealth equals the so called natural level				Imperfections in the labor market influence the development of the size of production
Long-term	Increase in the size of the production potential per capita as a result of changes in the intensity of the use of capital and the level of the technological development		Effects on international competing for mobile factors of production influence the development of tangible and human resources	Has influence on the development of the ability to use differences in the level of production of factors of production (through technological progress)	Has influence on the development of the process of structural changes and quality aspects of the competing process on the international scale

Table 1.1 Formation of various signs of international competitiveness of national economy of a given country in short-, medium-, and long-term perspectives, in accordance with model solutions. (Source: Heileman et al. 2006, p. 20)

1.2 Comparative Advantages and Competitive Advantages in Theoretical Aspects

To date, difficulties in understanding the nature of comparative advantages and the nature of competitive advantages have been observed. Therefore, it is worth organizing the relevant issues a little.

1.2.1 Comparative Advantages Under the Static Approach and Main Prerequisites of Their Occurrence

Under a static approach, the basic and unquestionable law of rational economic governance on an international scale is the principle of relative (comparative) costs formulated in 1817 by D. Ricardo, which constitutes a substantial part of his broader theory, commonly presented in many different textbooks of international commercial relations[4]. It is possible to consider it basic and unquestionable *inter alia* because of the fact that it constitutes an extension and at the same time a specific generalization of the principle of absolute advantages in international exchange, formulated in 1776 Smith. Furthermore, this principle—a remarkably important law of political economics—has constituted, until today, a starting point of specific reinterpretations which are, to be honest, nothing more than attempts at a slightly different approach to basic prerequisites of trade development and more broadly understood international economic exchange, and thus the sources of its existence.

Under the static approach at the macroeconomic level, the most important prerequisites for the development of international exchange of goods (products and services) and mobile production factors are believed to be the diversification of countries in terms of:

a. Level of technical and technological advancement;
b. Endowment of basic production factors (labor and capital);
c. Size and structure of demand, including in terms of various types of investment and consumer preferences, and inclinations of residents;
d. Economic volume and absorptiveness of internal markets and the related possibilities of achieving different economies of scale in the field of production and selling;
e. Shaping of the so-called market structures, that is, the number of economic entities and the intenseness of domestic competition between them;
f. Geographical location within the global economy and formation of transportation costs and costs of communication between partners (the so-called location and/or agglomeration advantages); and
g. A defined, foreign and international, economic policy pursued by competent domestic authorities (Deardorff 1980, 1982, 2003, 2005; Gupta 2007; Misala 2011a).

The diversification of countries in terms of the level of technical and technological advancement, and, more specifically, in terms of the productivity level of the applied production factors under conditions of free market and trade, is the most commonly indicated prerequisite for the development of the international exchange of products and services. Similarly, it is a source of the occurrence of statically considered comparative (relative) advantages, or lack thereof. After all, the occurrence of this factor stems directly from the deliberations of D. Ricardo (and also R. Torrens), if only the traditionally presented (i.e., in accordance with the classical authors' approach) principle of comparative costs were to be generalized and worded in a different way, that is, instead of taking into account the differences in production costs measured with labor outlays, take into account differences in labor

[4] There are many indications that the precursor in this area was another English economist, R. Torrens, who had published an interesting article a couple of years before D. Ricardo, but who had not received such recognition as D. Ricardo's breakthrough work, titled On the Principles of Political Economy and Taxation.

efficiency which condition the former. Allowing for differences in labor efficiency, the principle may be worded as follows: for each country, it is always profitable to specialize in production and export of those products in the case of which the country has a relative advantage in labor efficiency (in terms of the degree of technical and technological advancement) over the foreign countries, while it is unprofitable to specialize in production and export of those products in the case of which the country has no such relative advantage over the foreign countries (economic surroundings).

The principle of relative (comparative) costs and therefore also relative advantages in terms of technical and technological advancement and performance of production factors (their productivity) is, under the static approach, an important economic, technical, and balance law, which as a matter of fact cannot be called into question. This principle—an authentic and invaluable achievement of classical authors—indicates how to obtain an amount of a given product (good) while at the same time minimizing the expenditures or—which is equivalent—how to obtain a bigger product given the available expenditures.

In the light of the theoretical achievements to date, under the conditions of free market and trade, another and enormously important prerequisite for the occurrence of international business is the existence of differences between countries in terms of relative endowment of production factors, that is, labor and capital, and especially given additional assumptions of neo-classical economics concerning the lack of mobility of these factors on the international scale, and uniform production function, i.e., unchanging productivity of these factors. What is meant in such a case is the development of international business in accordance with the principle of resource abundance, which constitutes the core of the relevant theory, formulated in its final shape by E. Heckscher, B. Ohlin, and P. A. Samuelson (H–O–S). In accordance with the H–O–S principle, each country (or other economic entity) should specialize in production and export of products (goods and services) whose internal production (provision) requires more intense utilization of a relatively more abundant and thereby cheaper production factor (e.g., labor in Poland), while at the same time import products that require more intense application of a relatively less abundant, and thereby more expensive, factor (e.g., capital). The contents of this principle, which is easy to be noted, are nothing but a reinterpretation of the statically presented principle of comparative (relative) advantages, assuming many rigorous conditions mentioned above, in particular free market and trade and identicalness of production functions in the cooperating countries. On the other hand, E. Heckscher, B. Ohlin, and P. A. Samuelson explain additionally in their theory the origins of statically presented differences in costs and prices. They simply originate from different endowment of basic production factors, namely work and capital.

Under the static approach, the abundance principle also constitutes a basic element of the so-called neo-factor theories, i.e., theories in which a larger number of production factors and, at the same time, heterogeneity of human labor and capital, are taken into account.

The third prerequisite for the occurrence, under the conditions of free market and trade, of static comparative advantages of a given country is the diversification of countries in terms of size and structure of demand, or, more specifically, different preferences and inclinations of consumers and investors. It also takes place where we additionally face the same level of technical and technological advancement (with identical functions of their production) and identical relative endowment of production factors. Country *A* may have an absolute relative (comparative) advantage with regard to a product and/or products for which demand is relatively lower than that in the surrounding countries. Given the said conditions, prices of these products in this country are simply absolutely and relatively lower before the commencement of trade, that is, under the conditions of autarchy.

The fourth important prerequisite for the development of international trade in accordance with the statically presented principle of comparative advances is the diversification of countries in terms of economic size and absorptiveness of internal markets. On purely theoretical grounds, under the conditions of free market and trade, given identical production functions and identical endowment of resources, as well as identical preference functions, a so-called large country, which distinguishes itself with an absorptive internal market, has a comparative advantage over a country (surrounding) within the scope of a product and/or a number of products in the case of which it is possible to obtain growing economies of scale in the field of production and selling. Plainly speaking, the price of these goods in the country before trade commences is lower than in another country (surroundings) in the case of which the economies of scale do not occur.

The diversification of countries in terms of the so-called structure of their internal markets, i.e., the number of enterprises operating in them and the intenseness of cooperation, and at the same time, mutual competition between them, is usually considered to be the fifth prerequisite for the development of international trade in accordance with the statically presented principle of comparative advantages. It is assumed that these factors exert a significant, although diversified, influence on the formation of production and business costs in these countries and also on shaping of the so-called economies of scale. What matters here are not only growing economies of scale in the field of production and selling but also the formation of the so-called internal and external economies of scale of enterprises, the said notions having been introduced to the professional literature by Scitovsky (1958). According to him, the essence of internal economies of scale comes down to the following: as the production of enterprises in a given country grows, the average costs of these enterprises decrease and thereby they obtain a cost-and-price advantage over other enterprises in this country and/or enterprises from other countries, especially in the face of imperfect competition. The internal economies of scale, in turn, are understood nowadays as occurring because of the fact that, in a given enterprise or enterprises of a given region and/or country, relevant cuts in average costs and prices cause an increase in the level of their competitiveness when compared to other enterprises operating in a given industry in this region, country, and/or group of countries, and in the world as a whole.

Finally, the diversification of countries in terms of geographic location, shaping of costs of transportation of commodities, and costs of securing deliveries of components before the exchange commences is considered to be the sixth important prerequisite for the development of foreign and international trade in accordance with the static principle of comparable advantages. These aspects are discussed in great detail in the works of the representatives of the so-called traditional theory of business location (location theory) and its more developed versions in the form of the so-called new economic geography.

Even before the commencement of international exchange in each of the countries, when taking a decision on its commencement, the arrangement of static comparative advantages, or possibly the lack thereof, may be changed, even radically, by way of economic policy. For example, a one-off devaluation of national currency causes, almost automatically, a drop in domestic prices, immobile production factors (e.g., land and natural resources in it), and exported commodities, while revaluation (revalorization) causes, in a quasi-automatic manner (in practice with a certain delay), the opposite consequences. At this moment, however, on strictly theoretical grounds, we find ourselves at the stage of creation of dynamically understood comparative advantages, or, in other words, at the stage of transforming static advantages into dynamic ones, namely into comparative competitive advantages that may be created.

1.2.2 Transformation of Comparative Advantages into Competitive Advantages

In theoretical achievements to date, one may find many valuable clues concerning preservation of comparative advantages and creation of such new advantages, which obviously requires an appropriate economic policy. In the light of these achievements, all the above-listed prerequisites of static comparative advantage of a given country and/or group of countries may actually change more or less, and they do. Of course, without the intervention of economic politicians certain prerequisites evolve more slowly (e.g., society's disposition to saving, to inventions, and to innovativeness, or the so-called self-organizing, also in ethnic and cultural dimensions), while other prerequisites evolve faster (e.g., willingness to take risk or to comply with various supraregional, supranational, or even global standards). It is difficult to find a uniform interpretation formula for these skills and propensities in the context of those issues. Deardorff (2003) even proposes to consider them as relatively fixed values in the short- and medium-term perspective.

Possibilities of transforming static comparative advantages into dynamic competitive advantages exist *inter alia,* thanks to the active participation of a given country in broadly understood economic exchange, i.e., in foreign exchange of goods (products and services) and various production factors. In a given country *A,* the degree of technical and technological advancement may be increased (the production function may be changed favorably and the level of widely understood labor efficiency may be increased) by way of skillful specialization, or, more specifically, by focusing on the import of goods which are intense in terms of technology and on attempts to raise their importance in export. This issue has also been pointed

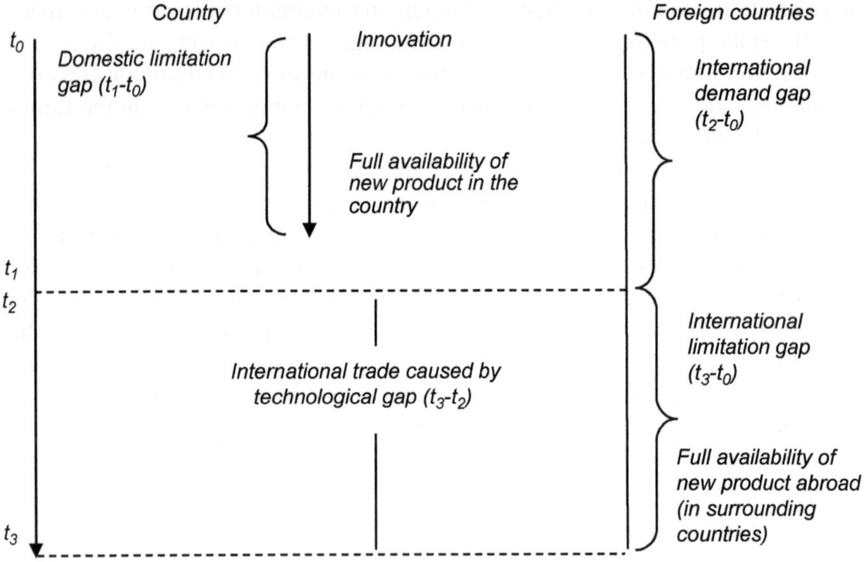

Fig. 1.3 International trade under conditions of technological gap between countries. (Source: Johns 1985, p. 112)

out, among others, by authors of neo-technological theories, i.e., technological gap theory, product life-cycle theory, and theory of economies of scale in production and selling (Fig. 1.3).

In the light of the authors of technological gap theory, an absolute international competitive advantage may be obtained by a country (more specifically, enterprises operating within its territory) that is distinguished by a high level of inventiveness and innovativeness, which is possible only in certain circumstances, and especially when relevant capital resources are available. Other countries may only imitate such a country more or less effectively.

The ability and tendency to market new products are also accentuated in another neo-technological theory, namely the product life-cycle theory. The author of this theory is Vernon (1966), but its main ideas were convincingly presented somewhat earlier by Posner (1961).

An extended version of the product life-cycle theory was presented by Hirsch (1997), who, when presenting the three product development stages, underscored not so much absolute and relative differences in the level of development of technical knowledge and technical advancement rate, as diversification of countries and regions in terms of endowment of various production factors, i.e., including capital. This author, on the basis of R. Vernon's idea, proposed a concept of life cycle of entire branches of economy. In his opinion, such branches are subject to typical transformations accompanied by changes in production methods and formation of demand, volume, and structure of sale. Of course, at each level of development of a region, or country, the possibilities of gaining benefits from the scale of production and selling are different.

The possibilities of transforming static comparative advantages into dynamic competitive advantages by way of development of foreign and international trade are also indicated by the authors of the so-called neo-factor theories. According to the authors and supporters of these theories, each country in the world has at its disposal, at a given moment *t*, specific resources of various production factors which should be and may be reasonably managed *inter alia* by way of relevant orientation of export (and also import) specialization structure. In their opinion, it is about "increasing"—in qualitative and quantitative terms—country's own resources, especially some of them, with the use of dynamic resource abundance principle, and creating in this way competitive advantages for the future. In this way, e.g., labor resources may be extended, as well as widely understood capital, technical knowledge, and the specific production factor in the form of natural environment.

In the conditions of free market and trade, a more effective way of strengthening static comparative advantages and creating dynamic competitive advantages through international exchange is constituted by direct international movement of mobile production factors which in theory may be complementary or serve as a substitution in relation to international movement of commodities and services. On the basis of a relevant orthodox theory, the occurrence of complementarity may be explained and proved in relation to a relatively small number of cases. Summing up, there are many more cases of relevant competitiveness (substitutability), which may be explained on the basis of contemporary heterodox theories (Purvis 1972; Svensson 1984; Neary 1995; Springer 2000; Misala 2011b).

On theoretical grounds, the first and most meaningful proof that it is possible to consolidate comparative advantages and/or create dynamic competitive advantages is T. M. Rybczyński's theorem (1955), based on the H–O–S theory, which says that an increase in the resource of a particular production factor causes, in the conditions of fixed prices of relative factors, an absolute and at the same time magnified growth in the production volume of a product (sector) that requires more intensive usage of this factor, and at the same time an absolute and magnified drop in the production of a product (sector) that requires more intensive usage of a factor whose resources are not changing.

The possibility of occurrence of the so-called magnification effect with all its implications is an important matter. The nature of the problem is that in a given country, the ratio of prices of an asset-intensive product (sector) and a labor-intensive product (sector) cannot change only where during the production of both products (also within the corresponding sectors) the ratio of labor to capital volume (*L*/*C* ratio) does not change, and at the same time the efficiency level of these production factors is constant. However, if, for example, capital resources are increasing, the absolute and relative production volume of a product (products) requiring relatively more intensive outlay of this production factor is also growing, and furthermore, in a magnified manner (due to the occurrence of somewhat excessive attractiveness of capital). Naturally, the situation in the case of the development of production of a product (sector) that is relatively more labor-intensive is quite the opposite.

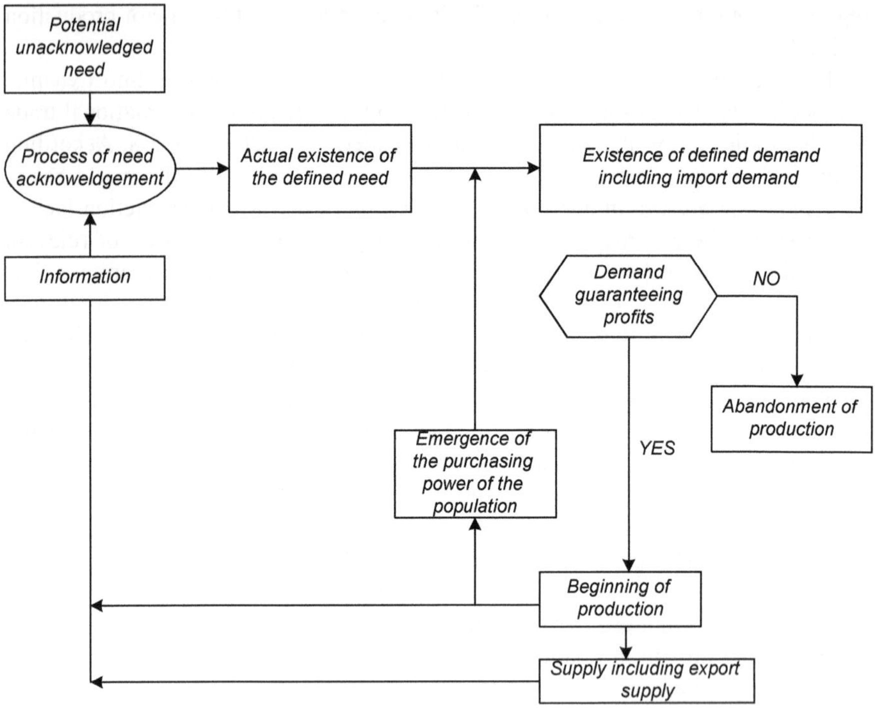

Fig. 1.4 Basic relations between supply and demand. (Source: Tesch 1980, p. 193, with author's own supplementation)

The impact of demand on the formation of comparative advantages and international trade has been already addressed by the classical authors of political economics (especially J. St. Mill—the creator of the famous principle of mutual demand), but abolition of the assumption that preferences of buyers (consumers and investors) are formed in the same way anywhere and always, and that these preferences do not have any influence on the development of this business and benefits flowing therefrom, formulated by Ohlin (1999), was a certain turning point. It has been noted with increased frequency that there are many significant relations between demand and supply (and in consequence, also in import and export of each of the countries), which may be utilized in the formation of dynamic competitive advantages (Fig. 1.4).

In the light of corresponding theoretical achievements, relatively large internal demand, complemented in a way by diversifying foreign demand, constitutes important prerequisites for utilization of the existing comparative advantages of a given country *A* and/or occurrence of possibilities of creation of new competitive advantages. In addition, these possibilities are usually enhanced with the growth of national income and individual income inside the country and in countries which are exchange partners, the more so as the demand for commodities and services may become diversified in many ways, adjustment to which is to the benefit of producers and exporters and to which they usually adjust, especially in the conditions of monopolistic competition. What is important here is that the very nature of each product (its basic features and functions) generates possibilities for diversification of demand therefor, and thus also a possibility of relevant adjustment on the part of foreign producers and exporters. Furthermore, in practice, product demand and supply may be diversified with the use of various marketing techniques.

For many years, the diversification of demand for various goods (and also of the corresponding responses on the part of supply and in international trade) has constituted a basis for intensive research resulting in the so-called theories (or rather hypotheses), sometimes referred to as demand-and-supply theories, which are also discussed in detail in Polish professional literature (Misala 2001, 2005, 2009; Czarny 2002; Rynarzewski and Zielińska-Głębocka 2006; Budnikowski 2006; Krugman and Obstfeld 2007; Bożyk 2008; Świerkocki 2011). However, this research clearly indicates the diversification of products as a new (when compared to classical and neo-classical ones) basis for the development of the contemporary international trade. Moreover, this diversification is divided into horizontal and vertical. The horizontal diversification is explained using two types of models, namely those referring to the concept of monopolistic competition by E. Chamberlin and those referring to the concept of asymmetry of buyer utility function by H. Hotelling. Models of the neo-Chamberlain-type trade and of the neo-Hotelling-type trade have many features in common. It is particularly important that the distance between the *perfect* variety of a particular product from the purchaser's perspective and the available version of the same product becomes considerably smaller as the international trade develops, especially in the case of trade between two countries at similar levels of economic development and, accordingly, also between countries with similar production and consumption structures.

In the light of research conducted so far, the vertical (qualitative) diversification of products (as well as product sets, subsets, and components) and the parallel trade between different countries (i.e., the parallel trade in goods and services) may be mainly explained in terms of the principle of resource abundance, and especially its dynamic interpretation. In fact, it is possible to observe the vertical (qualitative) diversification of traded products as well as the diversification of their price levels, caused by the fact that different countries use different combinations of various production factors. It is worth adding that their resources change at a relatively rapid pace, in terms of both time and space.

It is also possible to explain the vertical (qualitative) diversification of products and their parallel trade with the use of a model by Falvey and Kierzkowski (1987), according to which an increase in the income of societies of particular countries is accompanied by the increase in the so-called representative (domestic and foreign) demand for higher-quality goods, whose prices are also higher. As a result, the vertical intra-industry trade intensifies, as does the global trade.

The possibility of preserving trade comparative advantages as well as the possibility of creating new comparative and competitive advantages through trade development (particularly by the development of export) have already been suggested by Ohlin (1933). According to his nonformalized version of the resource abundance theory, even if two or more countries have the same resources and the same buyer preferences, the development of international trade is justified and may be beneficial. These benefits would be the increasing economies of scale with regard to production and sale, which may be achieved after foreign trade has been initiated. The above remarks form the so-called concept of absorptiveness of external markets, according to which larger and as institutionally integrated as possible supranational markets enable domestic enterprises to benefit as described above while offering a larger variety of products at lower prices to various buyers.

The increasing returns to scale are nowadays typically divided into static (short-term) and dynamic (medium- and long-term) returns. We may observe static returns when production batches and sales batches grow as a result of specialization and trade development, which leads to the reduction of unit production costs, thanks to the reduced frequency of switching production apparatus, increase in labor efficiency, distribution of the so-called fixed costs (e.g., design costs) over a larger number of units, as well as thanks to the gradual introduction of minor improvements to the production technique applied (production method improvement in relation to the production experience accumulation). Dynamic returns of scale of production and sales, on the other hand, may be observed when more radical production techniques are applied, which leads to, among other things, an improved production technology, increase in the production capacity or performance of the production equipment, etc.

Dynamic returns to scale usually accompany inventive, innovative, technological, and marketing activity of various companies. A significant factor of such activity is learning by doing, i.e., the process of broadening experience by repeating production processes, which has an adequate effect on the formation of the so-called learning curve, and therefore on the production and sales costs, including foreign sales costs. It is also important to learn how to gain new knowledge, learn to do, which obviously requires considerable investment in people (in other words, human resource development), but usually brings significant results, including the so-called external economies of scales, which are of interest to us. External economies of scale arise when, in a given enterprise or in enterprises from a particular region and/or country, appropriate reductions of usual costs and prices result in an increase in their competitiveness level in comparison to other enterprises functioning within the same industry sector, region, country, and/or group of countries, including the world as a whole. These observations indicate that human resources, innovativeness, and dynamic comparative advantages are all related to each other.

According to Scitovsky (1958), it is possible to distinguish between pure and pecuniary external economies of scale. On the level of enterprises, an increase in production and export usually results in an introduction of technological improvements, which in turn changes the relation between the expenditures and the results. In addition, it is possible to observe the effects of spreading particular knowledge among companies in the same country or region, which enhances their competitiveness level. These economies are therefore synonymous to those described before as dynamic returns, which result from learning by doing and from learning to do.

In accordance with theoretical achievements to date, we may observe pecuniary economies of scale virtually under the conditions of imperfect competition alone. The nature of such economies may come down to the following: enterprises change the structure of production taking into consideration, for example, the functioning in a given region, country, or group of countries, of an absorptive ready market for particular products or their components, which would also be characterized by sufficient purchasing power of the population, and, more precisely, an effect of the purchasing power of the population of particular regions or countries characterized by the focus on the so-called dynamic production areas, where employment generates sufficient revenues. In general, the key issue is to adopt particular adjustment measures based on the conditions of business operations development, in a broad sense, regarding both the demand and supply.

The formation of the so-called external economies of scale is inextricably connected to another possibility of consolidating the existing comparative advantages and/or creating new ones, namely the broadly understood organization and structure of national markets and the efficiency of their functioning, as first noted by Marshall (1919). He recognized the meaning of these factors and called them the "fourth agent of production," which was no less significant than land with natural resources in it, labor resources, and broadly understood capital resources. As Krugman and Obstfeld (2002, p. 198) note, "(…) Marshall claimed that there were three reasons for a group of companies (it should be added that such companies often operate in different countries—note by J. M.) to be more effective than a single company operating in isolation: possibility of maintaining specialized suppliers by the group of companies; labor market activation resulting from geographical concentration; increased diffusion of knowledge, also as a result of the geographical concentration. The same factors conducive to the external economies of scale do apply nowadays."

In the light of theoretical achievements to date, including the achievements of authors of the so-called new economic geography, another method of consolidating the already existing comparative advantages and/or creating new comparative advantages seems to be the active use of the international movement of mobile production factors, including domestic ones, i.e., formed within the territory of a particular country and called "advanced" by Porter (1990) and Hämäläinen (2003) (such as skilled labor force, loan and production capital, new technologies, including human communication technologies, brands, and trademarks). The aim should be—as suggested, among others, by Fujita et al. (1999)—to reduce the distance between a given country and its economic environment indirectly (through trade) and directly (through international exchange of production factors).

Today the theory of international movement of production factors is neither complete nor coherent. The reason for this situation is, *inter alia,* the fact that particular elements of the desired target theory, i.e., the theory of international movement of labor force, capital, and technical knowledge, not to mention the theory of rational environmental management, are still at the stage of development. We know, however, that individual parts of the overall theory are heterodox, complement each other to a greater or lesser extent, and, more importantly, contain several interesting elements, which are significant from our point of view.

All modern theoretical concepts of international movement of *labor force* share an assumption that the labor factor is not homogeneous, and that in fact the point is to explain problems related to migration of people of a certain age, qualifications, skills, dispositions, etc. Thus, the authors of these theories consistently assume that national and, above all, international labor markets are more or less segmented, and that regional and interregional (including international) differences in the level of wages can be observed. As stated by, *inter alia,* Siebert (1999, p. 51), on an international scale, "(…) in addition to differences in skills and human capital resources, employees in a particular country can, to a greater or lesser extent, use specific endowment of capital or local immobile labor force, which enables them to be more or less efficient. The differences in terms of such endowment should definitely be distinguished from the market segmentation caused by specific policy, e.g., specific decisions concerning immigration. The preference for a particular location can finally be significant enough to make people willing to accept lower wages." It is therefore important to competently manage domestic and foreign labor force in a given country.

The starting point for the macroeconomic approach to international labor movement are considerations based on the microeconomic theory of migration, in which it is assumed that migration decisions are the result of a process of optimization of the relevant benefits (profits) and diseconomies (losses). The theory created by Hicks (1932) and developed in the studies of Todaro (1969) and Harris and Todaro (1970) is known in the literature as the Harris–Todaro migration model. The model assumes that an important factor determining the change of residence and work is not the fact of obtaining employment itself, but rather a realistic prospect of getting a job with a higher salary. In other words, this is about a certain calculation of gains and losses of the migration from a dynamic point of view, and therefore—which is worth emphasizing—certain sequentiality of the search process with all its implications[5].

Contemporary theoreticians of international migration of labor force explicitly refer to the above-mentioned microeconomic considerations. They claim, among others, that the international differences in levels of economic development and wage rates (and therefore also the international variation in relative prices of production factors) are not the only causes of emigration and immigration of labor force between different countries and their groups. According to the theoreticians, international migration of labor force is related to the incidence of broadly defined costs and benefits. The costs include, above all, travel expenses, loss of existing revenue associated with the reallocation, expenses associated with finding new employment and with training, costs of risk of decisions made, and costs of separation from family and the former surroundings. As for the benefits, they include higher rates of wages and nominal income obtained in the country of emigration, better educational opportunities, possibility of improving the financial situation of children, etc.

In accordance with considerations that are of interest to us, concerning the theoretical approach to the impact of active participation of national economy of a country in the international movement of labor force on the development of its

[5] See for more detail, e.g., Straubhaar (2000); Rusangirsi (2003); Rynarzewski, Zielińska-Głębocka (2006); Siek (2009).

international competitiveness (including long-term competitive advantages), at least two aspects deserve attention. First, if a process of evaluation of dynamically presented gains and losses related to international migration and to sequentiality of those calculations by the citizens really takes place, it is possible to observe a specific accumulation of positive and negative effects of the phenomenon in time. Assuming that long-term effects of the relevant calculations should be positive, it is possible to conclude that emigration from a given country (including the so-called brain drain) does not necessarily lead to the deterioration of the international competitive ability and the international competitiveness of national economy of a given country. In fact, cumulative returns of better-educated and wealthier emigrants, which are sometimes referred to as the "snowball" of emigration and subsequent immigration (Either 1982; Ruffin 1984; Straubhaar 2000; Foders and Langhammer 2006), need to be taken into account. However, the situation can also be the opposite. The movement of foreign labor force may favor the increase in international competitive ability and international competitiveness of national economy of a given country and, in the long term, cause the erosion.

At this point, attention should be paid to another important aspect of the process of dynamic calculation of respective gains and losses by migrants. The authors of many relevant analyses indicate that this is when multidirectional migration movements overlap.

From the theoretical point of view, the phenomenon of overlapping of somewhat parallel migration movements between different countries can occur in its most intense form, when full freedom of international labor force migration is possible, that is, when individual calculations of persons of different age, qualifications, abilities, and skills can perfectly cross on an international scale. At least in some way it is very similar to specialization and inter-industry trade of goods and services (when the crucial reason for migration are differences in the level of wages) and to specialization and intra-industry trade of goods and services (i.e., when other reasons, such as the will or need to improve qualifications, are taken into account). On the basis of the theory, it can be proved that all these phenomena and processes are complementary and that the higher the level of economic development of collaborating countries, the more intense the phenomena and processes become. Moreover, in the light of contemporary theoretical analyses, including the results of relevant empirical analyses, there is barely any doubt that:

a. Direct and indirect international movements of labor force substitute or, eventually, complement each other;
b. Direct international movement of labor force is, in a way, naturally linked to direct and indirect international movement of capital and technical knowledge;
c. Direct international movement of labor force of different qualifications, abilities, skills, etc. can take place on the pattern of inter- and intra-industry trade of goods and services, further referred to as horizontal and vertical movement on international scale; and
d. It is possible to observe strict correlation between the horizontal and vertical (parallel and, so to say, crossing) international movement of labor force and the international movement of other mobile production factors.

Until now, a complete and coherent theory of international movement of *loan and production capital* has not yet been formulated. There have been and still are several theories and hypotheses concerning these issues. As regards the movement of loan capital, two theories have been created thus far (the so-called portfolio investment theory and the theory of investment portfolio diversification); each of them has some advantages and, at the same time, drawbacks[6].

While analyzing the effects of loan and production capital movement on the development of comparative advantages, it is important to pay attention primarily to theories and hypotheses of the development of foreign direct investment (FDI). The synthesis of the most important determinants of FDI development has been performed by Dunning (1981) in the so-called eclectic paradigm, in which the author presents the majority of benefits gained by the enterprises seeking to internationalize their production and trade through the export of capital. The concept of Investment Development Path (IDP; Dunning and Narula 1996) develops the aforementioned idea. According to this concept, the configuration of specific proprietary advantages, localizations, and internalizations changes together with the degree of economic development of the country or region that hosts the FDI. As a result, the net investment position of the country changes and its comparative advantages in the international trade evolve.

Many other contemporary economists have tried to formulate a model of the above-mentioned FDI types (including the movement of technical knowledge and, partially, of labor force as well) which would be compact and internally coherent with modern models of the international trade development. These attempts have been reflected, *inter alia,* in the model of Markusen et al. (1996). The authors clearly distinguish:

a. Horizontal FDIs and
b. Vertical FDIs.

According to J. R. Markusen, A. J. Venables, D. E. Konan, and K. H. Zhang, horizontal FDIs (i.e., relative investments realized by the so-called large transnational corporations (TNCs)) can develop and do so mainly between countries and regions of similar size and endowment of broadly understood natural resources, their final effect coming down to manufacturing a given product and/or a group of products in the country and abroad, as well as to the so-called parallel horizontal trade. On the other hand, according to reflections of the economists, vertical FDIs and the relevant movement of technical knowledge (as well as labor force migrations) develop mainly when a given country-investor (i.e., enterprises operating within its territory) visibly dominates in terms of endowment of the most important production factors nowadays, i.e., human capital and technical knowledge. Then, the transfer of production capital (which often takes place together with the transfer of people with adequate knowledge) is primarily aimed at—except for the still omnipotent desire for profit—the transfer of one or more production phases of the final product or, eventually, their larger group (including the production of certain sets, subsets,

[6] See for more detail Misala (1990, 2003, 2007); Södersten and Reed (1994).

or components), which results in, *inter alia,* the development of intra-industry trade (vertical and horizontal), together with all its consequences for the development of international competitive capacity, international competitiveness, and for the economic growth and development of particular regions—countries of the world and their groups, which will be discussed further on.

Until now, a complete and coherent theory of international movement of *technical knowledge,* in the form of technology embodied in goods and services and of disembodied technology, i.e., patents, licenses, and know how, has not been formulated either. It is known, however, that creating appropriate knowledge and skillful knowledge management through international exchange have a great significance from the point of view of development of competitive advantages and/or their lack in the international trade because, among others, the leaders in technical and technological progress always have, to a greater or lesser extent, first-mover advantages versus respective disadvantages at their disposal. It can be deduced from, *inter alia,* the model of the so-called flying geese paradigm by K. Akamatsu, according to which countries that dominate in the international competitive rivalry owe their position to the fact that they have cutting-edge technical knowledge at their disposal, which they can, to a greater or lesser degree, disseminate on an international scale by exporting it in a direct and/or indirect way. According to K. Akamatsu's model (which is called the flying geese paradigm due to the shape of graphs), countries that lag behind the world's leaders in terms of the degree of technical and technological advancement, technical culture, managerial skills, etc. initially meet their needs in these areas through adequate imports. Nevertheless, after meeting a number of conditions, they can become more or less effective competitors for the leading countries, basically in all of the above-mentioned domains. From this perspective, the degree of convergence in levels of economic, technical, technological, and system development of the partners is very important. The higher it is, the more favorable the situation generally becomes for the imitating countries[7].

The relations occurring in the process of international competitiveness of a given national economy through an active participation in the indirect (i.e., commercial) international turnover of a factor referred to as *the environment* may be also viewed in a synthetic way. From the dynamic point of view (in the development of the dynamic competitive advantages), it is possible to observe a certain fundamental trade-off between the development of the level of international competitiveness of a given economy and the development of quality of broadly understood environment of a given country. It becomes evident when the degree of effectiveness of the basic production factors is measured by their total factor productivity, and the quality of the environment is evaluated by such measures as the local emission of sulfur dioxide or carbon dioxide. The occurrence of the fundamental trade-off and the need for its resolution (more specifically, its costs constituting the reverse side of the development of both general production factors and the quality of the environment) determine the development of the international competitiveness of a given economy.

[7] See for more detail, e.g., Klodt (1989); Keller (2003); Deason (2009).

Table 1.2 Desired indirect effects and selected instruments aiming to achieve changes in the intensity and structure of comparative advantages. (Source: Deardoff (2003) and own supplementation)

	Desired indirect effects	Selected instruments and methods
1.	Raising technical and technological level and changing production functions	Budgetary expenditure on research and implementation work (including those in the military sphere), instruments for protection of intellectual property, indirect subsidies, and tax breaks
2.	Change in endowment of basic production factors	Encouragements to save (e.g., increased interest rates and tax breaks), tax allowances for investors, government purchases, and budgetary expenditures on development of economic and social infrastructure
3.	Change in demand size and structure	Budgetary expenditures for special educational programs (e.g., anti-smoking campaign), customs duties, indirect subsidies, and tax credits
4.	Increasing ability to achieve economies of scale in production and sales	Indirect tax credits and subsidies, government purchases, and top-down promotion of exports
5.	Development of flexible market structures	Strengthening of competition in market policy instruments, liberalization of foreign economic turnover, and use of market exchange rate
6.	Development of conditions to communicate with partners and of costs of transferring goods and production factors	Budgetary expenditures on development of appropriate infrastructure, indirect subsidies to exporters, and liberalization of functioning of economy, including foreign exchange

Finally, the last way of consolidating the existing comparative advantages and/or creating new ones is through application of various, broadly understood instruments of the economic policy, including instruments of a relevant foreign and international policy (Table 1.2).

Except for the use of certain instruments of discriminatory nature (e.g., direct subsidies), which will be discussed further on[8], all of the previously discussed ways and methods of consolidating comparative advantages and/or creating competitive advantages are theoretically justified. Furthermore, from a strictly theoretical point of view, the liberalization of the economy, including the broadly understood foreign exchange of a given country *A*, may result in an increase in its international competitive ability and pure international competitiveness in the sense described earlier (Fig. 1.5)[9].

[8] See the following sections, and in particular Sect. 1.3. of this chapter.

[9] Compare investigations contained in Sect. 1.1.2.

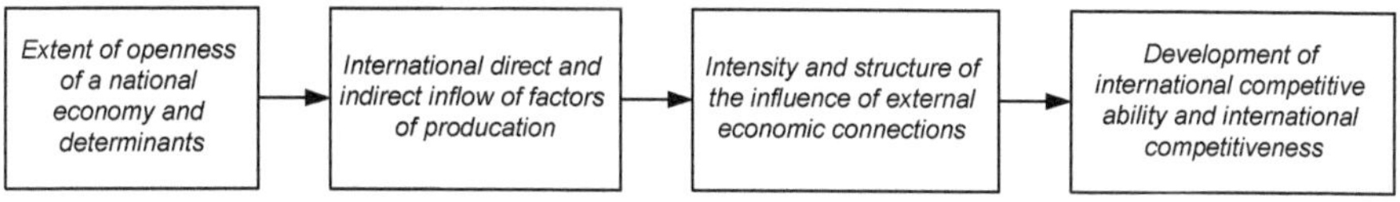

Fig. 1.5 Influence of external international exchange on shaping of international competitive ability and competitiveness of economy of a given country. (Source: Author's own study on the basis of Ozawa (1992, p. 43))

1.3 Theoretical Concepts of Developing International Competitiveness of a National Economy

Nowadays, we observe many concepts of developing the international competitive ability and international competitiveness of national economies of particular countries of the world and their groups, which are often understood as some kind of strategies of economic growth and development. An important part of these concepts are reflections concerning the use and consolidation of current comparative advantages, as well as the creation of new ones, often called the dynamic comparative advantages, which, to make it simpler, are further referred to as the development of comparative (relative) advantages, which, of course, is inherently an appropriate process. It is very often, however not always, accompanied by deliberations concerning possibilities and methods of a specific, top-down control of the process, including the possibility of using a variety of instruments of broadly defined economic policy and even of politics *sensu stricto*.

Various concepts of developing the international competitive ability and international competitiveness of national economies (including of course the concepts of developing comparative advantages) are divided in different ways and in accordance with different criteria (Reinert 1994; Köppen 1998; Reichel 2002; Plate 2004; Misala 2007, 2011; Lodha et al. 2007; Nijkomp and Siedschlag 2011). However, while rejecting the concept of autarky policy and different mixed concepts, such as the so-called "Japanese practice-based one," wrongly classified by Pitelis (2011) as theoretical, actually only four concepts, which evolve over time, deserve attention. They are as follows: the concept of locational competition and international competitiveness (germ. Standortwettbewerb), the concept of the so-called strategic trade policy, M. E. Porter's concept of international competitiveness complemented in the meantime by J. H. Dunning (the so-called diamond concept), and S. D. Gupta's concept, which the author himself calls the concept linking comparative to competitive advantages.

1.3.1 The Concept of Locational Competition and International Competitiveness

Among various concepts of shaping international competitiveness of national economy (its competitive ability and pure competitiveness) particular attention should

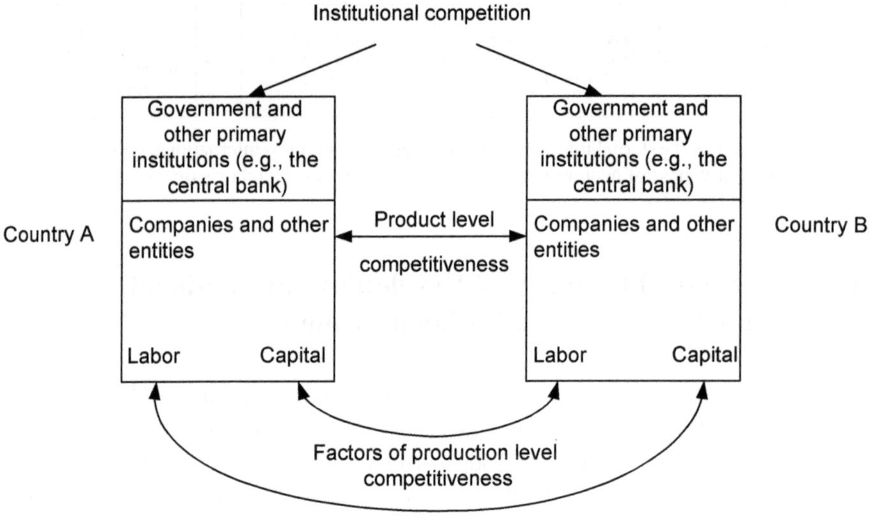

Fig. 1.6 Main elements of global competition according to the concept of locational competition. (Source: Siebert and Klodt (1999) and author's own supplementation)

be paid to the concept of locational competition and international competitiveness. It is stressed by, *inter alia,* Csaba (2005, p. 2), who recognizes the paradigm of locational competition as crucial to explain changes in the structure of the world's production and trade. The concept of locational competition refers not only to the development of costs and profits of governments which were modified and/or distorted by national policies. The essence of this concept comes down to three main elements. The first one is the so-called institutional competition. The second one is the creation of favorable conditions for the inflow of internationally mobile production factors, especially capital and technical knowledge, to a given country (region). Finally, the third element refers to the intensification of the competition mechanism between enterprises (Fig. 1.6).

The concept of locational competition and international competitiveness includes various elements of competition (on the international and inter-regional level, as well as between individual cities). According to Siebert (2000), there are three levels of this competition:

a. Competition between enterprises in markets for goods and services;
b. Competition between countries and regions for mobile production factors, i.e., mainly capital, skilled labor force, and technical knowledge; and
c. Competition between immobile and/or barely mobile production factors (e.g., land, raw materials, and broadly defined environment or unskilled labor force) through the indirect participation in the international exchange, and, specifically, through the international trade.

Product markets (for goods and services) constitute the area of competition for enterprises, which can increase their profits by being competitive in terms of prices and other values (quality) on an international scale. However, governments and other institutions of individual countries compete with each other as well. Their policy conduces, to a greater or lesser extent, to increase the attractiveness of a given country (region) to foreign (and, of course, domestic) mobile production factors, which in turn helps, to a greater or lesser extent, to improve the use of immobile labor force, domestic land, or raw materials. In consequence, electoral chances of politicians increase accordingly, as voters are trying to maximize their utility function in the sense that they take into account their situation on the labor market, the level of their income, etc. It is clear that the three levels of competition on an international scale are visibly linked to each other.

From the point of view of the created theory of international competitiveness of national economy, an important advantage of the concept of locational competition and international competitiveness is the possibility of extending the concept, in the sense of using various international exchange theories (theories of exchange of products and production factors), the basic assertions of which, as it is well known, may be empirically verified and compared on an international scale. It is indirectly indicated by Siebert (2006), who distinguishes six channels of the pure international competitiveness of national economy which can be examined. The author presents the following channels:

a. Movement of productive capital, together with reinvestments;
b. Movement of technical knowledge (including managerial knowledge), which, once gained and/or transferred, usually provides long-term effects that are properly spread out in time;
c. Movement of skilled labor force;
d. Short-term migration of residents;
e. Spread of the so-called demonstration effect through trade and international telecommunications links in a broad sense; and
f. Movement of portfolio investments influencing, *inter alia,* the development of productive capital resources and technical knowledge, as well as the exchange rate of relevant national currencies and financial resources of individual countries.

As stated by Siebert (2000 and 2006), while competing on an international scale in the institutional sphere for various production factors, the broadly defined government of a given country can use a range of instruments (tools) in order to increase the attractiveness to mobile production factors and/or to encourage them to remain in the country's territory. The point is only to skillfully use appropriate instruments. According to H. Siebert, especially the following instruments of economic policy shall be skillfully used:

a. Instruments favoring the increase of quantity and the improvement in quality of the so-called public goods (government, G);
b. Appropriately chosen tax rates (taxes, t); and
c. Appropriate institutional arrangements (institutional arrangements, In).

From the perspective of development of international competitiveness of national economy, it is very important that the broadly understood governments appropriately select and properly apply instruments that conduce to the increase in quantity and to the improvement in quality of different types of public goods (*G*), particularly the transport and telecommunications infrastructure as well as the education system in a broad sense. Furthermore, the appropriate choice of tax rates (*t*) is significant, as it influences, among other things, the development of the so-called tax base and the willingness to pay taxes. Eventually, the institutional competitiveness (*In*), understood as various constitutional provisions, principles of social negotiations, norms and procedures concerning modes of production and trade, ways of patenting and licensing, etc., as well as many informal norms and rules concerning the behavior of citizens, is very important. All the above-mentioned elements determine, to a large extent, the broadly understood location decisions made by residents and by owners of different types of enterprises and private institutions.

In view of the aforesaid conditions and the interdependence between them, an important task of economic authorities of every country, equated by H. Siebert with a broadly understood government, is to choose a certain optimal set of the above-mentioned instruments and solutions, i.e., an optimal instrument mix. The appropriate set should help to increase the domestic and international competitiveness on the one hand and constitute a resultant of other current preferences of the government on the other.

In accordance with the model considerations of Siebert (2006), governments use economic policy instrument mixes (*G, t,* and *In*) in order to optimize their functions (e.g., they aim at increasing national income *Y*, although they need to maintain a balanced state budget). At the same time, households are trying to maximize their utility function. Simultaneously, enterprises seek to maximize profits by using the production function, in accordance with which the size of global production *Y* depends on the amount of labor *L* and capital *K*, with capital resources being determined by public goods resources *G*, tax rate *t*, and appropriate institutional arrangements *In*. Therefore, the production function takes the form of $Y=f(L, K(G, t, In))$, in which *Y* may be interpreted as a supply vector. The vector has to be completed with benefits from trade and more broadly understood international exchange, which has already been mentioned[10]. In any case, it is clear that the vector can be extended by opening national economy. However, the opening (and, more precisely speaking, its forms, scale, and scope) has to be conducted taking into account the need for continuous and reasonable management of domestic and foreign production factors, in the presence of the phenomenon of international competition for the factors and the possibility of fast movement of mobile production factors on an international scale.

[10] Compare investigations contained in Sect. 1.2.3 of this chapter.

1.3.2 *The Concept of Strategic Trade Policy*

The essence of the strategic trade policy is the use of different instruments of economic policy (mainly trade, but also industrial and technological policies) to correct a variety of market distortions. Generally, it consists in using, in one's own business, deviations from V. Pareto optimum (the so-called first best solution) and/or, at a later stage, i.e., when appropriate measures of interventionism do not bring the desired results, in using corrective instruments called suboptimal and used as the so-called second best solution.

It is also possible to look at models constituting the concept of the so-called strategic trade policy from a different angle. In other words, they can be divided into models of implementation of the policy:

a. In monopolistic competition;
b. In oligopolistic competition;
c. In order to achieve higher production volumes and economies of scale; and
d. In order to achieve prices better than those of the partners.

Appropriate model considerations can easily be expressed in a more general way. The more or less explicit aim of these models is always to create intra-national conditions in order to capture as many benefits of international exchange as possible. According to Krugman and Venables (1987), benefits understood in this way can be achieved by means of appropriate actions of the government aiming at ensuring, in a given country, higher income ("economic rent") of production factors used for determined purposes, and not in other possible (alternative) but less cost-effective purposes, which may bring positive results for the society of the given country. The authors and the supporters of the so-called strategic foreign and international trade policy wish to achieve a higher profit rate in a given branch of industry, compared to branches of industry with a similar degree of risk, by using various instruments of foreign economic policy (as well as domestic policy instruments). An alternative goal can be to achieve higher wages in a given branch of industry than in other branches, in which employees with similar qualifications are employed. Therefore, if there are rents to be taken in a given branch or business or in a given industry, the strategic trade policy (e.g., proper construction of tariffs and skillful use of subsidies) can ensure a greater share of the rents (profits) of more effective branches or businesses to a given country and, eventually, increase the size and the level of national income. However, it involves many problems. First, the costs of protectionism. Second, the problem with identifying and picking up the priority branches, businesses, or industries. Third, the problem with the top-down control of the development of these branches in the long run and consequences of such a situation for international competitiveness. Fourth, the likelihood of corruption and social costs of this phenomenon.

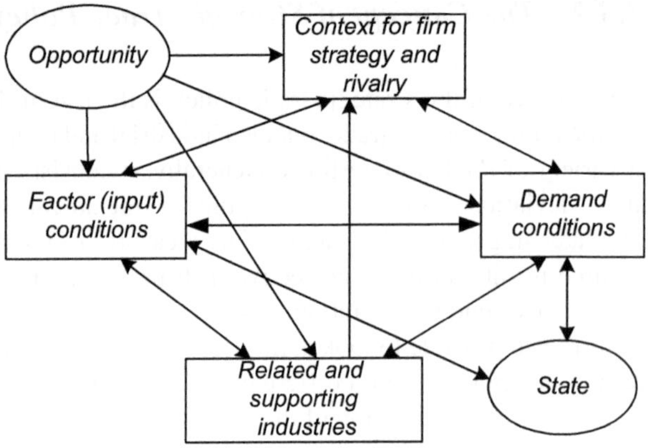

Fig. 1.7 M. E. Porter's diamond. (Source: Porter 2001, p. 207)

1.3.3 M. E. Porter's concept and J. H. Dunning's supplementations

The core of the concept of international competitiveness by Porter (1990) presented in 1990 is constituted by four groups of endogenous factors that determine the concept and are sometimes called Porter's diamond or rhombus. It refers to the following synthetic factors presented visually in the form of rhombus vertices:

a. conditions for development of basic mobile and immobile production factors (the so-called resource conditions and economic infrastructure);
b. conditions for development of domestic demand;
c. terms and conditions for operation of related and supporting industries; and
d. strategy and market structure, as well as competition between companies.

Moreover, M. E. Porter adds another two somewhat external factors called the opportunity (favorable circumstances) and the government (activity of relevant state institutions), which influence the development of the above-mentioned endogenous factors *sensu stricto*. In the end, the so-called "Porter's diamond" takes the form presented in Fig. 1.7.

According to M. E. Porter, the development of comparative advantages is highly influenced by the positive interaction between all four basic elements of the diamond that support each other, and which additionally can strengthen different kinds of opportunities understood as favorable circumstances of internal and external nature that have been used. Nevertheless, appropriate adjustment processes may also be supported by an appropriate activity of the state (broadly understood government), which, however, should be compliant with relevant international standards and principles. The activity may take the form of an active structural and regional policy (on different levels), as well as of the international economic policy.

Due to the relatively low "explanatory capacity of the Porter's diamond" as well as, but not only, limited possibilities of the so-called operationalization of the concept, the idea met with fairly widespread criticism. This is the reason why, inter

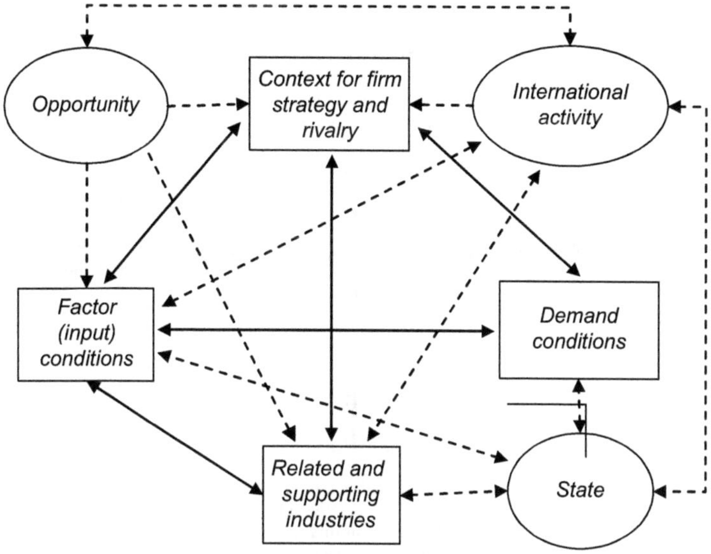

Fig. 1.8 Determinants of international competitive ability and competitiveness of countries (nations) by M. E. Porter, including the remarks of J. H. Dunning. (Source: Dunning 1992, p. 40)

alia, Dunning (1992) complemented the concept by adding to it the activity of large TNCs, the significance of which remains high (among other things, because of their impact on the size and structure of international movement of goods and production factors) and which usually shows an upward trend. As a result, "Porter's diamond" has taken the form presented in Fig. 1.8.

As stated by M. E. Porter and J. H. Dunning, it is the resultant of the effect of all the factors listed in Fig. 1.8, which is important in the process of development of dynamic comparative advantages of each country; however, the intensity of this influence can be smaller (dashed lines) or greater (solid lines). These factors (including the activity of the state) can support or weaken each other. According to the authors, the optimal situation occurs when competitive ability and international competitiveness are both positively determined by all factors. The difficulties in maintaining (erosion) competitive advantage (ability to compete and comparative advantages) arise when it results from the impact of a smaller number of factors or the presence of tensions and contradictions in their influence[11].

1.3.4 Linking Comparative and Competitive Advantages: The Concept by S. D. Gupta

Relatively recently, Gupta (2007) presented an interesting and compact concept of how the international competitiveness of national economies is shaped in line with

[11] See also Moon et al. (1998).

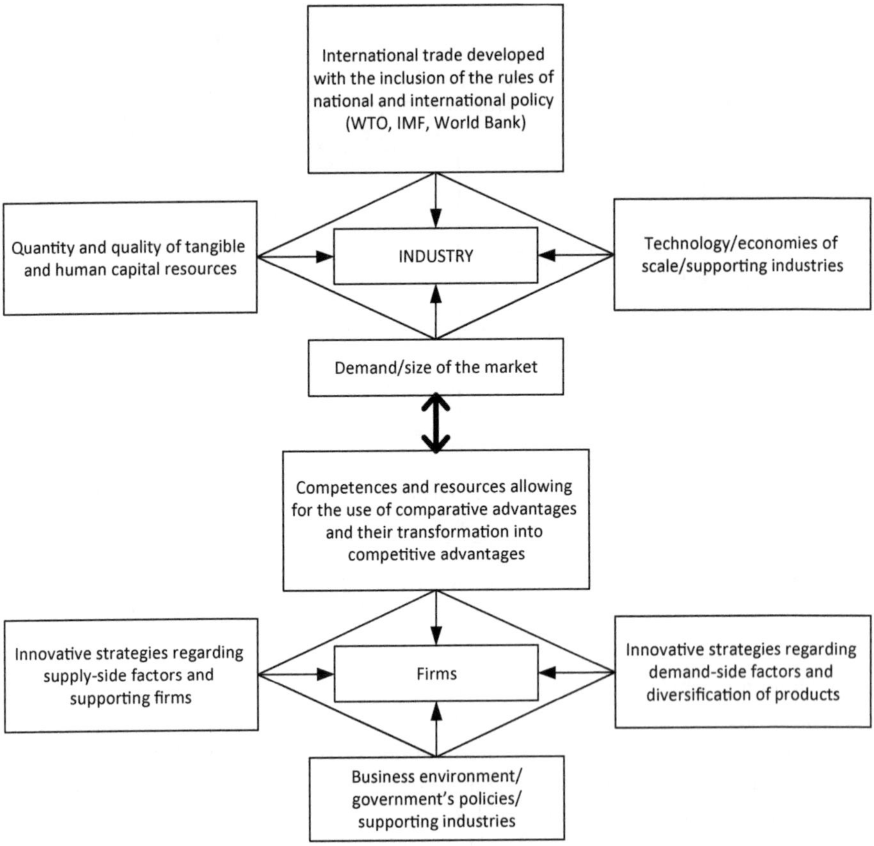

Fig. 1.9 Links between the concept of shaping international competitiveness in compliance with the principle of comparative advantages and the principle of competitive advantages according to S. D. Gupta. (Source: Gupta 2007, p. 34)

the dynamic principle of comparative advantages. Gupta assumed that, in dynamic terms, it is significant to ensure specific harmony of ventures at the microeconomic level (at the level of basic business entities, i.e., enterprises) and at the macroeconomic level. However, he only narrowed this concept down to industry in the broad sense[12] (Fig. 1.9).

Gupta's model (2007) contains a specification of the main determinants of parallel formation of comparative and competitive advantages in the given country in the form of two "diamonds" (rhombi), the links between which are evident. This means, among others, that the factors determining the formation of comparative advantages in one "diamond" influence the shaping of advantages in the other rhombus (i.e., competitive advantages) and vice versa. This might of course oc-

[12] A similar approach is displayed by K. Aiginger in a number of his works. However, Aiginger has yet to present an appropriate formalized model. Cf: Aiginger (1998, 2006, 2008, 2009).

cur only in certain systemic conditions. In such cases, according to Gupta (2007, p. 26): "(…) pursuing the forces related to competitive advantage in those goods and services in which a nation already has some comparative advantage offers a better promise for success in gaining competitive advantage for the nation. In this "double diamond" framework, it is also possible for forces of competitive advantage to further strengthen the operation of the forces of comparative advantage or even help create comparative advantage in goods and services, which did not exist in the light of classical theories of comparative advantage." Of course, certain conclusions for the domestic (national) economic policy follow from these statements. S. D. Gupta does not formulate explicit recommendations. However, he points out that national policies for infrastructure development, export promotion, education, or research and development (R&D) policies of exporting industries can be conducive to the creation and strengthening of comparative advantages. Industrial policy instruments, such as production subsidies, tax reliefs, and government procurement, are often employed to provide advantages to domestic industries. Moreover, the policies implemented by international organizations, such as the World Bank, the International Monetary Fund, or the World Trade Organization, can also constitute sources of comparative advantages or their absence in the case of certain industries in states influenced by these policies (Gupta 2007, pp. 26, 27).

1.3.5 Main Conclusions on the Theory

The above theoretical concepts of shaping international competitive capacity and international competitiveness of domestic economies contain certain common elements. The most significant common points are the drive to maintain previous dynamic comparative advantages (competitive advantages) and to create new ones that are sometimes referred to as dynamic competitive advantages. These concepts are also linked by the fact that their authors are rather certain that these advantages should develop in a situation where free competition reaches its maximum possible level on appropriate markets of basic economic entities. Economic policy can be used to shape these advantages in the area of goods production and service provision rather than in the area of their direct trading. Various economic policy instruments should be used when internal (intra-state) disruptions of the volume and structure of foreign trade occur and/or if they result from explicitly protectionist activity of the partner towards foreign partners.

At the same time, the discussed concepts contain many elements that allow for differentiating between them. Apart from the vastness and contents of these concepts, they differ in terms of the scope of issues covered by their authors. From this point of view, we can assume that the presented concepts can be divided into two groups. The first group encompasses the relatively narrow concept of strategic trade policy, while the second group contains relatively more universal concepts of M. E. Porter (completed by J. H. Dunning) and S. D. Gupta, especially the concept of locational competition and international competitiveness, which is also included in the more comprehensive model of the so-called social market economy of A. Müller-Armack.

An important issue that differentiates the concept of strategic trade policy focuses the attention of its authors mainly on tasks of the broadly understood government and its activity within the scope of foreign trade of goods and services. Thus, as a consequence—to continue the largely outdated deliberations of mercantilists and advocates of the so-called *beggar the neighborhood policy*—the authors of this concept suggest that various protectionist instruments of restricting imports and promoting exports are, in certain conditions, employed on purpose. Some, for instance, Krugman (1984), suggest that customs protection of the domestic market might be a factor conducive to the growth of exports of a national monopolist. However, others—e.g., Brander and Spencer (1985), or Eaton and Grossman (1986)—claim that in the case of an oligopoly, the application of customs and subventions, understood literally or also as taxes and production subsidies, may be effective, which is *nota bene* tantamount to equating foreign trade policy with industrial policy. However, by promoting harmful protectionism (sometimes even the purposeful application of quantitative import restrictions), in general they do not go beyond deliberations concerning international competitiveness capacity and international competitiveness, the effects of which are reflected by the current account balance of the given country, and, more precisely, the first and foremost in the balance of trade and the services trade balance. They certainly do not take into account the diverse factors of the regional economic integration theory and the more broadly understood theory of business activity location in the world.

From the perspective of the issue of international competitive capacity and international competitiveness of the domestic economy of the given country, the analyses of M. E. Porter and S. D. Gupta reach much further. This concerns especially the authors of the concept of locational competition and international competitiveness. According to them, nowadays one cannot view this issue solely or almost solely from the perspective of foreign trade in goods and services. Currently, it is necessary that the authorities of individual countries create conditions conducive to business. Authorities cater for a business-friendly environment by enabling business infrastructure development, deregulation of product and production factors' markets, citizens' qualifications level, and financial and social stability. This is aimed at "retaining" own production factors in the country and "attracting" other mobile factors, especially capital and technical knowledge, to the country on an international level. In the social market economy model and in its inherent part constituted by the locational competition discussed above, among other things, it is crucial—while ensuring the independence of the central bank—to provide a coherent and mutually complementary framework of norms and rules so as to guarantee that business activity is compliant with the market mechanism (law of demand and supply).

Another significant difference between the concept of strategic trade policy and the remaining theoretical assumptions is the inclusion of the realities of modern international economy from different authors' perspectives. Authors of the strategic trade policy formulate their concept on the basis of the so-called large countries. Such countries can exert influence on international markets of products and production factors, as well as on the development of institutional and instrumental infrastructure of modern international economy, including among others the principles of operation of the IMF, the World Bank, or the World Trade Organization. The starting point selected by the authors of other concepts discussed above is visibly divergent, as their theses are characterized by a higher degree of universality. In any case, their concepts seem to be better adapted to the realities of functioning in the modern international economy of the so-called small- and medium-sized countries, including Poland.

This is connected with another difference between the concepts discussed, mainly that the approaches of their authors to the issue of speed and the so-called scale of desired outcomes of employing appropriate economic policy instruments generally differ. Now, it is difficult to fight off the impression that the authors of the so-called strategic trade policy (thus the policy that is in definition limited to trade) expect that the planned outcomes will occur almost at once. Authors of the remaining concepts, especially S. D. Gupta and authors of locational competition and international competitiveness, seem to advocate the thesis that the shaping of competitive advantages consists of rather long-term and non-automatic processes. In their opinion, this concerns especially the process of long-term development of competitive advantages that are based on human capital and innovativeness. This is accompanied by the belief that it is crucial to skillfully combine the impact of market forces with the influence of broadly understood governments and specialized national and international organizations and institutions that also include the so-called large TNCs.

From a purely theoretical point of view, the authors of the discussed concepts concerning international competitiveness of domestic economies have varied opinions on the methods and measures that should be used to evaluate the effectiveness of utilizing their proposals, including evaluation methods and measures of shaping competitive advantages. To a greater or lesser degree (depending on the concept), the authors point out that it is necessary to perform such evaluations to diagnose the situation in individual countries. This can facilitate assessing the position of countries in the international competitive struggle and enable formulating the appropriate strategy for the future.

1.4 Methods of Analysis and Measures of Competitive Advantages

These days, we have access to numerous methods of analyzing how competitive advantages of a given country are shaped. On their basis, various measures of the intensity and structure of these advantages are constructed. In principle, these are *ex post* methods and measures, i.e., they allow for the evaluation of the development of the situation and events that took place in the past, as well as for the prognosis of appropriate indexes for the future.

1.4.1 Methods and Measures Based Directly on the Classic Principle of Comparative Advantages and Their Developments

Many analysis methods and measures of comparative advantage shaping in a given country refer directly to the achievements of D. Ricardo, i.e., to the classic principle of comparative (relative) advantages, as well as to its various interpretations in the form of the real costs principle, the alternative costs principle, and the Heckscher–Ohlin theorem[13].

This applies especially to the methods of analysis of these advantages presented in the following models:

a. Ricardian comparative advantage model of M. Guzek;
b. Comparative advantage model of T. Gries and C. Hentschel;
c. Comparative advantage model of F. Ezeala-Harrison; and
d. Comparative advantage model of E. Siggel.

The discussed models of analysis of the way in which comparative and competitive advantages will be shaped in the future contain many elements that can be classed as parts of the broadly understood theory of international inter-state competition. They explicitly refer to D. Ricardo's relative costs principle and its developments, especially dynamic ones. At the same time, they attempt to incorporate the so-called optimum formulated by V. Pareto, adherence to which allows the formulation of many significant conclusions about the national economic policy.

From the perspective of the needs of the economic policy, these models also have a significant feature in common—in contrast to Balassa's method or the intra-industry trade analysis—to a large extent, they constitute *ex ante* solutions. In any case, the results of the studies performed by means of the methods discussed can also be of practical significance in terms of the possibilities of correcting mistakes made by economic politicians in the given country in the past. However, it is difficult not to note that the authors of the discussed models focus mainly on how foreign trade in individual countries was shaped in the past and how it could be shaped in the future, as foreign trading always, to a greater or lesser extent, reflects the standing of their national economy. This is because there is a lot of truth in the claim that foreign trade of a given state (its intensity, structure, and foreign account balance) is the reflection of the general economic situation and applied systemic solutions. Nevertheless, the authors of the discussed models consider the meaning of the so-called systemic competitiveness to be somewhat secondary.

However, advocates of slightly simpler partial measures of international price competitiveness, including of course cost competitiveness, refer to the suggestions of the authors of the statically analyzed comparative costs principle and its reinterpretation[14]. They believe that, among others, the level of unit labor costs (ULCs), of

[13] More in Misala (2011a) and references cited therein.

[14] More in, among others, Wziątek-Kubiak (2003, 2006); Gorynia and Łażniewska (ed., 2009); Misala (2011a).

course against the backdrop of other comparative countries, constitutes a significant measure of this type of competitiveness. This measure can be presented in the form of the following formula:

$$ULC = \frac{W/L}{Q/L} = \frac{w}{a}, \qquad (1.1)$$

where

ULC	Unit labor costs;
W	Labor cost;
L	Labor;
Q	Quota;
w	Labor costs per employee; and
a	Quota per employee.

The way in which ULCs are shaped in the analyzed country indicates the labor costs incurred in the given period to manufacture a production unit. It is therefore a measure that is much more economically justified than the absolute labor cost level that for many various reasons (e.g., due to different labor market regulations in individual countries, different levels of economic development, and different levels of work effectiveness) is not thoroughly justified in terms of economy. Another measure referred to in the literature of the field as relative ULCs (RULCs) is much better justified. This measure can be presented as follows:

$$RULC = \frac{ULC_K}{ULC_Z} \cdot e, \qquad (1.2)$$

where

ULC_K	Unit labor cost in analyzed country “X”;
ULC_z	Unit labor cost in business environment countries; and
e	Nominal exchange rate.

Formula (1.2) is of determinist nature, as a certain cause yields particular effects and vice versa. What follows from this formula is that ULCs—a significant determinant of international competitiveness—rise, or fall, for a number of reasons in the given country. One of them is faster growth of (or decrease in) the level of salaries in comparison to foreign countries. Another reason is slower growth of (or decrease in) national labor productivity in comparison to foreign partners. The third reason is the appreciation (or depreciation) of the national currency of the analyzed country. RULC ratios are often interpreted in such a way that their decrease translates into an increase in international competitiveness of the given country, whereby its increase is tantamount to a drop in the level of this competitiveness (Wziątek-Kubiak 2003, 2006; Olczyk 2008). Such an approach, however, does not take into account any changes in the exchange rate level, not even of its nominal value.

Already in the 1960s, Bruno (1965), referring nota bene to the analyses of D. Ricardo and G. Haberler, presented a different method of approaching the analysis of the shaping of comparative advantages in international trade. M. Bruno outlined the importance of the international price competitiveness index, currently referred to as the domestic resource cost (DRC) index. Deliberations on this index were continued by many outstanding economists (including B. Balassa, D. M. Schydlowsky, A. Krueger, J. Bhagwati, T. N. Srinirasan, and J. Kotyński), but its essential features have remained essentially unchanged. Deardorff (2006) described this index as an *ex ante* measure of comparative advantages. This term is not fully correct. Generally, one should use the following general formula in various approaches (nationwide, industry, etc.):

$$DRC_i = \frac{WD}{K}, \tag{1.3}$$

where

DRC	Domestic resource costs;
WD	Added value;
K	Costs of production factors; and
i	Industry, branch.

Nowadays, various versions of formula (1.3) are used not only in reference to entire domestic economies but also in reference to particular industries or sectors[15].

The most important variations of this formula are:

a. The so-called *value-added* (*VA*) *ratio* that boils down to calculating the relation between added value and global production value for "*i*" in the following years;
b. The so-called *social profit ratio* (*SPR*) that boils down to calculating the relation between global production value and the value of outlays for individual branches, sectors, etc. in the following years.

For each "*i*" in the given period "*t*," the following close correlations occur between DRC, VA, and SPR:

$$\frac{1}{SPR} = 1 - VA \cdot (1 - DRC). \tag{1.4}$$

In reference to the dynamic Heckscher–Ohlin theorem, Schydlowsky (1984) presented DRC indexes in a slightly different manner. Schydlowsky used the term "domestic resource costs of foreign exchange" (DRCFE). In his opinion, in the case of a given country (but also a given branch, industry, or product, $i = 1 \ldots n$) in reference to the initial period "*t*," one should apply the following formula:

[15] Cf., among others, Kotyński (ed., 1988); Holmes et al. (1993); Lubiński et al. (1995); Wyżnikiewicz (1998); Wziątek-Kubiak (2003); Zielińska-Głębocka (ed., 2003); Frejtag-Mika (ed., 2006).

$$DRCFE = \frac{LH_{rs} \cdot \text{cost of labor} + KH_{rs} \cdot \text{ cost of capital}}{p^{CIF} - W \cdot p^{CIF}}, \tag{1.5}$$

where

LH_{rs} Labor per production unit;
KH_{rs} Capital per production unit;
p^{CIF} Average import costs per product unit; and
W Costs of materials used, calculated by taking into account the average prices of utilized production factors.

D. M. Schydlowsky was aware that also in market economy one should not tolerate the significant drawback of the Heckscher–Ohlin theorem, which is its static nature. This static nature involves failure to acknowledge the influence of changes in production costs in connection with large-scale production development, the influence of technical progress, or changes in consumer preferences. Schydlowsky proposed calculating the appropriate ratios for the given period by means of the following formula:

$$Dyn\ DRCFE = \sum_{t=0}^{n} \frac{1}{dt} \frac{LH_{rs} \cdot \text{cost of labor} + KH_{rs} \cdot \text{cost of capital}}{\text{net foreign exchange inflows}}. \tag{1.6}$$

In many countries, when analyzing cost-price competitiveness, the so-called relative unit value method is applied[16]. The following formula is used most often:

$$C_{ex/im} = \frac{\sum V_{ex\ i}}{\sum Q_{ex\ i}} \div \frac{\sum V_{im\ i}}{\sum Q_{im\ i}}, \tag{1.7}$$

where

$V_{ex\ i}$ Value of exports of a given country (e.g., Poland) to the selected market (e.g., EU member states or the global market);
$Q_{ex\ i}$ Volume of exports of a given country to the selected market;
$V_{im\ i}$ Value of imports of a given country from the selected market;
$Q_{im\ i}$ Volume of imports of a given country from the selected market; and
i Goods, groups of goods.

Another method of analyzing international cost-price competitiveness of the given country is the method that boils down to calculating and interpreting the so-called relative export prices (REPs). These prices are calculated by means of the following formula:

$$REP = \frac{\sum_{i} (P_i E_i \cdot W_i)}{P_k E_k}, \tag{1.8}$$

[16] More in, among others, Burzyński et al. (1997); Marczewski (2002); Misztal (2010).

where

P_k, P_i Index of export prices of country X and competitor countries;
E_k, E_i Nominal value of exchange rate of country X and competitor countries; and
W_i Significance ascribed to individual competitor countries.

Among the price (including cost) measures of international competitiveness, *terms of trade*, or, to be more precise, their variations, are most popular. There are various types of terms of trade (nominal, real, income, and factor-related). When measuring international competitiveness, the so-called nominal (price) terms of trade are used most often. This measure displays the relation between the prices obtained in the exports of the given country in the analyzed period and the prices of the goods imported in that period. If the price terms of trade (similar to other types of these measures) change over time, then this sheds light on the mutual advantage or the change in advantages of the parties involved in the trading process.

1.4.2 The Methods and Measures of Revealed Comparative Advantages

Balassa (1965) is the author of the revealed comparative (relative) advantages method. Balassa suggested that relevant ratios (the so-called relative comparative advantage or RCA_i ratios) should be calculated by means of the following formula:

$$RCA_i = \frac{X_{ij}}{X_j} \div \frac{X_i}{X}, \tag{1.9}$$

where

X_{ij} Value of exports of the "i" goods group of a given country "j" (e.g., Poland);
X_j Value of global exports of a given country "j" (e.g., Poland);
X_i Value of global exports of goods group "i" (e.g., fuels); and
X Value of global world exports.

Nowadays, certain reinterpretations of the original formula of Balassa (1965) are applied in most cases[17]. These reinterpretations are numerous, but the following one is used most frequently:

$$RCA = \ln\left(\frac{x_{ij}^K}{m_{ij}^K} \div \frac{X_j^K}{M_j^K}\right), \tag{1.10}$$

[17] More in, among others, Vollrath (1991); Hatzichronoglou (1996); Reichel (2002); Moenius (2006); Hoen and Oosterhaven (2006); Guordon (2006).

where

x_{ij}^{K} Exports of goods group "i" from country "K" to country or group of countries "j";
m_{ij}^{K} Imports of goods group "i" to country "K" from country or country group "j";
X_{j}^{K} Total exports of country "K" to country or country group "j";
M_{j}^{K} Total imports of country "K" from country or country group "j";
i Industry according to SITC;
K Analyzed country (e.g., Poland); and
j Other countries of the world, or member states of the enlarged European Union.

If the value of the RCA ratio is higher than zero, this indicates that the revealed relative advantage occurs and is intensive. RCA_i ratios whose value is below zero indicate that we are dealing with a disadvantage of lesser or greater intensity. Using the logarithmic form of the formula allows the maintaining of symmetry between positive and negative RCA_i ratios within the range oscillating around zero.

A significant advantage of this method, and mainly of the revealed comparative advantage ratio, is the possibility to adapt them to the analysis of appropriate streams of goods and services trade, i.e., to the international trade hypothesis. In particular, these are possibilities of using these indexes in reference to products with specified different absorptive capacity of production factors used in goods manufacturing and/or service provision (e.g., goods that are technologically intensive or goods that require relatively high expenditure, e.g., of human capital)[18].

The constantly developed method of revealed comparative advantages and appropriate partial indexes as well as the synthetic weighted means of partial ratios (revealed comparative advantage indexes, RCA indexes) are the current canon of studies on international competitive capacity in the area of international trade. However, the most important thing here is that always and at any moment, the specified structure of cost-price comparative advantages of the given country over foreign countries and the specific absence of such advantages (in the sense of existence of nonabsolute advantages) determine to a large extent the development of foreign trade in each of the countries and more broadly understood international trade. This is confirmed by the results of many empirical analyses (Deardorff 1984, 2007; Leamer and Levinsohn 1995).

Substantive justifications and attractiveness of the dynamic principle of comparative advantages, e.g., according to Dornbusch et al. (1977) or Schydlowsky (1984), are clear and obvious. However, if one takes a closer look at this principle, it is relatively easy to note that its application entails mainly the occurrence of the above-mentioned competitive advantage principle in practice. This is explicitly indicated by, among others, Guzek (2004) and Siggell (2006). In any case, as Siggell (2006, p. 138) stated, Balassa's measure reflects success in the domestic exports

[18] One can also come across attempts of using Balassa's concept in relation to services trading and exchange of technical knowledge. Cf., Weresa (2005); Kuźnar (2008); Misala (2010).

in relation to total international exports. However, the data on exports might be distorted due to subsidies and other types of stimuli. Their application influences the way in which international competitiveness is shaped, but does not provide an exhaustive explanation of real comparative advantages.

Comparative advantages in the interpretation of B. Balassa (in our opinion rather competitive advantages) are based on historical data that reflect the level of development of individual production fields of the given state in comparison with the level of development of individual fields on the international scale. The comparison of the so-called Ricardian approach and Balassa's approach reveals a few other problems. Most importantly, the so-called Ricardian approach assumes that the lack of export or low export volume in the case of a particular product or group of products do not necessarily have to be synonymous to the lack of possibilities for developing an appropriate comparative (also competitive) advantage in the future. However, according to the approach of B. Balassa and the advocates of the application of his method in formulating conclusions on economic policies, relative high advantages might be revealed (in the past and on a current basis), even in the case of low profitability of the given product or its exports, and even in the case of its nonexistence. In addition, the comparative (rather competitive) advantage optimality calculation performed according to B. Balassa does not correspond to Pareto's efficiency. At the same time, however, his results can be treated as an approximation of the capacity of the given country to compete in international trade and the basis for formulating appraisals of how this was shaped in the past. What unequivocally follows from the above theoretical achievements (as well as practical experience) is that formulating such appraisals should be supplemented by a reliable analysis of the intensity and structure of intra-industry trade of the specified country and/or group of countries.

1.5 Basic Formulas for the Analysis of the Intensity and the Structure of Intra-Industry Trade

The formula used most frequently to establish the intensity and structure of intra-industry trade is the formula developed by Grubel and Lloyd (1975), which is based on the following definition of this trade:

$$W_i = (X_i + M_i) - |X_i - M_i|, \tag{1.11}$$

where (X_i+M_i) expresses the global value of turnover in foreign "I" industries, i.e., total exports (X_i) and imports (M_i) in the specified period (most often in the specified year), whereas $|X_i-M_i|$ is the absolute value of the difference between exports and imports of the products of the given industry. In order to ensure the comparability of the data for various industries, it is worth taking a look at intra-industry trade against the backdrop of the total value of international trade in the given industry:

$$R_i = \frac{(X_i + M_i) - |X_i - M_i|}{(X_i + M_i)}. \tag{1.12}$$

The average level of intra-industry trade can therefore be calculated as follows:

$$R_S = \sum_{i=1}^{n} \frac{\{(X_i + M_i) - |X_i - M_i|\}}{(X_i + M_i)}. \tag{1.13}$$

Additionally, the weighted mean is used, whereby the measures adopted here are, respectively, shares of individual industries in the global exports of the analyzed country ($a_i = a_1, a_2, \ldots, a_n$) or in the global imports of that state ($b_i = b_1, b_2, \ldots, b_n$). As a result,

$$R_W = \sum_{i=1}^{n} \frac{\{(a_i X_i + b_i M_i) - |a_i X_i - b_i M_i|\}}{(a_i X_i + b_i M_i)}. \tag{1.14}$$

The value of the R_i ratios (expressing the intensity of intra-industry trade in individual industries) and the R_S and R_W ratios (expressing the average level of intra-industry trade in the total trade of the given country) is within the 0–1 range, whereby $R = 0$ means the lack of intra-industry trade and $R = 1$ means that trade is solely of intra-industry nature. Thus, the higher the value of R_i (or R_S or R_W), the larger the share of intra-industry trade in total international trade of the given industry or country. Of course, all the ratios mentioned above can be expressed as percentages. To that end, their value must be multiplied by 100.

The presented formulas are not free of faults. For instance, if ΣX_i differs from ΣM_i, i.e., when the foreign trade of the given country is experiencing deficit or surplus, then the average R reduces the significance of intra-industry trade in total trade (it will never be able to assume the value of 1 or 100). Taking this into account, H. Grubel and P. J. Lloyd suggest using the adjusted formula that is referred to as the adjusted mean R_C (in contrast to the ordinary mean R_S and weighted mean R_W). The adjusted formula is to take into account the distorting influence of instability of total trade of the analyzed country on the intensity of intra-industry trade and is as follows:

$$\overline{\overline{R}}_C = \frac{\sum_{i=1}^{n} (X_i + M_i) - \sum_{i=1}^{n} |X_i - M_i|}{\sum_{i=1}^{n} (X_i + M_i) - \left| \sum_{i=1}^{n} X_i - \sum_{i=1}^{n} M_i \right|}. \tag{1.15}$$

Aquino (1981) pointed out that the above formula for the intra-industry trade intensity ratio constitutes only a partial adjustment. In his opinion, the formula reduces the significance of intra-industry not because total trade is imbalanced, but because instable trade occurs in individual industries selected. Errors are made during the calculation of partial R_i measures. Therefore, A. Aquino suggests a different method for the calculation of intra-industry trade intensity at the level of domestic economy. This model is based on the assumption that unstable trade is evenly distributed between all selected industries (the so-called Aquino average, hereinafter referred to as R^A). According to this suggestion, one should first of all establish the hypothetical value of imports and vhat total trade is balanced:

$$X_i^n = X_i \cdot \frac{\frac{1}{2}\sum_{i=1}^{n}(X_i + M_i)}{\sum_{i=1}^{n} X_i}, \tag{1.16}$$

$$M_i^n = M_i \cdot \frac{\frac{1}{2}\sum_{i=1}^{n}(X_i + M_i)}{\sum_{i=1}^{n} M_i}, \tag{1.17}$$

where

X_i^n hypothetical value of "i" industry exports in a situation of balanced total trade and

M_i^n hypothetical value of "i" industry imports in a situation of balanced total trade.

If total trade is unstable, the ratios X_i^n differ from X_i ratios, and M_i^n ratios differ from M_i ratios. Usually, in contrast to the assumption of A. Aquino, the factors leading to instability are not distributed evenly between individual branches.

The calculation of X_i^n and M_i^n is only a step towards the calculation of the adjusted share of intra-industry trade. In the opinion of A. Aquino, this share should be calculated in accordance with the following formula:

$$R^A = \frac{\sum_{i=1}^{n}(X_i + M_i) - \sum_{i=1}^{n}|X_i - M_i|}{\sum_{i=1}^{n}(X_i + M_i)}. \tag{1.18}$$

Owing to the assumption that international trade balance is distributed evenly between individual industries, also the R^A average cannot be considered the perfect measurement of intra-industry trade intensity, i.e., its share in total international trade. Therefore, the discussion is still pending, and new definitions and methods of measuring intra-industry trade are being put forward.

In 1994, Brülhart (1994) drew attention to the fact that indexes calculated according to the formula of H. Grubel and P. J. Lloyd can somewhat artificially increase the volume of intra-industry trade when the increase in intra-industry trade contributes to the weakening of the trade imbalance in individual industries, branches, groups of goods, etc. As a result, he suggested that the so-called marginal intra-industry trade indexes should be calculated as follows:

$$MIIT_i = \frac{\Delta X_i + \Delta M_i - |\Delta X_i| + |\Delta M_i|}{|\Delta X_i| + |\Delta M_i|}, \tag{1.19}$$

where Δ stands for the increase in appropriate values between the analyzed periods. The value of $MIIT_i$ ratios varies also within the range from 0 to 1 (or 0–100 when expressed in percentages). Ordinary means and $MIIT_i$ index means can be calculated similarly, as in the case of standard indexes calculated according to the formula of H. Grubel and P. J. Lloyd.

One of the significant drawbacks of indexes calculated in accordance with H. Grubel and P. J. Lloyd's formula, but also M. Brülhart's formula, was the lack of possibility of separating horizontal intra-industry trade from vertical intra-industry trade. Among others, Greenaway et al. (1995) attempted to eliminate this setback. They suggested that it should be assumed that horizontal intra-industry trade (*HIIT*) meets the following criteria:

$$1-\alpha \le \frac{\text{export} \quad UV_{ijK}}{\text{import} \quad UV_{ijK}} \ge 1+\alpha, \tag{1.20}$$

on the other hand, it should be assumed that vertical intra-industry trade covers the following turnover:

$$\frac{\text{export } UV_{ijK}}{\text{import } UV_{ijK}} < 1-\alpha \quad or \quad \frac{\text{export } UV_{ijK}}{\text{import } UV_{ijK}} > 1+\alpha, \tag{1.21}$$

where

UV Unit value of a specified product;
α UV dispersion index;
i Product "i";
j Industry, branch, etc.; and
K Country "K".

According to D. Greenaway, R. Hine, and C. Milner, the so-called unit values of individual products should be taken as the criterion of horizontal and vertical intra-industry trade separation. Such reasoning is based on the assumption that it is the export–import price level that reflects quality differences, on which, in turn, vertical intra-industry trade is based. In the opinion of the above-mentioned economists, vertical intra-industry trade can be defined as trade, where, in the case of multi-digit (six-digit to be precise) disaggregation of data, the unit value of 1 kg of imports of a specified type of goods (on the Free On Board, FOB, basis) divided by the unit value of a kilogram of imports of these goods (on the Cost, Insurance and Freight, CIF, basis) exceeds the arbitrarily determined $\pm\alpha$ limit. Consequently, they go on to claim that if correctly calculated export unit values divided by import unit values do not exceed this limit, then we are dealing with horizontal intra-industry trade. Of course, also in the case of using the method and formulas proposed by D. Greenaway, R. Hine, and C. Milner, one can additionally calculate appropriate means, weighted averages, and variation coefficients and/or deviations of appropriate indexes.

A very similar method was presented by Fontagně et al. (1997). Their method is based on the assumption that intra-industry trade in the given line of business or category of products is not limited to the overlapping of a specified import and export amount or value. According to them, the decisive fact is that in the specified category of products or just one product, bilateral parallel movements take place. At the same time, they arbitrarily identify the selection criterion for these movements as the 10 % threshold, i.e., a smaller trade stream must constitute less than 10 % of the larger trade stream within the specified product category to allow for the calculation of the total of appropriate imports and exports to intra-industry trade. If this criterion is not met, then, in their opinion, one should speak of one-way trade and/or inter-industry trade. By determining the appropriate threshold, L. Fontagně, M. Freudenberg, and N. Peridy stray from the essence of intra-industry trade defined in accordance with the formula of H. Grubel and P. J. Lloyd, irrespective of the data disaggregation level. They believe this figure should be as high as possible. To be more precise, it should have eight digits.

Great significance is ascribed to the method of L. Fontagně, M. Freudenberg, and N. Peridy in connection with the differentiation of horizontal intra-industry trade from vertical intra-industry trade, as well as the key role of export and import prices within this scope. As regards these prices, the economists mentioned above adopt further selection criteria, namely, they assume that in the case of horizontal intra-industry trade, the relations between export and import prices of the given product category and/or group, subgroup, and components can fluctuate only within the range of 0.85–1.15; thus appropriate deviation cannot exceed the value of ± 15 %. If appropriate deviations are higher than ± 15 %, then in the opinion of L. Fontagně, M. Freudenberg, and N. Peridy, we are dealing with vertical intra-industry trade. According to them, these higher deviations are justified by the fact that, in the case of this type of intra-industry trade, it is these deviations that indicate quality differences between the listed products, as well as their groups, subgroups, and components.

A completely different method of calculating exclusively the intra-industry trade intensity (and not its structure) is presented by Glejser et al. (1979). In particular, they state that the formulas of B. Balassa, H. Grubel, and P. J. Lloyd and A. Aquino do not take into account, among others, the somewhat necessary specialization of the given country (group of countries) in imports and in exports. Moreover, they do not appropriately reflect the consequences of imbalance in foreign trade. In the opinion of H. Glejser, K. Goossens, and M. Vanden Eede, it is necessary to first establish the level of pro-import and pro-export specialization (import- and export-oriented) of individual branches and the entire economy of the analyzed country (or a group of countries), according to the following formulas:

$$a_i = \log\left\{\frac{X_i \div X}{X_{gi} \div X_g}\right\}, \tag{1.22}$$

$$b_i = \log\left\{\frac{M_i \div M}{M_{gi} \div M_g}\right\}, \tag{1.23}$$

and then

$$a = \frac{1}{2} \cdot \sum_{i=1}^{n} \log\left\{\frac{X_i \div X}{X_{gi} \div X_g}\right\} = \frac{1}{n}\sum_{i=1}^{n} \log(a_i), \tag{1.24}$$

$$b = \frac{1}{2} \cdot \sum_{i=1}^{n} \log\left\{\frac{M_i \div M}{M_{gi} \div M_g}\right\} = \frac{1}{n}\sum_{i=1}^{n} \log(b_i), \tag{1.25}$$

where

a_i	Ancillary index in reference to exports of industry “i”;
b_i	Ancillary index in reference to imports of industry “i”;
a	Average value of ancillary indexes in reference to global exports of the analyzed country;
b	Average value of ancillary indexes in reference to global imports of the analyzed country;
X	Value of global exports of the analyzed country (to international market, or to selected groups of countries);
M	Value of global imports of the analyzed country (from the whole world, or from selected groups of countries);
X_{gi}	Value of world exports or exports of selected groups of countries to the analyzed country in industry “i”;
M_{gi}	Value of world imports or imports of selected groups of countries from the analyzed country in industry “i”;

X_g Value of global exports or mutual exports within the selected group of countries; and

M_g Value of global imports or imports within the selected group of countries.

The calculation of indexes a_i and b_i in accordance with the above formulas is not yet a sufficient basis for the evaluation of the significance of intra-industry trade in global trade. In the opinion of H. Glejser, K. Goossens, and M. Vanden Eede, these indexes only constitute the basis for drawing conclusions on the meaning of intra-industry trade. They claim that in the case of intra-industry specialization advantage, the value of ratio $(X_i \div X)/(X_{gi} \div X_g)$ and ratio $(M_i \div M)/(M_{gi} \div M_g)$ is close to 1 for each branch. As indexes a_i and b_i are calculated by means of logarithms, the hypothetical means calculated on the basis of the presented formulas should assume values close to zero. Furthermore, in conditions of intra-trade specialization advantage, one can expect that

$$\frac{1}{n}\sum_{i=1}^{n}(a_i - a)^2 = A_{ai}^2, \tag{1.26}$$

$$\frac{1}{n}\sum_{i=1}^{n}(b_i - b)^2 = B_{bi}^2, \tag{1.27}$$

i.e., that the variation of indexes a_i and b_i will oscillate at low levels. Finally, H. Glejser, K. Goossens, and M. Vanden Eede suggest that A^2 and B^2 should be calculated in accordance with the above formulas. In their opinion, the less the variation, the more significant is the intra-industry trade, and vice versa.

1.6 Summary and Conclusions

The focal point of theoretical studies of broadly understood international competitiveness of national economy is the economic category referred to as comparative (relative) advantages, i.e., the existence of comparative advantages or their absence in foreign trade of the given country and/or in international trade of a specified group of countries. The occurrence of these advantages and/or absence thereof is, in the static approach, a general and universal principle of rational management at an international scale. The basic reasons for the occurrence of comparative advantages and/or disadvantages are considered to be as follows:

a. Differences across countries in terms of technical and technological advancement;
b. Differences across countries in terms of endowment of basic production factors;
c. Differences across countries in terms of demand size and structure;

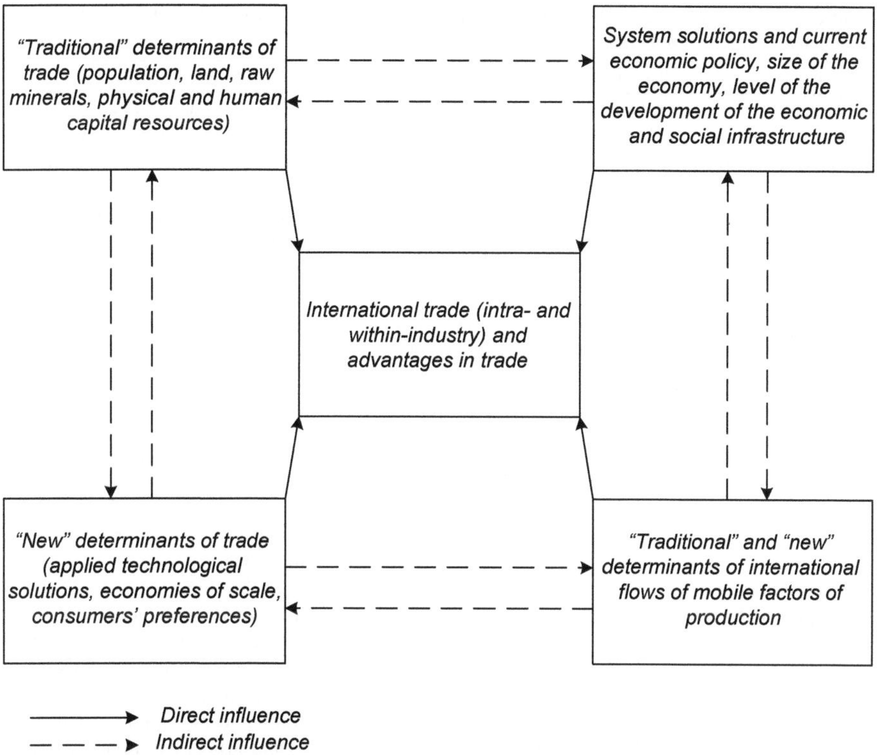

Fig. 1.10 Basic product features and functions and possibilities of preparing variants of demand, supply, and international trade structure. (Source: Own compilation)

d. Differences across countries in terms of economic size, absorptive capacity of internal markets, and possibilities of obtaining the so-called economies of scale in the area of production and sales;
e. Differences across countries in terms of broadly understood organization and structure of internal markets;
f. Differences across countries in terms of geographic location within the global economy, transport costs levels, costs of broadly understood communication between partners, etc.; and
g. Differences across countries in terms of broadly understood economic policy (its intermediate goals, instruments applied, etc.).

In each of these countries, it is possible to transform static comparative advantages (as well as their absence) into competitive advantages and/or create such new advantages (Fig. 1.10).

The main factors determining comparative and competitive advantages in foreign trade of the given country can be used to interpret the results of empirical analyses, as well as to program future developments, as well as the economic policy (Table 1.3).

Table 1.3 Types of factors determining comparative and competitive advantages of the given country according to selected criteria. (Source: Own compilation)

Criterion	Classification	Main factors and examples
1. Level of dependence on the analyzed country	1.1. Dependent factors (internal benefits)	Productivity understood as the relationship between results achieved and expenses incurred, effectiveness of domestic economy understood as the level of achievement of the goals set in real terms and in the institutional and instrumental area
	1.2. Independent factors (external advantages and disadvantages)	More or less favorable structure of external noneconomic conditions (e.g., political situations, assumed international obligations) and economic conditions (e.g., changes in the size and structure of import demand of partners or long-, medium- and short-term fluctuations of economic activity in the world)
2. Correlation with supply and demand	2.1. Supply factors	Level of development of economic infrastructure and level of its efficiency, effectiveness of utilizing basic production factors, and effects of using domestic and foreign technical knowledge
	2.2. Demand factors	Sizes and structure of domestic and foreign demand in dynamic terms
	2.3. Demand and supply factors	See points 2.1 and 2.2 Moreover, level of flexibility of home and foreign markets
3. Level of eclecticism		See points 2.1, 2.2, and 2.3
4. Correlation with real and institutional component	4.1. Factors that determine the real component	Efficacy of economic infrastructure and effectiveness of utilizing basic production factors, especially the so-called accumulated and created factors
	4.2. Factors that determine the institutional component (systemic competitiveness)	Transparency of rules and level of openness of the ownership system, efficacy and flexibility of the functioning of the economy and functional sub-systems headed by the broadly understood financial system, and applied macroeconomic stabilization policy and its effects

Table 1.3 (continued)

Criterion	Classification	Main factors and examples
5. Level of usefulness for economic politicians	5.1. Various divisions aimed mainly at providing basic information to economic politicians so that they are able to formulate appropriate economic development strategies	Level of macroeconomic stabilization, understood mainly as the effect of implemented fiscal and monetary policies, the pace and structure of technical progress, broadly understood effectiveness of the activity of basic business entities operating in the analyzed country

Strictly from a theoretical viewpoint, it is impossible to determine a priori the benefits from inter-industry labor division development in a precise manner. However, there is much in favor of the thesis that, when implementing the policy of transforming current comparative advantages into competitive advantages, it is particularly beneficial to prefer solutions conducive to maintaining international intra-industry trade labor division, especially in the case of vertical intra-trade. What is important here, among other things, is that the development of such a division of labor visibly stimulates inventiveness and innovativeness, and enables utilizing broadly understood economies of scale in the area of production and sales. These processes are coupled together, because—as the theoretical concepts presented above indicate—increased innovativeness and accumulation of human capital resources translate into the structure of competitive advantages in trade with foreign states in successive periods.

Chapter 2
Human Capital and Innovation—Basic Concepts, Measures, and Interdependencies

Ziemowit Czajkowski, Arkadiusz Michał Kowalski, Beata Michorowska and Marzenna Anna Weresa

With the acceleration of globalization processes, human capital stock and quality as well as invention and innovations have become particularly important. The greater mobility of capital on an international scale has led to a situation in which its allocation in a given country is increasingly determined by the quality of labor, in addition to such aspects as the price of labor, the tax system, and the quality of physical infrastructure and institutions. There are a few reasons for this. Production departments based on low-qualified labor force have been moved to areas where the labor costs are the lowest. The competition between the OECD (Organisation for Economic Co-operation and Development) countries and the dynamically developing countries, such as Chile, India, or Brazil, starts to concern the location of the technologically advanced production as well as allocation of research and development (R&D) works and R&D centers. The important elements of that competition are people, their knowledge, creativity, and the ability to convert these resources into innovation. For that reason, as the analysis of various theoretical approaches set out in the preceding section indicates, human capital and innovation are becoming increasingly critical for shaping the competitive advantage of countries and boosting benefits achieved in international exchange. The model approach to that kind of correlation as well as the empirical verification of its direction and strength requires supplementation of previous theoretical deliberations regarding the nature of international competitiveness with selected theoretical strands relating to human capital and innovation. This section is thus an attempt to integrate terms such as human

Z. Czajkowski (✉) · A. M. Kowalski (✉) · B. Michorowska · M. A. Weresa
World Economy Research Institute, Warsaw School of Economics,
Al. Niepodleglosci 162, 02-554 Warsaw, Poland
e-mail: zczajk@sgh.waw.pl

A. M. Kowalski
e-mail: arkadiusz.kowalski@sgh.waw.pl

B. Michorowska
e-mail: bz26456@sgh.waw.pl

M. A. Weresa
e-mail: mweres@sgh.waw.pl

M. A. Weresa (ed.), *Innovation, Human Capital and Trade Competitiveness*, Innovation, Technology, and Knowledge Management, DOI 10.1007/978-3-319-02072-3_2,

capital and innovation based on the concept of innovation systems, as well as to present and select measures adequate for analyzing these complex and overlapping phenomena.

2.1 The Term Human Capital

Ziemowit Czajkowski

The term *human capital*, even though introduced as early as in the seventeenth century by Petty (1690) and discussed, among others, by Smith (1776),[1] disappeared from the sphere of interest of economics, *inter alia*, after the critique by Marshall (1930), who held that "while human beings are incontestably capital from an abstract and mathematical point of view, it would be out of touch with the market place to treat them as capital in practical analyses."[2] The renaissance of research on human capital came in the 1960s, thanks to works by Mincer (1958), Schultz (1961), and Becker (1964).

It is also worth mentioning Machlup's studies on transformations of economies in highly developed countries, which resulted in the concept of a knowledge-based economy. This term is commonly used today but it had to wait until the 1990s to become popularized by Drucker (1992). The key to success in the knowledge-based economy is the human capital.

New theories on economic growth treat human capital as an important determinant of economic development. Such theories have been developed since the 1980s and started with the pioneer publications by Romer (1986), Lucas (1988), and other researchers (e.g., Jones and Manuelli 1990). An up-to-date review of research in the field of the theories about the growth and the role of human capital and innovation in that process can be found in studies by Stroombergen et al. (2002), Woessmann (2003), Le et al. (2005), Florczak (2007), Liberda and Maj (2009), and Weresa (2012).

The issue of valuation of people's skills and abilities has been covered by Adam Smith, Jean Baptiste Say, and many other economists over the last 400 years (Kiker 1966). The methods for estimation of human capital have evolved over the centuries, but its current definition as "the knowledge, skills, competences and other attributes embodied in individuals that are relevant to economic activity" (OECD 1998, p. 9) or "the knowledge, skills, competencies and attributes embodied in individuals that facilitate the creation of personal, social and economic well-being" (OECD 2001, p. 18) is not far from the definition by Adam Smith, who understood human capital as "all of the acquired and useful abilities of all of the inhabitants

[1] "A man educated at the expense of much labor and time to any of those employments which require extraordinary dexterity and skill, may be compared to one of those expensive machines. The work which he learns to perform, it must be expected, over and above the usual wages of common labor, will replace to him the whole expense of his education, with at least the ordinary profits of an equally valuable capital." (Smith 1776, p. 118).

[2] After Schultz (1961, p. 3).

of a country"[3] (Schultz 1961, p. 2). The above definitions make it impossible to directly measure the *stock* of human capital; the capital can only be estimated by means of measurable phenomena presented in statistics and proxies as well as in indexed combinations of those measures.

2.2 Methods of Human Capital Evaluation

Ziemowit Czajkowski

Although the availability of statistical data and development of mathematical methods had a significant impact on the way of estimating human capital, it is possible to distinguish two basic approaches to that issue. In one of them (the retrospective approach), the main focus is on establishing the *cost of production* of human capital which, depending on the study, may comprise living, health care, and education expenses or focus only on education spending. The originator of that method was Engel (1883). In the second (prospective) approach, the estimate of the human capital stock is based on the discounted stream of future income of workers in a given economy. The pioneer of that approach was Petty (1690). The approaches presented are well established in the theory of microeconomics. Unfortunately, due to the lack of relevant statistical data, it is not possible to apply those approaches in studies. The aforesaid problem has led to the development of the third approach related to the concept of the *cost of production* of human capital, the approach based on education level. Statistics on adult literacy rates, school enrollment rates, dropout rates, repetition rates, average years of schooling, and scores of competence tests (Program for International Student Assessment, PISA) were available for many countries and enabled both the comparative analyses (cross-sectional studies) and the analysis of trends and estimation of econometric models with the use of relevant time series. The last approach relied on the use of indexes, which were based on educational and demographic measures as well as detailed data from the labor market. The following subsections, based on Woessmann (2003) and Le et al. (2005), briefly present the main features of the approaches described:

1. Approach based on costs of production (retrospective approach),
2. Approach based on future income (prospective approach),
3. Approach based on education parameters (related to the retrospective approach),
4. Approaches using indexes (methods aggregating a number of variables) and *benchmarking* (competence tests, e.g., PISA, TIMSS, IALS).[4]

[3] "Adam Smith boldly included all of the acquired and useful abilities of all of the inhabitants of a country as a part of capital" (Schultz 1961, p. 2).

[4] Respectively: Program for International Student Assessment (PISA), International Adult Literacy Test (IALS) and Trends in International Mathematics and Science Study (TIMSS). Cf. further in this subsection.

2.2.1 Approach Based on Costs of Production

The originator of the approach based on production costs was Engel (1883). In this perspective, all expenses covering the cost of living and raising a human being from conception to the age of 25 should be treated as the *cost of production*. Engel proposed the following formula[5]:

$$c_{xi} = c_{0i} + xc_{0i} + \sum_{1}^{x} k_i c_{0i} = c_{0i}\left(1 + x + \frac{k_i x(x+1)}{2}\right) \quad (2.1)$$

where i is the social class (i = 1, 2, 3, respectively, for lower, middle, and upper class), x the age, $x < 26$, c_{0i} the costs incurred up to the point of birth, and $c_{0i} + k_i c_{0i}$ the annual cost of living.

From the viewpoint of the theory of investment, the weakness of the formula was that it ignored the time value of money. From the viewpoint of the theory of economics, it did not take into account the intangible investments crucial for the final value of human capital stock. Kendrick (1976) tried to tackle the said problem by assuming that tangible investments are the cost of living of a child up to the age of 14, while the intangible component comprises expenditures on health and safety, mobility, education, and opportunity costs of time spent on education.

The above approach, however, has significant flaws:

1. The cost of production of capital does not necessarily translate into its quality.
2. The value of capital is determined by the demand, not by the cost of production.
3. Components of expenditures and their prices are not recognized well enough to make their cost estimates useful, especially when the line between consumption and investment is not clear (e.g., Kendrick (1976) considered expenditures on necessities such as clothing and food to be investments, while Machlup (1962) regarded them as consumption).
4. The approach does not take into account the social costs of production of human capital (Dagum and Slottje 2000).
5. Variability of the appreciation/depreciation rate of human capital value (knowledge and skills become outdated in an abrupt manner as technology advances, but there is also the straight-line loss of competence due to forgetting)—there is no unequivocal position on that issue in the literature (see Kendrick 1976; Mincer 1958).
6. In macro-level studies, there is a problem of how to distinguish between the investments which have been completed and the investments which are still in progress—it is very difficult to isolate the part of spending on education which has already delivered graduates (Jorgenson and Fraumeni 1990).

[5] Quote after Kiker (1966, p. 483).

2.2.2 Income-Based Approach

The originator of a completely new scientific approach to measurements of human capital value was Farr (1853). He proposed to evaluate the human capital of a given individual by estimating the current value of future earnings net of living expenses and adjusted on the basis of a life table. That concept was further developed (Dublin and Lotka 1977).[6] The authors took into account also the employment rate (W. Farr's model assumed full employment conditions). According to their formula, the human capital of an individual at birth may be expressed as follows[7]:

$$V_0 = \sum_{x=0}^{\infty} \frac{P_{0,x}(E_x Y_x - C_x)}{(1+i)^x} \tag{2.2}$$

where i is the interest rate, $P_{0,x}$ is the probability (at birth) of living to age x, E_x is the employment rate at age x, Y_x is individual's annual earnings from age x to $x+1$, and C_x is the annual cost of living.

Formula (2.2) may be easily transformed to obtain the value of human capital at a given age a:

$$V_a = \frac{P_0}{P_a} \sum_{x=a}^{\infty} \frac{P_x(E_x Y_x - C_x)}{(1+i)^{x-a}} \tag{2.3}$$

Similarly, the total cost of living up to a given age a can be expressed as

$$C_a = \frac{1}{P_a} \sum_{x=a}^{a-1} \frac{P_x(C_x - E_x Y_x)}{(1+i)^{x-a}} \tag{2.4}$$

After combining formulas (2.2), (2.3), and (2.4), we have

$$C_a = V_a - \frac{(1+i)^a}{P_{0,a}} V_0 \tag{2.5}$$

Equation (2.5) may be interpreted in the following way: the cost of rearing a person up to a given age a equals the difference between his/her current value and the value at age a adjusted for the probability of living to age a.

Although the approach proposed by Dublin and Lotka was later developed and modified (*inter alia,* by the categorization of income based on education—cf. Mincer 1958), the above formulas provide sufficient discussion of the concept of prospective measurement of human capital.

One of the disputable areas is the approach to cost of living. Eisner (1989) considered subtracting the cost of living from income to be a mistake (which he justi-

[6] First edition in 1927.

[7] Quote after Kiker (1966, p. 484).

fied with the heterogeneity of the cost of living in a society and difficulty in distinguishing between investments in human capital and cost of living as well as of estimating the consumption of public goods). Other researchers (Graham and Webb 1979) held the view that net income (after deducting the cost of living) is a better estimate of the value of a given individual for others, while the total income better reflects the total social product (value for the society) of work done by a given person. Another problem considered in the literature was whether the prospective measurements should take into account the rise in real wages resulting from price adjustments which are an effect of economic growth (Houthakker 1959; Miller 1964). Characteristic features of the prospective approach include the following:

1. Valuation of human capital is expressed in market prices.
2. Valuation allows for the deprecation rate; hence, there is no need to determine it arbitrarily or to estimate it.
3. Data necessary to estimate human capital (life or mortality tables, wages, unemployment rate, and education structure) are available[8]; the only element which needs to be estimated is the term *structure of interest rates* (which can be traced, e.g., on the basis of treasury bond yield) used to discount income streams.

The main flaws of this approach are as follows:

1. The assumption that differences in wages truly reflect the differences as regards productivity (this is a critical assumption; if it is not met, the valuation model produces a wrong estimation of human capital stock; factors contributing to wages variation include, e.g., unemployment, which puts an employee in a worse negotiating position (wages do not keep up with productivity changes), local monopsony on the labor market, and redistribution of added value generated by lower-level workers for the benefit of employees at senior management levels of a company).
2. The measure is very sensitive to changes of the retirement age (change in number of financial flows) and of term *structure of interest rates* (change of discounting factors).

2.2.2.1 The Newest Studies Using Prospective Methods

The most important development of the method was proposed by Jorgenson and Fraumeni (1989, 1990). This model was used to estimate the human capital of the entire US population classified by the two sexes, 61 age groups, and 18 education types (0–17+ years of schooling) based on data collected from 2196 cohorts (Jorgenson and Fraumeni 1989). The basic model from the times of Engel was based on the assumption that a person at a given age x in n years will earn the same salary as the salary earned today by a person aged $x+n$ of the same sex and with the

[8] Data on wages in developing countries are very often unavailable or may contain an error in estimation. Until the mid-twentieth century, the data on wages were not a part of statistics of many developed countries.

same education. The proposed change involved the simplification of the discounting procedure of future income streams to the present value. The authors adopted a recursive formula in which the current value of the lifetime income stream ($L_y(s, a, e)$) obtained by an individual at a given age is the income from the current year plus the total lifetime income computed for the following year ($L_{y+1}(s, a+1, e)$), weighted by survival probabilities ($S_y(s, a+1)$).

Jorgenson and Fraumeni identified five stages of the human life cycle (Fraumeni 2011):

- Preschool: no school and no work (age 0–4),
- School and preproduction: school but no work (age 5–13),
- School and production: school and work (age 14–34),
- Production: work but not school (age 35–74),
- Postproduction: no school or work (age 75+).

By assumption, the discounted income for the last age group is zero. The formalized[9] formula for the production group is as follows:

$$L_y(s,a,e) = Y_y(s,a,e) + S_y(s,a+1)L_{y+1}(s,a+1,e)\frac{(1+g_y)}{(1+i_y)} \tag{2.6}$$

where s is the sex, a the age, e the education, and y the year. $Y_y(s,a,e)$ is the average income of a person of sex s at a given age a, with e education observed in year y, calculated as the product of average hourly rate and average number of hours worked, increased by *non-market* income, the estimation of the value of *non-commercial* activity (e.g., entertainment, sports, and self-education); $L_y(s,a+1,e)$ is the discounted per income stream of year y of a person of sex s at a given age a with education e; $S_y(s,a+1)$ is the probability of living up to age $a+1$ for a person of sex s at a given age a, observed in year y; g_y is the real growth rate of remuneration; and i_y is the real discount rate.

Another important contribution made by Jorgenson and Fraumeni was to include the potential value generated by individuals who receive formal education at a given moment (age category 14–34). Allowance made for the years of schooling affects the lifetime income of individuals from the second and the third group. The correct formula is then

$$\begin{aligned} L_y(s,a,e) = Y_y(s,a,e) + [&E_y(s,a,e)S_y(s,a+1)L_{y+1}(s,a+1,e+1) \\ &+ (1-E_y(s,a,e))S_y(s,a+1)L_{y+1}(s,a+1,e)]\frac{(1+g_y)}{(1+i_y)} \end{aligned} \tag{2.7}$$

where $E_y(s, a, e)$ is the school enrollment rate for a given set (s, a, e) in a given year.

The most serious objection to the approach proposed by Jorgenson and Fraumeni comes from the work of Dagum and Slottje (2000)—the approach does not take

[9] Designation changed. Full description of the method for all age groups is contained in Fraumeni, (2011, p. 3–5).

into consideration the differences in capabilities as well as the impact of gender and of environmental and social conditions on productivity achieved as a result of education. Nonformal schooling is not considered, nor are the differences between various types/fields of study. Another objection concerned the rough approximation of productivity during the time of schooling. Yet another objection was that the employment rate was not considered: the stock of human capital computed using formulas (2.6) and (2.7) will not change whether the employment rate is 100 % or, e.g., 50 % (Conrad 1992).

2.2.2.2 Index Measures Based on the Prospective Approach

The concept of index was proposed by Mulligan and Sala-i-Martin (1995). It involved the approximation of human capital using an index obtained by dividing the average salary in economy by the average salary of an unqualified worker.[10] The reason for the adoption of such a method was that the salary incorporates not only the skills and abilities of a given worker but also the physical capital available to him or her. This means that unqualified workers tend to receive higher wage rates in countries providing better technical conditions (due to differences in the productivity of unqualified labor depending on the stock of production capital used per one worker). It would lead to wrong estimation of the stock of human capital. Another reason for the adoption of the aforesaid method was the differentiation between the quality and material scope of education with regard to both time and location. When combined with the first factor, the cross-sectional studies for the same level of education may give different wage levels. Mulligan and Sala-i-Martin (1995) propose a method to reduce the impact of technical conditions of unqualified labor on the human capital index in the following way:

$$h_i(t) = \frac{\int_o^\infty w_i(t,s)\eta_i(t,s)ds}{w_i(t,0)} \tag{2.8}$$

where $w_i(t,s)$ is the wage of a worker after s years of schooling in state i, $\eta_i(t,s)$ is the percentage of population receiving education for s years in state i, and $w_i(t,0)$ is the wage of an uneducated worker in state i.

In practice, depending on time series available, discrete forms of formula (2.8) are used (cf. Florczak 2007, p. 120):

$$h_i = \sum_e \frac{w_{i,e}L_{i,e}}{w_{i,0}}, \tag{2.9}$$

[10] Hence, it is not a standardized 'non-*numéraire*' index; the *numéraire* is a wage of an unqualified worker.

where $w_{i,e}$ is the wage of a worker with level of education e in state i, $w_{i,0}$ is the average wage of a worker with minimum level of education (basic or none education) in state i, and $L_{i,e}$ is the percentage of labor force with e education in state i.

The advantages of this approach include:

1. Elimination of the impact of the accumulated physical capital on higher productivity of unqualified labor,
2. The possibility of capturing the impact of years of schooling (but not of its quality) on the wage variability,
3. Allowing for interstate differences in the stock of the human capital of individuals with the same years of schooling.

The approach proposed by the authors is not free from serious drawbacks (Wachtel 1997):

1. It relies on the assumption that an unqualified worker has the same stock of human capital irrespective of the country and time of observations (it does not apply to the modification with formula (2.9)).
2. The identical number of years of schooling in different countries and in different periods does not result in the identical growth of human capital; using the wage of an unqualified[11] worker (with no educational at all) in the denominator may be insufficient to obtain an index that can be compared among different countries.
3. It does not take into account the differences in wages that are not the result of a different value (marginal productivity) of human capital (the wage structure does not necessarily correspond to the structure of value added generated).
4. It ignores the contribution to human capital by factors other than formal schooling (*inter alia,* nonformal schooling, learning-by-doing, and vocational trainings, as well as health, which is inconsistent with models proposed, e.g., in Mincer (1974)).

The method proposed by Jorgenson and Fraumeni (1990) was further developed by Jeong (2002), who used the average wage in industry as the denominator in the formula. The change was justified with the observation that the nature of work in that sector is least diversified across countries. Individuals with no education constitute too small a percentage of the population in developed countries to use data for that group as a basis for international comparisons.

2.2.3 Approach Based on Educational Parameters

Studies using an approach in which educational parameters are applied to estimate human capital are based on the following variables:

[11] The wage of an unqualified worker is estimated with the use of a power function with an exponent which is the constant from Mincer's wage regression, calculated separately for each country in each year of the studied period of time.

1. Adult literacy rates,
2. School enrollment rate,
3. Average years of schooling, including the following adjustments:
   a. Dropout rates,
   b. Repetition rates,
4. Competence test scores (e.g., PISA[12]).

The above indicators are used as indirect measures of educational investments which are considered to be at the center of human capital formation. At the beginning, the measures referred to above were used only as variables stratifying the population in research on market effectiveness of educational investments (cf. e.g., Mincer 1958).

Estimation of the stock of human capital using educational indicators is a convenient method considering the availability of statistical data; however, it does capture the most important aspect from the viewpoint of economics: how productively is the generated capital used and to what extent do the outlays translate into the real growth of human capital (effectiveness and quality of education).[13]

2.2.3.1 Adult Literacy Rates

The adult literacy rate is defined as the percentage of the population aged 15+ who can "read and write a simple statement on his or her everyday life."[14] The indicator was used as a control variable in early studies on growth, which took into account the human capital factor (e.g., Romer 1989; Azariadis and Drazen 1990).

The main flaws of the indicator include the following:

1. The definition of adult literacy rate is not consistently applied, which results in discrepancies in international statistics, and the data are difficult to compare.
2. It misses out the *advanced* skills being the key components of human capital (from the point of view of their impact on workers' productivity) in developed countries where education is universal.

2.2.3.2 School Enrollment Rates

The school enrollment rate is defined as[15]:

---

[12] http://www.pisa.oecd.org/

[13] See Woessmann (2003), p. 243. By definition, the quality of education is taken into account in the approach based on tests' scores.

[14] Definition after UNESCO (1993).

[15] Depending on the country and customary or legal regulations, 'schooling obligation' might be only a matter of customary habits and might be not supported by any legal sanction.

$$\text{School enrollment rate}_i = \frac{\text{Number of students enrolled at a given level } i}{\text{Population of the age group subject to compulsory education at a given level } i} \qquad (2.10)$$

The indicator was used as an estimation of human capital in growth models, *inter alia* in studies (Barro 1991; Levine and Renelt 1992; Mankiw et al. 1992; Gemmell 1996).

The indicator has certain flaws, such as:

1. It is an estimation of investments in human capital being done at a given time (and of future growth of the stock of human capital) and not an estimation of the stock of human capital available for productive applications in national economy nor an estimation of current changes of human capital.
2. Investments implemented at a given time do not necessarily translate into a growth of accumulated human capital in the future (graduates may not become a part of the labor force in a given national economy).
3. Changes in the stock of human capital are the result of a balance between qualifications of people who exit and who enter the labor force; hence, school enrollment rates do not take into account the full scope of changes in human capital.
4. If the indicator is used in studies on economic growth, there might be an important problem related to reverse causality: a high level of school enrollment rates, according to studies (Wolff 2000; Judson 2002), is the result of the increased productivity and not its cause.

2.2.3.3 Average Years of Schooling

Measuring the stock of human capital using the average years of schooling requires an assumption (cf. Wachtel 1997) that there is a relation between the years of schooling and the formation of human capital. The main flaw of this method is that the relevant time series are not available: the average years of schooling are not recorded and thus must be computed using one of the following:

1. Census (/survey) estimation,
2. Extrapolation,
3. Perpetual inventory method.

The next part of this subsection discusses the main drawback of this method.

Average years of schooling: the survey/census-based estimation

The pioneer work on estimation of average years of schooling was the publication by Psacharopoulos and Arriagada (1986),[16] where calculations were made according to the following formula (cf. Nehru et al. 1993, p. 2):

$$L = \sum_i l_i S_i \tag{2.11}$$

where l_i is the percentage of labor force with i level of education, S_i is the years of schooling typical of i level of education, and i is one of the following: no education, incomplete primary education, primary education, incomplete secondary education, secondary education, and tertiary education (no category of *incomplete tertiary education*).

The main advantage of the above approach was the availability of the time series l_i for 66 countries (and the possibility of calculating the corresponding values for 33 other countries on the basis of other time series). The problems associated with applications of this method included:

1. Different number of years typical of a given education level across countries,
2. No information on the typical number of years of schooling for *incomplete primary* and *incomplete secondary* education (Nehru et al. 1993) as well as different definitions of the term *labor force* across countries (l),
3. Omission of trainings at the workplace and vocational trainings (Kyriacou 1991), and
4. Neglecting the issue of the quality of education (Kyriacou 1991).

When discussing the methodology proposed by Psacharopoulos and Arriagada (1986), it is important to mention its best known and most commonly applied development being prepared since 1993 by Barro and Lee (1993, 1996, 2000, 2010). In the most recent studies (Barro and Lee 2010), the estimations of which were published in the database of the World Bank,[17] formula (2.10) was modified in the following way:

$$s_t = \sum_{a=1}^{A} l_t^g s_t^a, \tag{2.12}$$

$$s_t^a = \sum_j h_{j,t}^a Dur_{j,t}^a, \tag{2.13}$$

[16] Cf. also Psacharopoulos and Arriagada (1992), where authors repeated the studies using updated data.

[17] http://databank.worldbank.org/ddp/home.do, Education Statistics base, series with the prefix "Barro-Lee."

where t is the year, l_t^g the percentage of g category in the population aged 15+, observed in year t, a the age group (indexed ranges), a∈ {(1; 15–19), (2; 20–24), …, (13; 75+)}, s_t^a the years of schooling at age category a, $h_{j,t}^a$ the percentage of the group g that completed education at the level j ∈ {pri, sec, ter}, and $Dur_{j,t}^a$ the length of schooling at the level j in years.

The basic difference in relation to formula (2.10) was the use of data from life tables to adjust the data on school enrollment rates in given years. The adjustment was made using the variable $h_{j,t}^a$ (cf. Barro and Lee 1993, p. 14 and next).

Average years of schooling: the extrapolation method

The method was proposed in the work by Kyriacou (1991) in order to fill in the gaps in the time series for L [18]. It relied on the assumption that the parameters α_i in model (2.13) are stable across time and do not vary across countries:

$$H75 = \alpha_0 + \alpha_1 PRIM60 + \alpha_2 SEC70 + \alpha_3 HIGH70 \tag{2.14}$$

where *H*75 denotes the average years of schooling computed in Psacharopoulos and Arriagada (1986) for 1975, *PRIM*60 is the enrollment ratio for primary education in 1960, *SEC*70 is the enrollment ratio for secondary education in 1970, and *HIGH*70 is the enrollment ratio for tertiary education in 1970.

The model required an additional assumption that there are no variations as regards the length of schooling at a given schooling level and the non-promotion and dropout rate. The above method of estimating human capital allowed expanding the data set computed in Psacharopoulos and Arriagada (1986) to include the total of 133 countries in the years 1965, 1970, 1980, and 1985. The results obtained were of limited predictive value due to the fact that the assumption as regards the stability of α_i parameters was not satisfied.

Average years of schooling: the perpetual inventory method

The perpetual inventory method to estimate the stock of human capital based on data on school enrollment rates was used for the first time in studies by Lau et al. (1991)[19] and Lau et al. (1991). The authors defined the *educational capital stock* as the total number of person-school years of the working-age population. The method consisted of summing up the number of enrollments for each age group at all education levels and adjusting the obtained result for each year for the survival probability based on life tables. In studies (e.g., Nehru et al. 1993), the formula of perpetual inventory was adjusted for data on nonpromotion and dropout rates. Having available the number of enrollees in each year and at each education level P, we

[18] Kyraciou denotes L as $\overline{S}(MEAN)$.

[19] Quote after Nehru et al. (1993).

may develop the following index for an education level taking *k* years of schooling by indexing the current grade with *g*:

$$S_{P,T} = \sum_{T-64+6}^{T-15+6} \sum_{g=1}^{k} E^{*}_{P,g,T-g-1}, \tag{2.15}$$

where

$$E^{*}_{P,g,t} = E_{P,g,t}(1 - r_{P,g,t} - d_{P,g,t}), \tag{2.16}$$

with $E_{P,g,t}$ being the school enrollment rate *gross* in grade *g* at a given education level *P*, observed in year *t*, $r_{P,g,t}$ the repetition rate[20] in grade *g* at a given education level *P*, observed in year *t*, $d_{P,g,t}$ the percentage of permanent dropouts in grade *g* at level *P*, observed in year *t*, and $E^{*}_{P,g,t}$ being the school enrollment rate *net*.

The index (2.14) does not take into account the depreciation of the stock due to mortality. The modified version of the index would need to include an additional factor, $\theta_{P,g,T}$, the probability of living to year *T* by a person who completed *g* years of schooling at a given education level *P*. In the case when formal education at a given level *P* at age results from the data on schooling obligation in a given country, it is possible to use the life tables. In the case of the tertiary level of education, such an approach introduces a systematic error, which cannot be avoided due to the lack of relevant data.[21] The adjusted index, $S^{\theta}_{P,T}$, would be as follows:

$$S^{\theta}_{P,T} = \sum_{T-64+6}^{T-15+6} \sum_{g=1}^{k} \theta_{P,g,T} E^{*}_{P,g,T-g-1}. \tag{2.17}$$

The use of this index poses very serious problems. Due to gaps in the time series and the lack of data going back far enough, the results obtained with the use of this method were, to a large extent, based on time series extrapolation and were affected by error, although more complete real data (going back to 1930, and for some countries to 1902) were used in studies (e.g., Nehru et al. 1993).

To sum up, despite the substantial advantage over the previous education-based measures, the average years of schooling as a measure of the stock of human capital is liable to errors from a variety of causes, including both the simplifications made and unavoidable errors resulting from extrapolation as well as contradictions between assumptions on which the measure itself is based and empirical observations of economic processes.

[20] Problems connected with records and calculation of this share are discussed in further detail in Nehru et al. (1993, p. 5).

[21] All quoted studies assumed the continuity of education and mortality from the general population θ_a,characteristic of age *a* at which—in accordance with the law or social norms—an individual should attend grade *g* at level *P*.

1. If *average years of schooling* was to be the measure of the human capital stock, and not an estimation of outlays, then it should take into account the variations of the human capital growth rate for different grades and education levels. The first year of schooling, as described in the above methods, raises human capital by the same amount as the first year of study at a polytechnic. In the literature; however, diminishing returns to education measured with income were observed (Mincer 1958, 1974; Freeman 1981; Psacharopoulos 1994), or even the impact of education was negated (Arrow 1973).
2. The assumption that the quality of education (and, as a consequence, the amount of human capital stock generated during 1 year) is homogeneous across time and countries is at odds with empirical data (e.g., scores of the Program for International Student Assessment (PISA)[22], International Adult Literacy Test (IALS)[23], Trends in International Mathematics and Science Study (TIMSS)[24]); there were attempts to make relevant adjustments in further studies on this issue (Gundlach 1995).
3. There is no differentiation between various education categories; the approach assumes full mobility among different professions of graduates from any field of study, hence the homogeneity of human capital. No such phenomenon was observed (Jorgenson 1995a, b).
4. It ignores all investments on human capital formation other than those related to formal education (vocational trainings, courses, learning by doing process, and health investments).

2.2.3.4 Competence Test Scores

Another education-based method to estimate the quality of human capital involves international competence tests. They include both literacy tests and the mathematics and science tests. The most important international institutions that carry out periodic tests of competence and skills of employees and students are the OECD and the National Center for Education Statistics in the U.S. (NCES,[25] the major governmental center that collects and analyzes data related to education).

The problem of the quality of schooling (and, consequently, the quality of human capital stock) was initially ignored in the analysis of the stock of human capital. In the growth models where the stock of human capital was estimated based on average years of schooling (e.g., Barro and Lee 1993, 1996, 2000, 2010), an important problem was that the quality of education varied across countries. The problem was approached using, *inter alia,* such measures as educational spending per student (Barro and Lee 1996) and student–teacher ratio (Barro 1991). In studies (e.g., Barro and Lee 2000), it was proposed that international competence tests (IALS

[22] http://www.pisa.oecd.org/

[23] http://www.statcan.gc.ca/bsolc/olc-cel/olc-cel?catno=89-588-X

[24] http://nces.ed.gov/timss/

[25] *The National Center for Education Statistics*, a unit of the U.S. Education Department.

or TIMSS) should be used as the measure of the quality of human capital, while highlighting the significant discrepancies between the results of both tests. In other works on the same issue (Hanushek and Kimko 2000), the indexes used were based on data from six studies carried out over 30 years by IEA[26] (four studies) and IAEP[27] (two studies), the aim of which was to take into account all information coming from available tests.

Competence tests such as PISA, Progress in International Reading Literacy Study (PIRLS), and TIMSS as potential measures of the human capital quality have one significant drawback: they provide information on future workers but not on the human capital available now in a given economy.

2.2.4 Approach Based on Indexes and the Benchmarking Method

Discussed below are selected indicators (simple or complex) that have no numéraire or no direct designations in the theory of economy, which limits their interpretability and cognitive value. The key features of this approach include:

1. standardization of the observed variable (combined with the loss of numéraire and economic interpretation), the aim of which is to obtain the predetermined variation range and comparability of a given index as well as to make it possible to organize the observed objects into categories based on the calculated value of the index and
2. in the case of aggregation of a number of variables, multiplying the vector of the standardized variables by the vector of weights adopted by experts or obtained using a selected statistical method (e.g., principal component method); the approach, however, results in distortion of the observed variability of the data, which lost their economic interpretability as early as at the first stage.

Due to the above properties, the index-based approach does not provide an answer to the question of how the socioeconomic parameter varies in reality. Information about a better score of a given country does not indicate how that score is better or what specifically is better.

Despite the problems mentioned above, there are also a number of arguments in favor of the use of indexes. Messinis and Ahmed (2009) point out the following reasons:

[26] International Association for the Evaluation of Educational Achievement, http://www.iea.nl. Assessments of mathematical competences were conducted in the years 1963–1964 (11 countries) and 1980–1982 (17 countries). Assessments of competences in science were conducted in 1966–1973 (17 countries) and 1983–1986 (23 countries). Cf., e.g., Hanushek and Kim (1995).

[27] International Assessment of Educational Progress, an organization that conducts international competence tests based on the NAEP methodology.

1. heterogeneity and time-varying returns to education (Psacharopoulos and Patrinos 2002; Hartog and Oosterbeek 2007),
2. noncognitive skills (Heckman et al. 2006; Flossmann et al. 2007),
3. obsolescence of skills (Alders 2005; Görlich and de Grip 2007; Pfeiffer and Reuß 2008), and
4. skill–job mismatch (Cheng and Ghulam 2007; Korpi and Tåhlin 2009).

Furthermore, the latent-factor-estimation approach is an effective strategy in dealing with measurement errors and endogeneity (Temple 1999; Durlauf et al. 2005; Heckman et al. 2006).

The following subsections present some of the most important index-based methods (Woessmann 2000; Messinis and Ahmed 2009; Ederer 2006; Ederer et al. 2007; Anger 2007). Detailed discussion of the Lisbon Council's *Human Capital Index* (Ederer 2006; Ederer et al. 2007) will not be possible due to the fact the calculation methodology used has not been made publicly available.[28]

2.2.4.1 Index Based on Educational Quality Measures

Hanushek and Kimko (2000), based on the studies confirming the importance of international test scores in mathematics and science in determining the productivity and income (Bishop 1992), developed two indexes to measure human capital (Hanushek and Kimko 2000, p. 1187). The first one, QL1, was obtained by multiplicative transformation of series of observations so as to achieve the global mean of 50. It should be noted that the method relied on the assumption that the intertemporal mean in performance is constant across all countries and that the set of countries is a random sample from a population with distribution stable in time. The second index, QL2, set aside the assumption that the intertemporal mean in performance is constant. Additional data collected by the National Assessment of Educational Progress (NAEP) were utilized. It was assumed that the change in the average global level of performance for international tests corresponds to the trend observed in NAEP tests for USA. The measure of quality of schooling for each country was the weighted average of all tests available for a given country. The weights used were the normalized inverse of the country-specific standard error for each test.

QL1 and QL2 indexes—as well as the studies on which they are based—are not a precise measure of schooling quality as they miss out the significant impact of innate abilities. This flaw, however, is not that important if the object of comparison is the quality of human capital. Another issue is how to use observations about students' competence level to infer about competences of human capital available in economy; in order to do that additional assumptions are required (e.g., regarding consistent quality of schooling over time).

[28] It was not possible to obtain the technical documentation of the methodology from the authors.

Indexes proposed in the work by Hanushek and Kimko (2000) were used to develop an intensive[29] measure of human capital stock based on Mincer's (cf. Mincer 1974, p. 8–23) return to education function. In accordance with previous studies (Bils and Klenow 2000, p. 1161) Wößmann (2000) assumed that

$$H^M = e^{\phi(s)}L \Leftrightarrow \frac{H^M}{L} \equiv h^M = e^{\phi(s)}, \tag{2.18}$$

where

$\phi(s)$ is the relative effectiveness of labor unit with s years of schooling versus labor unit with no education, r the rate of return to education, $\frac{d\phi(s)}{ds} = r$ the condition imposed by Mincer: the derivative of the function $\phi(s)$ must be equal to the rate of return to education r,

H^M the human capital stock in economy according to Mincer,

h^M the human capital stock per one worker, and

L the working population in a given economy.

The education quality index was introduced by Wößmann into the $\phi(s)$ function in the following manner:

$$\phi(s) = \sum_a r_a Q_i s_{ai}, \tag{2.19}$$

where r_a is the world average rate of return to education at level a, series after Psacharopoulos (1994), Q_i is $QL1_i/QL1_{USA}$ or $QL2_i/QL2_{USA}$, where i is the index for a given country, and s_{ai} is average years of schooling at level a in country i, series after Barro and Lee (2000).

When incorporating (2.19) into (2.17), we obtain

$$H_i^M = \exp\left(\sum_a r_a Q_i s_{ai}\right) L_i \Leftrightarrow h_i^M = \exp\left(\sum_a r_a Q_i s_{ai}\right). \tag{2.20}$$

The advantages of measure (2.19) include the following:

1. There is no maximum limit to the quality of human capital.
2. There is a maximum limit to the amount of human capital that can be accumulated through education (s_{ai}).
3. The measure captures a number of aspects of human capital in one single variable.
4. It has economic interpretability.

An important drawback of this measure is that it is strongly dependent on the availability and quality of the input data, the result of which is the lack of possibility of performing the required calculations or the low reliability of results obtained.

[29] Per capita.

2.2.4.2 Index Based on the Latent Factor

This method is a pioneer approach to human capital measurement. Its authors (Messinis and Ahmed 2009) followed the example of the concept developed for measurement of patent quality (Lanjouw and Schankerman 2004). They captured in a single index both the quality of education and the market valuation of skills. In devising the latent variable index, they used the common factor analysis:

$$I_{k,\,jt} = \mu_k + \lambda_k h_{jt}^S + e_{k,\,jt}, \tag{2.21}$$

where $I_{k,jt}$ is the natural logarithm of k indicator for j country in year t, μ_k is the average value of k indicator, λ_k is thecommon factor loading for k indicator, h_{jt}^S is the common factor, and $e_{k,jt}$ is the idiosyncratic error.

The common factor h_{jt}^S is an unobserved characteristic of human capital in a given country. The following indicators[30] were used for the calculation:

- The TIMSS test score delayed by two periods,
- Scientific publications in mathematics and science per capita,
- Book publications in the field of mathematics and science (theoretical works and applied science) per capita,
- Equipment (value of machines and devices) per capita.

The obtained index (h_{jt}^S) was compared with other measures of human capital stock used in the literature on the theory of growth: average years of schooling, TIMSS test scores (and TIMSS delayed by 2 years), and intelligence quotient (IQ) index defined in Lynn and Vanhanen (2002). According to the results obtained by the authors, the new index seems to be more closely linked to both gross domestic product (GDP) dynamics and GDP level than the indexes used so far.

2.2.4.3 The Lisbon Council Human Capital Index

This index can be found in a number of studies (Ederer 2006; Ederer et al. 2007). Although the calculation methodology was not disclosed, it is worth presenting this index because the authors take account of aspects which are not captured in the traditional approach to human capital analysis. The authors adopt a retrospective approach to the measurement of the stock of human capital. The costs of its production comprise:

1. Parental education: speech, social competences, and system of values, expressed as the opportunity cost of lost earnings of a parent,
2. Preschool and school education, expressed in salaries of tutors and spending on educational facilities,
3. Tertiary education, expressed in costs incurred (tuition fees or wages of academic teachers as well as spending on educational facilities),

[30] The full list of indicators is contained in the supplement A to the study (Messinis, Ahmed 2009).

4. Formal and informal adult education of adults (vocational trainings, courses, and self-education in the spare time), measured at opportunity costs of lost earnings,
5. Learning by doing, measured at opportunity costs.

Categories 1 and 5 seem impossible to estimate precisely enough to give reliable results when used in international comparisons.

The authors distinguish four areas of human capital analysis:

1. Human capital endowment measures the cost of all types of education and training per one employee (in accordance with the above list), taking depreciation into account.
2. Human capital utilization evaluates how much of a country's human capital stock is actually deployed (this indicator differs from the employment ratio in that there is *human capital* of employed people in the numerator and the entire human capital of a country's population in the denominator).
3. Human capital productivity measures the relation of GDP to the human capital employed in the economy.
4. Demography and employment—the index of the forecasted number of people employed in the national economy of a given country in year 2030,[31] taking into consideration the demographic and migration trends observed.

Due to the lack of any data on the method of calculating the index, it is not possible to discuss it in more detail. It should be noted, however, that the areas 2, 3, and 4 have not been taken into account in the human capital measures, as they concern the interaction between other variables and human capital. Nevertheless, the impacts are especially important for the analysis of efficiency of the economic policy, and for preparing the development strategy.

2.2.5 Human Capital Measurement: Summary

The measurement of human capital, despite the availability of appropriate theoretical models, presents a serious problem due to the unavailability of sufficiently detailed data, the lack of records on specific areas of economic activity at state statistical offices, or too short time series. The insolubleness of the problem forced researchers to use indirect measures of variable theoretical and statistical quality. The indirect measures of human capital presented in the text are the effect of previously described problems and of gaps in time series and methodology which is not internationally heterogeneous. The said problems concern more the cost-based approach.

The most common approach from among those analyzed above is the set of measures proposed by Barro and Lee (2010), which has been systematically updated and published by the World Bank.[32] The said approach—as indicated above—is not free from drawbacks. Another method is the approach proposed by Jorgenson and

[31] In 2035 (Ederer et al. 2007).

[32] http://databank.worldbank.org/ddp/home.do, Education Statistics base, series with the prefix "Barro-Lee".

Fraumeni (1989), which was adopted as a statistical standard by OECD (Fraumeni 2011) in the framework of "The OECD Human Capital Project" (Mira and Liu 2010).

The analysis of human capital impact on shaping long-term competitive advantage cannot be based on one measure only. It seems that studies should incorporate a set of measures (The Lisbon Council Human Capital Index). When assessing human endowment, the most convenient method will be to use the measure, average years of schooling, developed by Barro and Lee or to make one's own calculations using formula (2.9). For the purpose of the analysis of human capital utilization, the following set of indexes might be used: employment rate and unemployment rate with special focus on individuals at the ages of 15–30 years. Human capital productivity can be measured by labor productivity (GDP per hours worked in a given economy). Another aspect of comparative studies should also be the forecasted number of the working population in 2030, which is possible to estimate with the use of population forecasts, especially taking into consideration the fact that the aging of societies and shrinking of the working-age population are important problems facing most of the European countries.

2.3 Innovativeness: Basic Definitions and Economic Interpretation

Arkadiusz Michał Kowalski, Beata Michorowska

According to a number of economic theories, including the new theory of growth, the new economic geography, or the concept of knowledge-based economy, innovativeness is one of the key factors of competitiveness.[33] The aim of this subsection is to present the essence of innovation and innovativeness of economy in the context of the economic dimension of these concepts.

The basic term that should be a starting point in analyses of innovativeness of economy is innovation. Innovation is the implementation—in business practices, in workplace organization, or in the firm's external relations—of a new or significantly improved (OECD/Eurostat 2008, p. 48) product, process, marketing method, and new organizational method, which are discussed as follows:

- Product (a good or service) (*product innovation*): changes in the functional characteristics or intended use of a product;
- Process (*process innovation*): changes in the production or delivery method in terms of techniques, equipment, and/or software intended to decrease unit costs, to increase quality, or to produce or deliver new or significantly improved products;
- Marketing method (*marketing innovation*): changes in product design or packaging, product placement, and product promotion or pricing aimed at better addressing customer needs, opening up new markets, or newly positioning a firm's product on the market, with the objective of increasing the firm's sales;

[33] Cf. theories analyzed in Sect. 1.

- New organizational method (*organizational innovation*): changes in workplace organization or in the firm's external relations intended to increase a firm's performance by reducing administrative costs or transaction costs.

Adoption of the above definition means that the minimum requirement for innovation is that the product, process, marketing method, or organizational method must be new (or at least significantly improved) to the firm. Innovation may therefore consist of the implementation of a single significant change or of a series of smaller incremental changes that together constitute a significant change. A common feature of all kinds of innovation is that they have been implemented.

By definition, all innovations contain a degree of novelty. There are three levels of the novelty of innovations: new to the firm, new to the market, and new to the world.

Innovations are typically considered as an element of characteristics of an enterprise. It is possible, however, to perceive innovativeness from the macroeconomic perspective. Innovativeness of economy is usually defined as "the ability of a given economy to create innovation, where *ex ante* is the potential to create innovations and *ex post* is the total effect of innovative activities of enterprises functioning in the economy in a given period of time" (Weresa 2002, p. 14). The implications of this definition are the two concepts: *capacity to innovate* and *innovative position.* The first term denotes the extent to which a given country is capable of creating and commercializing new ideas, while the second means the effect of the creativity of a nation combined with financial resources in a given economic and institutional environment (Weresa 2012, p. 32). The prevailing opinion in the economic literature is that the innovativeness of economies is an evolving process, where the ability to create and introduce changes is the function of knowledge and experiences gained earlier (Soete 1981; Usher 1982; Weresa 2002, 2012).

When looking for interdependencies between innovativeness and gaining long-term competitive advantages in international trade, one cannot forget about the relatively new phenomenon in global economy, globalization of innovative activities, referred to as the techno-globalism. It includes activities aimed at creating new ideas, their application, and international diffusion (Zorska 1998). Globalization of innovative activities is considered to be one of the recent phenomena in the global economy, having impact on the international competitiveness of countries, regions, industries, and enterprises.[34]

2.4 Methods for Measuring Innovation

Arkadiusz Michał Kowalski, Beata Michorowska

[34] For more on this subject, see Weresa (2012).

2.4.1 New Innovation Indexes in the Context of Evolution of Innovation Process Models

Until the mid-twentieth century, the issue of measuring innovativeness was not broadly addressed and there were multiple reasons for that, both subjective and objective. The objective reasons were associated with the fact that, in the past, no clear linkages were perceived between scientific and inventive activities as well as between inventiveness and production. The subjective reasons were that there was no research methodology formed, which is required for the analysis of the dynamic process of scientific and technical progress. The models of economic growth created until that time assumed nonmeasurability of this factor or it was considered to be exogenous (Fiedor 1979, p. 21).

The evolution of methods to measure innovativeness and of the indexes used for that purpose was determined by the changing nature of innovation, finding its reflection in subsequent generations of innovative process models. In the past, according to the linear model presented by Schumpeter, the process of development of innovation happened in a linear fashion, starting from R&D works, going through successive stages, up to the practical application of knowledge in business activity. The model prioritizes research and scientific activity, not paying that much attention to the stage of commercialization of results of R&D works. Now, it is widely recognized that this stage is the *bottleneck* of innovative processes. Therefore, while in the 1960s the basic indicator used to measure innovativeness was the level of spending on R&D, at present equally important elements are the measures of the economic effects of innovation, such as export of mid-tech and high-tech industrial products, sale of innovation for the market and for the firm, as well as income from patents and licenses from abroad. The views on the factors determining innovations in economy evolved too. In accordance with assumptions of the modern model of innovation processes, innovations are treated as the product of interactions between people and organizations and their environment. In this concept, the basic element that contributes to an innovation process is no longer R&D activity, which is the source of the so-called technological push, but the opposite phenomenon of the innovation pull, i.e., focusing R&D activities on specific fields in order to meet market requirements.

In modern business activity, innovations start to be perceived not as specific projects implemented by firms but as a constant process occurring in business entities at the level of creating new products (goods and services) and processes and better adjustment of organizational structure. The new trends in innovativeness concern both the process of creating products and the changes taking place in the structure of an enterprise and its surrounding. Another argument in favor of the departure from the linear model depicting the typical process of routine production is the modern feature of the innovation process, interactivity, which is connected with the interdependence of different stages of the process and the existence of a number of feedbacks. The form of cooperation in modern business activity is networking, which is an important element of modern innovation models. Enterprise networking involves establishing and developing corporate relationships as a result of exchange of knowledge, experience, and best practices as well as mutual support and building a network of contacts. Thanks to the utilization of rare resources and diffusion of innovation, networks allow obtaining collective competitive advantage. Networking allows increasing the extent of specialization of individual entities through focusing on key capabilities, which are utilized in a coordinated manner. Such processes of cooperation enable smart, joint utilization of network resources and knowledge potential (Kowalski 2011a). In connection with the growing role of networking in modern innovation models, a number of indexes were developed in the last decade to measure innovativeness such as commercialization of scientific research or innovative (Small or Medium-sized Enterprises) SMEs cooperating with other entities. Examples of new innovativeness measures that were developed along with subsequent models of innovation processes are presented in Table 2.1.

The evolution of views on innovative processes determined the development of new indexes and expanded the scale of innovation measurements. It affected a number of various factors such as new organizational structures of R&D and innovative activities as well as modern concepts and policy instruments (Smits and Kuhlmann 2004; Mytelka and Smith 2002). The increasing role of innovation policy, which is one of the newest areas in the economic policy, was also important. Active implementation of the innovation policy required determination of what should be measured (e.g., public sector spending on R&D and government grants for innovative enterprises) and in what terms (e.g., impact on the competitiveness of economy), while the results of measurements influenced the priorities of public actions (what and how should be supported with public funds (Cervantes 2007).

2.4.2 Economy Innovativeness Measures

The basis for developing a set of indexes for the purpose of studying the interdependencies between innovativeness and competitive advantages of countries in international trade were the global theoretical and empirical achievements as regards the methodology used to evaluate the innovativeness of economies.

The most important criterion used when selecting the set of measures for the study was the ability to capture the complex nature of innovative processes. Therefore, the indexes adopted are intended to measure various elements of innovativeness, starting from the conditions that drive innovation (enablers, e.g., public spending in R&D in GDP), through firm activities, up to outputs of the innovative activities (e.g., share of high-tech product export in total amount of export).

The indexes were divided into three groups in accordance with the methodology of calculating the Summary Innovation Index (SII) used by the European Commission in assessing the level of innovativeness of individual European Union member states (e.g., European Commission 2011, 2012): enablers, firm activities, and outputs. Moreover, using the methodology of the World Economic Forum, innovation

Table 2.1 New innovation indexes in the context of evolution of innovation process models

Type of innovation model	Period	Innovation indexes
Linear model of technological push	1950s and 1960s	Spending on R&D
Linear model of innovative pull	1970s	As above and, additionally: statistics regarding patents, technology balance of payments
Coupling innovation model	1980s	As above and, additionally: high-tech products, bibliometric studies, human resources statistics, innovation surveys
Systemic innovation model	1990s	As above and, additionally: innovation surveys in manufacturing sector, innovation described in technical literature, budget support for innovative activities, investments in intangible assets, ICT indexes, productivity, high-risk capital
Integrated, networked innovation model	2000s	As above and, additionally: innovation surveys in all sectors, technological innovation (product, process), nontechnological innovations (organizational, marketing), tax incentives, science–economy cooperation, commercialization of scientific research; cooperation among enterprises (e.g., within clusters), internationalization and techno-globalization, development of new sectors (e.g., biotechnology and nanotechnology), statistics regarding knowledge-based economy, intellectual capital, creativity indexes, disaggregated government budget appropriations for R&D (GBAORD) data

Based on: European Commission, *Innovation and enterprise creation: Statistics and indicators, Proceedings of the conference held at Sophia Antipolis, 23 and 24 November 2000*, Innovation papers No 18, Office for Official Publications of the European Communities, Luxemburg 2001; Górzyński, *Przegląd wskaźników monitorowania systemów wspierania innowacyjności w krajach UE i wybranych krajach pozaeuropejskich—wnioski i rekomendacje dla Polski [Review of indexes intended for monitoring systems supporting innovation in EU states and in selected non-European countries—conclusions and recommendations for Poland]*, Warsaw 2005, as modified and supplemented.

indicators were supplemented with indexes related to technological advancement. The most important innovativeness measures are listed in Table 2.2.

Due to the complicated nature of innovativeness, the measurement approaches and methodologies presented in the theory and applied in practice evolve constantly. As no single universal method to measure that phenomenon has been established, a number of various approaches are used for this purpose.

Both the theoretical models and the method for developing international indexes to measure competitiveness and innovativeness point out the need for multifaceted measurement of innovativeness. For that reason, the set of selected indexes includes both those which refer to: innovation enablers, connected with expenditure in economy and available resources allowing us to carry out innovative activities; firm activities, demonstrating firms' activities toward development through innovation; and indexes showing the outputs of the innovative activities.

The presented set of indexes will allow for international comparisons.

Table 2.2 The most important innovativeness measures. (Source: Own study)

No.	Index
Enablers	
	Gross domestic expenditure on R&D (GERD) as % of GDP
	Government expenditure on R&D (GOVERD) as % of GDP
	Percentage of expenditure on R&D financed by the government
	Percentage of expenditure on R&D financed by other domestic sources
	Percentage of expenditure on R&D financed from abroad
	Household with access to the Internet (as % of all households)
	Human resources in science and technology as a share of the labor force
Firm activities	
	Venture capital investments as % GDP
	Business expenditure on R&D (BERD) as % of GDP
	Percentage of expenditure on R&D financed by industry
Outputs	
	Employment in medium–high and high-tech manufacturing as % of total workforce
	Employment in medium–high and high-tech manufacturing as % of total workforce
	Employment in knowledge-intensive services as % of total workforce
	Employment in knowledge-intensive services as % of total workforce
	Number of patent applications per $1 billion of R&D expenditure
	Number of patent applications per million of inhabitants
	High-tech product export as % of total export

2.5 Summary and Conclusions: Human Capital and Knowledge as Elements of National Innovation Systems

Marzenna Anna Weresa

Concluding from the analyses made in the preceding subsection, innovativeness is a very complex phenomenon and the plurality of stages of creation and implementation of innovation as well as feedbacks between them incline researchers to look at the innovation process from a broader and systematic perspective. Such an approach is based on the so-called interactive innovation model of Kline and Rosenberg (1986, p. 25) and the coupling model of Rothwell and Zegveld (1985); their evolution and further transformations led to the formation and development of the concept of national innovation systems. The concept originates from two economic trends: the evolutionary trend describing the technological change as a gradual process connected with achievements made in the preceding period of time (cf., e.g., Nelson and Winter 1977) and the trend concentrating mainly on innovations being the outcome of interactive learning processes (cf., e.g., Freeman 1987; Lundvall 1992).

Critical analysis of the literature on various interpretations of the national innovation system allowed formulating a broad definition, capturing the key aspects of its characteristics.[35] The national innovation system can be understood as a set of four interrelated groups of phenomena. The said phenomena include:

[35] The analysis and its results as well as empirical studies related to this issue are the subject of a separate monograph—cf. Weresa (2012).

1. The set and structure of entities coming from a given country (but operating also globally) that participate in the creation of new knowledge, its diffusion (home and abroad) and commercialization, as well those that take part in the creation of innovations and their application (utilization, use);
2. A set of institutions providing the environment for the development of science, technology, and entrepreneurship as well as for their change over time;
3. Mutual relations (interactions) of the aforesaid entities and institutions;
4. Knowledge resources accumulated within a given economy (Weresa 2012, p. 23).

Such a broad approach combines the two trends of theoretical studies on national innovation systems and might be useful in the analysis carried out in this monograph, concerning the relationship between the human capital and innovation and the formation of competitive advantages of countries in international trade. The concept of the national innovation system integrates the formation of new knowledge and its commercialization (so creation and implementation of innovation) with the key aspects characteristic of human capital stock, making reference to the processes of interactive learning. The central point of the learning processes is a human being or—in a broader sense— human resources and their development through the system of education, labor market, and learning by doing and also various interactions within the innovation process (Freeman 1987; Dosi et al. 1988; Lundvall 1992; Metcalfe 1995; Edquist 1997). As noted by Lundvall (2007, p. 21), interactive learning processes—being an element of the national innovation system—refer both to the acquisition of knowledge and to the shaping of skills, which perfectly corresponds to the definition of human capital of OECD quoted in Sect. 2.1., according to which human capital means "the knowledge, skills, competencies and attributes embodied in individuals that facilitate the creation of personal, social and economic well-being" (OECD 2001, p. 18). Moreover, as indicated in numerous empirical studies, investments in human capital are one of the most important innovativeness factors, determining the level of absorptions of national and foreign technical solutions (Fiedor and Czaja 2003).

Summing up, the assessment of relationships between the human capital and innovativeness and the competitiveness of countries, which finds its expression in competitive advantages on foreign markets, will be carried out in the empirical part of this monograph with the use of the concept of the national innovation system. The said concept integrates human capital materialized in entities within the system and knowledge resources accumulated in the system, institutions serving the development of science, technology, and education, as well as regulations concerning the running of business activity, forming an entire set of related elements and events, the effect of which is the formation of product, process, marketing, or organizational innovation. The essence of the national system of innovation understood in that way is the interactions of private and public enterprises, individual inventors, as well as universities and various government agendas, aimed at the development of science and technology in a given country (Niosi et al. 1993). Mutual interactions among all elements of the innovation system concern technological, legal, commercial, financial, and social issues; yet, the most important relations take place in the field of developing, protecting, financing, or regulating the new knowledge and technology and their introduction to the market as innovations.[36]

Such a combined approach to innovativeness and human capital within the national innovation system finds its justification not only in the theories of innovation systems but also in the methodology of measurement of innovation and human capital. Concluding from the preceding subsections, numerous human capital measures (e.g., human resources in science and technology, school enrollment rates, employment in high-tech sectors, indexes based on patents, etc.) describe both human capital and innovativeness of a given economy.

The above statement, however, requires an additional comment. Namely, when using the concept of the national innovation system as the framework for further analysis, one should remember that, despite its universal character, it is not possible to take account of all aspects of innovativeness and human capital.

First, not all innovations have the economic dimension, which does not exclude their impact on the economic sphere. For example, social innovations (which comprise changes in education systems, state administration, etc.)—although not directly connected with business activity—may have a direct influence on the conditions in which the activity is run. Standard innovativeness measures allow grasping the significance of that group of new solutions to a limited extent, which makes it impossible to take greater account of that area in this study.

Second, human capital in its broad term—in addition to its characteristics related to education and knowledge (including science and technology)—incorporates also certain aspects connected with demography and health of the society (Schultz 1972, p. 4). These issues are very broad, making the analysis of their entire complex spectrum impossible, and a far-reaching extrapolation of that issue goes beyond the objectives of this study. The proposed narrowing of the term *human capital* to aspects connected with knowledge and education is a common practice in studies of this type (cf., e.g., Lucas 1988, p. 17; Mankiw et al. 1992, p. 418).

Third, not all phenomena may be quantified and described in a set of statistical data. Incompleteness of data and gaps in time series are yet another limitation to the scope of further analysis.

The drawbacks of the methodology indicated above—unavailability of certain statistical data and serious delays in providing those data, which limits the possibility of capturing new trends—will be taken into account when interpreting the outcome of the empirical study.

[36] Discussed in more detail: Weresa (2012).

Chapter 3
Concept of National Innovation System and International Competitiveness—A Theoretical Approach

Marzenna Anna Weresa

The aim of this chapter is to identify, on the basis of the theory, the relationship between human capital and innovativeness and international competitiveness of the economy. The previous parts of the monograph determined the analytical framework that will be used to conduct this evaluation.

First, it follows from the deliberation presented in Chap. 1 that one of the main symptoms of international competitiveness is competitive advantages in foreign trade (which reflect the ability to sell goods and services to foreign markets). Second, Chap. 2 proves that human resources available to the economy, their quality, and the innovativeness of the economy can—assuming some simplification—be jointly captured in the concept of the national innovation system (NIS). Third, several theoretical and empirical works confirm that innovativeness, human capital, and competitive advantages in foreign trade are strongly influenced by other types of international business trading, particularly by the foreign direct investments (FDIs)[1]. They may cause changes (both positive and negative) in the competitiveness of countries in which foreign capital was invested (Dunning 1994; Markusen and Venables 1997; Weresa 2001; Lipsey 2002, 2006; Keller 2009; Radulescu and Serbanescu 2012; Kuntluru et al. 2012). That is the reason why this aspect was included in the spectrum of the analyzed questions.

The basic concept of competitiveness used in this chapter is Porter's competitive advantage of nations, expanded to include aspects of international production relations exposed on the theoretical basis by Dunning[2].

The macroeconomic approach toward the relationship between the innovation system and competitiveness refers to the broad definition of competitiveness connected with changes in the level of prosperity.

[1] Mentioned in Chap. 1, as well as in, e.g. Weresa 2002.

[2] Mentioned in Chap. 1.

M. A. Weresa (✉)
World Economy Research Institute, Warsaw School of Economics,
Al. Niepodleglosci 162, 02-554 Warsaw, Poland
e-mail: mweres@sgh.waw.pl

M. A. Weresa (ed.), *Innovation, Human Capital and Trade Competitiveness*, Innovation, Technology, and Knowledge Management, DOI 10.1007/978-3-319-02072-3_3,

3.1 Innovation—The Main Determinant of Comparative Advantages of Nations

The existence of a relationship between innovativeness and competitiveness of countries, regions, and companies was confirmed both on the theoretical basis and in numerous empirical works (Porter 1990; Lubiński et al. 1995; Bossak and Bieńkowski 2001; Cantwell 2005; Castellaci 2007; Bieńkowski et al. 2010; Rynarzewski 2011; Di Bello and Andreta 2012). It is possible to sum them up with a general statement that the conditions for the functioning of the NIS, which determine countries' innovativeness, can influence the competitiveness in different ways in both the short and long term. The relationship between the NIS and competitiveness is illustrated in the form of a model in Fig. 3.1.

The first conclusion that arises after the analysis of Fig. 3.1 relates to the complexity of innovation processes and the existence of multidimensional relationships between individual elements forming the NIS, which have already been emphasized above. It is possible to distinguish four principal groups of components of the NIS:

1. system entities (i.e., companies, research units, organizations supporting the knowledge transfer, users of innovations, administrative units shaping the innovation policy, etc.);
2. system institutions (i.e., regulations, standards, and customs);
3. reciprocal relations and links between entities and institutions (illustrated by the arrows in Fig. 3.1); and
4. resources of knowledge accumulated in the system, as well as processes of its transfer and absorption[3].

The entities (actors) of the NIS are primarily companies that introduce innovations to the market of their structure (e.g., networks and industrial clusters). Another group of entities includes research units, educational institutions, and other entities (e.g., individual inventors, academics, and students—this group constitutes the core of the human capital of the system and appears in Fig. 3.1 as the "science system"). Other entities of the NIS are the organizations that intermediate in the transfer and diffusion of knowledge to business, and the organizations that form the innovation policy at the central and local levels. The newest works on innovation systems consider also the users of innovations (consumers) to be the system entities. It corresponds with new approaches concerning the creation of innovations by users (the concept of the so-called user-driven innovation). The significance of the last of the mentioned groups was emphasized by Rogers (2003), who had concluded, while examining the diffusion of innovations, that consumers were able to find new uses of a given innovation, which would make it a new innovation.

The second group of NIS components—institutions—relates to the conditions of conducting scientific and research activity, as well as innovative activity. Furthermore, the group includes legal norms, customs, standards and technical norms,

[3] More on the subject in Weresa (2012, pp. 37–39).

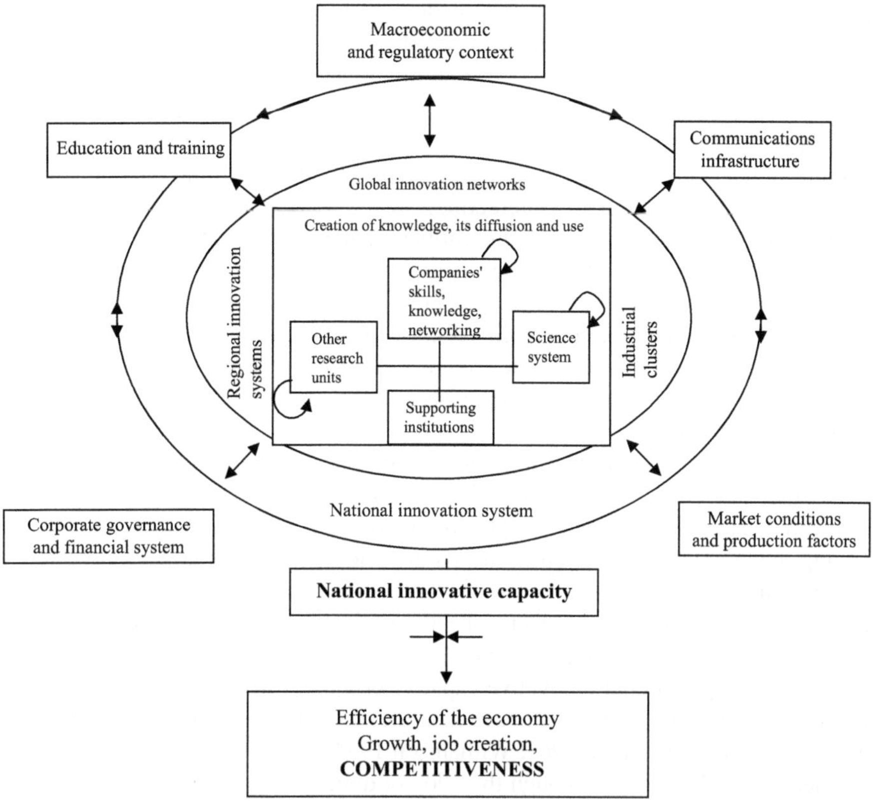

Fig. 3.1 National innovation system and its relationship with competitiveness. (Source: OECD (1999), *Managing National Innovation Systems*, Paris with additions and highlights introduced by the author.)

research and innovation funding instruments, and innovative policy tools. Regulations related to the intellectual property protection and commercialization of knowledge are especially important. The NIS institutions comprise also regulations in the domain of the education system, including rules for the financing of education, accreditation of educational institutions, etc. Another group of NIS institutions contains legal norms that are directly related to entrepreneurship, e.g., market competition regulations (in Fig. 3.1, these elements are schematically presented in the form of rectangles partially placed outside the system, but having a common part with the NIS).

The third group of components of the NIS includes interactions between all elements of the NIS—in Fig. 3.1, the senses of the arrows show that the interaction of all these elements, their relationships, and feedback influence the efficiency of the whole economy and, thus, economic growth and competitiveness.

Finally, the fourth group of the NIS contains knowledge resources accumulated in the economy, which are presented in Fig. 3.1 as “companies’ skills, knowledge, and networking" and are also partially included in the category of "science system" as knowledge and skills embodied in people. Codified knowledge is reflected in scientific and technological specialization of a given country, usually measured by patent indexes. Uncodified knowledge is a dynamic resource; so the way in which it is created, transferred, and absorbed is very important. Learning processes are the essence of the concept of the innovation system, including learning by doing, learning by failure, and learning by interacting.

The synthetic presentation of the entities of the NIS introduced above allows their confrontation with the competitiveness factors analyzed in Chap. 1 of this monograph. They match one of the categories of factors which determine the development of international competitiveness of national economy called the "ability to invent and innovate", extended by knowledge resources (mentioned in Fig. 1.4 in Chap. 1). It confirms the existence of a direct relationship between the NIS and competitiveness.

Another observation concerning the relation between NIS and competitiveness that can be made on the basis of Fig. 3.1 relates to the category called the "national innovative capacity". It seems that the innovative capacity is the key element influencing the creation of knowledge and its diffusion and use, as it is a kind of contribution transformed into the outcome, i.e., into innovations. It is therefore worth taking a closer look at the concept of national innovative capacity in connection with the theories on the competitiveness.

The notion of national innovative capacity originates from the empirical studies conducted in the 1960s for industries such as the production of iron, steel, oil refining, and coal mining (Mansfield 1963; Williamson 1964). These approaches relate to the mesoeconomic level and, although they do not directly define the notion of innovative capacity, they indicate its determinants, which include, e.g., the structure of the industry or specific characteristics of companies (resources and scale of operation) determining competitive advantages. Nevertheless, it is important, from the perspective of the analyzed topic, to indicate the relationship between innovativeness and competitiveness at the micro- and mesoeconomic levels.

The macroeconomic approach to the question of innovative capacity observed from the national perspective occurs in the literature in various forms: from the narrow approach that is focused on the patent activity and its impact on economic growth (Suarez-Villa 1990), through the statistical and dynamic approach analyzed from the perspective of the planned economy (Balcerowicz 1980), to the wide, multifaceted approach integrating the macro- and microeconomic aspects of research on innovativeness (Stern et al. 2000, 2002).

The widest approach defines the national innovative capacity as a long-term ability to create and commercialize a stream of new and previously unknown ideas, i.e., innovations (Stern et al. 2000, p. 10). Thus, this definition emphasizes three issues: the uniqueness of innovation on a global scale, the continuous development of the stream of new solutions, and the need for their practical use. Factors that determine national innovative capacity include tangible and intangible resources of a nation, investments making their use and improvement possible, and innovation policy tools (Stern et al. 2002, p. 900).

In the context of the analysis of the relationships between human capital, innovativeness, and competitiveness, the bases of the concept of national innovative capacity that clearly refer to the theory of competitiveness are very important. Three theoretical pillars mentioned by Stern et al. (2000, p. 2–10) are:

- new theory of growth (Romer 1990),
- concepts of the NIS (in particular, the evolutionary trend of technical changes (Nelson 1993)), and
- Porter's (1990) theory of competitive advantages of nations.

From these three trends, factors that can be included in the specificity of a given territory (country), conditioning the emergence of innovations in this—and no other—territory, were selected. Importantly, mutual links between these factors and the efficiency of their interactions were highlighted (as in the case of the concept of NIS). According to Stern et al. (2000, p. 11–13), the most important determinants of the national innovative capacity can be narrowed down to:

1. common innovation infrastructure for the economy, i.e., human and financial resources, institutions, in particular, and the innovation policy;
2. specific environment for the creation of innovations in industrial clusters; and
3. quality of links between (1) and (2), understood as the strength and the direction of their interaction.

These factors are schematically presented in Fig. 3.2 (Stern et al. 2002, p. 907):

This fragment requires a reference to the concept of competitive advantages of nations characterized in Chap. 1 and elaborated by Porter (the so-called Porter's "diamond") (Porter 1990, p. 71; 2008, p. 183). The concept points out four main groups of factors determining competitive advantages of nations, which interact with each other and, by these means, form the diamond of competitive advantage [4]. These are:

- production factors (resources), both mobile and immobile (such as human resources, scientific base, and technical knowledge), and conditions for their development,
- domestic demand (including the size of the demand, its structure, and dynamics of changes),
- relative and supporting sectors (which may favor or impede the creation of new ideas and innovations), and
- companies' strategies and structure of the market, as well as competition between companies (competition conditions).

The above-mentioned determinants of the national competitive advantage create the environment of competition between companies. Its most significant feature, which determines the economic success of a country, is the formation of stimuli to create and implement innovations. It is favored by fast accumulation of unique, country-specific resources and abilities, which may be based on a relatively bigger

[4] Mentioned in Chap. 1, Fig. 1.7.

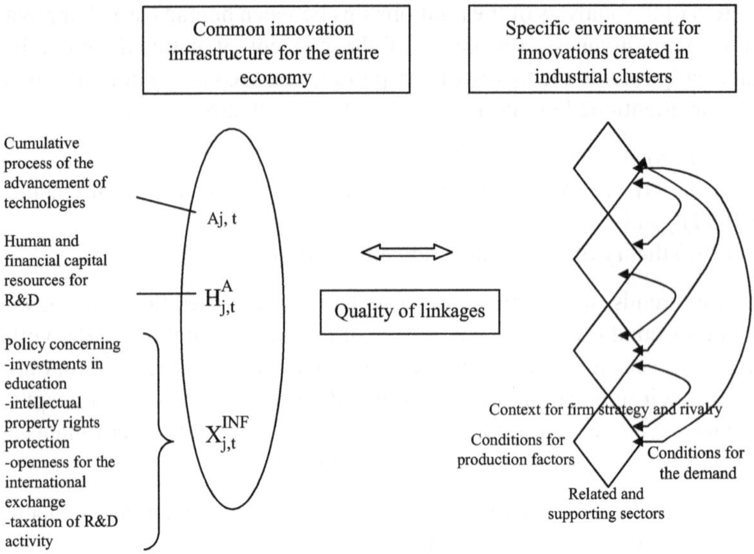

Fig. 3.2 National innovative capacity according to Stern et al. (Source: Stern et al. 2002, p. 906)

effort and engagement in business activity than in other territories. Domestic companies gain competitive advantages also when they have a wider access to information and a better recognition of the needs of the market, in the case of both demand and supply. The key element of Porter's diamond is, however, the motivation to undertake innovative activity, which may originate from any of the four groups of factors determining competitiveness (i.e., from any apex of the so-called "diamond") or their interaction. Innovations determine the improvement in the already-owned comparative advantages or an acquisition of such advantages in new domains (Porter 2008, p. 183).

In Fig. 3.2, the model of Porter's diamond was used to characterize industrial clusters (the right side of Fig. 3.2). According to Porter, the clusters are precisely the elements that form the special environment for the creation of innovations, as they form unique relationships, facilitate the sharing of knowledge and the formation of technological spillover effects, and strengthen cooperation as well as competition (Porter 2008, p. 221). Thus, they indicate the national innovative capacity. The clusters, which (similar to entire economies) can be analyzed with the use of the diamond, appear in Porter's model of national competitive advantages in the category of "context for firm strategy and rivalry".

This synthetic analysis of the concept of the national innovative capacity shows its indissoluble link with Porter's concept of competitive advantages and is another evidence that, in theory, the thesis about the existence of the relationship between human capital, innovations, and competitiveness is fair.

These deliberations lead to another question: in what way does Porter's diamond of competitive advantages of nations correspond with the concept of the NIS? It is possible to find at least a couple of common points in both theories.

First of all, the common core of Porter's diamond and NIS is innovativeness. New solutions, the result of all interactions between different elements of the innovation system, are also the main axis of Porter's diamond, around which competitive advantages of a country are concentrated. The implementation of innovations is indispensable for the development of all four groups of factors determining competitive advantages of nations. Innovations condition the development of resources of a given country and the effectiveness of their use is defined as the added value per labor or capital unit. Moreover, new solutions can constitute an element of formation of demand and its growth, which helps to increase the scale of production, to obtain additional benefits, and to gain competitive advantages in international markets or to strengthen the existing ones (mentioned in Chap. 1).

Second, a common point can be observed in relation to production factors, which are explicitly enumerated by Porter as factors of competitiveness and which also occur in the concept of the NIS as its entities (including human resources) and knowledge accumulated in the system. It is worth mentioning that, as it was highlighted in Chap. 1 of the present monograph, production factors (their resources and prices) are also—according to classical economics—one of the bases of the international trade and, thus, a determinant of competitive advantages (or their lack) in foreign markets.

Third, another shared point of both analyzed concepts is institutions, interpreted in studies on NISs as a set of regulations and norms, as well as procedures that ensure the compliance with these "rules of the game" (North 1984, pp. 5–6). Regulations, legal norms, and informal customs are also, to a small extent, contained in each of the tops of Porter's competitiveness diamond and, in particular, they determine the companies' strategies, the market structure, and the competition (the method and strength of the competition). It is the domestic conditions and contexts that determine how, and how efficiently, a company may be set up or liquidated (barriers to entry and to exit), the organizational structure of companies, management policies and practices, and competitive conditions, as well as stimuli to innovations (Porter 2008, p. 194). While comparing the institutional dimension of the NIS with Porter's diamond, one should note that the entire institutional order that constitutes the innovation system (norms and regulations concerning research and development (R&D), science, education, entrepreneurship, and innovation policy) is included in Porter's diamond. On the other hand, although the institutional component of the innovation system is principally limited to factors related to the development of science, knowledge, technology, education, and entrepreneurship, its institutional dimension is a little bit wider than the concept of competitiveness of countries, since, besides institutions themselves, it comprises also their interactions with entities that form the system. As a result, the evaluation of competitiveness of countries conducted with the use of the concept of the NIS can also include the sphere of influence of government policy (particularly, science and innovation policy) on the innovativeness of companies (Solleiro and Castanon 2005, p. 1066).

All in all, the common points of the concept of the innovation system and Porter's diamond of competitive advantages are: the availability and quality of resources (human capital and technology) and the way in which they are used, institutions, and, in particular, changes introduced to these elements in a way that would result in gaining advantage in the innovative activity and in improving innovations. If domestic companies were to introduce novelties on an international scale, they can, by getting ahead of global trends, use this advantage in their expansion into the international markets. The export of goods and services contributes to the dissemination of not only the material goods, but also the system of values of a given nation, which reinforces the already possessed competitive advantages (Porter 2008)[5]. This observation refers to the third group of mutual interdependences between the innovation system and competitiveness, i.e., internationalization.

3.2 Innovative Activity of Transnational Corporations as a Factor of Competitiveness of Countries, Regions, and Industries

Another element that connects the innovation system with competitiveness is related to the increasing degree of internationalization not only of business activity but also of research and education. Globalization, which manifests itself in the strengthening of interdependence in the global economy, the expansion of transnational corporations (TNCs), and the increase in the number of scientific and technological links, influences competitiveness of economies. Studies on innovation systems indicate progressive internationalization of the R&D field (Soete 2010; Weresa 2012). As discussed in Chap. 1, international economic cooperation and, in particular, trade links and investments are the main signs of the international competitive position of countries understood from the statistical point of view (mentioned in Fig. 1.4 in Chap. 1). On the other hand, the theory shows that it is possible to use FDIs and foreign trade to consolidate comparative advantages in statistical terms and to create new advantages of this type. It is referred to in the Rybczyński theorem (1955), neo-technological theories of international trade (Posner 1961; Vernon 1966), Dunning's eclectic paradigm (1988), learning concepts leading to obtain economies of scale (Scitovsky 1958), as well as in the achievements of the new economic geography (Krugman 1990; Fujita et al. 1999).

The achievements of the theory of international trade mentioned above show that the technological potential of a country conditions the country's specialization in the international trade and shapes the flow of direct investments, which means, according to the adopted definition, that it determines the competitive position of the country. The technological potential is considered by the literature to be a component of the NIS or even the entire economic system itself, with the assumption that innovations are limited to their technological dimension (Narula and Wakelin 1996, pp. 373–387). The uneven distribution of the technological change in space and time determines the directions of the international trade and starts the adjustment processes within a given country and between individual economies (Soete 1990).

[5] This topic is discussed in more detail in Weresa (2012), Chap. 4.

Adjustments within a country consist in specializing in those industries which present comparative advantages in the technological development and are more innovative. Thanks to specialization, these industries become more efficient and develop in a more dynamic way than in countries where specialization is absent. It results in the deepening of preexisting cross-sectoral gaps and, consequently, the intensification of specialization in the international trade and changes in the share of particular countries in global exports. Shifts in the attractiveness of countries and industries to the inflow of FDI constitute the second effect of this process. This is how the technology (or, to put it more broadly, the accumulated knowledge), which is an element of the NIS, determines the competitive position of a country (understood in static terms, as the ability to sell and the ability to attract—mentioned in Chap. 1).

However, this static approach has to be complemented by the second aspect of the issue, namely by the ability to shape comparative advantages through trade and through the influx of FDIs. The two-way correlation between competitiveness of an economy and its links with the external environment means that in the long term foreign exchange of goods and services influences the level of innovativeness and the possessed competitive advantages (Weresa 2002). It is illustrated by the extended model of Porter's competitiveness diamond, which was complemented by Dunning with the activity of TNCs, which is considered to be one of the factors that influences the dynamic competitive advantages (mentioned in Fig. 1.16 in Chap. 1). The most recent scientific papers on this topic formulate foundations of the theory of FDI made by the knowledge-based TNCs (Narula and Guimon 2009; Pitelis 2011). It is a reinterpretation of Dunning's eclectic paradigm elaborated on the basis of innovation systems, taking into account the consequences of the activities of TNCs on the competitiveness of countries, regions, and industries. The new approach to Dunning's eclectic paradigm pays more attention to interactions between companies and competitors, suppliers, and customers. The deliberations were extended by transnational interactions, i.e., those within the TNCs, between their branches located in different countries, as well as links with foreign suppliers of knowledge (universities, laboratories, and companies from other countries). They are favored by an increasing mobility of capital, technical knowledge, and people (Narula and Guimon 2009, p. 4). At the heart of Dunning's complemented concept are interactive learning processes (Pitelis 2011, p. 226), which directly relate to the basis of the concept of innovation systems. The new approach toward Dunning's paradigm refers to Penrose's (1959) theory of growth of the company based on learning processes. The main argument states that the creation of knowledge within a company takes place due to dynamic interactions between the internal resources and the external/market opportunities (including those offered by the global environment). The interactions favor the acquisition of additional competitive advantages through learning processes (Pitelis 2011, pp. 227–230).

This element of interaction between companies, their suppliers, and customers also plays a crucial role in Porter's diamond of competitive advantages, although it does not widely mention interactions with the external environment. Consequently, the comments on the extension of the diamond made by Dunning, which have already been mentioned above, are a valuable complement to the concept.

As it was indicated above, foreign investments and the activities of TNCs were distinguished by Dunning as factors affecting competitiveness for two main reasons: 1) The first one is due to their impact on innovations conditioning resources and the effectiveness of their use. 2) The second is due to the fact that the location of direct investments made by TNCs is a source of innovativeness and competitiveness of economies, as it performs the following functions:

- creational, involving the direct innovative activity of entities with foreign capital in the local market,
- informational, by facilitating the flow of information about new products and manufacturing techniques and their assimilation,
- educational (e.g., by learning to learn, learning by watching, and learning by doing and by using),
- motivational, by motivating local companies to search for better solutions or imitations, and
- increasing efficiency, by influencing the efficiency of utilization of resources[6].

During the development of knowledge-based economy, the most important functions of TNCs in stimulating innovativeness and competitiveness of economies are the first three of them, namely creational (creating new knowledge), informational (accelerating the information transfer), and educational. The following functions are performed through the influence that the entities with foreign capital have on the supply or demand in the country in which the investments are located.

Direct investments of TNCs can constitute an innovation stimulus to supply, e.g., in a situation in which they launch new, unknown products and create a demand for them. Furthermore, companies with foreign capital, which are established as a result of the inflow of investments (e.g., subsidiary companies of TNCs), can directly create innovations based on the results of R&D of their parent companies. The second function of direct investments as the innovation stimulus is to respond to the preexisting unmet demand.

Foreign investors influence the creation and diffusion of innovations in the country in which the investments are made at all stages of the innovation process: invention, innovation, and diffusion–imitation. Therefore, potentially, at least three groups of positive effects may occur:

- the capital may be transferred together with solutions used in the investor's country and, thus, diffusion of innovation on an international scale may take place;
- companies with foreign capital may create innovations in the host country, using, *inter alia*, R&D results of the parent company; and
- ideas introduced by foreign investors may be imitated by the local companies or may constitute an inspiration to create new solutions (spillover effects).

Sometimes, however, investments of TNCs may have a negative impact on the innovativeness of the host country (thus limiting the country's competitiveness at the same time), e.g., in a situation in which the inflow of capital limits the technologi-

[6] This topic is discussed in more detail in Weresa (2002).

cal specialization of the country by directing it to the branches with a low share of added value. Furthermore, a phenomenon of substitution of the investment, in the sphere of domestic R&D by foreign capital, may take place. In the long term, it would have a rather negative impact on innovativeness due to the subordination of the technological development of the country to preferences of foreign investors.

Most of the theoretical concepts explaining the phenomenon of direct investments assume that the investor has an advantage in the domain of manufacturing technologies and marketing or organization of work[7]. The advantage can be used also in a direct way—by stimulating the creation of innovations in companies with foreign capital operating in the country in which the investments are located. In processes of creation of new solutions in TNCs, it is possible to use the R&D results of the parent company (unknown in the local market). Moreover, assuming that the investor has higher technological potential at its disposal in a given domain, as a result of the accumulation of technology achieved in prior periods, the innovations created in subsidiary companies can be unique even on an international scale (Cantwell 1994).

In addition to the direct diffusion of innovations from the investor's country to the host country and the creation of innovation by companies with foreign capital, direct investments influence innovativeness also in an indirect way by stimulating reactions of companies operating in the investor's local environment in the host country (so-called spillover effects).

Innovations of domestic companies may be forced by the activity of foreign investors due to a demonstration of new ideas, as a result of competition and cooperation (horizontal or vertical) or as a consequence of training of the investor's employees.

The demonstration effect occurs in the case in which the technological level of the investor is higher than that of the local companies. The imitation of innovations by local companies is favored by the direct contact with national partners and the movement of labor force between companies.

Innovativeness is also influenced by an increase in competition caused by the activity of a company with foreign capital (Dunning 1988). Competition forces local companies to search for new solutions to improve operational efficiency. This effect has an impact on the domestic companies operating within the same sector and, thus, it has an intra-industry scope (Markusem and Venables 1997, pp. 15–16).

Therefore, the competition results in: first, a shortening of the innovation development cycle, as the fight for market share intensifies the innovative activity of companies, and second, an erosion of benefits that would be gained by the monopolist if there was a monopoly, caused by competitors, that is, the increasing number of companies, which can benefit from innovations. In addition, imitators appear, who are willing to make an effort to innovate.

In some cases, however, very strong foreign competition may have a negative impact on innovativeness—e.g., in a situation in which it would lead to the withdrawal of local companies from the market and strengthening of the monopoly of one supplier (Markusem and Venables 1997, p. 12).

[7] More on the subject in Zorska (1998).

Another group of stimuli connected with the activity of TNCs, which can encourage innovativeness and, therefore, competitiveness of the economy, are the links between the investor and local suppliers and customers. This refers to a situation in which a company established with foreign capital obtains raw materials and components in the local market or is a supplier of components for local customers. Consequences for the process of creation and diffusion of innovations are, in this case, of inter-industry nature. Moreover, the presence of foreign capital may have educational effects—through training in new technologies, offered by local subcontractors and suppliers to the employees.

Innovational effects of all groups of stimuli are usually closely related. Positive results of the activities of TNCs for companies, operating in the local environment of the country in which investments are located, can therefore come down to the imitation (imitative innovations), some minor improvements of the imitated solution (imitative and adapted innovations), or the inspiration for completely new ideas (creative innovations). The intensity of the impact largely depends on whether and to what extent the innovation, used by the investor, is original and is unknown, i.e., whether it is new on a global scale or in the country, industry, or just in the company. The second element, which is important from the viewpoint of the size and pace of the innovation resonance, is the way of entering the market, i.e., whether it is done together with a local partner or a subsidiary of a foreign company is established by the company only.

Positive effects of the expansion of innovations transferred via foreign investments in the local environment may, however, be limited by an active policy of TNCs that are interested in maintaining the technological advantage for as long as possible. Methods of such active influence include a restricted access of certain subsidiary companies to certain research results, technologies, or markets. The occurrence of the effect of competition may also be limited by the use of transfer pricing. Furthermore, the strength of the spillover effects depends also on the factors attributed to the host country. If the technological gap between the investor's country and the host country is too large, positive effects may not occur at all (Knell 2000). In such conditions, direct investments may also entail other negative effects, such as limitations of domestic R&D and of innovations of local companies (in particular, if the market position of the foreign investor is privileged due to subsidies or tax holidays), which impede innovativeness of the host country. Thus, the final result of the inflow of foreign capital is, as far as innovativeness is concerned, resultant of these opposing effects.

Educational foreign investments are of great importance to a knowledge-based economy, as education is considered to be the medium of a knowledge-intensive economy. They become visible in the process of creation and diffusion of innovation principally in the field of supply and include two spheres:

- sphere of production, in which the adaptation and imitation of innovations take place, on the basis of which potential improvement or creation of new solutions may be effected (e.g., learning by watching and learning by doing) and
- sphere of R&D, whose educational efforts lead to gaining knowledge on the ways of learning and increasing the capacity of the economy to create innovations (learning to learn).

In both cases, the company, while making educational efforts, analyzes the costs of creating knowledge and juxtaposes them against the potential benefits resulting from the introduction of a new or improved product or manufacturing technology[8].

As far as the sphere of production is concerned, the educational process takes place by means of observation (watching) of the actions of other entities in the market (learning by watching), by action or use (learning by doing or learning by using), or by interaction between new and previously known elements (learning by interacting).

Learning by watching other entities combines the above-mentioned effects of competition and imitation. In order to improve their market position, companies imitate solutions used by their competitors. In an extreme situation, it can consist in, e.g., buying a new product abroad, meticulously examining its construction, and creating a true copy (reverse engineering). Solutions used in companies with foreign capital are observed and are imitated in a similar way, which favors the spread of new ideas in the market and, afterwards, their further improvement. The process of improving an observed innovation constitutes another innovation.

Learning by watching may also relate to the demand side of the innovation process, when innovations are noticed, e.g., in foreign markets by customers, and result in a creation of the demand for specific goods or services in the country. It causes a launch of an innovation stimulus to the demand.

A direct educational effect, which may accompany foreign investments, is the training of employees conducted by foreign investors, as well as marketing and promotional activity, which, besides its other functions, provides information on new products and thus, to some extent, educates consumers (but also potential competitors).

The second sphere, besides the production, in which foreign investments of TNCs can result in educational effects, is R&D. The process of learning to learn is multilayered and complex and it includes:

- direct R&D activity conducted in a host country by companies with foreign capital, which influences the broadening of knowledge of local employees involved in the activity,
- the use of research achievements of the parent company by a subsidiary company, which broadens the general knowledge of employees of the subsidiary company,
- direct import of R&D results and their use in production, and
- spread of R&D results imported or gained by companies with foreign capital due to the flow of labor force between companies and branches.

As a result of these processes, skills of people employed in local R&D centers develop and can later be used to create other new innovations.

[8] It is possible to present educational efforts in the form of a formal model. For example, in the case of FDIs, a model of the efficiency of learning of domestic companies from companies with foreign capital was proposed by, *inter alia*, S. Das, *Externalities and Technology Transfer through Multinational Corporations: A Theoretical Analysis*, in: E.K.Y. Chen, (ed.),: *Transnational Corporations and Technology Transfer to Developing Countries*, Routledge, London 1994, pp. 99–110.

The foregoing educational effects of the activities of TNCs have a crucial meaning in the knowledge-based economy, as information, its creation, and the pace of transfer are crucial in building added value and developing competitive advantages.

3.3 Foreign Trade and Innovativeness of Companies, Industries, and Countries

J. Dunning, like M. Porter, considers innovations to be the most important factor of competitiveness of countries. The author also notices that not only the technological and investment cooperation with foreign partners, but also the foreign trade may constitute a stimulus to introduce new solutions. Foreign trade favors an increase in effectiveness, thanks to the development of specialization of a country in those domains in which the country is relatively more efficient. Moreover, the presence in international markets implies the necessity to fight off international competition. The fact of competing provides motivation for searching and using in practice the results of research and R&D, new concepts, ideas, and inventions and also encourages the improvement and development of existing technologies, the implementation of new solutions in organization and management, and the development of infrastructure (Dunning 1994). In other words, cooperation with foreign countries may influence a country's innovativeness by determining its competitiveness in dynamic terms, by encouraging innovations, and by facilitating the diffusion of new solutions. The foreign trade, as well as direct investments, can shape the innovation system of an economy by fostering the imitation or adaptation of foreign ideas or by inspiring the creation of entirely new solutions. The conditions of operating and competing in foreign markets as well as the presence of foreign entities in the domestic market may stimulate national companies to seek additional strengths in order to decrease the costs of manufacturing goods or to improve their utility. Therefore, the economic cooperation with foreign countries may be a stimulus to improve products and production processes and to introduce innovations in the field of technologies and organization and marketing. Furthermore, the innovativeness of countries is also affected by the increase in common transfer of research activity to countries in which the costs of such activity are the lowest. Thus, new entities conducting R&D activity appear in the NIS. This is the way in which foreign investments and trade directly influence innovativeness and competitiveness of countries (Narula 2003). At the same time, the economic cooperation with foreign countries plays an important role in the innovation diffusion processes. Foreign trade, similar to direct investments, is a medium of knowledge transfer. This is the way in which a flow of new ideas between countries takes place. Then, after a certain period of time, new solutions are always imitated (Weresa 2009). It induces changes in resources of the accumulated knowledge available in the country and, thus, influences the NIS. The causes and consequences of the process are characterized in Chap. 1.

3.4 Impact of Innovations on Competitive Advantages of Companies

The analysis of innovations from the perspective of a company should take into account the fact that the notion of innovativeness refers to all areas of the company's activities. According to the generally applicable definition, the innovativeness relates to technologies, business organization, and marketing and distribution.

As it was proven above on the basis of the theory, innovative activities are an indispensable condition for gaining competitive advantages of nations. However, competitiveness understood from the macro perspective (i.e., that of the national economy) consists of the ability of companies to introduce innovations and to improve the technological level (Porter 1990). It means that, from the macroeconomic viewpoint, besides the relationship between innovativeness and competitiveness, it is worth analyzing the significance of innovations in the development of the competitive position of companies.

One of the tools used in the literature to evaluate the relationship between innovativeness and competitive advantages of companies is the value chain (Porter 2006, p. 213). In relation to the technological dimension of innovativeness, it can be assumed that the competitiveness of companies depends on the technologies used in the company, in both the core and auxiliary activities. Therefore, innovations created with the use of these technologies have a direct or an indirect impact on the competitiveness of companies.

A technological change (innovation) influences competitive advantages of companies if it implies a change in the relative cost position of a company or if it leads to the achievement of new technological advantages over its competitors. These advantages do not have to originate from a groundbreaking innovation. They may be an effect of a skillful combination of technologies, which were already known. A technological innovation forms competitive advantages of a company by means of (Porter 2006, pp. 219–220):

- influence on a decrease in costs and/or an increase in the diversification of the company and the achievement of the position of the leader in technology,
- improvement in the cost structure and positive impact on the degree of differentiation of the company,
- initial benefits (investor's benefits) due to the fact that the company is the first to introduce an innovation and thus, gains, for example, an exclusive right to the learning processes, favorable access to factors of production, etc.), and
- improvement of the general structure of the branch.

The competitiveness of companies depends on the technological strategies developed by companies in order to gain comparative advantages. It comes down to choosing the technology to be used by a company, positioning in the field of technology, and its potential licensing if new generations of this technology are developed.

There are two possible types of technological strategies—cost leadership or differentiation strategy (Porter 2006, pp. 225–240). The choice of strategy is connected with the market position of a company. If the company aspires to be the leader in technology, it usually seeks to diversify its product. Another possible method is to imitate the technology used by leaders. The mechanism of imitation and creation of competitive advantages on its basis is explained in more detail in the so-called neo-technological theories (mentioned in Chap. 1). By choosing such a strategy, a company builds its competitive advantage on relatively low production costs.

The third strategic decision of companies relating to technology concerns the question of whether and when to license the technologies owned. Usually, licensing is a consequence of the emergence of a new generation technology in the market, but sometimes it may cause a loss of the competitive advantage of the company.

Furthermore, an important role in the diffusion of technology in a knowledge-based economy is played by the use of the Internet and other modern methods of remote communication. The rapid increase in the mobility of human capital observed in the present century is also significant. These factors have considerably accelerated the innovation diffusion process. It brings positive effects for the entire global economy, but may also jeopardize innovators, who want to use the new solution exclusively as long as possible. This is the reason why not only the systematically conducted R&D activity, which aims at elaborating new generations of technologies, but also the protection of information about novelties against competitors have a prominent place among innovative strategies (Weresa 2009). Both the foregoing factors—the R&D activity and regulations concerning the intellectual property protection—are included in NISs. Thus, it is another sphere, in which the relationship between the competitiveness of companies and the operation of innovation systems can be observed.

The fact that the competitiveness of companies is conditioned by the efficiency of the NIS is confirmed not only by the theory, but also by the results of empirical studies. The research on factors determining the competitiveness of companies allows the distinguishing of some groups of factors, among which the most important are (Atzei et al. 1999, p. 745):

- R&D activities,
- technologies used in production,
- education, skills, and creativity of employees,
- links between companies and research centers,
- efficiency of the commercialization of R&D,
- company's ability to restructure,
- methods of diversification of activity,
- global management and marketing skills,
- financial resources, and
- ability to influence the regulatory sphere of the economy.

The above-mentioned factors can be assigned to particular elements of the NIS. Technologies, education, creativity, and skills of employees are significant for building comparative advantages, as they form the knowledge accumulated in the innovation system. Links between companies and research centers, often highlighted as a competitiveness factor, as well as other various global links (e.g., marketing or management links) and the ability of companies to influence the environment (including the formation of regulations), constitute the core of innovations systems. Interactions within the system condition the efficiency of its operation. These convergences indicate a strong relationship between innovativeness and the competitive position of a company.

The analysis conducted above leads to the conclusion that there is a relationship between the functioning of innovation systems and competitive advantages of companies operating within the system.

3.5 Innovations and Level of Prosperity

While analyzing the links between innovation systems and the competitiveness of countries, it is worth taking a closer look at the notion of socioeconomic prosperity. Many definitions of competitiveness consider the level of prosperity to be its reflection[9]. Therefore, a question arises—whether and in what way innovations translate into changes in the prosperity level. Figure 3.3 presents a simple sequence of transforming human creativity and invention into socioeconomic prosperity. This simplified model shows that the creative effort of inventors may influence the increase in the prosperity level via labor or goods market. If a new solution (process innovation) leads to the growth of labor productivity, then the consumers may gain considerably modified and cheaper goods and/or services.

Creativity and invention may be turned into socioeconomic prosperity also if they are used to develop new products. If an innovative product is accepted by the consumers, it increases their contentment and, as a consequence, their prosperity. Furthermore, consumers and their creativity may also cause an increase in the prosperity. It happens in a situation where the consumers propose new uses of a product (so-called user-driven innovation), while producers are willing to apply the ideas of the consumers in order to fill the existing market niche (von Hippel 2005).

The model presented above refers principally to the economic dimension of the relationship between the innovativeness and the prosperity, as the translation of innovations into prosperity takes place mainly due to the increase in productivity. Nevertheless, an increase in prosperity can also happen by means of innovations implemented in the social sphere, which influence, e.g., the quality of the natural environment or health care (Swann 2009, p. 237).

Today, a very significant element of the relationship between innovativeness and prosperity is the natural environment. This factor is also referred to in the concepts of competitiveness (mentioned in, for example, the concept of locational competition discussed in Chap. 1). Technologies and innovations do not always favor environmental protection (and sometimes can even cause environmental degradation). It is even possible to notice a discrepancy (trade-off) between the eagerness to improve international competitiveness of a given economy and the evolution of quality of

[9] Mentioned in deliberations concerning the definition of competitiveness in Chap. 1.

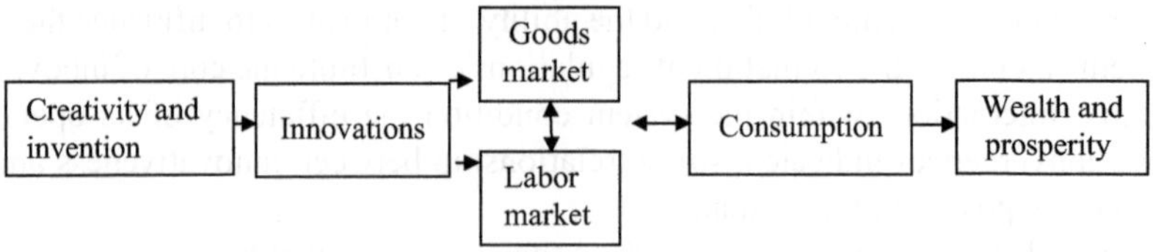

Fig. 3.3 Role of innovations in creating prosperity. (Source: Study, supplements and adaptation based on: Swann, *The Economics of Innovation: An Introduction*, Edward Elgar, Cheltenham, Northampton 2009, p. 236)

the broadly understood natural environment. It consists in the fact that broadly understood costs of environmental protection may cause a decrease in the price competitiveness of products. On the other hand, an increase in the quality of the natural environment implies an increase in socioeconomic prosperity.

3.6 Stages of Competitive Development and Taxonomy of NISs

According to the concept of Porter (1990, p. 546), the strength of the impact of innovations on competitive advantages of nations is closely related to the level of the economic development of a given country. The main factor that is necessary to compete in the underdeveloped countries is the availability of natural resources. The development of competitive advantages in foreign markets based on a relative abundance of resources is explained in a theory formulated in the twentieth century by Heckscher et al. (H–O–S). The abundance principle, understood in statistical terms, constitutes also a basic element of the so-called neo-factor theories[10]. However, competitive advantages based on resources are time-limited. The depletion of resources or an increase in costs weakens the advantages in resources. Then, an indispensable condition that has to be met so that new competitive advantages can develop is an increase in investment in the processing industry. According to Porter's concept, due to an increase in the income per capita, countries pass to the second stage of the competitive development based on investments. The import of technologies from abroad as well as their adaptation to local needs constitute the basis for the improvement in competitiveness. An increase in investments and a gradual accumulation of capital become an entry barrier for the competitors. However, at the time of strong market volatility and fast technological progress, the import of technologies becomes, after a certain period of time, insufficient to maintain the competitive advantage, especially if the income per capita is gradually rising. Then, the innovativeness of local entities becomes the method for developing competitive advantages. The stage of the development based on innovations relies on the innovative activity of countries, activity, which not only develops imported

[10] Mentioned in Chap. 1.

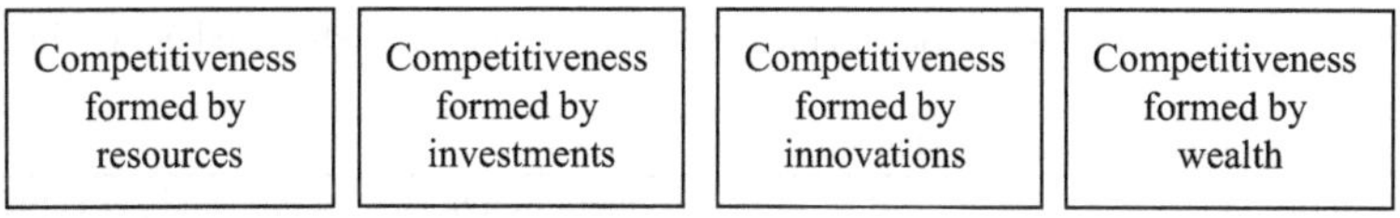

Fig. 3.4 Stages of competitive development. (Source: Based on M. Porter, *The Competitive Advantage of Nations,* New York, The Free Press 1990, p. 546)

technological solutions, but also creates entirely new solutions, which are unique on a global scale. The profile of the international specialization of a given country shifts toward goods that are technologically advanced and require expertise. In order to maintain a competitive advantage in these domains, it is necessary to continue to invest in research and to translate the results of the research into innovations. Simultaneously, it is possible to observe a further increase in prosperity, which is reflected in the increasing income per inhabitant. It translates into an increase in the costs of production, R&D, and innovative activity. As a result, the cost competitiveness is lost. At this stage of development, the competitiveness of a country is shaped by the accumulated wealth of the nation (Fig. 3.4).

The classification of stages of competitive development elaborated by Porter has been adopted by the World Economic Forum (WEF) to create the *Global Competitiveness Report* (WEF 2012). These studies assumed that, for that moment, no country had achieved the last/fourth stage and, as a result, three basic stages of competitive development were distinguished:

1. factor-driven economies,
2. efficiency-driven economies, and
3. innovation-driven economies.

Furthermore, the report assumed that transitional stages could occur between the basic stages (WEF 2012).

According to Porter's evolutionary development of competitiveness and to its modified version by the WEF, a country can be competitive at each stage if it elaborates its own, unique system of skills, links, and communication, which, in a sense, corresponds with the concept of the NIS, in which links play the key role. Evolution of the key factors of competitiveness within the stages of competitive development, from the resources, through investments and innovations, to the accumulated wealth, is closely related to the development of education. Links related to production, technology, science, and culture existing in the economy are also significant. Changes in those links are often supported by the economic policy and other country-specific institutions (Ainamo 1997). The foregoing elements constitute common points of the concept of competitiveness and the concept of the NIS.

Taking into account the above-mentioned thesis, it is worth comparing Porter's classification with the taxonomy of NISs. A multilevel, developed taxonomy of the NIS was elaborated, with the use of the hierarchical cluster analysis, by Godinho et al. (2003, 2006). The authors distinguished eight principal dimensions of innovation systems, to which they assigned indicators describing innovativeness and human capital. The NIS features included in this taxonomy and the corresponding indicators are presented in Table 3.1.

Table 3.1 NIS classification criteria and their indicators. (Source: Study based on: Godinho et al. 2006, p. 11–12)

Features of national innovation systems	Dimensions
Market conditions	Overall GDP size income and per capita, and population density
Institutional conditions	Income inequalities, life expectancy, demographic structure, and corruption index
Tangible and intangible investment	Expenditures on R&D and education per capita and as a percentage of GDP and rate of investment
Theoretical and applied knowledge	Percentage of population with secondary and higher education, percentage of science students, researchers as a percentage of labor force, and number of scientific papers per capita
Economic structure	Share of technologically advanced industries in exports and in the added value and revenue of companies conducting R&D activity on a global scale in proportion to the GDP
Links between the economy and the environment	Balance of foreign trade and direct investments in relation to the GDP and broadband Internet connections
Diffusion of knowledge	Internet access, density of mobile networks, and number of ISO 9000 and ISO 1400 certificates per inhabitant
Innovation	Number of patents and trademarks per capita

The NIS features and their indicators were used to empirically analyze 69 economies of different sizes, populations, and levels of economic development. The group of countries included in the classification represents about 87.4 % of the global population (Godinho et al. 2006, p. 13). This is the reason why the proposed taxonomy can be considered comprehensive and one that covers almost the entire global economy.

The classification of NISs was conducted according to the method of the hierarchical cluster analysis. The method makes use of the distance between features describing the analyzed innovation systems in multidimensional space in order to determine dense clusters—groups of countries of similar parameters. On this basis, two main types of NISs were distinguished: (1) developed innovation systems and (2) developing innovation systems. Within each of these NIS types, three smaller subtypes were distinguished. Some of them were subdivided into kinds that were not named by the researchers but were marked with symbols. Hong Kong was classified as a separate NIS type. As a result, the proposed taxonomy of NISs is composed of 15 country groups as illustrated in Table 3.2.

The taxonomy will be used in the econometric analysis and case studies of selected countries.

The choice of the countries, which are subject to case studies, ensures the diversity of the research sample. A purposeful selection has been used, so that the following two criteria are met:

Table 3.2 Taxonomy of national innovation systems operating in the global economy—a multidimensional approach. (Source: Based on Godinho et al. 2003 and 2006 as well as Godinho 2007)

NIS type	Subtype	Kind	Countries representing particular NIS types
T.O.	Hong Kong		
T.1.	T. 1.1.	Ireland, the Netherlands, Switzerland, Finland, Sweden, and Singapore	
Developed innovation systems	Dynamic innovation systems		
	T.1.2.	T.1.2.1.	Germany, Great Britain, France, Italy, South Korea, and Taiwan
	Performing innovation systems	T.1.2.2.	the United States and Japan
	T.1.3.	T.1.2.3.	Canada, Norway, Australia, Austria, New Zealand, and Spain
	Unevenly developed innovation systems	Denmark, Belgium, and Luxembourg	
T.2.	T.2.1.	T.2.1.1.	Portugal, Greece, Poland, Hungary, the Czech Republic, and Slovenia
Developing innovation systems	Catching-up innovation systems	T.2.1.2.	Malaysia and Malta
		T.2.1.3.	Latvia, Estonia, Lithuania, Slovakia, and Ukraine
	T.2.2.	T.2.2.1.	Russia
	Hesitating innovation systems	T.2.2.2.	China, Brazil, South Africa, Thailand, Argentina, India, and Mexico
		T.2.2.3.	Turkey, Colombia, Bulgaria, Indonesia, the Philippines, Peru, and Romania
		T.2.2.4	Egypt, Cyprus, Chile, and Venezuela
	T.2.3.	T.2.3.1.	Algeria, Iran, Vietnam, Morocco, and Bangladesh
	Unformed innovation systems	T.2.3.2.	Pakistan, Kenya, Ethiopia, Tanzania, Sudan, Nigeria, Congo, and Myanmar

1. the main types of NISs be represented and
2. countries placed on different levels of competitive development determined by Porter's classification be represented.

From the group of the so-called dynamic NIS, three countries were chosen to be analyzed in the form of case studies—Switzerland, Finland, and Ireland. They were classified by the WEF as innovation-driven development countries, occupying the highest positions in the ranking of competitiveness (1st, 3rd, and 27th place, correspondingly) (WEF 2012, p. 10 and p. 13). Countries with a stable NIS analyzed in detail by means of case studies were Germany, the U.S., Austria, and Spain. These were also countries classified as innovation-driven by the WEF, but they occupy different positions in the ranking of competitiveness (6th, 7th, 16th, and 36th place, correspondingly). The third group of countries characterized in detail in terms of the impact of human capital, innovations, and competitiveness includes two countries with the catching-up NIS—Poland and Hungary. Both these countries are in a transitional stage between the efficiency-driven and the innovation-driven development. They occupy further places in the ranking of competitiveness (41st and 60th place, respectively). The last group of case studies relates to countries with unbalanced NIS. These are China, Mexico, and Russia, which occupy the 29th, 53rd, and 67th places, respectively, in the ranking of competitiveness. Furthermore, these countries are at different stages of competitive development—China is still at the efficiency-driven stage, while Russia and Mexico represent a transition stage between the efficiency-driven and innovation-driven development (WEF 2012, p. 10 and p. 13). The more detailed analysis of the case studies will be preceded by an econometric analysis covering all countries in the world (subject to availability of statistical data), which is the subject of Chap. 4.

3.7 Conclusion

To sum up, the relationship between innovation systems and competitiveness can be observed in at least three dimensions—technological, institutional, and international. Certain elements of innovation systems—technology, innovations, human capital, and institutions—constitute factors of competitiveness of economies. The actors of systems of innovation, i.e., companies, research units, and administration bodies, influence the factors of competitiveness in both direct and indirect ways. Economic links with foreign countries are not included in innovation systems, but influence them, constituting factors of competitiveness of economies in dynamic terms and a manifestation of competitive position (in static terms) at the same time. Innovations are transferred from one country to another. Additionally, entities from several different countries undertake joint efforts to create new knowledge or to use it in an innovative way. All these processes influence the competitiveness of countries and companies. Investments in R&D as well as an appropriate system of education constitute the source of improvement in competitiveness by influencing the development of absorption capacities, i.e., the ability to recognize and to use the knowledge derived from both internal and external sources. The establishment of links between different organizations, the creation of cooperation networks of institutions from different countries, as well as the development of the ability to absorb new knowledge, which has become accessible thanks to such links, are therefore crucial.

The foregoing deliberations lead to the conclusion that the innovation system and competitiveness have a theoretically common area defined by the technology, by the operation of institutions, which are necessary for the development and transfer and diffusion of the technology, and by a number of areas of interaction, which include, *inter alia*, the relationships between business entities and foreign countries.

Chapter 4
Human Capital and Innovations As Determinants of Competitiveness

Ziemowit Czajkowski

This analysis is aimed at identifying the main determinants of international competitiveness of an economy and at designing measures thereof and verifying a hypothesis concerning the diversification of influence of individual determinants with respect to groups of countries singled out on the basis of characteristic features of their innovation systems. The issue of identification of the above mentioned determinants and diversification of their influence is of high importance in terms of verification of efficiency and purposefulness of allocation of resources among fixed assets, human capital, and research and development (R&D)—and thus, for both the purposes of assessing the economic policy pursued and designing future development strategies.

4.1 Model Assumptions

Econometric modeling is a technique rarely used in the field of analysis of competitiveness of national economy. The reasons for such a state of affairs are lack of a generally accepted method of measuring competitiveness (see Gomułka and Czajkowski 2010) and lack of possibility to express most elements of competitiveness analysis schemes (see Czajkowski 2010a) in the form of mathematical models. The works of Fagerberg (1988; Fagerberg et al. 2007) are fundamental studies that apply model approach at the macroeconomic level. An interesting concept in the field of dependent variables, introduced in these works, was linking the dependent variable with the relative levels of independent variables (with international environment as a reference group). The proposed model had also a drawback, as competitiveness of national economy was identified with its relative growth rate. Such formulation of the topic makes it impossible to analyze relationships between expenses and resources of a given national economy and its success in exports. On the basis of the

Z. Czajkowski (✉)
World Economy Research Institute, Warsaw School of Economics,
Al. Niepodleglosci 162, 02-554 Warsaw, Poland
e-mail: zczajk@sgh.waw.pl

M. A. Weresa (ed.), *Innovation, Human Capital and Trade Competitiveness*, Innovation, Technology, and Knowledge Management, DOI 10.1007/978-3-319-02072-3_4,

concepts of J. Fagerberg, an alternative model was developed (Czajkowski 2010b) defining the benchmark of competitiveness in accordance with Trabold (1995) as the ability to sell combined with the ability to adjust. Relative levels of price competitiveness (an indirect benchmark—unit labor costs (ULC)), technological competitiveness (patents, R&D expenses, investments in fixed assets, and percentage of R&D workers in the general number of workers), investment competitiveness (foreign direct investments, FDIs), and relative level of human capital (an indirect benchmark—percentage of people with higher education in the total number of the employed) have been used as independent variables. The aggregation of independent variables into indexes has been abandoned in the model in order to retain sound diagnostic and analytical properties (clear interpretation of parameters). A fixed-effects model has been adopted. This study develops and modifies the approach applied in Czajkowski (2010b). The number of countries used in the model shall be extended, longer time series used, and a division into groups of countries in accordance with innovation system taxonomy proposed in Godinho et al. (2005) shall be introduced. For each of the groups, a separate fixed-effects model will be estimated in order to verify the hypothesis on diversification of factors determining the competitive success depending on the characteristics of the innovative activity of entities operating in individual national economies.

4.2 The Design of the Model and Measurement of Variables

First, this subchapter presents a theoretical outline of the model—in the form of a scheme of correlations between the benchmark of competitiveness and generally defined independent variables. Then, taxonomy shall be presented, which was used to divide countries according to characteristics of their innovation systems (which determines their competitive strategy—and as a result, the strength of impact of individual competitiveness determinants is differentiated). In the following part, empirical interdependence between the dependent variable and independent variables is analyzed, and specific economic strategies shall be defined and used for the purposes of calculation of indirect benchmarks for independent variables. This stage of the study shall be conducted for unmodified time series (countries for which there are gaps in time series shall be excluded from it).

When compared to the version presented in Czajkowski (2010b), the basis model design has not been changed. The independent variables were selected on the basis of the approach proposed by Schumpeter (1950) and its implementation in Fagerberg's model (Fagerberg 1988; Fagerberg et al. 2007). The relative growth rate of the share in global foreign trade, in accordance with the adopted concept, depends on technological factors, production potential, and price competitiveness. Each of these aspects is taken into account in the proposed model. The set of competitiveness factors outlined above, proposed by J. Fagerberg, is supplemented with the foreign market penetration capacity. The capacity is an important aspect of competitiveness defined as the ability to produce goods and services attractive in international markets (ability to sell) and the ability to adjust to changes taking place in the environment (ability to adjust)[1]. The competitiveness factor described above comprises, among other things, the brand and the trust of business partners. They may, according to the research by Baroncelli et al. (2005) and Fink et al. (2005), significantly hinder entities from a given country in utilizing their comparative advantages—or significantly decrease benefits gained from foreign trade. The final form of the model scheme is as follows:

$$S_C = f(A,T,R,Q,P), \tag{4.1}$$

where

S_C is the benchmark of competitiveness of national economy, measured with relative export success,
A is the relative foreign market penetration capacity,
T is the relative level of technological advancement,
R is the relative R&D potential,
Q is the production potential, and
P is the relative price competitiveness.

In the discussed model, the dependent variable (S_C) is the relationship of the growth rate of exports of country *C* to the growth rate of foreign trade of the reference group (this benchmark shall be discussed in detail further). Independent variables are relationships of selected parameters of the economy of country *C* to the average levels of these parameters recorded in the reference group. Deviations from the average value, considered to be the "zero level," in accordance with the adopted concept of the model, constitute competitive advantages of the country or the environment in particular areas.

Interactions between variables:

1. Price competitiveness P
   - makes it possible to increase foreign market penetration capacity A;
   - a drop in price competitiveness may be compensated with an increase in the relative foreign market penetration capacity (brand and established business contacts);
2. Relative level of technological advancement T
   - makes it possible to maintain or increase relative foreign market penetration capacity A (patents $\Rightarrow$ new product, higher quality), despite decreasing price competitiveness P (salary growth);
   - or by process innovations and decrease of production prices—contributes to the increase of price competitiveness P;
   - growth of technological advancement level T results in growth of production potential Q and R&D potential R;

[1] See Chap. 1.

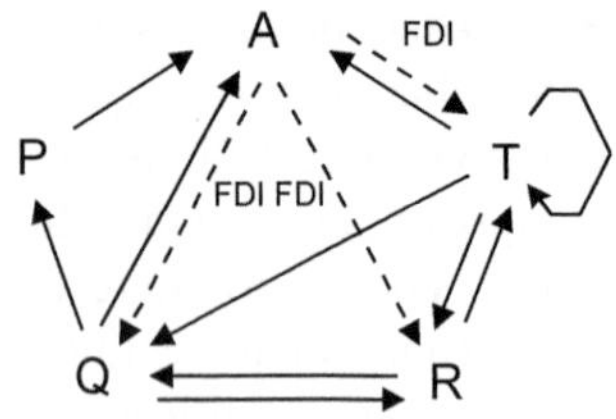

Fig. 4.1 Interactions between independent variables

3. Relative R&D potential R
   - influences the changes of relative level of technological advancement T
   - and production potential Q;
4. Relative production potential Q
   - through gains of scale and production technology (embodied in machinery park), as well as human capital (productivity resulting from quality of employees) it influences price competitiveness P;
   - by way of product quality—it influences relative foreign market penetration capacity A;
5. Relative foreign market penetration capacity A, in the case when it is a result of FDIs, exerts influence, through the spillover effect, on T, R, and Q (dashed arrows on Fig. 4.1).

Delayed interactions between individual independent variables are shown in Fig. 4.1. These interactions shall not be modeled and their occurrence is one of the reasons for the application of a multiplicative-exponential form of the model:

$$S_C = kP^{\alpha_1} T^{\alpha_2} R^{\alpha_3} Q^{\alpha_4} A^{\alpha_5}, \tag{4.2}$$

where $\alpha_i > 0, i = 1, 2,..., 5$.

Transmission mechanisms are discussed further in detail. Due to the abstract character of variables, it was necessary to develop methods of approximation of S_c, A, T, R, Q, and P variables.

4.2.1 Benchmark of International Competitiveness of Economy

The literature on the subject—as it was indicated in Chap. 1 hereof—presents many theories of competitiveness of national economy.

In this chapter, an approach focused on the result of competition shall be applied. As a variable approximating international competitiveness of economy of a given country, relative export growth rate shall be used (see Czajkowski 2010b, pp. 71–77). If a given country increases its exports more slowly than it stems from the growth ratio of international trade, it may be assumed that enterprises located in this country are unable to retain their share in the market (insufficient production capacities in conditions of growing demand, unacceptable combination of price, and quality for the purchaser), or to enter newly created markets.

The adopted method combines the ability to sell and the ability to adjust, as mentioned by Trabold (1995)[2], in a conceptual area, as well as the approach of "revealed competitive advantages" proposed by Balassa (1965). The adopted approach has a disadvantage consisting in not taking into account the aspect of quality of life in a country that is successful in terms of exports[3], because exports may be forced by, e.g., foreign debt (see e.g., Hatsopoulos et al. 1988)—and reduce significantly the standards of living of residents of a given country.

The resultant benchmark of competitiveness based on relative growth rate is expressed by the formula (4.3)[4]

$$S_{C,t} = \frac{\dfrac{X_{C,t}}{X_{C,t-1}}}{\dfrac{X^{C}_{W,t}}{X^{C}_{W,t-1}}}, \tag{4.3}$$

$$X^{C}_{W,t} = \sum_{\substack{c \in O \\ c \neq C}} X_{c,t}$$

where $X_{c,t}$ is the total exports of country C in time t, $X^{C}_{W,t}$ is the total exports of all countries from the reference group (with country C excluded) in time t, O is the set of indexes of all countries (or the reference group), and c is the summation index.

The formula relates the relative growth rate for a given country to the reference group for which the exports are summed up. For each country, S_c is calculated with the exclusion of this country from summation, so that the export growth rate of a given country, if it constitutes a substantial part of the whole international trade, does not distort the S_c indicator.

The index form of the indicator (4.3) adopted in the formula, instead of an incremental one, stems from the necessity to avoid division by zero in the case when there is no change in export volume for the reference group in a given period. Moreover, it allows avoiding the problem connected with different signs of the numerator and the denominator as well as excessively high values obtained for global trade growth rates (with country C excluded) close to 0. The indicator is a measure concentrated around 1, whereby values above 1 mean above-average export growth rate, while lower values mean problems with coping with international competitiveness. In addition, the indicator maintains economic interpretation after its logarithm has been computed.

However, the construction of the proposed indicator is not free from disadvantages:

- It does not answer the question on export profitability;
- It does not provide information whether the quantitative success is a result of

[2] see also Reiljan and Kulu (2002).

[3] The quality-of-life aspect is an important part of many definitions of competitiveness, see e.g., Aiginger (1998, 2006a, 2006b) and Czajkowski (2010a).

[4] See Czajkowski (2010b, pp. 73, formula 15).

  - an advantage of a given country in the form of technological superiority of a product and/or production process,
  - an underestimation of the relative (when compared to other countries) remuneration of some of production factors,
  - or price dumping in combination with state subsidies;
- Under-average export growth rate may be a result of an external shock that affected economies of main partners, and not a result of insufficient expenditures or incorrect allocation thereof;
- It ignores the structure of exports
  - which causes the countries that export low-technology products to be assessed as equally competitive as countries exporting Information and Communications Technology (ICT) products and services,
  - countries specializing in exports of products and services for which there was a rapid growth recorded in global demand are treated in the same way as countries specializing in less dynamic sectors of production and services,
  - export success observed with the use of the indicator may be a result of a change in terms of trade, and not the activities of enterprises or state institutions of a given country;
- It ignores geographical structure of exports and the impact of integration blocks;
- It does not provide information about the quality of life in a country deemed to be competitive (which is considered to be a big disadvantage by many researchers (see Aiginger 1998, 2006a, b); if capital flows due to means of production held outside of a given country constitute a significant element of dwellers' revenues, the indicator shall show a competitive shortage despite the fact that the quality of life in a given country may be significantly higher than in countries classified as more competitive;
- As data enabling calculation of the indicator solely for the trade inside the group were not available, values of the indicator were calculated on the basis of exports performed also to countries from outside of the reference group—thus, the indicator accounts for the global export success of country C in comparison to the growth rate of exports of the reference group.

Bearing in mind the disadvantages of the indicator described above, it should be underscored that in the case of analyses conducted at the level of international economies, it is necessary to apply simplifications, and the proposed indicator makes it possible to introduce the quantitative measurement of relations between expenditures (and accumulated expenditures), and the results of competition, to the competitiveness analysis.

4.2.2 *Relative Price Competitiveness P*

Relative price competitiveness P comprises, among other things, the differences in growth rate of salaries, labor force productivity, and deviations of the real exchange rate from the equilibrium rate. The relative price competitiveness may be reduced to relationships between prices of the exchanged products and services, expressed in a common currency. As an intermediate measure of competitiveness, relative cost of labor may be used and calculated in accordance with the formula

$$P_t = \frac{ULC_{C,t}}{ULC_{W,t}} = \frac{ULC_{C,t}}{\sum\limits_{\substack{c \in O \\ c \neq C}} \frac{EMP_{c,t}}{EMP_{W,t}} ULC_{c,t}}, \tag{4.4}$$

where $ULC_{C,t}$ is the unit labor cost[5] in country C in year t, $EMP_{c,t}$ is the employment in country c in year t, $EMP_{W,t}$ is the total employment for all the countries (or the reference group), with the exclusion of C, whereby other symbols have the same meaning as in (4.3).

With the use of weights it is possible to take into account the differences in quantities of products and services that are possibly available at the prices stemming from $ULC_{C,t}$, and, as a result, also the average level of prices of products and services exported by enterprises of a given country related to the average price on international markets (which is given as the average of $ULC_{C,t}$ for the environment, weighted with the size of employment of other market participants' economies). The assessment of the parameter in the case of variable P in the estimated model shall allow answering the question to what extent economies of a given country group base their export strategies on the labor price.

4.2.3 Relative Level of Technological Advancement T

The level of technological development T reached by entities operating in a given country is understood as a capacity to create products that are unique in terms of technology used or the innovative usage of technological solutions. A side effect of reaching a given level of technological development is the creation of grounds for further development. From the perspective of the capacity to deliver products and services with specific performance characteristics, the relative level of technological advancement depends on both accumulated knowledge and technology embodied in the fixed assets. Since it is impossible even to roughly estimate the amount of knowledge materialized in machinery and devices, it shall be assumed for the purposes of further analyses that T is independent of the fixed assets, or the growth rate of their development.

Technological and product competitiveness is determined, among other things, by technological knowledge resources available in the economy. It should be noted that as a result of technology diffusion (see e.g., Metcalfe 1988), specific solutions cease to play a significant role as competitive factors over time due to the fact that

[5] Various proxy benchmarks of ULCs and labor productivity will be used as $ULC_{C,t}$.

they are generally used. As an intermediate measure of T, the cumulative number of patents submitted by residents of a given country in n years preceding the moment of observation shall be used, in accordance with formula (4.5):

$$B_{C,t} = \frac{\dfrac{Pat_{C,t}}{Y_{C,t}}}{\dfrac{Pat_{W,t}}{Y_{W,t}}} = \frac{\dfrac{Pat_{C,t}}{Y_{C,t}}}{\dfrac{1}{Y_{W,t}} \sum\limits_{\substack{c \in O \\ c \neq C}} Pat_{c,t}} \tag{4.5}$$

$Pat_{c,t} = \sum\limits_{i=0}^{n} Patents_{c,t-i}$, where $Pat_{C,t}$ and $Pat_{W,t}$ are, respectively, the number of patents submitted in country C and in the reference group in the period $(t-n,t)$, whereby other symbols are defined as in (4.4). Other factors that comprise the relative level of technological advancement are covered in other parts, as staff qualifications and technology materialized in the fixed assets. Due to the above, the following identity is assumed: $T = B_{C,t}$.

4.2.4 Relative Research and Development Potential R

The relative R&D potential reflects the capacity of a given economy to produce new technologies and obtain existing ones—through the purchase of patents and fixed assets, spillover effects, diffusion, and imitation (see Fagerberg 1988; Fagerberg et al. 2007). To absorb technologies, an economy needs to have an adequate number of R&D staff, adequate amounts of funds earmarked for R&D, and the relevant infrastructure (R&D centers and laboratories with technical backup). The effects of this factor are threefold: low R&D potential leads to the decrease of relative technological potential T, which advances over time (the number of technologies produced locally and absorption capacity for external solutions go down, unless there are sufficiently strong spillover effects, generated, *inter alia*, because of FDIs). This effect is transmitted through T to Q, and, in addition, R&D development staff working to satisfy the needs of industry increase the production potential (e.g., via process innovations). This direct relationship is shown in Fig. 4.1 as Q ← R. The symbols are defined in the following way: $SG_{C,t}$ and $SG_{W,t}$ are the accumulated expenses for R&D: (GERD) in the period $(t-n,t)$.

Analogically as in the case of formula (4.5), we shall obtain

$$G_{C,t} = \frac{\dfrac{SG_{C,t}}{Y_{C,t}}}{\dfrac{SG_{W,t}}{Y_{W,t}}} = \frac{\dfrac{SG_{C,t}}{Y_{C,t}}}{\dfrac{1}{Y_{W,t}} \sum\limits_{c \in O} SG_{c,t}}, \tag{4.6}$$

where

$$SG_{c,t} = \sum_{i=0}^{n} GERD_{c,t-i},$$

and

$$E_{C,t} = \frac{\frac{RDE_{C,t}}{EMP_{C,t}}}{\frac{RDE_{W,t}}{EMP_{W,t}}} = \frac{\frac{RDE_{C,t}}{EMP_{C,t}}}{\frac{1}{EMP_{W,t}} \sum_{\substack{c \in O \\ c \neq C}} RDE_{c,t}} \tag{4.7}$$

where $RDE_{C,t}$ is the number of R&D workers in country C in year t, $RDE_{W,t}$ is the number of R&D employees in year t in countries of the reference group, and EMP is the total employment.

As a result, R shall assume the following form, which corresponds to the logic of the Cobb–Douglas production function:

$$R_{C,t} = E_{C,t}^{\gamma_1} G_{C,t}^{\gamma_2}, \tag{4.8}$$

where $\gamma_i > 0, i = 1, 2$.

4.2.5 Relative Production Potential Q

The technological supremacy of a given economy is not a sufficient condition for obtaining a higher share in the international market. It is also necessary to ensure relevant production capacities Q. We shall assume that the production capacities are comprised of fixed assets (machinery park and infrastructure) present in the economy, as well as skilled labor force resources L.

4.2.6 Fixed Assets

From the point of view of the international technological competitiveness—related to products and processes—of enterprises, the most important part of a machinery park is its most recent part. One should expect that intense modernization of fixed assets shall contribute to the improvement of price and quality competitiveness of produced goods and services. In consequence, a good approximation of fixed assets for the most advanced part of economy is constituted by accumulated investments (gross fixed capital formation, GFCF) from n prior periods. A problem may occur here, consisting in the fact that the pace of utilization of fixed assets differs across industries. Investment expenditures shall be calculated in relation to the gross domestic product (GDP), in accordance with formula (4.9; see formula 4.13):

$$K_{C,t} = \frac{\dfrac{KI_{C,t}}{Y_{C,t}}}{\dfrac{KI_{W,t}}{Y_{W,t}}} = \frac{\dfrac{KI_{C,t}}{Y_{C,t}}}{\dfrac{1}{Y_{W,t}} \sum\limits_{\substack{c \in O \\ c \neq C}} KI_{c,t}}, \tag{4.9}$$

where $KI_{c,t} = \sum\limits_{i=1}^{n} GFCF_{c,t-i}$.

4.2.6.1 Human Capital

Usually, due to the availability of statistical data, the following variables are used to measure human capital:[6]

- average years of schooling,
- average years of schooling at tertiary level, and
- percentage of employees with higher education in the total of the employed.

When trying to find a variable that reflects resource L we encounter two problems: the first one is the unavailability of sufficiently detailed statistical data (gaps connected with periodicity of the studied level and structure of education of the population, short time series) related to education, and the second one is mismatch of the education structure and the actual needs of the economy. A country that has a high percentage of workers with higher education may have serious shortages within the scope of highly skilled technical staff and technical personnel.

Ultimately, due to the problems indicated above, given the lack of data on persons with secondary and higher technical education in countries of groups covered by the study, an index calculated in accordance with the following formula will be used:[7]

$$L_{C,t} = \frac{\dfrac{TERT_{C,t}}{EMP_{C,t}}}{\dfrac{TERT_{W,t}}{EMP_{W,t}}} = \frac{\dfrac{TERT_{C,t}}{EMP_{C,t}}}{\dfrac{1}{EMP_{W,t}} \sum\limits_{\substack{c \in O \\ c \neq C}} TERT_{c,t}}, \tag{4.10}$$

where *TERT* is the number of people with higher education and *EMP* is the people employed in total.

[6] The indexes listed are the simplest intermediate benchmarks. This issue is discussed in more detail in Chap. 2.

[7] See Czajkowski (2010b).

Alternative methods of human capital measurement, based on the series developed by Barro and Lee (2010), shall be used for the purposes of a test analysis between S_C index and human capital. The relevant index shall be calculated as

$$H_{C,t} = \frac{h^l_{C,t}}{\sum\limits_{\substack{c \subset O \\ c \neq C}} p_{c,t} h^l_{c,t} / \sum\limits_{\substack{c \in O \\ c \neq C}} p_{c,t}}, \qquad (4.11)$$

where $h^l_{c,t}$ is the average number of years of education at a given level/ in country c in year t, $p_{c,t}$ is the population in country c in year t.

As a result, the formula approximating Q may be, analogically with (4.8), as follows:

$$Q_t = K_t^{\omega_1} L_t^{\omega_2}, \qquad (4.12)$$

where $\omega_i > 0, i = 1, 2$.

4.2.7 *Relative Foreign Market Penetration Capacity*

The relative foreign market penetration capacity of enterprises located in a given country is influenced by:

1. brands and trademarks that are recognizable internationally,
2. unique technology contained in a product, guaranteed with patents,
3. unique quality guaranteed by the technology used, and
4. direct access to the market as a consequence of completed FDIs.

It is worth stressing that FDIs, if they are oriented towards export production, are usually connected with one or more of factors 1–3. Furthermore, FDIs are conducive to spillover effects, so they influence, with a delay, variables T and R.

The inclusion of FDIs in the model is connected with several issues. FDIs are characterized by very high volatility over time, and often more than a half of flows registered in a given year are constituted by a single investment. In addition, production capacities are launched with a certain delay, and gradually. Spillover effects also occur with a certain delay—and they occur over a certain period of time as well. Only the access to markets is obtained in a leap, which nevertheless—for the above reasons—does not necessarily result in a leap in exports growth. In effect, a good proxy for the influence of FDIs is their accumulated values from n prior periods. In this way, we shall also obtain a certain smoothening of time series, which is important considering the high variance observed for FDIs, especially in the case of smaller economies. However, the estimation of the optimum n value is extremely difficult—due to high diversification of delays for which an investment completed in a given year influences the increase of exports and other independent variables. It is also difficult to determine influence depreciation coefficients for individual factors (Gaussian curve seems most appropriate to model this phenomenon). Taking the above into consideration, the relative extent of saturation of the economy with FDIs may be expressed with formula (4.13)

$$F_{C,t} = \frac{\frac{FDS_{C,t}}{Y_{C,t}}}{\frac{FDS_{W,t}}{Y_{W,t}}} = \frac{\frac{FDS_{C,t}}{Y_{C,t}}}{\frac{1}{Y_{W,t}} \sum_{\substack{c \in O \\ c \neq C}} FDI_{c,t}}, \tag{4.13}$$

where $FDS_{c,t} = \sum_{i=1}^{n} FDI_{c,t-i}$ is the accumulated FDI flows in n periods preceding moment t.

As for now, international data containing statistics of trademarks and brands submitted for registration make it possible to analyze relationships between them and the S_c index. An adequate measure for the relative resource of trademarks restricted by entities from a given country takes an analogical form as in the case of patents (formula 4.5):

$$M_{C,t} = \frac{\frac{ZT_{C,t}}{Y_{C,t}}}{\frac{ZT_{W,t}}{Y_{W,t}}} = \frac{\frac{ZT_{C,t}}{Y_{C,t}}}{\frac{1}{Y_{W,t}} \sum_{\substack{c \in O \\ c \neq C}} ZT_{c,t}} \tag{4.14}$$

$$ZT_{c,t} = \sum_{i=0}^{n} TrademarkApplications_{c,t-i},$$

where $ZT_{C,t}$ and $ZT_{W,t}$ are, respectively, the number of trademark applications submitted by residents in country C and in the reference group in the period $(t-n,t)$, whereby other symbols have the same meaning as in (4.4).

What is worth stressing here is that there are certain difficulties concerning the allowing for F and B in the formula, as for highly advanced countries with high income of the population the inflow of FDIs is not usually aimed at export production but rather at delivering products and services to the local market (avoiding possible anti-import barriers); also, spillover effects are not likely to occur.

Bearing in mind that the above-mentioned factors taken into account through patents T are responsible for part of the foreign market penetration capacity, they may be introduced to the model as

$$A_t = F_t^{\beta_1} M_t^{\beta_2}, \tag{4.15}$$

where $\beta_i > 0$, $i = 1, 2$.

4.3 Final Form of the Model

Finally, after inserting equations describing A,T,R,Q,P into formula (4.2), we shall obtain

$$S_{C,t} = k P_t^{\alpha_1} B_t^{\alpha_2} E_t^{\gamma_1} G_t^{\gamma_2} K_t^{\omega_1} L_t^{\omega_2} F_t^{\beta_1} M_t^{\beta_2}, \tag{4.16}$$

where $\alpha_i > 0,\ \beta_i > 0,\ \gamma_i > 0,\ \omega_i > 0, i = 1, 2.$

After computing logarithms for both sides (16) we obtain a linear model with regard to the parameters,

$$\begin{aligned}\ln(S_{C,t}) = k + \alpha_1 \ln(P_{C,t}) + \alpha_1 \ln(B_{C,t}) + \gamma_1 \ln(E_{C,t}) + \gamma_2 \ln(G_{C,t}) \\ + \omega_1 \ln(K_{C,t}) + \omega_2 \ln(L_{C,t}) + \beta_1 \ln(F_{C,t}) + \beta_2 \ln(M_{C,t})\end{aligned} \tag{4.17}$$

The model in form (4.17) has very favorable interpretation qualities

$$\begin{aligned}&(\ln(X_{C,t}) - \ln(X_{C,t-1})) - (\ln(X_{W,t}) - \ln(X_{W,t-1})) \\ &\quad = k + \alpha_1 \ln(P_{C,t}) + \alpha_1 \ln(B_{C,t}) + \gamma_1 \ln(E_{C,t}) + \gamma_2 \ln(G_{C,t}) \\ &\quad\ + \omega_1 \ln(K_{C,t}) + \omega_2 \ln(L_{C,t}) + \beta_1 \ln(F_{C,t}) + \beta_2 \ln(M_{C,t}),\end{aligned} \tag{4.18}$$

whereby $\frac{\Delta X_{C,}}{X_{C,t-1}} - \frac{\Delta X_{W,t}}{X_{W,t-1}} \approx (\ln(X_{C,t}) - \ln(X_{C,t-1})) - (\ln(X_{W,t}) - \ln(X_{W,t-1}))$.

The expression on the left is, as it may be seen, the difference in the pace of export growth of country C and the reference group. Estimations of parameters connected with variables shall, in such a situation, have the same interpretations as in the case of a level-log model. For instance, the interpretation of α_1 is as follows: an increase of value P by 1 % causes a change in the difference of pace of export growth in relation to the environment by $\alpha_1/100$.

4.4 Method and Criteria for Country Grouping

The results obtained in the study of Czajkowski (2010b) indicated a necessity of taking into account the heterogeneity of countries in the model, as well as allowing the diversification of influence directions of individual independent variables for different countries—a fixed (within) effects model turned out to be insufficient. Since the available time series were too short, it was impossible to make estimations for each of the countries individually (too low number of degrees of freedom). Therefore, it was necessary to group countries.

Many grouping methods are applied in econometric research. One such method was proposed by Hansen (1996, 1999, 2000; panel modeling with thresholds and

Table 4.1 Taxonomy of national innovation systems. (Source: Based on Godinho et al. 2003, 2006, as well as Godinho 2007)

NIS type	Subtype	Kind	Countries representing particular NIS types
T.0.	Hong Kong		
T.1. Developed innovation systems	T. 1.1. Dynamic innovation systems	Ireland, the Netherlands, Switzerland, Finland, Sweden, Singapore	
	T.1.2. Performing innovation systems	T.1.2.1.	Germany, Great Britain, France, Italy, South Korea, Taiwan
		T.1.2.2.	the United States, Japan
		T.1.2.3.	Canada, Norway, Australia, Austria, New Zealand, Spain
	T.1.3. Unevenly developed innovation systems	Denmark, Belgium, Luxembourg	
T.2. Developing innovation systems	T.2.1. Catching-up innovation systems	T.2.1.1.	Portugal, Greece, Poland, Hungary, the Czech Republic, Slovenia
		T.2.1.2.	Malaysia, Malta
		T.2.1.3.	Latvia, Estonia, Lithuania, Slovakia, Ukraine
	T.2.2. Hesitating innovation systems	T.2.2.1.	Russia
		T.2.2.2.	China, Brazil, South Africa, Thailand, Argentina, India, Mexico
		T.2.2.3.	Turkey, Colombia, Bulgaria, Indonesia, the Philippines, Peru, Romania
		T.2.2.4	Egypt, Cyprus, Chile, Venezuela
	T.2.3. Unformed innovation systems	T.2.3.1.	Algeria, Iran, Vietnam, Morocco, Bangladesh
		T.2.3.2.	Pakistan, Kenya, Ethiopia, Tanzania, Sudan, Nigeria, Congo, Myanmar

division of sample); grouping takes place during estimation of the model in order to maximize the fitting. However, this method does not permit using multidimensional variables in sample division. Similarities of national economy innovation systems are difficult to identify with the use of a single variable. In order to obtain relatively reliable grouping, justified in terms of the subject, the taxonomy obtained with the use of hierarchical cluster method in the study of Godinho et al. (2005) (Table 4.1.) was used [8].

[8] This topic is discussed in more detail in Chap. 3.

Due to the lack of data it was impossible to include the group T.2.3., as well as some other countries from other groups (these were marked out with underlining in Table 4.1), into account. Group T.1.3. had been removed from the study because it was impossible to apply a panel model for the available time series (insufficient number of observations). In total, data have been gathered enabling the comparison of 40 countries.

4.5 Empirical Verification of the Model

This subchapter presents relationships between the S_c index and actual data, the sources of which (with specific data series and formula in which it were used) are listed in Table 4.2. For the purpose of comparisons, average logarithmic values of the dependent variable and independent variables were used, recorded in the years 2000–2010. As a result of the logarithmic transformation, the average level of competitiveness and competitiveness factors is shifted from 1 to 0. Due to the size of group T.1.1. (five countries), the regression line drawn on the charts below for this group is of purely informative character. For the same reasons, regression lines for other groups should also be treated with great care. The analysis below is focused on initial examination whether there are "trajectories" common for countries belonging to respective groups, and to what extent deviations from these trajectories may occur (given sufficiently large deviations, the estimation of a common trajectory will not be possible for such small samples that we deal with in individual groups, but it will be good to initially analyze data on a higher level of aggregation). Additionally, charts of S_c dispersion in relation to individual dependent variables were prepared and split into groups as indicated in Table 4.1 (see appendix, Figs. 4.15–4.18).

Due to the low availability of data concerning unit labor costs,[9] it was necessary to construct an alternative measure for ULC, with similar interpretation possibilities. With the use of time series published by the World Bank and Eurostat, it is possible to replace the unit labor cost ($ULC_{C,t}$) in formula (4.4) with the following expression:

$$LC_{c,t} = \frac{w^h_{C,t}}{Y^{EMP}_{c,t}} = w^h_{C,t} / \frac{Y_{C,t}}{EMP_{C,t}}, \qquad (4.19)$$

where $w^h_{c,t}$ is the average hourly labor cost in industry (taking into account social insurance and other benefits),[10] $EMP_{C,t}$ is the employment in country C in year t,

[9] Data published by OECD did not include a major part of the studied countries, while other databases—e.g., EU KLEMS— cover too limited scope of years.

[10] Eurostat data: *Hourly labour costs – Nace Rev. 1.1 [lc_an_costh]*, Bureau of Labour Statistics (BLS) series *Hourly compensation costs in manufacturing, U.S. dollars, 1996–2010.*

Table 4.2 Data used in the model

Formula	Variable	Data[a]	Notes
(4.19)	$w^h_{c,t}$	Total labor costs (including social insurance). Eurostat: *Hourly labor costs—Nace Rev. 1.1 [lc_an_costh]*, Bureau of Labour Statistics (BLS) series *Hourly compensation costs in manufacturing, USD, 1996–2010,* additions—see notes	Data for China (Banister and Cook 2011)(BLS); Data for Chile: International Labour Organisation (ILO)[b], dane dla Indii (Sincavage et al. 2010) (BLS); data for Russia: Federal State Statistics Service[c] corrected by social insurance; data for Ukraine: State Statistics Service of Ukraine[d], corrected by social insurance; data for Turkey: Turkish Statistical Institute[e]
(4.19)	$Y^{EMP}_{c,t}$	GDP per employee; calculated on the basis of the World Bank data	(series: NY.GDP.MKTP.CD[f] employment series $EMP_{C,t}$)
(4.4), (4.10)	$EMP_{C,t}$	calculated on the basis of the World Bank data	$EMP_{C,t}$=(1-SL.UEM.TOTL.ZS[g]/100 [unemployment])*SL.TLF.TOTL.IN[h]
(4.5)	$Patents_{c,t}$	Patent applications by residents; the World Bank data (series IP.PAT.RESD)	Missing data supplemented using Cran-R package, the functions aregImpute from Hmisc and imputation from longitudinalData[i]. Source: OECD data, series *Patent grants at the USPTO* and *Patent applications to the EPO*
(4.3)	$X_{C,t}$, $X_{W,t}$	World Bank, series: exports in fixed prices (USD, year 2000), NE.EXP.GNFS.KD	Missing data supplemented with the World Bank data; exports in current prices: NE.EXP.GNFS.CD.
(4.13)	$FDI_{c,t-i}$	World Bank, series BX.KLT.DINV.WD.GD.ZS (FDI as % of GDP) in real values; fixed prices in USD (year 2000)	$FDS_{C,t}$ calculated for n=4.
(4.5), (4.6), (4.13)	$Y_{C,t}$, $Y_{W,t}$	World Bank, values in USD in current prices and in PPP standard; series NY.GDP.MKTP.CD	Missing values supplemented using the formula R package with imputation formula from longitudinal Data.
(4.7)	$RDE_{C,t}$	Number of researchers (in full time equivalent); compiled from the World Bank data SP.POP.SCIE.RD.P6 as well as OECD and Eurostat databases	Missing values supplemented using the formula R on the basis on Eurostat data (*Total R&D personnel by sectors of performance, occupation and sex [rd_p_persocc]* i OECD)
(4.6)	$GERD_{C,t-i}$	Values in USD, fixed prices and PPP in USD from the year 2000; World Bank, series: GB.XPD.RSDV.GD.ZS * NY.GDP.MKTP.KDOECD.	Missing values taken from Eurostat data (*Total intramural R&D expenditure (GERD) by sectors of performance [rd_e_gerdtot]*). $SG_{C,t}$ and $SG_{W,t}$ computed assuming n=4

Table 4.2 (continued)

Formula	Variable	Data[a]	Notes
(4.9)	$GFCF_{C,t-i}$	Values in USD, fixed prices and PPP in USD from the year 2000; World Bank, series *Gross fixed capital formation (constant 2000 USD), NE.GDI.FTOT.ZS.*	$KI_{c,t}$ calculated for $n=4$
(4.10)	$TERT_{C,t}$	Calculated on the basis of the World Bank data—series Barro–Lee percentage of population with higher education (BAR.TER.SCHL.15UP), labor force with higher education (SL.TLF.TERT.ZS)	Corrected using Eurostat data (*Employees by sex, age and highest level of education attained (1 000) [lfsa_eegaed]*) and OECD (*Graduates by field of education, Level of education: 905160: Tertiary-type A and advanced research programs*).

[a] Missing data supplemented using the Hmisc and longitudinal Data package R (formulas aregImpute and impute). R is statistical and econometrical package available at: http://cran.r-project.org/
[b] http://laborsta.ilo.org/data_topic_E.html
[c] http://www.gks.ru/bgd/regl/b11_12/IssWWW.exe/stg/d01/07-07.htm
[d] http://www.ukrstat.gov.ua/operativ/operativ2007/rp/ean/ean_e/osp_rik07_e.htm, http://www.ukrstat.gov.ua/operativ/operativ2005/gdn/prc_rik/prc_e/osp_e.html
[e] http://www.turkstat.gov.tr/PrelstatistikTablo.do?istab_id=1328i, http://www.turkstat.gov.tr/PrelstatistikTablo.do?istab_id=1292
[f] GDP in USD, current prices, series http://databank.worldbank.org/ddp/home.do, *World Development Indicators & Global Development Finance*
[g] Unemployment rate, series in the World Bank database
[h] Labor force (below the age of 15) series, as above

and $Y_{c,t}^{EMP}$ is the GDP per one employed person. As a result, expression (4.4) shall be as in the following formula:

$$P_{C,t}^{w,Y} = \frac{\dfrac{w_{C,t}^{h}}{Y_{c,t}^{EMP}}}{\dfrac{w_{W,t}^{h}}{Y_{W,t}^{EMP}}} = \frac{w_{C,t}^{h} \Big/ \dfrac{Y_{C,t}}{EMP_{C,t}}}{\left(\sum\limits_{\substack{c\in O\\c\neq C}} EMP_{c,t} w_{c,t}^{h} \Big/ \sum\limits_{\substack{c\in O\\c\neq C}} EMP_{c,t}\right) \Big/ \left(\sum\limits_{\substack{c\in O\\c\neq C}} Y_{c,t} \Big/ \sum\limits_{\substack{c\in O\\c\neq C}} EMP_{c,t}\right)} \tag{4.20}$$

Are there any correlations between the price competitiveness index defined in such a way and the competitiveness index? In Fig. 4.2, country groups selected from Table 4.1 are presented. Intuitively, the growth of $P_{C,t}^{w,Y}$ is accompanied by a drop in the relative export growth rate. What draws attention here is the fact that countries with the highest unit labor costs do not show the lowest S_c values. The rank of India as the second country after Germany with the highest unit labor costs may be somehow puzzling—but it may be explained in terms of low labor efficiency approximated by

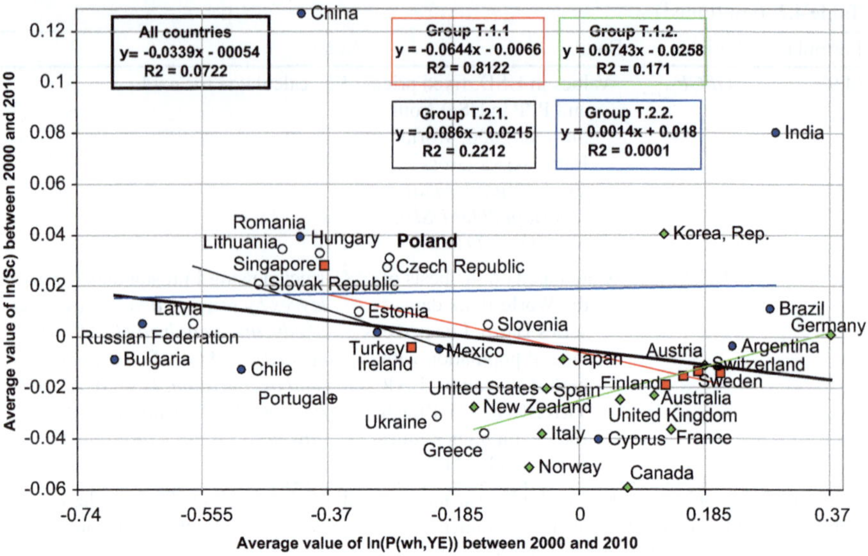

Fig. 4.2 Relations between $P_{C,t}^{w,Y}$ and relative dynamics of exports (S_c), average values between 2000 and 2010 (Symbol $X_{c,\,t}$ shall be used interchangeably with X(C, t) in description of axes of the chart; estimations of directional coefficients and curve adjustment parameters are of informative character; data used to prepare the chart come from the sources listed in Table 4.2. (Source: own work)

the use of $Y_{c,t}^{EMP}$. The analysis of dispersion for individual groups indicates that the systematics of countries proposed in Table 4.1. is not connected with the phenomenon that is decisive for the weak correlation visible in Fig. 4.2. (there is no relationship between the features of the innovation system and the influence of unit costs onto the export success). Only four countries with unit labor costs above average record relative export growth rate which is also above average (including India, for which the estimation may be significantly biased, which is described below).

The initial hourly labor cost shall be used as a supplementary measure:

$$P_{C,t}^{w,E} = \frac{w_{C,t}^{h}}{w_{W,t}^{h}} = \frac{w_{C,t}^{h}}{\sum\limits_{\substack{c\in O\\ c\neq C}} EMP_{c,t} w_{c,t}^{h} \,/ \sum\limits_{\substack{c\in O\\ c\neq C}} EMP_{c,t}}, \tag{4.21}$$

where $w_{W,t}^{h}$ is the average global hourly labor cost in industry with other symbols defined as above.

The measure ignores the labor productivity aspect, and it should be added that the index (4.20) underestimates the labor productivity in export industries—namely industries that shape the values of the S_c index. The issue of labor effectiveness in export industries was examined, *inter alia*, in Wagner (2005). In addition, in the case of countries with low nominal labor costs,the ownership structure of capital

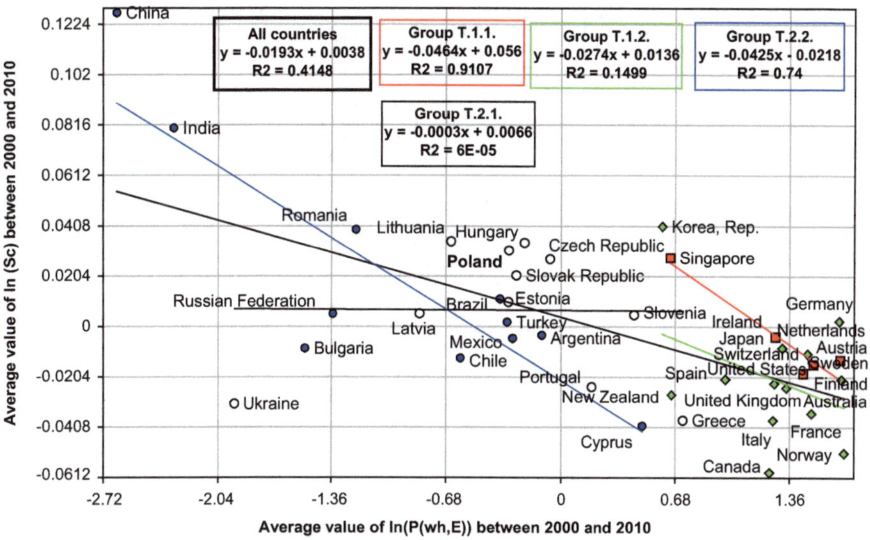

Fig. 4.3 Relative price competitiveness $(P_{C,t}^{w,E})$ and relative dynamics of exports (S_c), average values between 2000 and 2010. (Source: own work)

in exporting enterprises is important, due to the transfer of technologies and work organization. In order to verify results observed for India, Argentina, and Brazil, which are inconsistent with expectations, a chart was prepared that presents the relationship between S_c and $P_{C,t}^{w,E}$.

The change in the slope of lines (aside from T.1.2.) in Fig. 4.3 is mainly an effect of the change of scale on the horizontal axis. Once again, it may be seen that the systematics introduced on the basis of examination of innovation systems does not put countries in any proper order in terms of the role of labor cost factor in the export success. Countries of groups T.1.1. and T.1.2. show, in accordance with Fig. 4.3, over-average relative labor costs—and relative export growth rates that are lower than average. Countries with the lowest relative labor costs—India and China—are leading ones in terms of S_c index values. For the whole sample, there is a very weak negative correlation between the export success and both indexes of price competitiveness: $P_{C,t}^{w,Y}$ and $P_{C,t}^{w,E}$.

4.5.1 Relative Level of Technological Advancement T

For the purposes of examination of the relationship between $B_{c,t}$ and S_c, $n=4$ (see expression (4.5); application of higher n values did not lead to increasing the strength of the correlation, and lower values (1–3) decreased its strength) was assumed for calculation of $B_{c,t}$. The analysis of empirical data allows the ascertaining that there is large diversification of theoretical influence of the relative technological

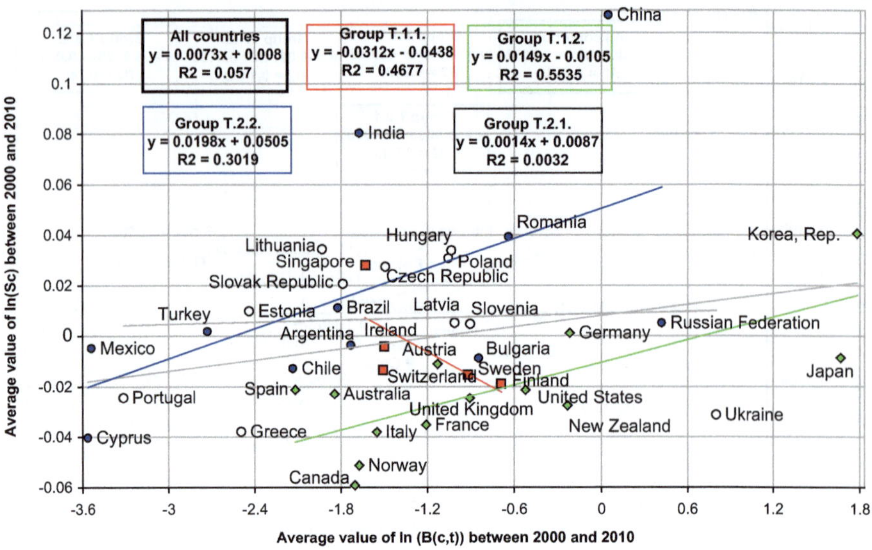

Fig. 4.4 Relative technological competitiveness ($B(C, t)$) and relative dynamics of exports (S_c), average values between 2000 and 2010. (Source: own work)

advancement measured with the relative "saturation" of the GDP with patents (4.5) on the export success. One may indicate a group for which patent applications are not connected with changes of S_c (T.2.1.), a group for which export success is not negatively correlated with patent activity (T.1.1.), and two groups for which the direction of the correlation is in line with expectations (Fig. 4.4). Among the countries that show the highest relative innovativeness measured with $B_{c,t}$, only South Korea records a value of $S_{c,t}$ that significantly departs from the average. All relationships are very weak.

4.5.2 Relative Research and Development Potential R

In accordance with the assumptions of the model, the relative R&D potential is determined by two factors. One of these factors is constituted by relative fixed assets resources, used for the purposes of the R&D activity, approximated with relative accumulated expenditures for R&D in prior periods ($G_{C,t}$, formula 4.6). The other one is the human capital available in the field of R&D, approximated with the relative percentage of scientists ($E_{C,t}$, formula 4.7; Fig. 4.6).

Trends observed for individual country groups are similar as in the case of the relative technological advancement level approximated with the relative patent activity. Among the countries with above-average expenses for R&D, only Germany, North Korea, and Singapore show greater export growth rate (Fig. 4.5).

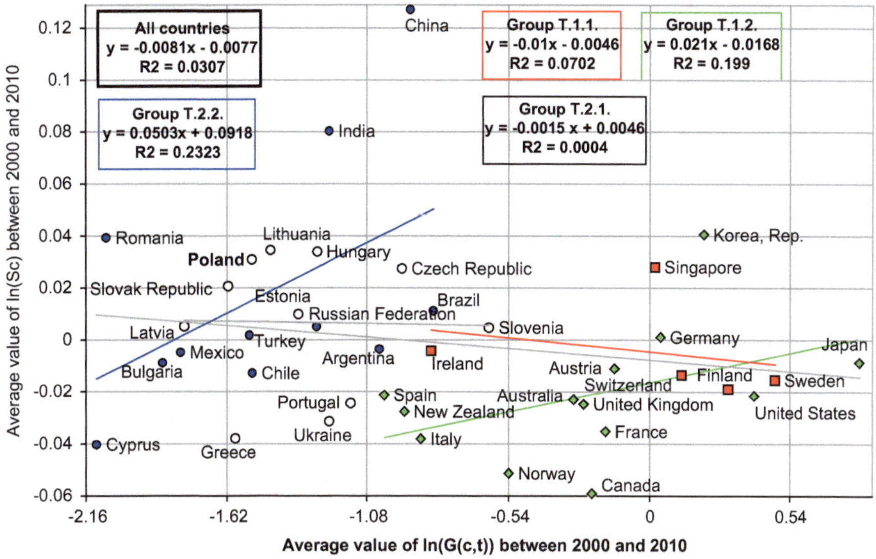

Fig. 4.5 Correlation between competitiveness measured with the average Sc value and relative accumulated expenses for research and development ($G_{C,\,t}$). (Source: own work)

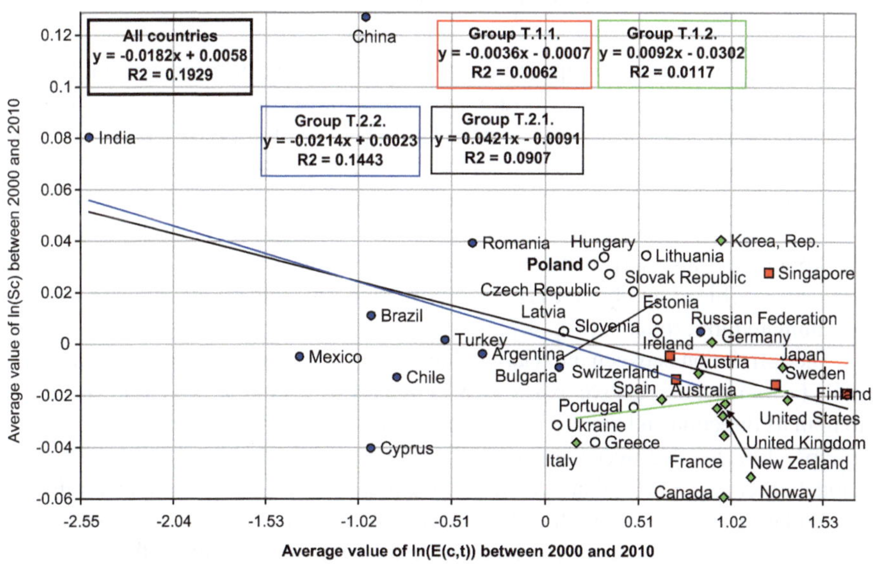

Fig. 4.6 Correlation between competitiveness measured with the average S_c value and the relative share of scientific workers in total employment ($E_{C,\,t}$). (Source: own work)

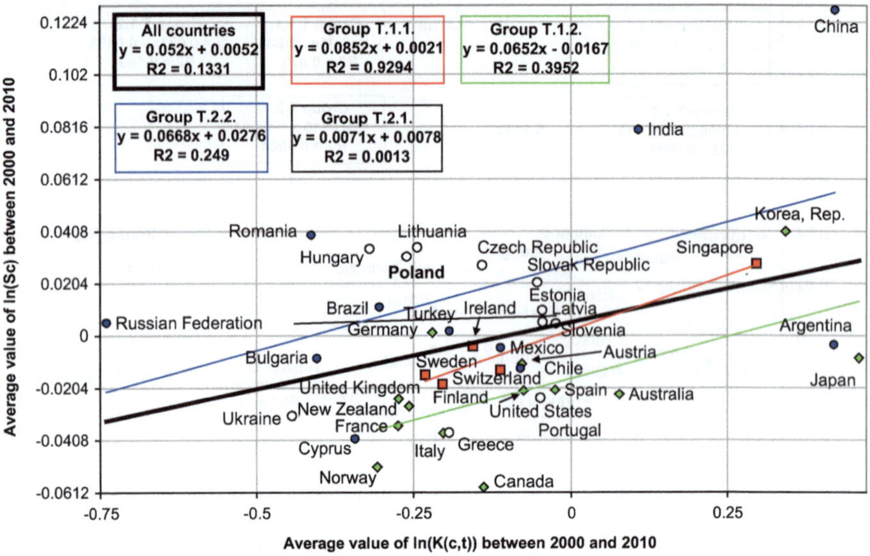

Fig. 4.7 Correlation between competitiveness measured with the average Sc value and the relative share of advanced production assets in economy ($K_{C,t}$). (Source: own work)

4.5.3 Relative Production Potential Q

For all country groups, except for T.2.1., there is a weak positive correlation between the relative export growth rate and accumulated investments in fixed assets (formula (4.9)). The virtual lack of correlation between the average relative export growth rate in 2000–2010 and development of the production potential is a counter-intuitive observation. Japan, which is modernizing its fixed assets in the most intense way, reaches a growth rate which is lower than average, and the result observed for Argentina is also surprising. Among highly developed countries, only Korea and Singapore achieve an above-average export pace given an above-average level of accumulated investments (Fig. 4.7).

The correlation between the relative share of labor force with higher education (formula (4.10)) and the relative export growth rate presented in Fig. 4.8 is surprising due to both its direction and the fact that the negative correlation is the case for all country groups. It means that over the period of 11 years, relatively higher levels of human capital observed in individual national economies do not seem to bring about the expected benefits measured with export success. Among the countries of group T.1.2., only South Korea achieved a relative export growth rate which is higher than that in the reference group. What also draws attention is the high relative level of human capital in the Russian Federation and Ukraine. These observations may be a result of inconsistencies in the methodology of recording of education levels in individual countries. In Fig. 4.9, another benchmark of the relative level of human capital in economy was applied (see formula 4.11)—the relative average number of years of education on the

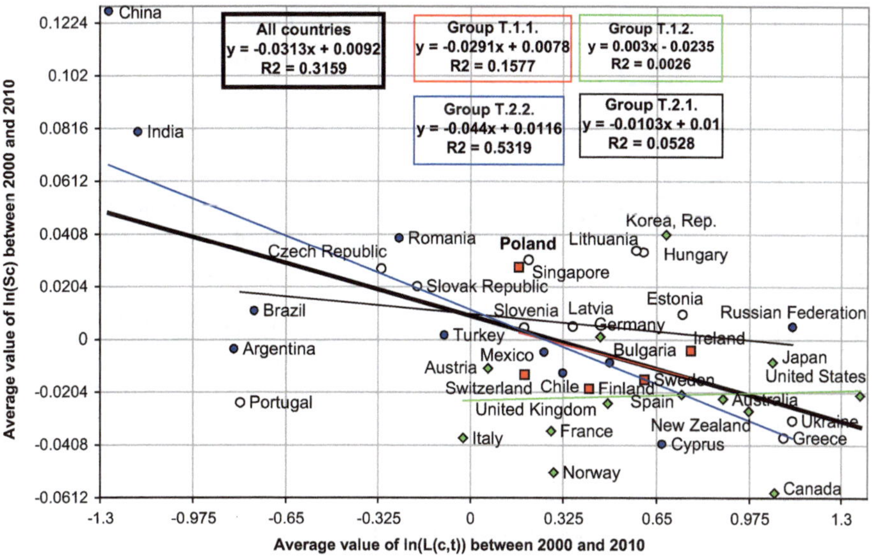

Fig. 4.8 Correlation between competitiveness measured with the average Sc value and the relative share of workers with higher education in total employment ($L_{C,t}$)

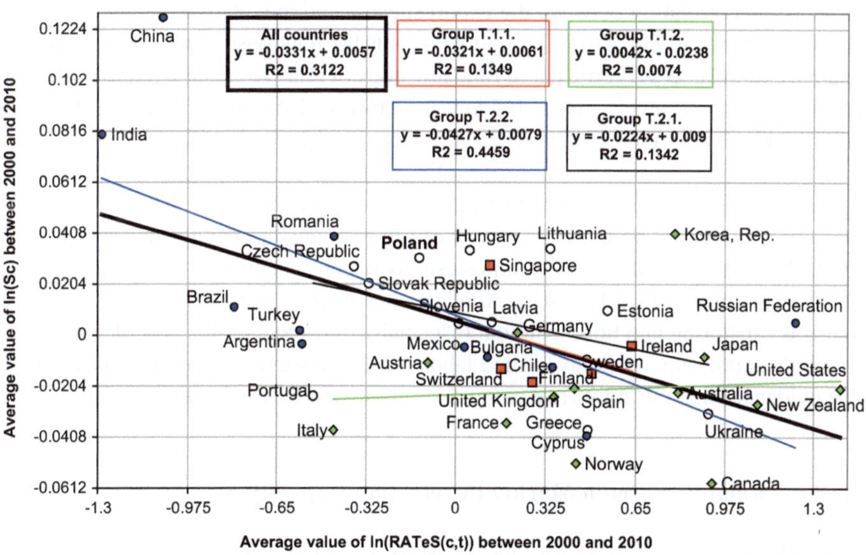

Fig. 4.9 Correlation between competitiveness measured with the average S_c value and the relative average number of years of education on the tertiary level ($\text{RATeS}_{C,t}$)

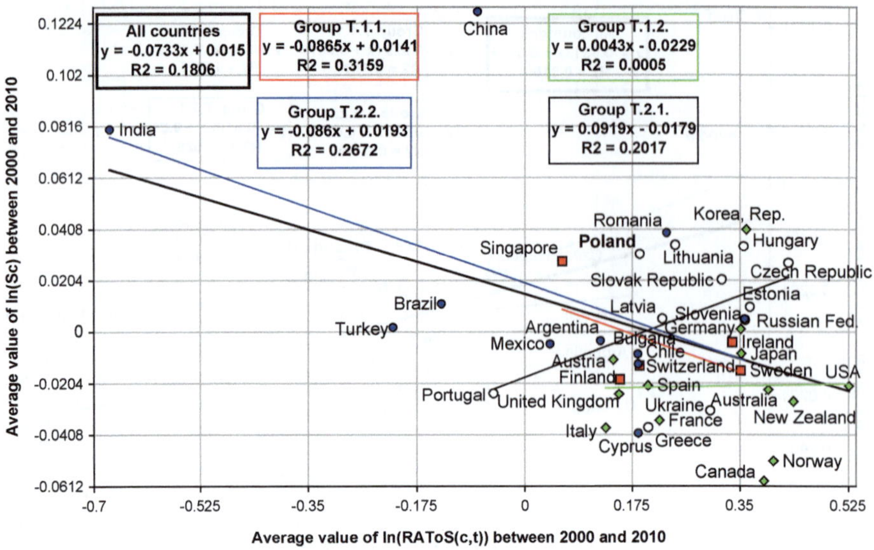

Fig. 4.10 Correlation between competitiveness measured with the average S_c value and the relative average number of years of education on all levels ($RAToS_{C,t}$)

higher level,[11] observations are analogical to those for Fig. 4.8. The last chart that presents the correlation between the export success and the relative human capital resources is shown in Fig. 4.10. The figure presents the relationship between an average relative export success (vertical axis) and the index of relative average number of years of education on all levels, calculated in accordance with formula (4.11) for the series published by Barro and Lee (2010)[12]. As opposed to the trends observed in Figs. 4.8–4.10 it is possible to observe, in relation to group T.2.1., a weak positive correlation between the export success and the relative level of human capital. The observed coincidence should not be understood as a causal relationship. Higher levels of human capital resources in national economies usually occur concurrently with higher levels of nominal salaries (see Fig. 4.3). Another phenomenon that explains the observed trends is offshoring and hollowing out (especially in respect of this part of production which does not require highly skilled labor force). However, Portugal is an example that evidences that the above reasoning does not necessarily exhaust the issue.

4.5.3.1 Relative Foreign Market Penetration Capacity *A*

According to expectations formulated earlier, a weak negative correlation between export success and the relative level of FDIs in GDP (formula 4.13) may be obser-

[11] Barro–Lee series were used: *Average years of tertiary schooling, age 15+, total*, available under signature BAR.TER.SCHL.15UP in the World Bank data base (http://databank.worldbank.org/ddp/home.do data collection for Education Statistics).

[12] Barro–Lee series *Average years of total schooling, age 15+, total,* series signature BAR.SCHL.15UP source as above.

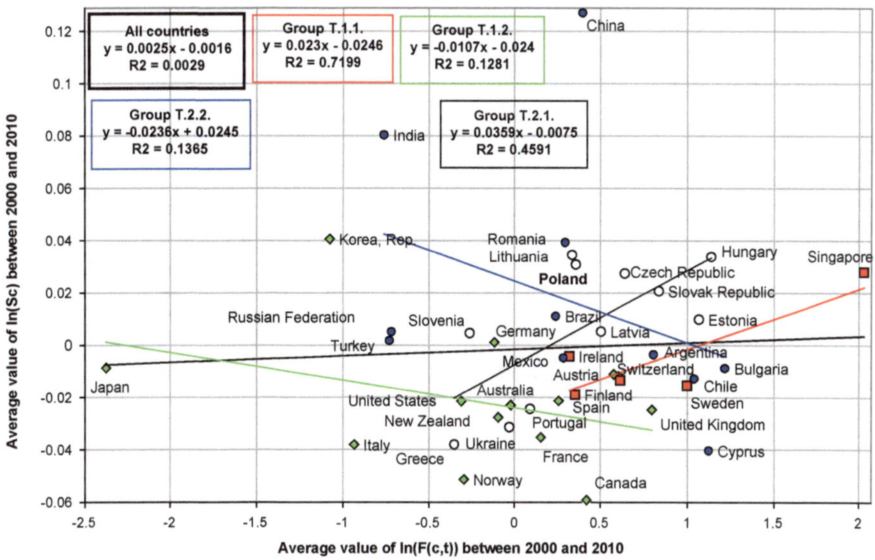

Fig. 4.11 Correlation between competitiveness measured with the average S_c value and the relative share of FDIs in GDP ($F_{C,t}$). (Source: own work)

ved for highly developed countries from group T.1.2. A trend observed for group T.2.1. is also in line with expectations (Fig. 4.11).

For all country groups, except for T.2.1., there is a weak positive correlation between the relative accumulated number of registered trademarks (formula 4.14) and the relative pace of export growth. What is surprising is that high values of $M_{C,t}$ were recorded for China and India. Among countries with high $M_{C,t}$ values, countries classified as "developing innovation systems" (group T.2., Table 4.1) dominate. The only country not belonging to T.2. that records above-average relative exports growth rate and at the same time a high value of $M_{C,t}$ is South Korea (Fig. 4.12).

4.6 Statistical Verification of the Model

The statistical analysis of data on the level of groups and countries revealed diversification of average values and range of variability of $ln(S_c)$ both at the level of country groups distinguished in Table 4.1 (Fig. 4.13) and at the level of individual countries (a relevant account of countries sorted by groups and alphabetically is presented in Fig. 4.14). What is distinctive here is strong intragroup diversification—which means that there must be factors that influence the dependent variable which are not connected with the characteristics of the innovation system that is not the only factor determining foreign trade competitiveness. The data structure indicates that an ordinary least squares (OLS)-type model (a pooled model, without individual (group), or time effects) will not be adequate.

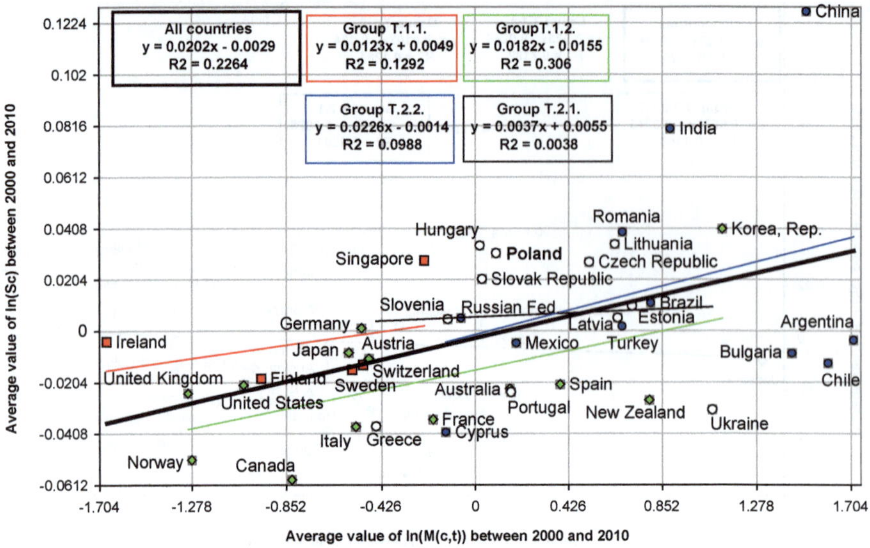

Fig. 4.12 Correlation between competitiveness measured with the average Sc value and the relative saturation of economy with registered trademarks ($M_{C,\ t}$). (Source: own work)

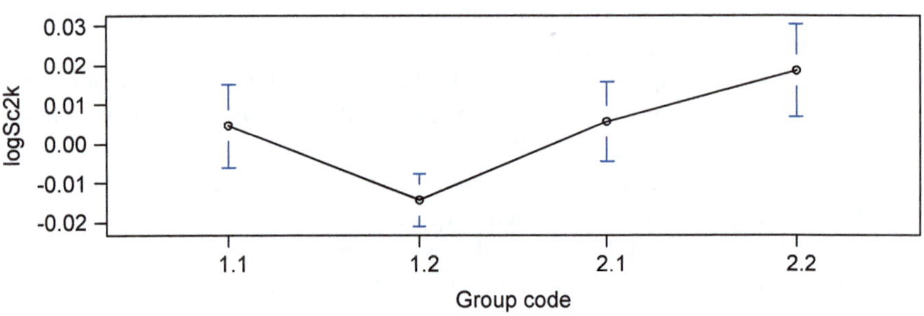

Fig. 4.13 Heterogeneity of $ln(S_c)$ observed on the level of groups T.1.1–T.2.2 (data: 1995–2010). Average values circles) and 95 % confidence intervals (*vertical lines*). (Source: own work)

4.6.1 Relationships Between Independent Variables

Prior to the estimation of a model, it is reasonable to analyze the structure of relationships between respective time series (Table 4.3).

The analysis of data contained in Table 4.3. allows ascertaining that at the level of all countries there are only weak correlations between the dependent variable and the independent variables (≥0.2; correlations <0.2 are considered a sign of lack of correlation between variables). It should be taken into consideration however that the results concern the whole and strongly heterogenic population. Directions of relationships for variables connected with R&D potential (columns/rows Ect and

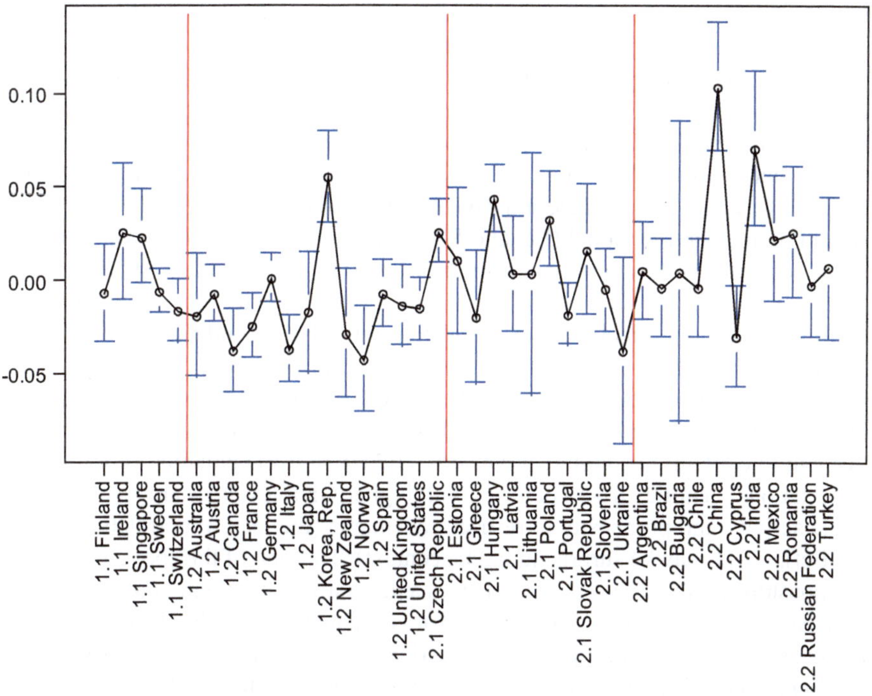

Fig. 4.14 Heterogeneity of $ln(S_c)$ observed at the level of individual countries (data: 1995–2010). Average values (*circles*) and 95 % confidence intervals (*vertical lines*; source: own work)

Table 4.3 Matrix of correlations between variables, all groups T.*i.j* together

ln(·) \ ln(·)	Sc2k	P(wh,YE)	~~P(wh,E)~~	Bct	Fct	Gct	Ect	Kct	Lct	~~RATeS~~	~~RAToS~~	Mct
Sc2k	1.00	-0.10	-0.28	0.09	0.02	-0.06	-0.22	0.23	-0.24	-0.25	-0.18	0.23
P(wh,YE)	-0.10	1.00	**0.52**	-0.04	-0.12	0.47	0.02	0.05	-0.09	-0.07	-0.18	-0.25
~~P(wh,E)~~	-0.28	**0.52**	1.00	-0.11	-0.13	**0.59**	**0.62**	-0.12	0.40	0.41	0.40	**-0.71**
Bct	0.09	-0.04	-0.11	1.00	-0.13	**0.50**	0.42	0.15	0.26	0.31	0.37	0.06
Fct	0.02	-0.12	-0.13	-0.13	1.00	-0.14	-0.06	0.00	-0.08	-0.09	0.02	0.20
Gct	-0.06	0.47	**0.59**	**0.50**	-0.14	1.00	**0.66**	0.28	0.24	0.33	0.29	-0.46
Ect	-0.22	0.02	**0.62**	0.42	-0.06	**0.66**	1.00	-0.09	**0.59**	**0.64**	**0.71**	**-0.53**
Kct	0.23	0.05	-0.12	0.15	0.00	0.28	-0.09	1.00	-0.26	-0.19	-0.12	0.28
Lct	-0.24	-0.09	0.40	0.26	-0.08	0.24	**0.59**	-0.26	1.00	**0.95**	**0.71**	-0.34
~~RATeS~~	-0.25	-0.07	0.41	0.31	-0.09	0.33	**0.64**	-0.19	**0.95**	1.00	**0.75**	-0.33
~~RAToS~~	-0.18	-0.18	0.40	0.37	0.02	0.29	**0.71**	-0.12	**0.71**	**0.75**	1.00	-0.26
Mct	0.23	-0.25	**-0.71**	0.06	0.20	-0.46	**-0.53**	0.28	-0.34	-0.33	-0.26	1.00

Gct) are counterintuitive (see Figs. 4.5 and 4.6). The situation is analogical in the case of variables describing human capital (Lct, RATeS, RAToS)—the results were probably influenced by values observed for India and China. Among other variables, lack of relative FDIs inflow (Fct) and very weak influence of relative level of technological advancement (Bct) may be disturbing.

Another problem that should be taken into account is very strong correlation between variables Lct, RATeS, and RAToS. It means that only one of these variables should be used in the model (Lct and RATeS are virtually identical, while RAToS in the case of less developed countries may be of greater importance than RATeS). Finally, Lct variables shall be used in the model due to better economical interpretability (despite marginally lower correlation with the dependent variable). Other interesting correlations between variables are [P(wh, E); {Mct; Gct; Ect}] (−0.71; 0.59; 0.62—respectively). The strong negative correlation between the average relative labor price in industry and the relative number of trademark applications is difficult to explain. The high relative correlation with R&D potential is in line with expectations—entities from countries with higher GDP levels usually show higher R&D activity and propensity to restrict trademarks (the lack of correlation [P(wh, E); Bct]: –0.11 may be surprising in this context). The first of the above-mentioned relationships (correlation −0.71) indicates a necessity of elimination of P(wh, E) or Mct. Since the price competitiveness factor is accounted for in P(wh, YE), we have to choose P(wh, E). Other possible reductions of the set of independent variables shall take place as part of the analysis after estimation of the model. Among other correlations, what is quite obvious is the relationship between both variables that approximate relative price competitiveness [P(wh, YE); P(wh, E)] (0.52), and the variables connected with innovative activity [Ect; Gct] (0.66) [Gct; Bct] (0.5). However, these correlations are not strong enough to serve as a basis for an a priori decision on elimination of any variables from the model, as in the case of the following three: Lct, RATeS, and RAToS.

After the elimination of variables in accordance with indications contained in Table 4.3, the intragroup correlation between independent variables and S_c was examined, and the results are given in Table 4.4. The values of correlations significantly differ between the groups; however, in the case of groups T.1.1. and T2.1., a major part of correlation estimates showed high values of significance level thresholds[13] (*P*-Value). Results contained in this table indicate the validity of the division into groups. For each of the groups, the analysis of correlations between independent variables was conducted once again, and as a result, logMct variable was eliminated for group T.1.1., due to its strong correlation with logGtc (0.745, given *P*-Value of 2.22e-15). Values of correlations after the elimination of logMct variable and including the Netherlands, which had been removed earlier to group T.1.1. due to lack of observations for $M_{c,t}$, are presented in Table 4.4.

[13] To obtain *P*-values, rcorr function of Hmisc library of R package was used.

Table 4.4 Correlations between the logarithms of independent variables and the logarithm of dependent variable, observed at the level of individual groups. (Source: own work)

Group	Value	Ln P(wh, YE)	lnBct	lnFct	lnGct	lnEct	lnKct	lnLct	lnMct
T.1.1.	*corr.*	*−0.36831*	*0.03886*	*0.02040*	*−0.21271*	*−0.12511*	*0.18034*	*0.09530*	*0.02315*
	P-Value	0.00078	0.73221	0.85744	0.05818	0.26883	0.10942	0.40042	0.83847
T.1.1.[a]	*corr.*[a]	*−0.37262*	*0.08256*	*0.01329*	*−0.16878*	*−0.06415*	*0.20172*	*0.07656*	–
	P-Value[a]	0.00019	0.42386	0.89775	0.10020	0.53466	0.04874	0.45847	–
T.1.2.	*corr.*	*0.11106*	*0.32164*	*−0.09981*	*0.19927*	*−0.05011*	*0.34327*	*0.02154*	*0.32743*
	P-Value	0.11026	0.00000	0.15147	0.00391	0.47223	0.00000	0.75746	0.00000
T.2.1.	*corr.*	*−0.17533*	*0.01957*	*0.04926*	*0.01103*	*−0.01177*	*0.04465*	*−0.08178*	*0.05832*
	P-Value	0.01994	0.79658	0.51617	0.88446	0.87679	0.55624	0.28056	0.44195
T.2.2.	*corr.*	*0.04682*	*0.17798*	*−0.06615*	*0.19229*	*−0.21390*	*0.33153*	*−0.32670*	*0.18731*
	P-Value	0.53722	0.01811	0.38307	0.01057	0.00437	0.00001	0.00001	0.01280

[a] After elimination of logMct, the Netherlands, that had been removed due to lack of data for M(c, t), were included into the group once again

4.6.2 Panel Model—Selection of the Model Type and the Estimator

Heterogeneity of population due to the characteristics of observed $S_{c,t}$ values and the diversification of the strength of relationship between $S_{c,t}$ and potential independent variables indicates that it is reasonable to make separate estimates of parameters of independent variables for each of the groups. The simplest solution was selected, consisting in the estimation of four separate panel models with the initial form of[14]

$$y_{i,t} = \alpha_i + \tau_t + \beta_{i,t}^T x_{i,t} + u_{i,t}, \tag{4.22}$$

where $i = 1,2,...,n$ is the number of the country, $t = 1,2,...,T$ is the year index, and $u_{i,t}$ is the random variable, $E(u_{i,t}) = 0$. The model (4.22) may be reduced to (4.23), if parameters are homogenous within the sample:

$$y_{i,t} = \alpha + \beta^T x_{i,t} + u_{i,t}. \tag{4.23}$$

During the estimation of the equation of linear regression for panel data it is assumed that there are differences between the values of constant α_i (individual effects[15]), τ_t (time effects), or different values of parameter β (different pace of changes of $y_{i,t}$ for individual groups i of the panel—or moments of observation t). In order to select a relevant form of the model, verification procedures were conducted for each of T.*i.j.* groups.

The first null hypothesis is the lack of diversification of α, τ, and β in terms of i and t (a model without individual and time effects (a pooled model):[16]

$$H_0 : \forall(i = 1, \dots, n; \ \ t = 1, \dots, T) \ \ \alpha_i = \alpha, \tau_t = 0, \beta_{i,t} = \beta \tag{4.24}$$

against the alternative hypothesis of variable coefficients model with individual effects

$$H_1 : \forall\left(t = 1, \dots, T; i = 1, \dots, n\right)\left(\tau_t = 0, \beta_{i,t} = \beta_i\right) \wedge \exists(i)\left(\alpha_i \neq \alpha, \beta_i \neq \beta\right) \tag{4.25}$$

Second null hypothesis is the occurrence of individual effects α_i and no diversification of τ due to t (no time effects) and β due to i and t (model with fixed effects (*FE*)/least squares dummy variable (*LSDV*)):

$$H_0 : \forall(i = 1, \dots, n; \ \ t = 1, \dots, T) \ \ \alpha_i \neq \alpha, \tau_t = 0, \beta_{i,t} = \beta \tag{4.26}$$

against the alternative hypothesis (4.25).

---

[14] Sometimes, instead of $\alpha_i + \tau_t$, the following notation is applied: $\alpha_{i,t}$.

[15] The term individual effects is in the literature used interchangeably with *group effects* in the literature of the subject.

[16] $\alpha + \tau$ is reduced in this model to $\alpha' \equiv \alpha$.

Table 4.5 Comparison of results of testing hypotheses (4.24) and (4.26) with the Chow F test as well as hypotheses related to time and two-way fixed effects tested with F test[a]). (Source: own work)

	Hypothesis	Test	Gr.	F	df1	df2	*P*-value
1.	$H_0{:}\forall(i,t)\ \{\alpha_i+\tau_t=\alpha \wedge \beta_{i,t}=\beta\}$ (neither individual nor time fixed effects present (*pooled*) (4.23))	Chow F	T.1.1.	1.70680	40	48	0.03846
			T.1.2.	1.37140	108	91	*0.06064*
			T.2.1.	1.54790	90	77	0.02502
			T.2.2.	1.06300	90	77	*0.39310*
2.	$H_0{:}\forall(i,t)\ \{\alpha_i \neq \alpha \wedge \tau_t=0 \wedge \beta_{i,t}=\beta\}$ (*individual fixed effects/LSDV*)	Chow F	T.1.1.	1.33450	35	48	0.17510
			T.1.2.	1.17780	96	91	*0.21590*
			T.2.1.	1.33460	80	77	0.10210
			T.2.2.	1.09330	80	77	*0.34740*
3.	$H_0{:}\forall(i,t)\ \{\alpha_i+\tau_t=\alpha \wedge \beta_{i,t}=\beta\}$ $H_1{:}\forall(i,t)\{\tau_t=0 \wedge \beta_{i,t}=\beta\}\wedge\exists(i)\ \{\alpha_i \neq \alpha\}$	F for significance of fixed effects (α_i)	T.1.2.	2.6754	12	187	0.002425
			T.2.2.	0.7829	10	157	*0.6452*
4.	$H_0{:}\forall(i,t)\ \{\alpha_i+\tau_t=\alpha \wedge \beta_{i,t}=\beta\}$ $H_1{:}\forall(i,t)\ \{\alpha_i=0 \wedge \beta_{i,t}=\beta\} \wedge \exists(t)\ \{\tau_t \neq \tau\}$	F for significance of time effects (τ_t)	T.1.1.	0.67620	15	73	*0.79920*
			T.1.2.	1.95780	15	184	0.02041
			T.2.1.	2.23650	15	152	0.00735
			T.2.2.	0.95660	15	152	*0.50380*
5.	$H_0{:}\forall(i,t)\ \{\alpha_i+\tau_t=\alpha \wedge \beta_{i,t}=\beta\}$ $H_1{:}\forall(i,t)\ \{\beta_{i,t}=\beta\} \wedge \exists(i,t)\ \{\alpha_i \neq \alpha \wedge \tau_t \neq \tau\}$ (*two-way fixed effects*)	F for significance of fixed effects both individual and time (α_i i τ_t)	T.1.1.	1.72560	20	68	*0.05023*
			T.1.2.	2.16890	27	172	0.00156
			T.2.1.	2.42820	25	142	0.00058
			T.2.2.	0.91880	25	142	*0.57950*

[a]The pFtest function from PLM library of R package has been used for the purposes of implementation of the test. See Croissant and Millo (2008), p. 22

To verify hypotheses (4.24 and 4.26), for models without effects (pooled models) and models with individual effects (a within model with effects for countries) the Chow test was applied[17]; the results are listed in Table 4.5. The first of the hypotheses (i.e., 4.24) should be rejected for the significance level of 0.05 in the case of groups T.1.1. and T.2.1. For the second one (i.e., 4.26), there are no grounds for rejection at the significance level of 0.1 in the case of all groups. It means that in the case of groups T.1.2. and T.2.2., the form of the model (whether it is a pooled one or LSDV) should also be verified. Test F was used to measure the significance of fixed effects in the LSDV model[18], and the test result is contained in row 3, Table 4.5. The test result for group T.1.2. suggests the necessity of rejecting H_0 (*P*-value $0.0024 \Rightarrow$ *FE* model), while T.2.2. does not provide any grounds to reject H_0 on the lack of occurrence of individual effects, which means that a pooled model shall be the appropriate one (the result may be surprising in the context of the observed heterogeneity of group T.2.2. in Fig. 4.14, but nevertheless the parameters of the pooled model are better than the fixed-effects model). In addition, the occurrence of time

[17] To implement the Chow test, pooltest function from PLM library of R package was used. See op.cit. Croissant, Millo (2008), pp. 20.

[18] pFtest function from PLM library of R package was used, see Croissant, Millo (2008), pp. 21.

Table 4.6 Model form preferred on the test basis. (Source: own work)

Group	Model
T.1.1.	Individual fixed effects/LSDV
T.1.2.	Two-way effects
T.2.1.	Two-way effects
T.2.2.	Pooled (no fixed effects)

effects was examined (Table 4.5., row 4). The test showed the occurrence of time effects in the case of groups T.1.2. and T.2.1. It is unfavorable because it means that a certain part of variability of dependent variable S_c was not captured in the selected set of independent variables. The last row of Table 4.5. contains results of tests for the hypothesis on the joint occurrence of time and individual effects; its result may not be analyzed irrespective of other tests (tests that verify the occurrence of time and individual effects separately). The models preferred for respective country groups in accordance with test results (Table 4.5.) are summarized in Table 4.6.

The next stage of the process of selection of the model form is to verify whether the random-effects (RE) model would not have better properties for the data set under examination. Models with fixed effects are preferred, if the sample (almost) exhausts the population (as in this study), but they are sensitive to autocorrelation and heteroskedasticity. The test of Lagrange multipliers showed that for all groups there are no grounds to reject the hypothesis on the lack of individual random effects (Table 4.7). For groups T.1.2. and T.2.1., the hypothesis on non-occurrence of random time effects had to be rejected. In the case of T.2.1, the test showed additionally that there are no grounds to reject the hypothesis on the concurrent occurrence of individual and time effects (a result that is inconsistent with the test for individual effects).

Random effects occurrence test results (Table 4.7.) should be additionally verified (with the exclusion of group T.2.2., for which the hypothesis on the occurrence of random effects was rejected) with the Hausman test (1978) that verifies the occurrence of correlation between errors and independent variables (H_0: no correlation ⇒ random and fixed estimators are equal, unbiased, and consistent, but the random one is effective and, thus, also the preferred one).

During the verification of the random model with individual effects and (/or) time ones it turned out that it was not possible to compute each of random-type estimators for every country group T.*i.j*. (errors for Swamy–Arora estimator (swar) and Nerlove estimator (nerlove)). Test results indicate that models with random effects should be used (exceptions: the fixed-effects model is preferred in the case of the Wallace–Hussein estimator (walhus) for model T.1.1 and the Swamy–Arora estimator (swar) for model T.2.1). For each of the groups in Table 4.8, the random-effects model that uses the Amemiya estimator proved to be a better one than the fixed-effects model.

Table 4.7 Significance of random components: Breusch–Pagan test[a] of Lagrange multipliers. (Source: own work)

Model	Pooled or random individual effects			Pooled or random time effects			Pooled or random two-way effects		
H_0, H_1	H_0:Var(α_i)=0			H_0:Var(τ_t)=0			H_0:Var(τ_t)=0 ∧ Var(α_i)		
	H_1: Var(α_i) ≠ 0			H_1: Var(τ_t) ≠ 0			H_1: Var(τ_t) ≠ 0 ∧ Var(α_i) ≠ 0		
Group	chisq	df	*P*-value	chisq	df	p-value	chisq	df	*P*-value
T.1.1.	0.02240	1	0.88110	0.89270	1	0.34470	0.91500	2	0.63280
T.1.2.	0.36580	1	0.54530	4.48020	1	*0.03429*	4.84600	2	0.08866
T.2.1.	2.55860	1	0.10970	5.91960	1	*0.01497*	8.47830	2	*0.01442*
T.2.2.	3.29810	1	0.06936	0.02540	1	0.87330	3.32360	2	0.18980

[a]The test was implemented in plm library of R package with plm test function. See Croissant and Millo (2008), p. 21; the authors did not emphasize that the test is designated to examine the occurrence of random effects in terms of the pooled model

Table 4.8 Hausman test—verification of selection preferences as regards random effects model (H_0) vs. fixed effects model. (Source: own work)

Group	Model	Estimator[a]	chisq	df	*P*-value
T.1.1.	Individual	swar	–	–	–
		walhus	20.5722	7	0.0045
		amemiya	3.9725	7	0.7829
		nerlove	2.6386	7	0.9163
T.1.2.	Two-way	swar	10.352	8	0.2412
		walhus	13.876	8	0.0851
		amemiya	3.818	8	0.8732
		nerlove	–	–	–
T.2.1.	Two-way	swar	64.3497	8	6.488e-11
		walhus	8.8025	8	0.3592
		amemiya	3.5826	8	0.8927
		nerlove	–	–	–

[a]See Croissant and Millo (2008), p. 12

Table 4.9 Autocorrelation in the idiosyncratic error component, Baltagi–Li and Wooldridge test. (Source: own work)

Group	Test	chisq	df	*P*-value
T.1.1.	Baltagi–Li[a]	6.50670	1	*0.01075*
	Wooldridge[b]	13.8274	–	*0.0002004*
T.1.2.	Baltagi–Li	0.08930	1	0.76510
	Wooldridge	0.3163	–	0.5738
T.2.1.	Baltagi–Li	0.13710	1	0.71120
	Wooldridge	0.2275	–	0.6333
T.2.2.	Wooldridge	0.46510	–	0.49520

[a]Function pbltest, plm library for R, see Croissant and Millo (2008), p. 25
[b]Function pbltest, plm library for R, see Croissant and Millo (2008), pp. 26-27

4.6.3 *Diagnostics of the Model and Adjustment*

In panel research for long time series, the occurrence of autocorrelation of the random element should be verified. In the case when short time series are used, this test may be skipped[19]. To examine the autocorrelation, the Wooldridge test was used (which is relevant for short data series), designed to diagnose autocorrelation in fixed-effects models[20], as well as the Baltagi–Li test designed to examine autocorrelation in random models[21]. Due to the selection of the random models, it was not necessary to apply the Wooldridge test for the first three groups. In the case of group T.1.1, both tests showed (Table 4.9) that the zero hypothesis on no autocorrelation should be rejected. It means that for the model for T.1.1, OLS estimators are

[19] In this study $t=16$, which is an intermediate value between *short* and *long* series.

[20] More specifically: the existence of serial correlation in the idiosyncratic error component.

[21] More specifically: the existence of AR(1) or MA(1) processes in the idiosyncratic error component.

Table 4.10 Heteroskedasticity—results of Berusch–Pagan[a] tests. (Source: own work)

Group	BP	df	*P*-value
T.1.1.	25.49990	12	0.01262
T.1.2.	271.44270	35	<2.2e-16
T.2.1.	129.48540	33	2.276e-13
T.2.2.	67.42730	8	1.595e-11

[a] The Berusch–Pagan test is implemented in bptest function of lmtest library of R package

inefficient and it is recommended to use a feasible generalized least squares (FGLS) or a panel-corrected standard errors (PCSE) estimator.

Another problem that may occur in panel data is heteroskedasticity. To verify the hypothesis on the occurrence of heteroskedasticity in the model, the Breusch–Pagan test was employed. The test verifies the zero hypothesis on the homoskedasticity. This hypothesis, in accordance with the results contained in Table 4.10, should be rejected for all groups.

In Table 4.11, characteristics of models selected for the purposes of estimation for individual groups are gathered, as well as the problems diagnosed for these models. The last column contains an account of selected methods of elimination of the observed problems, recognized in the literature on the subject.

In accordance with the results of the research of Beck and Katz (1995), the FGLS model included in Table 4.11, proposed by Parks (1967), underestimates standard errors for panel data[22] when the number of periods (T) is not significantly higher than the number of studied objects (N). This phenomenon occurs, according to the authors' findings, even for $(T/N)>3$. As a result of the underestimation of standard errors, statistically insignificant estimations of parameters were considered to be significant. In connection with the described reservations and short time horizon ($T=16$) which causes the T/N ratio in the examined models not to exceed 3, the FGLS method shall not be applied. Results obtained with the use of the Arellano estimator shall be compared with estimations obtained with the use of the panel-corrected standard errors (PCSE) method.

4.6.4 Model Estimation

The models for groups T.1.1., T.1.2., and T.2.1. were estimated in accordance with the guidelines described in Table 4.11. Next, as the serial correlation in the idiosyncratic error component (T.1.1) and heteroskedasticity (in all models) were diagnosed, estimations of errors and threshold significance levels were adjusted with the use of the covariance matrix estimator, proposed by Arellano (1987), which is immune to heteroskedasticity and to the serial correlation of errors[23]. In Table 4.12, results of estimations of parameters and selected parts of reports for individual mo-

[22] Time series cross section, TSCS.

[23] The estimator is available in lmtest library of R package, through of vcovHC function, see Croissant, Millo (2008), pp. 31.

Table 4.11 Account of models, problems, and possible solutions. (Source: own work.)

Group	Model	Problem	Selected solutions suggested in literature
T.1.1.	Individual random effects, Amemyia estimator	Autocorrelation, heteroskedasticity	Model: PCSE[a], FGLS[b] or Arrelano estimator (heteroskedasticity and autocorrelation robust covariance matrix estimator preferred to White's)
T.1.2.	Two-way random effects, Amemyia estimator	Heteroskedasticity	Model: PCSE, FGLS, heteroskedasticity and autocorrelation robust covariance matrix estimator[c] (Arellano or White)
T.2.1.	Two-way random effects, Amemyia estimator	Heteroskedasticity	Model: PCSE, FGLS, heteroskedasticity and autocorrelation robust covariance matrix estimator (Arellano or White)
T.2.2.	No fixed effects (pooled)	Heteroskedasticity	Model: PCSE, FGLS, heteroskedasticity and autocorrelation robust covariance matrix estimator (Arellano or White)

[a]Panel corrected standard errors, see Beck and Katz (1995)

[b]Feasible generalised least squares, see Parks (1967)

[c]Robust covariance matrix estimation, see Croissant and Millo (2008), pp. 28. and Kleiber and Zeileis (2008), pp. 111

Table 4.12 Account of results of model estimates for individual groups (the full report printout with model estimation is contained in printouts in 1, 2, 3, and 4 in the appendix). (Source: own work)

Group: T.1.1.		RCM[a] estimator: Arellano			Original (pre-RCM)	Group: T.1.2.		RCM estimator: Arellano			Original (pre-RCM)
Variable	Estimate	Std. Err.	*t* value	Pr(>|t|)	Pr(>|t|)	Variable	Estimate	Std. Err.	*t* value	Pr(>|t|)	Pr(>|t|)
logPwhYE	−0.14273	0.037415	−3.8147	0.0002528***[b]	0.0131819*	logPwhYE	−0.11454	0.05785	−1.98	0.049085*	0.04663*
logBct	0.083275	0.031483	2.6451	0.0096720**	0.0001321***	logBct	−0.00812	0.00736	−1.1036	0.271085	0.51518
logFct	−0.00198	0.000862	−2.2931	0.0242240*	0.0566536.	logFct	−0.00268	0.005491	−0.4885	0.625762	0.68632
logGct	0.049027	0.02685	1.826	0.0712480.	0.145077	logGct	0.060294	0.020796	2.8993	0.004160**	0.02273*
logEct	−0.12809	0.045979	−2.7858	0.0065382**	0.0004411***	logEct	−0.03501	0.024517	−1.4282	0.154813	0.17933
logKct	−0.02164	0.024217	−0.8934	0.3740561	0.573994	logKct	0.022312	0.025149	0.8872	0.376046	0.52255
logLct	−0.08405	0.038722	−2.1705	0.0326593*	0.110027	logLct	−0.03016	0.019132	−1.5766	0.116484	0.25941
logMct	–	–	–	–	–	logMct	0.026765	0.009042	2.96	0.003449**	0.03213*
R-Squared:		0.28145	Adj. R-Squared:		0.258	R-Squared:		0.11661	Adj. R-Squared:		0.11157
F-statistic: 4.92422 on 7 and 88 DF, P-value: 9.9769e-05						F-statistic: 3.2837 on 8 and 199 DF, P-value: 0.0015297					

Group: T.2.1.		RCM estimator: Arellano			Original (pre-RCM)	Group: T.2.2.		RCM estimator: Arellano			Original (pre-RCM)
Variable	Estimate	Std. Err.	*t* value	Pr(>|t|)	Pr(>|t|)	Variable	Estimate	Std. Err.	*t* value	Pr(>|t|)	Pr(>|t|)
logPwhYE	−0.1311	0.052493	−2.4975	0.01348*	0.01273*	logPwhYE	−0.01537	0.019658	−0.7817	0.43547	0.48876
logBct	0.004462	0.008557	0.5215	0.60274	0.77662	logBct	0.022512	0.004645	4.8469	2.849e-06***	0.00185**
logFct	−0.00598	0.006396	−0.9356	0.35082	0.28084	logFct	−0.00602	0.005674	−1.0612	0.29011	0.46842
logGct	−0.00585	0.028563	−0.205	0.83784	0.84533	logGct	−0.01415	0.008729	−1.6213	0.10684	0.45243
logEct	−0.00875	0.018339	−0.4773	0.6338	0.79617	logEct	−0.0322	0.005414	−5.9471	1.553e-08***	0.00771**
logKct	−0.04394	0.042556	−1.0325	0.30331	0.2027	logKct	0.05807	0.019862	2.9237	0.00394**	0.02029*
logLct	0.007584	0.018169	0.4174	0.67692	0.85228	logLct	−0.00184	0.007211	−0.2545	0.7994	0.88595
logMct	0.029044	0.012474	2.3284	0.02109*	0.16123	logMct	−0.00222	0.007172	−0.3094	0.75742	0.84679
R-Squared:		0.053788	Adj. R-Squared:		0.051037	R-Squared:		0.22113	Adj. R-Squared:		0.20983
F-statistic: 1.18665 on 8 and 167 DF, P-value: 0.30989						F-statistic: 5.92682 on 8 and 167 DF, P-value: 1.1067e-06					

[a]Robust Correlation Matrix

[b]Signif. codes: *** $p < 0.001$; ** $p < 0.01$; * $p < 0.05$; . $p < 0.1$

dels (R^2, F statistics for the Wald test) are presented. In addition, in Table 4.13, results obtained with the use of the RCM Arellano estimator are presented along with estimations of parameters, standard errors, and threshold significance levels obtained with the use of the PCSE model[24].

The estimation results presented in Table 4.12 show a relatively good matching of the models adopted for groups T.1.1. and T.2.2. (R^2 0.28 and 0.22, respectively), and fairly poorer matching in the case of the model for group T.1.2. (R^2=0.117, the result of the Wald test does not allow the adoption of the null hypothesis on no correlation between the dependent variable and all independent variables[25]). The model estimated for group T.2.1. was negatively verified with the use of the Wald test (no grounds for rejection of the above-mentioned zero hypothesis). The observed diversification of the model form verification results (Wald test), as well as the quality of the model's matching (R^2), confirms that the analysis should be conducted separately for different country groups; however, the negative verification for group T2.1. indicates a possibility of erroneous model specification (omission of important independent variables)—or erroneous specification of the systematics contained in Table 4.1. In this study, the foregoing shall not be determined.

Before the commencement of the analysis of significance and interpretation of the estimated values, the multicollinearity of independent variables was examined in the models estimated for individual groups (Table 4.14). It is considered that the level of the variance inflation factor (VIF) coefficient that indicates the occurrence of the problem of multicollinearity is 10.

Multicollinearity was not ascertained only in the case of T.2.2. group. The multicollinearity in a model results in the overestimation of parameter errors and underestimation of statistics t (and so, the risk of considering statistically significant independent variables to be insignificant increases). In addition, multicollinear independent variables make it difficult to interpret estimates of their parameters in the model due to extreme sensitiveness of such valuations to adding or removing observations (e.g., shortening or lengthening of time series). Possible solutions to this problem include reduction of the set of independent variables, ridge regression, or principal component regression.

Parameter estimates and assessment of their significance differ considerably between individual T.*i.j.* groups. Estimates of errors and threshold significance levels are different for models adjusted with the Arellano estimator and their equivalents estimated with the use of the PCSE model.

4.6.4.1 Interpretation of Parameter Estimates for Group T.1.1.

Models estimated for group T.1.1. (Ireland, the Netherlands, Switzerland, Finland, Sweden, and Singapore, see Table 4.1) showed statistical significance for the achievement of the export success (in each of the models) of the following factors:

---

[24] The implementation in R package is discussed in Bailey, Katz (2011).

[25] H_0 of the Wald test has the following form: $\forall(i)\ \beta_i=0$.

Table 4.13 Estimations of parameters: results derived via Arellano RCM and PCSE. (Source: own work)

Group: T.1.1.	RCM estimator: Arellano				PCSE (Listing 5)[a]			
Variable	Estimate	Std. Err.	*t* value	Pr(>|t|)	Estimate	PCSE	*t* value	Pr(>|t|)
logPwhYE	−0.1427	0.0374	−3.8147	0.0002528***b	−0.1187	0.0893	−1.3283	0.1877268
logBct	0.0833	0.0315	2.6451	0.0096720**	0.0962	0.0202	4.7608	0.0000081***
logFct	−0.0020	0.0009	−2.2931	0.0242240*	−0.0023	0.0012	−1.8701	0.064997.
logGct	0.0490	0.0269	1.8260	0.0712480.	0.0625	0.0320	1.9546	0.0540027.
logEct	−0.1281	0.0460	−2.7858	0.0065382**	−0.1524	0.0427	−3.5685	0.0005999***
logKct	−0.0216	0.0242	−0.8934	0.3741	−0.0398	0.0350	−1.1366	0.258979
logLct	−0.0841	0.0387	−2.1705	0.0326593*	−0.1136	0.0740	−1.5355	0.1284635
Group: T.1.2.	RCM estimator: Arellano				PCSE (Listing 6)			
Variable	Estimate	Std. Err.	*t* value	Pr(>|t|)	Estimate	PCSE	*t* value	Pr(>|t|)
logPwhYE	−0.1145	0.0579	−1.9800	0.049085*	−0.1747	0.0675	−2.5864	0.0105263*
logBct	−0.0081	0.0074	−1.1036	0.2711	−0.0241	0.0273	−0.8846	0.3776091
logFct	−0.0027	0.0055	−0.4885	0.6258	0.0002	0.0088	0.0217	0.9827137
logGct	0.0603	0.0208	2.8993	0.004160**	0.0676	0.0302	2.2358	0.0266544*
logEct	−0.0350	0.0245	−1.4282	0.1548	−0.0185	0.0305	−0.6083	0.5437685
logKct	0.0223	0.0251	0.8872	0.3760	0.0433	0.0377	1.1484	0.2523893
logLct	−0.0302	0.0191	−1.5766	0.1165	−0.0731	0.0388	−1.8842	0.0612266.
logMct	0.0268	0.0090	2.9600	0.003449**	0.0189	0.0210	0.9006	0.3690417
Group: T.2.1.	RCM estimator: Arellano				PCSE (Listing 7)			
Variable	Estimate	Std. Err.	*t* value	Pr(>|t|)	Estimate	PCSE	*t* value	Pr(>|t|)
logPwhYE	−0.1311	0.0525	−2.4975	0.01348*	−0.1264	0.0720	−1.7551	0.0813992.
logBct	0.0045	0.0086	0.5215	0.6027	0.0144	0.0280	0.5137	0.6082918

Table 4.13 (continued)

Group: T.2.1.	RCM estimator: Arellano				PCSE (Listing 7)			
Variable	Estimate	Std. Err.	*t* value	Pr(>|t|)	Estimate	PCSE	*t* value	Pr(>|t|)
logFct	−0.0060	0.0064	−0.9356	0.3508	−0.0050	0.0061	−0.8271	0.4095542
logGct	−0.0059	0.0286	−0.2050	0.8378	−0.0226	0.0373	−0.6054	0.5458756
logEct	−0.0088	0.0183	−0.4773	0.6338	0.0096	0.0364	0.2647	0.7916081
logKct	−0.0439	0.0426	−1.0325	0.3033	−0.0732	0.0377	−1.9391	0.0544667.
logLct	0.0076	0.0182	0.4174	0.6769	0.0981	0.0830	1.1823	0.2390723
logMct	0.0290	0.0125	2.3284	0.02109*	0.0301	0.0311	0.9657	0.3358469

Group: T.2.2.	RCM estimator: Arellano				PCSE (Listing 8)			
Variable	Estimate	Std. Err.	*t* value	Pr(>|t|)	Estimate	PCSE	*t* value	Pr(>|t|)
logPwhYE	−0.0154	0.0197	−0.7817	0.4355	−0.0154	0.0203	−0.7559	0.4507501
logBct	0.0225	0.0046	4.8469	2.849e-06***	0.0225	0.0081	2.7964	0.0057743**
logFct	−0.0060	0.0057	−1.0612	0.2901	−0.0060	0.0066	−0.9154	0.3613157
logGct	−0.0142	0.0087	−1.6213	0.1068	−0.0142	0.0143	−0.9874	0.3248869
logEct	−0.0322	0.0054	−5.9471	1.553e-08***	−0.0322	0.0099	−3.2470	0.0014097**
logKct	0.0581	0.0199	2.9237	0.00394**	0.0581	0.0198	2.9306	0.0038567**
logLct	−0.0018	0.0072	−0.2545	0.7994	−0.0018	0.0147	−0.1248	0.900837
logMct	−0.0022	0.0072	−0.3094	0.7574	−0.0022	0.0134	−0.1658	0.8685

[a] Printouts can be found in the Appendix to this chapter
[b] Signif. codes: *** $p < 0.001$; ** $p < 0.01$; * $p < 0.05$; . $p < 0.1$

Table 4.14 Multicollinearity of independent variables, generalized VIF coefficient.[a] (Source: own work)

Variable	T.1.1.	T.1.2.	T.2.1.	T.2.2.
logPwhYE	27.9576	13.9836	4.9163	2.7321
logBct	13.8810	118.2780	34.7251	3.4327
logFct	1.5332	8.8980	2.5572	1.6019
logGct	18.3292	51.1832	8.3397	3.0123
logEct	25.9136	12.2012	4.8170	3.8224
logKct	6.5223	14.0945	3.7495	2.6931
logLct	28.6251	72.5726	113.5194	3.4336
logMct	–	30.2326	9.0919	1.9677

[a] Calculated with the use of VIF function from car library of R package.

- relative level of technological advancement ($B_{c,t}$), whereby a growth of $B_{c,t}$ value by 1% leads to a growth in export, which is faster by 0.000833 (PCSE: 0.000437) than in countries of the reference group (we divide the parameter estimates by 100); when converting them, for convenience reasons, into percentage points (p.p.)—we obtain 0.0833 p.p.[26] (0.0437 p.p.);
- $F_{c,t}$ (a component of the relative foreign market penetration capacity A); growth $F_{c,t}$ by 1 % translates into export growth rate lower by 0.002 p.p. in relation to the reference group (PCSE: 0.004 p.p.) (estimation without practical significance);
- $G_{c,t}$ (a component of the relative R&D potential R, relative accumulated R&D expenses); growth by 1 % translates into export growth pace higher by 0.049 p.p. (PCSE: 0.0091 p.p.) than in the countries of the reference group; and
- $E_{c,t}$, a growth by 1% translates into export growth rate lower than 0.1281 p.p. (PCSE: 0.0506 p.p.) in the countries of the reference group; it means that, against expectations, countries with over-average percentage share of scientific workers in the total workers (which means an over-average R&D potential measured with the engaged labor force) do not show competitive advantage.

In addition, solely for the Arellano RCM estimator, the parameter estimates for the following factors were considered to be statistically significant:

- relative unit labor costs (indirect measure $P_{C,t}^{w,Y}$, see formula (4.20))—growth of $P_{C,t}^{w,Y}$ by 1 % translates into export growth rate which is lower by 0.1427 p.p. when compared to the growth rate observed for the environment (which is in line with expectations) and
- relative human capital $L_{c,t}$ measured with the proportion of the number of people with higher education in the total employment—growth of $L_{c,t}$ by 1 % translates into export growth rate which is lower by 0.0841 p.p. than in the countries of the reference group (observation incompliant with the expectations).

[26] In this case, due to the manner of construction of $S_{c,t}$ and the interpretation of the logarithm of this variable, this is admissible; see comment to formula (4.17).

4.6.4.2 Interpretation of Parameter Estimates for Group T.1.2.

Models estimated for group T.1.2. (Germany, Great Britain, France, Italy, South Korea, Taiwan, USA, Japan, Canada, Norway, Australia, Austria, New Zealand, and Spain, see Table 4.1) showed statistical significance for the achievement of the export success (in each of the models) of the following factors:

- $G_{c,t}$ (a component of relative research and development potential *R*, relative accumulated R&D expenses); growth of $G_{c,t}$ by 1 % translates into export growth pace by 0.0603 p.p. (PCSE: 0.0041 p.p.) than in the countries of the reference group; and
- $M_{c,t}$ (the relative trademark resource, a component of relative foreign market penetration capacity *A*); growth of $M_{c,t}$ by 1 % translates into export growth rate lower by 0.0268 p.p. in relation to the reference group (PCSE: 0.0214 p.p.) (estimation without practical significance);

In addition, solely for the Arellano RCM estimator, parameter estimates for unit labor costs were indicated as statistically significant (indirect benchmark $P_{C,t}^{w,Y}$, see formula (4.20))—a growth of $P_{C,t}^{w,Y}$ by 1 % translates into export growth rate lower by 0.1145 p.p. when compared to the growth rate observed for the environment (which is in line with expectations).

One should bear in mind that very strong multicollinearity of independent variables occurred in the model.

4.6.4.3 Interpretation of Parameter Estimates for Group T.2.1.

The generalized least squares (GLS) model estimated for group T.2.1. (Portugal, Greece, Poland, Hungary, Czech Republic, Slovenia, Latvia, Estonia, Lithuania, Slovakia, and Ukraine) was negatively verified with the use of the Wald test. Further, the value of R^2 coefficient disqualifies the model. The above may be caused by excessive heterogeneity of the sample, or, in other words, diversification of the strength and direction of influence of individual competitiveness factors, which give 0 as a resultant of estimates for β_i.

The parameter estimates showed that in these countries, unit labor cost $P_{C,t}^{w,Y}$ is of statistical significance for the achievement of the export success (in each of the models); a growth by 1 % translates here into export growth pace by 0.1311 p.p. lower (PCSE: 0.1264 p.p.) than in the countries of the reference group.

In addition, solely for the Arellano RCM estimator, the estimation of $M_{c,t}$ (formula (4.14)), the relative trademark resource, a component of relative foreign market penetration capacity *A*) was indicated as statistically important; a growth of $M_{c,t}$ by 1 % translates into export growth rate lower by 0.029 p.p. in relation to the reference group. In the model estimated with the PCSE method, a direction of influence of relative accumulated investments in fixed assets $K_{c,t}$ (formula (4.9), a component of relative productive potential *Q*) was observed, which was inconsistent with the expectations: it seemed that an increase by 1 % of the variable should result in the export growth pace lower by 0.0732 p.p.

4.6.4.4 Interpretation of Parameter Estimates for Group T.2.2.

For the last group (Russia, China, Brazil, Argentina, India, Mexico, Turkey, Bulgaria, Romania, Egypt, Cyprus, Chile, and Venezuela), a pooling model was estimated—without any time or individual effects. The model showed a relatively high value of the coefficient of determination (for panel data), the parameter estimates for the Arellano RCM estimator and the PSCE model are almost identical, and both methods indicated the same set of statistically significant independent variables. In addition, no multicollinearity of independent variables (understood as VIF value > 10, see Table 4.14) occurred in the model.

The following variables influencing the relative export growth rate were indicated as statistically significant:

- $B_{c,t}$ (formula (4.5)), a proxy benchmark of the relative level of technological advancement Q); growth of $B_{c,t}$ by 1 % translates into export growth pace higher by 0.0225 p.p. (PCSE: 0.0225 p.p.) than in the countries of the reference group;
- $E_{c,t}$ (formula (4.7)), the relative percentage of scientific workers in the total of the employed, a partial benchmark of the relative research and development potential R); growth of $E_{c,t}$ *by* 1 % translates into export growth rate lower by 0.0322 p.p. in relation to the reference group (PCSE: 0.0322 p.p.) (the estimation inconsistent with the expectations); and
- $K_{c,t}$ (formula (4.9), the relative accumulated investments in fixed assets, a component of the relative production potential Q); growth of $K_{c,t}$ *by* 1 % translates into export growth rate lower by 0.0581 p.p. in relation to the reference group (PCSE: (0.0581 p.p.).

4.7 Summary and Conclusions

The study confirmed the thesis on the diversification of the strength and direction of influence of competitiveness factors depending on the characteristics of the economy of a given country. In this study, a hypothesis was verified that the key characteristic that determines the significance of individual factors of competitiveness of countries in foreign trade is the type of the national innovation system (NIS).

The quality of the econometric models estimated for stable innovation systems (group T.1.2.), and catching-up innovation systems (group T.2.1.) does not allow to formulate conclusions with regard to the importance of innovation and human capital for competitiveness in foreign trade. The above may be interpreted in two ways. In the case of the two said groups, the characteristics of the innovation system are not the only criterion defining the crucial set of competitiveness determinants, or on the second level of typology, the differences between countries, important from the point of view of the role played by competitiveness factors, are not captured.

However, conclusions related to the importance of innovation and human capital for competitive advantages in foreign trade may be expressed for dynamical NISs (T.1.1.) and nonbalanced NISs (T.2.2.). The econometric study implies that in the case of dynamical NISs (T.1.1.), the relative level of technological advancement is an important determinant of competitiveness benchmarked with export success. In addition, high positive estimation of the parameter for relative accumulated R&D expenses seems to confirm this hypothesis. Furthermore, the model indicates that the accumulation of human capital is not sufficient to maintain relative export growth rate at a level guarantying a nondecreasing share in the global trade (i.e., it does not prevent the displacing from international markets). It may be assumed that the above results from the fact that a high level of human capital is accompanied by a high relative unit labor cost, which is not fully balanced by other competitiveness factors. Countries with high human capital levels have more serious problems with retaining the share in international markets. It may be assumed that it results from the fact that a high level of human capital is accompanied by a high labor cost which is not fully balanced by higher work effectiveness. These observations should be compared with the reverse trends observed for average values of relative export growth rate and relative unit labor costs in group T.1.2. (Fig. 4.2.), and the neutrality of indirect benchmarks of human capital for this group (Figs. 4.8–4.10).

For nonbalanced NISs (T.2.2.), the interpretation of parameter estimates and estimates of critical significance level allows ascertaining that the success drive for exports of these countries (e.g., improvement of competitiveness in foreign trade) is the extension and modernization of production capacities. The hypothesis on the importance of technological innovations for development of competitive advantages in trade is therefore indirectly confirmed for countries with low unit labor costs as the main success drives for exports are knowledge and technology materialized in the fixed assets. The estimation of the parameter of the relative level of technological advancement indicates that these countries pass from the stage of copying to the stage of improvement of technologies obtained thanks to diffusion. However, there are no grounds to accept the hypothesis on the importance of human capital for international competitiveness of these countries (logLct is statistically insignificant).

Moreover, it should be noted that strong multicollinearity was discovered for independent variables (with the exclusion of group T.2.2.), which may lead to instability of parameter assessments (i.e. potentially significant changes in the direction and strength of influence of independent variables on the dependent variable may occur given even a slight change of the time series under consideration). This issue is related to the nature of the studied macroeconomic phenomena, which implies a need to apply the ridge regression or the principal component regression[27].

The above conclusions that stem from econometric modeling indicate the direction and methodology of further research on the subject of factors of competitiveness.

---

[27] See Maddala, 1992, pp. 269–295.

Appendix

Model estimates in accordance with test results per effect type (see Table 4.5 and 4.11), error values and significance levels before adjustment

Printout 1. Model T.1.1—with random individual effects, estimation with the Amemiya estimator, prior to the adjustment of errors in and significance levels of parameter estimations (signif. codes for all printouts: *** $p < 0.001$; ** $p < 0.01$; * $p < 0.05$; . $p < 0.1$).

```
Balanced Panel: n=6, T=16, N=96
Effects:
                    var std.dev share
idiosyncratic 0.00140 0.03742 0.569
individual    0.00106 0.03256 0.431
theta:   0.7239
Residuals :
    Min.   1st Qu.   Median   3rd Qu.      Max.
-0.10400 -0.02160  0.00668  0.01950  0.11400
Coefficients :
                  Estimate Std. Error t-value  Pr(>|t|)
(Intercept)      0.2777227  0.0634692  4.3757 3.316e-05 ***
logP_wh_YERel -0.1427269  0.0564113 -2.5301 0.0131819 *
logBct           0.0832748  0.0208245  3.9989 0.0001321 ***
logFct          -0.0019769  0.0010236 -1.9314 0.0566536 .
logGct           0.0490266  0.0333472  1.4702 0.1450773
logEct          -0.1280899  0.0350726 -3.6521 0.0004411 ***
logKct          -0.0216368  0.0383436 -0.5643 0.5739935
logLct          -0.0840485  0.0520626 -1.6144 0.1100265
Total Sum of Squares:    0.16992
Residual Sum of Squares: 0.1221
R-Squared       :  0.28145
      Adj. R-Squared :  0.258
F-statistic: 4.92422 on 7 and 88 DF, p-value: 9.9769e-05
```

Printout 2. Model T.1.2—with random bidirectional effects, estimation with the Amemiya estimator, prior to the adjustment of errors in and significance levels of parameter estimations.

```
Balanced Panel: n=13, T=16, N=208
Effects:
                    var   std.dev share
idiosyncratic 0.0015702 0.0396258 0.505
individual    0.0014262 0.0377653 0.459
time          0.0001112 0.0105433 0.036
theta  : 0.7463 (id) 0.2784 (time) 0.2712 (total)
Residuals :
    Min.  1st Qu.   Median  3rd Qu.     Max.
-0.17300 -0.02190 -0.00227  0.02110  0.18700
Coefficients :
                Estimate Std. Error t-value Pr(>|t|)
(Intercept)    0.0578620  0.0288868  2.0031  0.04653 *
logP_wh_YERel -0.1145407  0.0572093 -2.0021  0.04663 *
logBct        -0.0081230  0.0124594 -0.6520  0.51518
logFct        -0.0026820  0.0066313 -0.4044  0.68632
logGct         0.0602941  0.0262634  2.2957  0.02273 *
logEct        -0.0350137  0.0259829 -1.3476  0.17933
logKct         0.0223117  0.0348321  0.6406  0.52255
logLct        -0.0301632  0.0266689 -1.1310  0.25941
logMct         0.0267650  0.0124032  2.1579  0.03213 *
Total Sum of Squares:    0.35544
Residual Sum of Squares: 0.31399
R-Squared      :  0.11661
      Adj. R-Squared :  0.11157
F-statistic: 3.2837 on 8 and 199 DF, p-value: 0.0015297
```

Printout 3. Model T.2.1—with random bidirectional effects, estimation with the Amemiya estimator, prior to the adjustment of errors in and significance levels of parameter estimations.

```
Balanced Panel: n=11, T=16, N=176
Effects:
                    var   std.dev share
idiosyncratic 0.0036020 0.0600163 0.337
individual    0.0066368 0.0814667 0.621
time          0.0004452 0.0210990 0.042
theta  : 0.8189 (id) 0.349 (time) 0.3451 (total)
Residuals :
    Min.  1st Qu.   Median  3rd Qu.     Max.
-0.21200 -0.03340 -0.00187  0.03460  0.17000
Coefficients :
                Estimate Std. Error t-value Pr(>|t|)
(Intercept)   -0.0493097  0.0565948 -0.8713  0.38485
logP_wh_YERel -0.1311017  0.0520592 -2.5183  0.01273 *
logBct         0.0044619  0.0157004  0.2842  0.77662
logFct        -0.0059841  0.0055309 -1.0819  0.28084
logGct        -0.0058546  0.0299652 -0.1954  0.84533
logEct        -0.0087523  0.0338286 -0.2587  0.79617
logKct        -0.0439406  0.0343579 -1.2789  0.20270
logLct         0.0075840  0.0406642  0.1865  0.85228
logMct         0.0290442  0.0206398  1.4072  0.16123
Total Sum of Squares:    0.63707
Residual Sum of Squares: 0.60281
R-Squared      :  0.053788
      Adj. R-Squared :  0.051037
F-statistic: 1.18665 on 8 and 167 DF, p-value: 0.30989
```

Printout 4. Model T.2.2—without random or constant individual or time (pooled) effects, prior to the adjustment of errors in and significance levels of parameter estimations.

```
Balanced Panel: n=11, T=16, N=176
Residuals :
    Min.  1st Qu.   Median  3rd Qu.     Max.
-0.29600 -0.03760  0.00443  0.03340  0.31600
Coefficients :
                 Estimate Std. Error t-value Pr(>|t|)
(Intercept)     0.0178683  0.0253516  0.7048  0.48190
logP_wh_YERel -0.0153678  0.0221495 -0.6938  0.48876
logBct          0.0225120  0.0071153  3.1639  0.00185 **
logFct         -0.0060219  0.0082865 -0.7267  0.46842
logGct         -0.0141517  0.0187906 -0.7531  0.45243
logEct         -0.0321956  0.0119368 -2.6972  0.00771 **
logKct          0.0580697  0.0247814  2.3433  0.02029 *
logLct         -0.0018353  0.0127759 -0.1437  0.88595
logMct         -0.0022189  0.0114660 -0.1935  0.84679
Total Sum of Squares:    1.0973
Residual Sum of Squares: 0.85467
R-Squared      :  0.22113
      Adj. R-Squared :  0.20983
F-statistic: 5.92682 on 8 and 167 DF, p-value: 1.1067e-06
```

Estimation of parameters by the use of the PCSE model

The estimated PCSE models were modified in relation to the models estimated above to correspond to them in the best possible way.

Printout 5. Model T.1.1. In order to take into account the individual effects, a PCSE model was estimated with binary variables for countries (an equivalent of the fixed-effects model with individual effects; it is not possible to obtain an equivalent of the random effects model with the use of PCSE).

```
Estimate        PCSE    t value      Pr(>|t|)
(Intercept)                       0.346872635 0.067396423  5.1467514 1.747135e-06
logP_wh_YERel                    -0.118661999 0.089335138 -1.3282791 1.877268e-01
logBct                            0.096173385 0.020201041  4.7608132 8.069660e-06
logFct                           -0.002264260 0.001210778 -1.8700879 6.499698e-02
logGct                            0.062522416 0.031988096  1.9545526 5.400263e-02
logEct                           -0.152386677 0.042703619 -3.5684722 5.998238e-04
logKct                           -0.039767670 0.034988460 -1.1365939 2.589790e-01
logLct                           -0.113559418 0.073955792 -1.5355040 1.284635e-01
factor(Country)Ireland            0.003623109 0.054041410  0.0670432 9.467085e-01
factor(Country)Netherlands        0.009111429 0.056994042  0.1598663 8.733748e-01
factor(Country)Singapore          0.014681770 0.062029959  0.2366884 8.134817e-01
factor(Country)Sweden            -0.035325297 0.032115337 -1.0999510 2.745336e-01
factor(Country)Switzerland -0.081620824 0.036225682 -2.2531204 2.688923e-02
# Valid Obs = 96; # Missing Obs = 0; Degrees of Freedom = 83.
```

Printout 6. Model T.1.2. In order to take into account the individual effects, a PCSE model was estimated with binary variables for countries and years (an equivalent of the bidirectional fixed-effects model; it is not possible to obtain an equivalent of the random effects model with the use of PCSE).

	Estimate	PCSE	t value	Pr(>|t|)
(Intercept)	5.733494e-02	0.056461831	1.0154637689	3.113111e-01
logP_wh_YERel	-1.746765e-01	0.067536916	-2.5863858036	**1.052622e-02**
logBct	-2.411150e-02	0.027257056	-0.8845966300	3.776091e-01
logFct	1.912068e-04	0.008812047	0.0216983409	9.827137e-01
logGct	6.757560e-02	0.030224993	2.2357522893	**2.665433e-02**
logEct	-1.852719e-02	0.030455628	-0.6083339411	5.437685e-01
logKct	4.325667e-02	0.037666197	1.1484214497	2.523893e-01
logLct	-7.311986e-02	0.038807073	-1.8841889562	**6.122660e-02**
logMct	1.891281e-02	0.020999416	0.9006350415	3.690417e-01
factor(Country)Austria	-1.045163e-02	0.040921417	-0.2554073579	7.987137e-01
factor(Country)Canada	6.669305e-03	0.030811689	0.2164537195	8.288906e-01
factor(Country)France	-2.021761e-02	0.032922399	-0.6140989253	5.399614e-01
factor(Country)Germany	7.457513e-02	0.060535084	1.2319324463	2.196558e-01
factor(Country)Italy	-6.165652e-02	0.042019094	-1.4673453113	1.441083e-01
factor(Country)Japan	5.912931e-05	0.092585952	0.0006386423	9.994912e-01
factor(Country)Korea, Rep.	7.296183e-02	0.074937877	0.9736308309	3.316072e-01
factor(Country)New Zealand	3.357414e-02	0.052164423	0.6436213719	5.206782e-01
factor(Country)Norway	-3.812421e-02	0.040166272	-0.9491597185	3.438711e-01
factor(Country)Spain	1.184442e-03	0.027748420	0.0426850255	9.660021e-01
factor(Country)United Kingdom	1.724156e-02	0.057895969	0.2978024381	7.662134e-01
factor(Country)United States	3.271133e-02	0.062452724	0.5237774783	6.011076e-01
factor(Year)1996	1.317599e-02	0.001901695	6.9285535654	8.163221e-11
factor(Year)1997	1.535212e-02	0.002436992	6.2996169850	2.415030e-09
factor(Year)1998	9.996369e-03	0.003017180	3.3131501239	1.124663e-03
factor(Year)1999	5.876986e-03	0.003235694	1.8162986539	7.106568e-02
factor(Year)2000	-1.986386e-02	0.004686239	-4.2387632670	3.659825e-05
factor(Year)2001	8.764290e-03	0.005464037	1.6039953102	1.105497e-01
factor(Year)2002	-7.952507e-03	0.004898859	-1.6233385332	1.063484e-01
factor(Year)2003	-1.916805e-02	0.005120808	-3.7431698730	2.474999e-04
factor(Year)2004	-1.924473e-02	0.006305374	-3.0521161263	2.633148e-03
factor(Year)2005	-1.510533e-02	0.007105189	-2.1259568660	3.493380e-02
factor(Year)2006	-1.222896e-02	0.008151934	-1.5001297015	1.354142e-01
factor(Year)2007	-6.133435e-03	0.009280096	-0.6609236292	5.095456e-01
factor(Year)2008	-8.611606e-03	0.009623121	-0.8948870228	3.720981e-01
factor(Year)2009	2.412464e-02	0.009754080	2.4732867340	1.435881e-02
factor(Year)2010	-2.851390e-02	0.010477087	-2.7215484288	7.166305e-03

Valid Obs = 208; # Missing Obs = 0; Degrees of Freedom = 172.

Printout 7. T.2.1. In order to take into account the individual effects, a PCSE model was estimated with binary variables for countries and years (an equivalent of the bidirectional fixed-effects model; it is not possible to obtain an equivalent of the random-effects model with the use of PCSE).

```
                                    Estimate        PCSE    t value     Pr(>|t|)
(Intercept)                     -0.042441776 0.057123002 -0.7429892 4.587156e-01
logP_wh_YERel                   -0.126395415 0.072016145 -1.7550983 8.139911e-02
logBct                           0.014374175 0.027984125  0.5136546 6.082918e-01
logFct                          -0.005021021 0.006070452 -0.8271246 4.095542e-01
logGct                          -0.022558487 0.037261751 -0.6054060 5.458756e-01
logEct                           0.009648229 0.036446753  0.2647212 7.916081e-01
logKct                          -0.073151109 0.037723391 -1.9391445 5.446666e-02
logLct                           0.098095030 0.082971373  1.1822756 2.390723e-01
logMct                           0.030076775 0.031145747  0.9656784 3.358469e-01
factor(Country)Estonia          -0.109467570 0.101251594 -1.0811442 2.814657e-01
factor(Country)Greece           -0.140445473 0.105448715 -1.3318842 1.850321e-01
factor(Country)Hungary          -0.093885172 0.067775446 -1.3852387 1.681517e-01
factor(Country)Latvia           -0.145280141 0.058883183 -2.4672603 1.480413e-02
factor(Country)Lithuania        -0.134661376 0.078151040 -1.7230913 8.704883e-02
factor(Country)Poland           -0.055617636 0.045307388 -1.2275622 2.216425e-01
factor(Country)Portugal          0.027572810 0.060907551  0.4526994 6.514561e-01
factor(Country)Slovak Republic -0.030924845 0.029363780 -1.0531630 2.940551e-01
factor(Country)Slovenia         -0.040489527 0.053838236 -0.7520589 4.532604e-01
factor(Country)Ukraine          -0.267717657 0.129923602 -2.0605775 4.116844e-02
factor(Year)1996                 0.052922231 0.006894686  7.6757997 2.413100e-12
factor(Year)1997                 0.049350654 0.011591520  4.2574790 3.738428e-05
factor(Year)1998                 0.085071199 0.011600015  7.3337149 1.564420e-11
factor(Year)1999                 0.002782088 0.014688192  0.1894098 8.500421e-01
factor(Year)2000                 0.066794603 0.017318160  3.8569110 1.735524e-04
factor(Year)2001                 0.091092514 0.016964060  5.3697353 3.142541e-07
factor(Year)2002                 0.032746389 0.017090653  1.9160409 5.736944e-02
factor(Year)2003                 0.042097251 0.019361763  2.1742468 3.134326e-02
factor(Year)2004                 0.028299103 0.019902968  1.4218534 1.572615e-01
factor(Year)2005                 0.029592569 0.019870676  1.4892583 1.386370e-01
factor(Year)2006                 0.033424957 0.020928295  1.5971180 1.124624e-01
factor(Year)2007                 0.041869822 0.022871264  1.8306737 6.924576e-02
factor(Year)2008                 0.042009039 0.024343677  1.7256653 8.658295e-02
factor(Year)2009                -0.006618455 0.021458882 -0.3084250 7.582110e-01
factor(Year)2010                 0.017075317 0.020174345  0.8463877 3.987606e-01
# Valid Obs = 176; # Missing Obs = 0; Degrees of Freedom = 142.
```

Printout 8. T.2.2. A model without binary variables (an equivalent of the pooled model).

```
Estimate         PCSE     t value      Pr(>|t|)
(Intercept)     0.017868335 0.022813392  0.7832389 0.434596201
logP_wh_YERel -0.015367843 0.020329445 -0.7559401 0.450750068
logBct          0.022512040 0.008050414  2.7963830 0.005774240
logFct         -0.006021934 0.006578667 -0.9153730 0.361315677
logGct         -0.014151698 0.014332636 -0.9873758 0.324886853
logEct         -0.032195577 0.009915527 -3.2469859 0.001409678
logKct          0.058069712 0.019814731  2.9306334 0.003856618
logLct         -0.001835315 0.014706785 -0.1247937 0.900836941
logMct         -0.002218918 0.013381566 -0.1658190 0.868499941
```

Dispersion charts for ln(S_c) in relation to natural logarithms of independent variables per groups

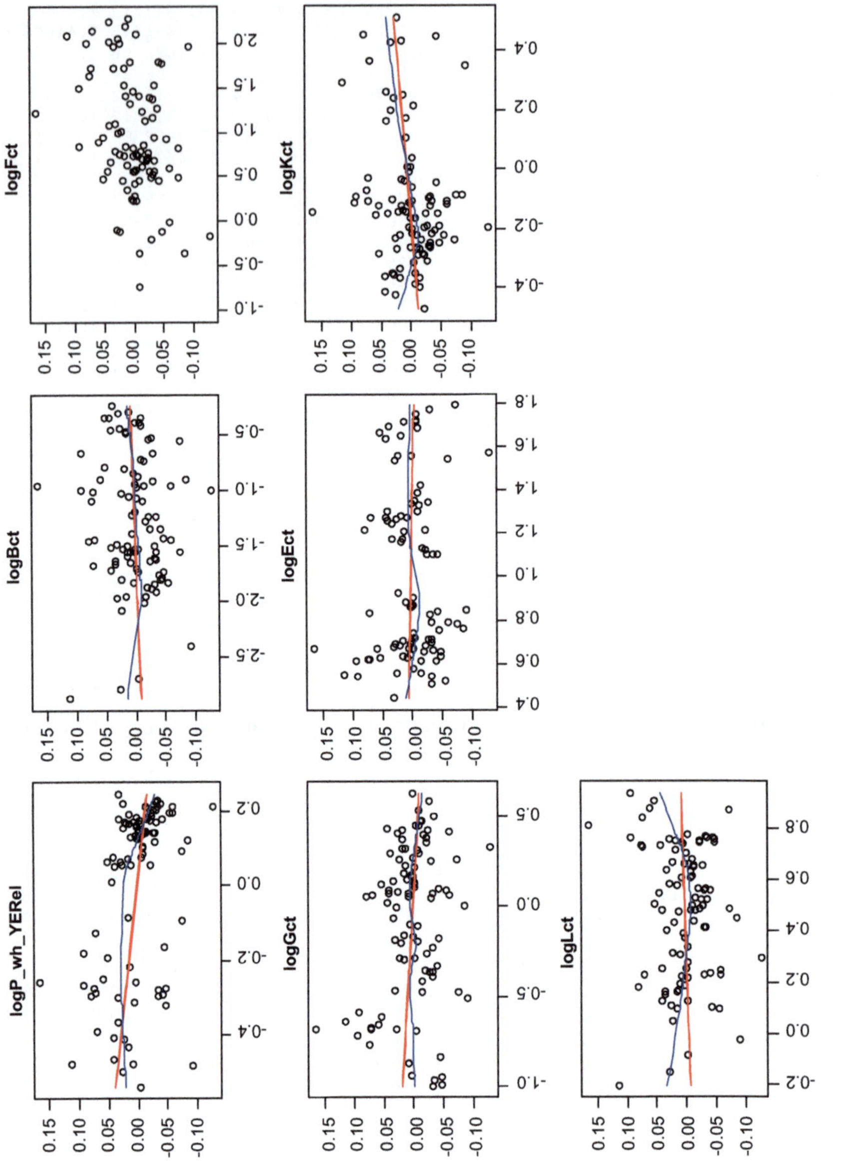

Fig. 4.15 Group T.1.1. (after including the Netherlands and elimination of independent variable M(c, t), see Table. 4.4; source: own work)

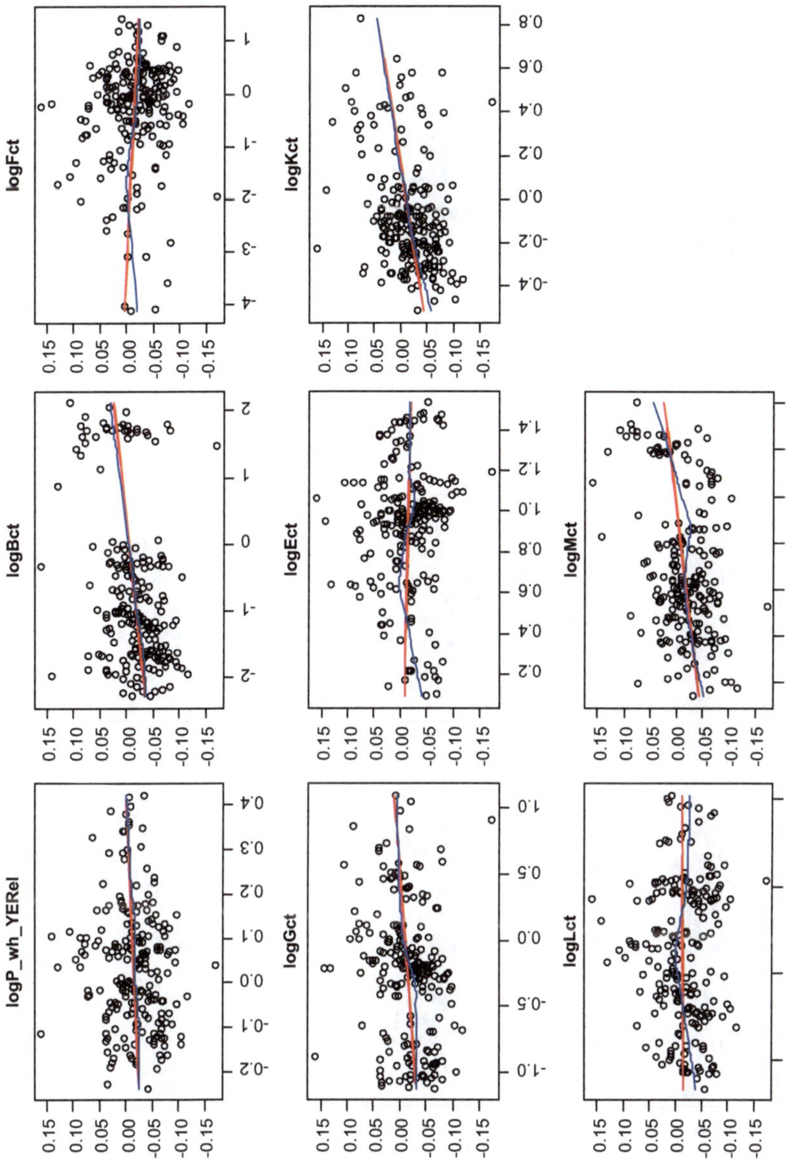

Fig. 4.16 Group T.1.2. (Source: own work)

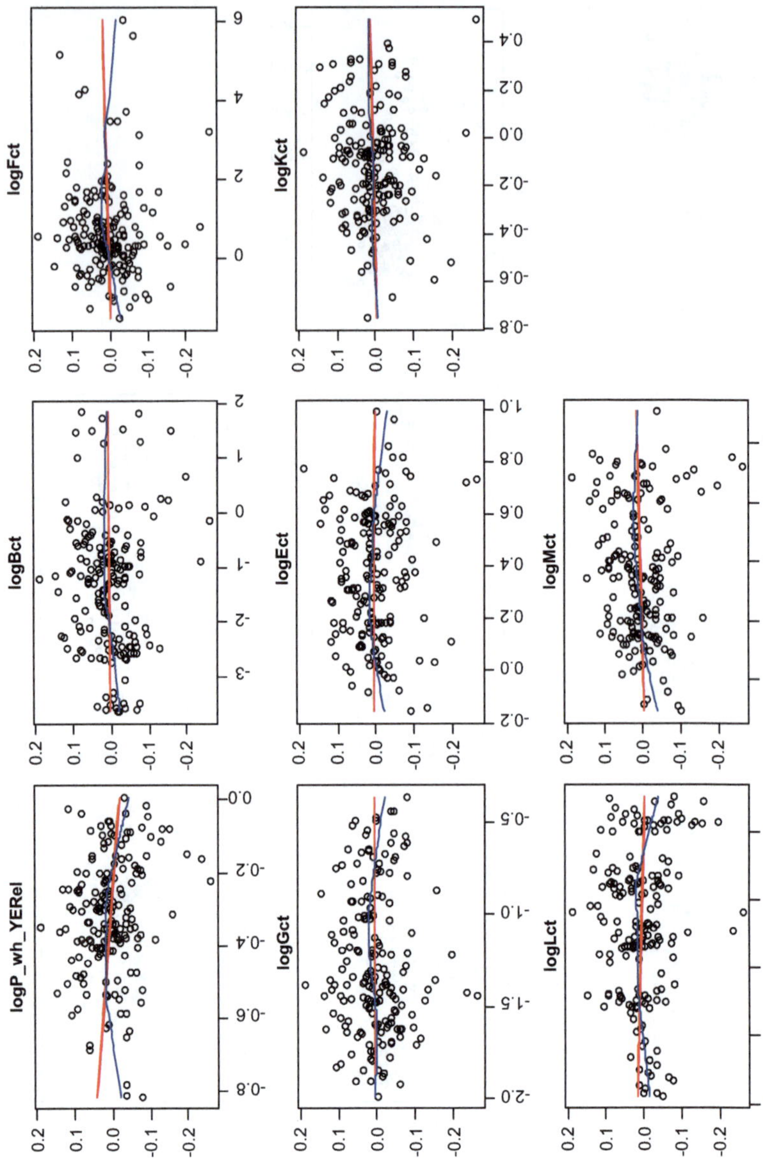

Fig. 4.17 Group T.2.1. (Source: own work)

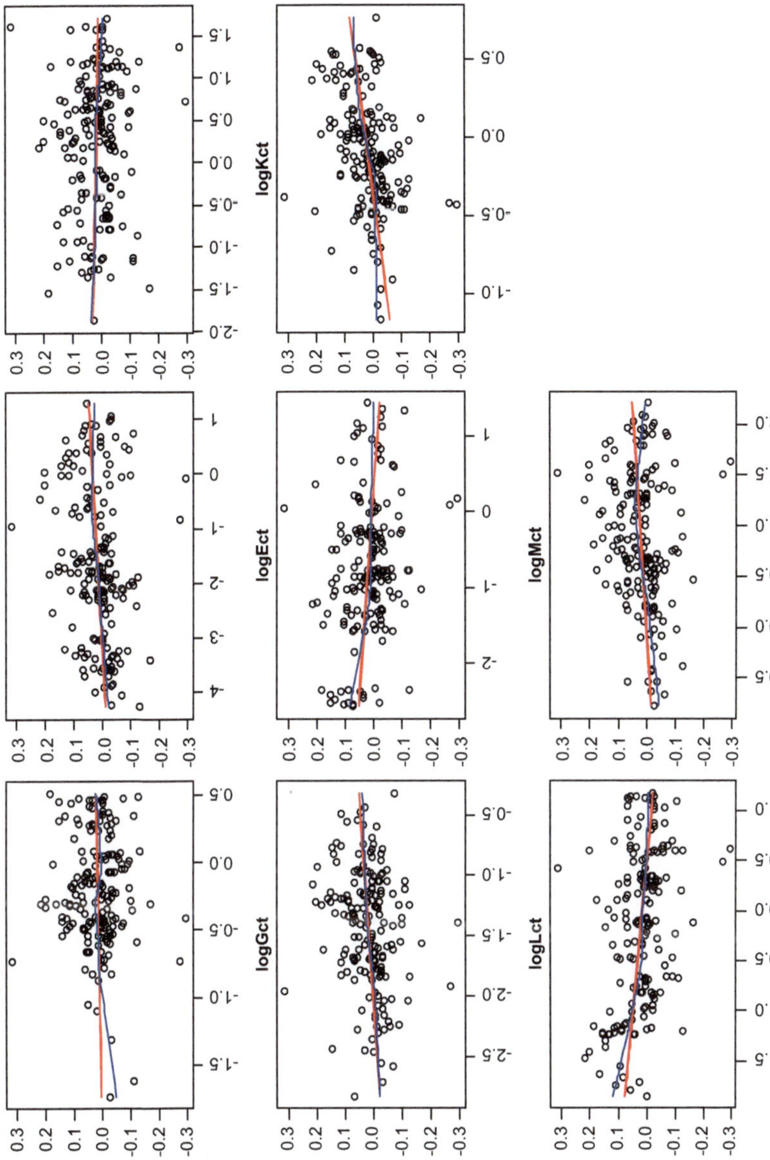

Fig. 4.18 Group T.2.2. (Source: own work)

Part II
Changes in Competitive Advantages in Foreign Trade and National Innovation Systems' Development: Comparing Countries in Europe, North America and Asia

Chapter 5
International Competitiveness of Countries with Dynamic Innovation Systems

Ireneusz Bil, Krzysztof Falkowski and Arkadiusz Michał Kowalski

The group of countries known for their dynamic innovation systems includes five European countries (Ireland, the Netherlands, Switzerland, Finland, and Sweden) as well as Singapore—a small and very competitive Asian country. Dynamic innovation systems are characterized by a high innovative ability and an innovation position above average. The typical qualities of their functioning are a drive to develop knowledge and human capital resources, significant internationalization of business and research and development (R&D) sphere, high flexibility of institutions, as well as dynamic network connections between main entities of the innovation system. Within the group of a dynamic national innovation system (NIS), it is possible to distinguish two different approaches toward using those strengths in order to develop competitiveness. Finland and Switzerland focus on using and developing their own resources mainly on the basis of the internal factors, while treating the internationalization as an accompanying element. A similar approach can be observed in Switzerland, a country ranked among the most competitive economies but not a member of the European Union (EU). By contrast, Ireland and Singapore place emphasis on internationalization as the main factor of the development of internal resources[1] (Weresa 2012).

In order to better illustrate those differences between various dynamic NIS, the following sections contain the analysis, in the form of case studies, of the international competitiveness of three economies which represent the dynamic model of an

[1] A comprehensive evaluation of various innovation systems, along with an analysis of their characteristics, may be found in Weresa (2012).

I. Bil (✉) · K. Falkowski (✉) · A. M. Kowalski (✉)
World Economy Research Institute, Warsaw School of Economics, Al. Niepodleglosci 162, 02-554 Warsaw, Poland
e-mail: ibil@fae.pl

K. Falkowski
e-mail: kfalkow@sgh.waw.pl

A. M. Kowalski
e-mail: arkadiusz.kowalski@sgh.waw.pl

M. A. Weresa (ed.), *Innovation, Human Capital and Trade Competitiveness*, Innovation, Technology, and Knowledge Management, DOI 10.1007/978-3-319-02072-3_5,

innovation system, but utilize different strategies of using innovations and human capital in order to form their competitiveness in international trade, namely Finland, Ireland, and Switzerland.

5.1 Case Study: Finland

Krzysztof Falkowski

Nowadays, Finland is one of the most innovative and competitive countries in the world. In 2012, it was ranked the fifth most innovative European country (following Switzerland, Sweden, Denmark, and Germany) (EC 2012). The Finnish economy is also classified among the world's leading countries in terms of competitiveness (according to The Global Competitiveness Report 2011-12, it is ranked third, following Switzerland and Singapore) (WEF 2012, p. 13). On the other hand, it is a relatively small country, considering its area (338,100 sq. km—66th place in the world ranking) and population (5.3 million inhabitants—112th place in the world ranking).

Therefore, a question arises in relation not to Finland's competitiveness on the international level but to its underlying causes. This section will, in particular, focus on the role of human resources (in terms of quantity and quality), as well as on innovations (both technological and organizational).

5.1.1 Evaluation of Finland's Competitiveness in Foreign Trade Between 1995 and 2010

While analyzing the turnover of Finland's foreign trade between the years 1995 and 2010, it is possible to distinguish three characteristic periods. During the first period, from 1995 to 2001, the value of Finland's foreign trade remained at a constant level, running from US$ 69.9 billion in 1995 to US$ 79.4 billion in 2000. In the second period, from 2002 to 2008 (in other words, from the introduction of the single currency, the euro, to the outbreak of the world economic crisis), Finland's foreign trade turnover increased almost 2.5 times, from US$ 77.9 billion in 2002 to US$ 189.1 billion in 2008. The third characteristic period in Finland's foreign trade falls between 2009 and 2010, when, due to the global recession, the value of foreign trade turnover decreased, respectively, to US$ 123.7 billion in 2009 and US$ 131.7 billion in 2010.

For the entire analyzed period between 1995 and 2010, the exports were more significant to Finland's total foreign trade turnover than the imports (Fig. 5.1). The pace of changes was similar for both export and import. Only once, in 2010, the value of Finnish imports was higher than the value of the exported Finnish goods and services.

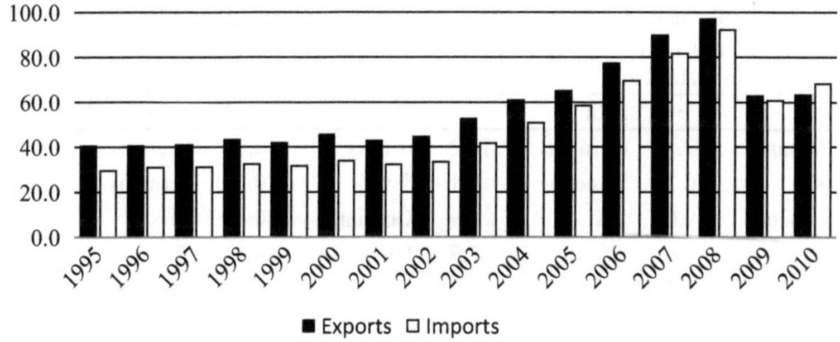

Fig. 5.1 Value of Finland's exports and imports between 1995 and 2010 (in US$ billion). (Source: Own study on the basis of data from the United Nations Commodity Trade Statistics Database)

During the entire period from 1995 to 2010, Finland's trade with foreign countries was characterized by an active balance of sales turnover, with the exception of 2010, when the foreign trade deficit amounted to US$ 4.9 billion. However, a significant decrease in the active balance of Finland's foreign trade turnover was observed in 2008 (a decrease by 42.7% in relation to the corresponding balance in 2007), and particularly in 2009 (a decrease by 55.3% in relation to 2008 and by as much as 74.4% in relation to 2007). By all means, it was a direct effect of the world economic crisis.

When analyzing the geographical structure of Finnish exports between 1995 and 2010, it is important to stress the fixed position of main markets for Finnish goods. In 1995, 2000, 2005, and 2010, virtually the same countries constituted the group of Finland's five main export partners. Traditionally, Finland's most important export partners were Germany, Sweden, and the United Kingdom, and—from the second half of the first decade of the twenty-first century—also Russia (whose share in total Finnish exports amounted only to 4.7% in 1995, while in 2005 and 2010—to almost 10%) (Table 5.1).

It should also be emphasized that in 2010 the combined share of Finland's five main export partners constituted almost a half of the total value of goods and services sold abroad (precisely 48.39%).

In the analyzed period of 1995–2010, the geographical structure of Finnish imports did not significantly differ from the geographical structure of the exports. The same countries, although in a slightly different order, were the main foreign markets for the Finnish economy's supply of goods and services. For years, three countries have been the main import partners for Finland, namely Germany, Russia, and Sweden.

For many years, Germany had been Finland's leading foreign supplier of goods. Its share in total Finnish imports was very high and ranged from 15.5% in 1995 to 13.3% in 2010. The fact that Germany lost the leading position to Russia is primarily due to the increase in prices of energy raw materials imported by Finland from this country. Because of Finland's energy imports from Russia, the country was traditionally placed, through the entire analyzed period, among Finland's most im-

Table 5.1 Finland's main export partners in chosen years of the period from 1995 to 2010 (% share in total exports). (Source: Own study on the basis of data from the United Nations Commodity Trade Statistics Database)

	1995		2000		2005		2010	
1.	Germany	13.1	Germany	12.2	Russia	10.8	Sweden	12.5
2.	UK	10.2	Sweden	9.1	Sweden	10.7	Germany	11.0
3.	Sweden	9.0	UK	8.9	Germany	10.4	Russia	9.8
4.	US	6.6	US	7.7	UK	6.6	US	7.7
5.	Russia	4.7	France	5.0	US	5.7	The Netherlands	7.4

Table 5.2 Finland's main import partners in chosen years of the period from 1995 to 2010 (% share in total imports). (Source: Own study on the basis of data from the United Nations Commodity Trade Statistics Database)

	1995		2000		2005		2010	
1.	Germany	15.5	Germany	14.2	Germany	14.9	Russia	17.8
2.	Sweden	11.6	Sweden	10.3	Russia	13.9	Germany	13.3
3.	UK	8.2	Russia	9.3	Sweden	10.5	Sweden	10.1
4.	US	7.2	US	7.1	China	5.9	China	7.3
5.	Russia	7.2	UK	6.5	UK	4.4	The Netherlands	5.4

portant import partners, while its share was steadily rising (in 1995 it amounted to 7.2 %, in 2000—to 9.3 %, in 2005 it was already 13.9 %, while in 2005 it amounted to as much as 17.8 %) (Table 5.2).

While analyzing Finland's imports in the 1995–2010 period, it is important to stress the increase in the significance of China. From the middle of the first decade of the twenty-first century, China has been gradually increasing its share in Finnish imports, being one of Finland's five biggest import partners. The largest part of Finland's exports in the years 1995–2010 were manufactured goods classified by material (group 6 of the Standard International Trade Classification (SITC)), as well as machinery, equipment, and transport equipment (group 7 of the SITC classification). These two groups accounted for as much as 76.68 % of the value of total exports in 2000. In the following years, this share decreased to 74.57 % in 2005 and to 64.65 % in 2010 (Table 5.3).

In the 1990s, the biggest role in Finnish exports was played by the goods of group 6 of the SITC classification (including, among others, metals and metallurgical products). However, as soon as in the first decade of the twenty-first century, and especially in its first half, the clear leader in this respect were the goods of group 7 of the SITC classification (machinery, equipment, and transport equipment), the share of which in total Finland's export turnover in 2000 amounted to 45.35 %, while in 2005 it was 44.1 %. Within this goods group, the sale of electronic and optical equipment abroad was the most significant. After 2005, the importance of the foregoing group decreased considerably. In 2010, the goods from this group accounted for only 32.55 % of total Finnish exports (which meant a decrease by as much as 11.55

Table 5.3 Structure of Finland's exports and imports in chosen years of the 1995–2010 period by SITC groups (%). (Source: Own study on the basis of data from the United Nations Commodity Trade Statistics Database)

SITC group		Exports				Imports			
		1995	2000	2005	2010	1995	2000	2005	2010
0	Food and live animals	1.98	1.39	1.55	2.08	4.88	4.28	4.10	5.53
1	Beverages and tobacco	0.29	0.17	0.18	0.26	0.65	0.70	0.88	0.96
2	Crude materials, inedible, except fuels	8.91	6.85	5.71	6.80	7.31	6.61	7.66	8.59
3	Mineral fuels, lubricants, and related materials	1.89	3.47	4.44	8.23	8.83	11.91	13.76	18.50
4	Animal and vegetable oils, fats, and waxes	0.08	0.08	0.11	0.19	0.19	0.10	0.10	0.59
5	Chemicals and related products	5.93	5.87	7.39	10.71	12.21	10.42	11.24	11.36
6	Manufactured goods classified chiefly by material	38.88	31.33	30.47	32.10	15.45	11.77	12.08	12.77
7	Machinery and transport equipment	35.35	45.35	44.10	32.55	38.48	41.89	38.65	28.37
8	Miscellaneous manufactured articles	5.72	4.94	5.20	5.45	10.18	9.90	9.67	9.99
9	Commodities and transactions not classified elsewhere in the SITC	0.97	0.55	0.87	1.62	1.82	2.41	1.86	3.34
	Overall	100.0	100.0	100.0	100.0	100.0	100.0	100.0	100.0

percentage points). The above-mentioned situation was caused, first, by an increase in the competition in the international market for telecommunications equipment and the weakening position of Finnish Nokia, as well as by the global economic crisis, which adversely affected the demand for consumer electronic goods.

Changes related to the second of the above-mentioned goods groups, which are incredibly important for Finland's exports, i.e., the goods of group 6 of the SITC classification (manufactured goods classified according to material), were not that significant in the 1995–2010 period. Although in 1995 they accounted for 38.88 % of Finnish exports, afterward their corresponding share remained at a lower but relatively constant level (31.33 % in 2000, 30.47 % in 2005, and 32.1 % in 2010).

The third group of goods in terms of the share in Finnish exports in 2010 were the goods of group 5 of the SITC classification (chemicals and related products), the share of which amounted to 10.71 %. It should be stressed that the share of goods of the group had almost doubled, as it amounted to 5.93 % in 1995. A spectacular growth of an analogical share had been observed in the case of the goods of group 3 of the SITC classification (mineral fuels, lubricants, and related materials), the share of which in total Finland's exports had quadrupled (in 1995 it amounted to 1.89 %, while in 2010 it was already 8.23 %). Nevertheless, the significance of the two foregoing goods groups in Finnish exports was not as big as that of the goods of groups 6 and 7 of the SITC classification.

The goods structure of Finnish imports was, however, much more diversified in the 1995–2010 period than the analogous structure of Finnish exports discussed above. Nevertheless, it is possible to indicate a strong leader, i.e., the goods group of the SITC classification with the largest share in Finland's total imports. The group includes machinery, equipment, and transport equipment (group 7 of the SITC classification). Throughout the analyzed period of 1995–2010, the share of this goods group in total Finnish imports decreased by as much as 10.11 percentage points (from 38.48 % in 1995 to the level of 28.37 % in 2010). However, over the whole period, the share considerably exceeded the corresponding shares of other goods groups.

Other goods groups of the SITC classification with a share of over 10 % in Finnish imports in the period 1995–2010 were group 3 (mineral fuels, lubricants, and related materials), group 6 (manufactured goods classified by material), and group 5 (chemicals and related products).

The biggest increase in the significance of Finnish imports throughout the analyzed period was observed in the case of the goods of group 3. The share of mineral fuels, lubricants, and related materials increased from the level of 8.83 % in 1995 to 18.5 % in 2010 (an increase of 9.67 percentage points). Such a significant growth in the share of this goods group was due to an increase in world prices of fuels imported by Finland mainly from Russia.

In summary, while analyzing the changes in the goods structure of Finland's foreign trade in the years 1995–2010 according to the goods groups of the SITC classification, it should be emphasized that Finnish exports were strongly dominated by highly processed goods, such as manufactured goods classified by material (group 6 of the SITC classification) as well as machinery, equipment, and transport equipment (group 7 of the SITC classification). In the case of the import of goods

Table 5.4 Structure of Finland's exports and imports in chosen years of the 1995–2010 period according to the OECD classification of manufacturing industries based on the technology level. (Source: Own study on the basis of data from the United Nations Commodity Trade Statistics Database)

Industries	Exports				Imports			
	Total share in exports (%)				Total share in imports (%)			
	1995	2000	2005	2010	1995	2000	2005	2010
High technology	11.21	17.90	16.07	9.54	12.48	15.18	14.01	9.87
Medium-high technology	24.14	24.13	27.54	29.13	38.97	39.28	39.96	38.43
Medium-low technology	42.49	39.15	40.39	46.58	36.07	33.65	34.36	37.95
Low technology	22.16	18.82	16.00	14.75	12.48	11.89	11.67	13.75

to Finland throughout the examined period, a significant decrease in the importance of machinery and transport equipment (group 7 of the SITC classification), as well as a growing significance of mineral fuels, lubricants, and related materials (group 3 of the SITC classification), was observed (Table 5.3).

From the perspective of international competitiveness of a given economy, it is a good idea to take a closer look at changes in the goods structure in its foreign trade, including the technological factor. In accordance with the OECD classification (OECD, Organisation of Economic Co-operation and Development), high-technology, medium-high-technology, medium-low-technology, and low-technology industries are distinguished (Table 5.4).

In the 1995–2010 period, Finnish exports were strongly dominated by medium-low-technology goods, in terms of both the value of their foreign sales and of their share in Finnish exports. The value of export of the medium-low-technology goods throughout the whole period almost doubled (in 1995 it amounted to US$ 34.5 billion, while in 2010 it was as much as US$ 65.4 billion). With regard to the share of this goods group in Finnish exports, it increased in the analogical period from 42.49 % in 1995 to 46.58 % in 2010 (an increase of 4.09 percentage points).

The second position in Finland's exports was traditionally occupied by medium-high-technology goods. In the case of this goods group, a growth in both the value of turnover in exports (an increase of US$ 21.3 billion, from the level of US$ 19.6 billion in 1995 to US$ 40.9 billion in 2010) and of the share in the turnover (an increase of 4.99 percentage points, from 24.14 % in 1995 to the level of 29.13 % in 2010) was observed in the analyzed period.

In the case of other categories of manufactured goods, i.e., those of high technology and low technology, a decrease in their share in Finnish exports was observed at the same time, although the value of foreign sales of these goods slightly increased in both cases. In this context, a considerable decrease in the significance of high-technology goods in Finland's exports has to be pointed out. While in 2000 the share of the group in total Finnish exports amounted to 17.9 %, 10 years later (in 2010) it was only 9.54 %. The foregoing fact explains a considerable decrease in the significance of the goods of group 7 of the SITC classification (i.e., machinery, equipment, and transport equipment), which was measured by the share of the goods in Finland's export (explained above) and was as high as 12.8 percentage points (from 45.35 % in 2000 to 32.55 % in 2010).

In the case of Finnish imports in the analyzed 1995–2010 period, medium-high- and medium-low-technology goods played the most significant role. Their share in Finland's total imports was stable throughout the examined period. In the case of medium-high-technology goods, it amounted to 38.97 % in 1995, 39.28 % in 2000, 39.96 % in 2005, and 38.43 % in 2010. At the same time, analogical shares of medium-low-technology goods were as follows: 36.07, 33.65, 34.36, and 37.95 %.

Just as in the case of Finnish exports, in the years 1995–2010, a visible decrease in the share of high-technology goods in imports could be observed (a decrease of 5.31 percentage points, in comparison with the share of these goods in imports in 2010 and 2000). The share of low-technology goods in Finland's total imports remained, however, at the same level, ranging from 12.48 % in 1995 to 13.75 % in 2010.

In connection with an increase in the total turnover of Finnish imports (comparing the level of 2010 to that of 1995), the value of individual goods groups distinguished according to the degree of technological advancement also increased over the analyzed period. The value of Finland's imports of each of the groups, except for high-technology goods, almost doubled in the examined period. Imports of high-technology goods also increased their value, but to a lesser extent (an increase of 42.5 %).

All in all, in the years 1995–2010, the biggest role in Finnish exports was played by medium-low-technology goods, while in imports by medium-high and medium-low-technology goods. Taking into account the dynamics and trend of changes in the significance of individual goods groups in the Finnish foreign trade, it should be noted that, in the case of exports, the key role of the above-mentioned medium-low-technology goods was strengthening. As for imports, medium-high-technology and medium-low-technology goods maintained their position throughout the entire analyzed period. In the second half of the twenty-first century, a decrease in the significance of high-technology goods could be observed in the case of both Finnish exports and imports.

While analyzing the goods structure of Finland's foreign trade in the 1995–2010 period from the perspective of the significance of goods with different extents of use of tangible and intangible expenses in the production process, it is important to highlight that Finnish exports are traditionally dominated by capital-intensive and mainstream goods (Table 5.5). In 2010, the share of capital-intensive goods in exports amounted to 40.68 %, while in 1995 it was 32.97 %, which indicates an increase of 7.71 percentage points. Incidentally, in the case of all other goods groups distinguished on the basis of the extent of use of tangible and intangible expenses in the production process, a decrease in the share in Finnish exports was observed in the years 1995–2010. The biggest decrease concerned labor-intensive and technology-driven goods (3.84 and 2.99 percentage points, respectively).

Furthermore, with a growth in the share of capital-intensive goods in Finnish exports in the years 1995–2010, the value of the exports had also been gradually increasing, from US$ 18 billion in 1995 to US$ 36 billion in 2010. In the case of other goods groups distinguished according to the same Austrian Institute of Economic Research (WIFO) methodology, although an increase in the value of exports was noted, it was considerably smaller than that of the export of capital-intensive goods.

Table 5.5 Structure of Finland's exports and imports in chosen years of the period from 1995 to 2010 according to the WIFO classification of manufacturing industries based on the extent of use of tangible and intangible expenses in the production process. (Source: Own study on the basis of data from the United Nations Commodity Trade Statistics Database)

Industries	Exports				Imports			
	Total share in exports (%)				Total share in imports (%)			
	1995	2000	2005	2010	1995	2000	2005	2010
Basic	37.18	31.66	32.64	36.72	25.08	22.25	21.98	25.72
Labor intensive	12.09	9.72	8.01	8.25	10.79	10.98	11.24	12.10
Capital intensive	32.97	29.15	31.94	40.68	22.86	19.65	21.29	25.55
Marketing based	3.48	3.35	3.02	3.05	10.79	10.40	10.05	11.07
Research based	14.28	26.12	24.39	11.30	30.48	36.72	35.44	25.56

Taking into account the extent of use of tangible and intangible expenses in the production process, the goods structure of Finnish imports in the 1995–2010 period was much more miscellaneous. The largest share in Finnish imports during the analyzed period belonged to mainstream goods, capital-intensive goods, and technology-driven goods. As for the significance of the foregoing goods groups for Finnish imports throughout the examined period, some characteristic changes could have been observed. Despite some fluctuations, the share of mainstream goods and capital-intensive goods increased slightly (comparing the status of 2010 to that of 1995). The share of technology-driven goods, on the other hand, decreased by 4.92 percentage points.

In the case of the two other goods categories, i.e., labor-intensive goods and marketing-driven goods, the share of which in Finnish exports was not as significant as that of the three above-mentioned goods groups (as it only slightly exceeded 10% in both the first and second case), a slight increase was noted in the years 1995–2010.

To sum up, in the years 1995–2010, capital-intensive and mainstream goods played the main role in Finland's exports. An increase in the significance of capital-intensive goods in that period took place, however, at the expense of a decrease in the importance of labor-intensive goods. It was a direct consequence of the adopted strategy of development of the economy of Finland. As for Finnish imports, the largest (comparable) shares were attributed to mainstream goods, capital-intensive goods, and technology-driven goods. It has to be noted that the significance of technology-driven goods was decreasing.

Another approach to the analysis of the goods structure of Finland's foreign trade takes into account the classification of the skill level of labor force necessary to produce the goods that are subject to foreign trade, in the case of both exports and imports (Table 5.6). In the years 1995–2010, the main role in trade was played by the goods produced by highly skilled, blue-collar workers. Their share in Finnish exports was very high throughout the whole analyzed period—in 2010 it amounted to 46.58%, while in 1995 it was 42.54%.

The second most important goods group in Finnish exports in the 1995–2010 period, taking into consideration the skill level of the labor force, are goods pro-

Table 5.6 Structure of Finland's exports and imports in chosen years of the 1995–2010 period according to the classification of manufacturing industries based on the skill level of labor force. (Source: Own study on the basis of data from the United Nations Commodity Trade Statistics Database)

Industries based on:	Exports				Imports			
	Total share in exports (%)				Total share in imports (%)			
	1995	2000	2005	2010	1995	2000	2005	2010
highly skilled, white-collar workers	11.22	17.92	16.07	9.54	12.48	15.18	14.01	9.87
semiskilled, white-collar workers	24.17	24.16	27.54	29.13	38.97	39.28	39.96	38.43
highly skilled, blue-collar workers	42.54	39.08	40.39	46.58	36.07	33.65	34.36	37.95
low-skilled, blue-collar workers	22.07	18.84	16.00	14.75	12.48	11.89	11.67	13.75

duced by semiskilled, white-collar workers. The share of the goods group had also increased in the analogous period (from 24.17% in 1995 to 29.13% in 2010), just like the total value of their sales in foreign markets (from the level of US$ 19.6 billion in 1995 to US$ 40.9 billion in 2010). An increase in the importance of goods manufactured by highly skilled, blue-collar workers and semiskilled, white-collar workers, which took place in Finland in the years 1995–2010, included a reduction in the share of the goods manufactured with the help of highly skilled, white-collar workers and low-skilled, blue-collar workers.

As for the import of goods to Finland in the analyzed period, it was dominated by goods manufactured by semiskilled, white-collar workers and highly skilled, blue-collar workers. In 2010, they constituted 38.43 and 37.95% of Finnish imports, respectively.

In the case of Finnish imports, just as in the case of exports, the significance of goods produced by highly qualified, white-collar workers (measured by their share in imports) in the analyzed period was the lowest in comparison with the other groups (9.87% in 2010). As in the case of exports, the group's share decreased in the second half of the first decade of the twenty-first century.

All in all, taking into consideration the origin of the goods that were subject to Finland's foreign trade in the examined period of 1995–2010, the analysis of industries based on the skill level of the labor force shows that Finland's exports were dominated by goods manufactured by highly skilled, blue-collar workers, while the main role in imports was played by goods produced by semiskilled, white-collar workers and highly skilled, blue-collar workers.

The foregoing analysis of the goods structure of Finland's foreign trade in the years 1995–2010 leads to the following general conclusions:

1. Exports of Finnish goods to foreign markets were dominated by medium-low-technology goods, capital-intensive goods, and mainstream goods manufactured by highly skilled, blue-collar workers. It was proven by the fact that the largest part of Finland's exports were manufactured goods classified by material (group 6 of the SITC classification), as well as machinery, equipment, and transport equipment (group 7 of the SITC classification).

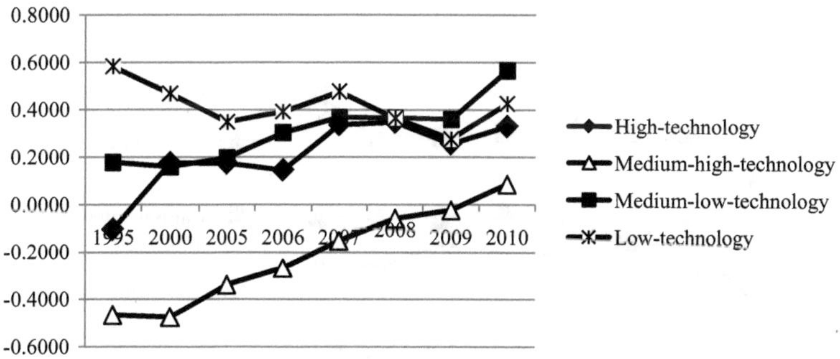

Fig. 5.2 Development of comparative advantages (RCA) in Finland's foreign trade in chosen years of the 1995–2010 period according to the OECD classification of manufacturing industries based on the technology level. (Source: Own study on the basis of data from the United Nations Commodity Trade Statistics Database)

2. As for the import of goods to Finland, the dominant role was played by medium-high- and medium-low-technology goods, capital-intensive goods, mainstream goods, and technology-driven goods produced by semiskilled, white-collar workers and highly skilled, blue-collar workers. The accuracy of the above-mentioned data was confirmed by the fact that the biggest share in Finland's imports belonged to machinery, equipment, and transport equipment (group 7 of the SITC classification), mineral fuels, lubricants, and related materials (group 3 of the SITC classification), as well as to manufactured goods classified by material (group 6 of the SITC classification).

While assessing the level of Finland's competitiveness in foreign trade, one should pay attention to the level of so-called revealed comparative advantages (RCAs).[2]

Throughout the whole 1995–2010 period, Finland recorded comparative advantages in the trade in low-technology goods (mainly paper products, wood, and furniture), medium-low-technology goods (mainly construction and repair of ships), and high-technology goods (mainly aircraft and information and communication technology (ICT) products). In 2010, Finland also gained comparative advantage in the trade of medium-high-technology goods. In the case of the last of the foregoing goods groups, its international competitiveness significantly improved in the years 1995–2010. It was mainly related to the improvement in competitiveness of Finnish electronic and nonelectronic devices (Fig. 5.2).

Taking into consideration the goods groups distinguished on the basis of the extent of use of tangible and intangible expenses in the production process, it is important to notice that, in the 1995–2010 period, comparative advantages occurred in Finland's foreign trade in relation to capital-intensive, mainstream, and labor-intensive goods.

[2] Mentioned in Chap. 1, formula (11) and its interpretation

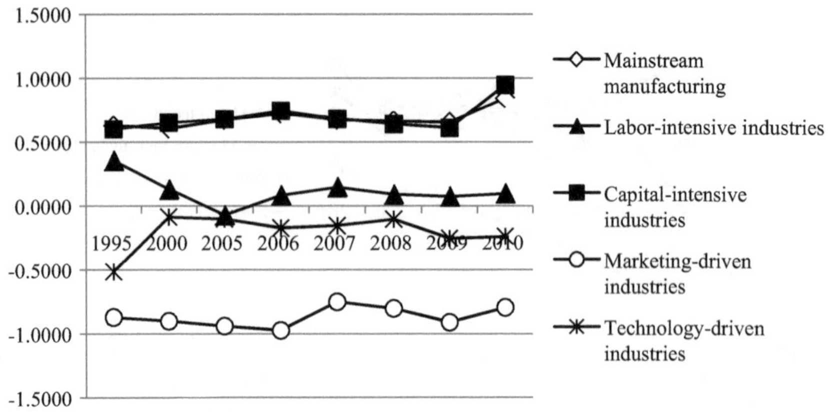

Fig. 5.3 Development of comparative advantages (RCA) in Finland's foreign trade in chosen years of the period from 1995 to 2010 according to the WIFO classification of manufacturing industries based on the extent of use of tangible and intangible expenses in the production process. (Source: Own study on the basis of data from the United Nations Commodity Trade Statistics Database)

In the case of capital-intensive goods, in the analyzed period, Finland's largest comparative advantages were noted in wood pulp and cartons trade, as well as in refined and petrochemical products trade. As for mainstream products, in the years 1995–2010, articles made of paper, electric motors, generators and transformers, pipes, glass and glassware, and weapons and ammunition were the most competitive goods in the international market. Among labor-intensive goods, in the examined period, the most competitive were Finnish ships and boats, heating appliances, wood and sawmill products, and fur products.

In the case of other goods groups (i.e., marketing-driven and technology-based), the RCA index took on negative values in the entire 1995–2010 period, which means that no comparative advantages were recorded. The biggest shortage of those advantages was observed in marketing-driven goods trade (Fig. 5.3).

In the years 1995–2010, the Finnish trade in goods groups distinguished on the basis of the skill level of the labor force needed for their production was characterized by comparative advantages related to goods manufactured by semiskilled blue-collar workers, low-skilled blue-collar workers, and highly skilled white-collar workers. The Finnish goods group that had no comparative advantages throughout the entire analyzed period was the group of goods produced by semiskilled, white-collar workers. However, an exceptionally strong growth rate of the gradual improvement in competitiveness of products from this group made it possible for comparative advantages of the Finnish trade in these goods in 2010 to emerge (Fig. 5.4).

All in all, the Finnish economy is very competitive on the international stage, with the highest competitiveness in 1995 in mainstream and capital-intensive, low-technology goods produced with the help of low-skilled, blue-collar workers. On

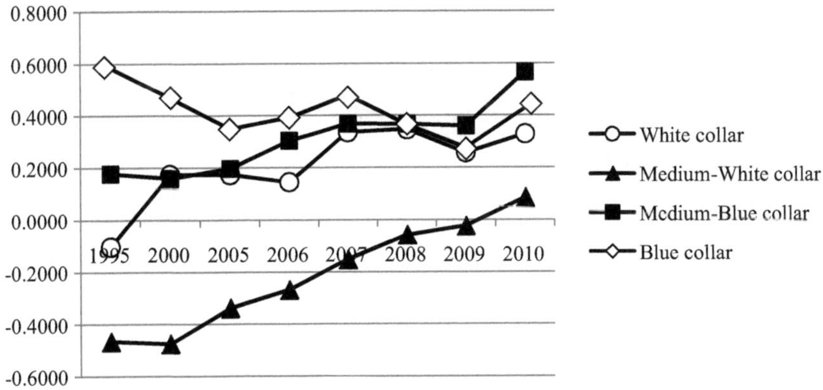

Fig. 5.4 Development of comparative advantages (RCA) in Finland's foreign trade in chosen years of the 1995–2010 period according to the WIFO classification of manufacturing industries based on the skill level of labor force. (Source: Own study on the basis of data from the United Nations Commodity Trade Statistics Database)

the other hand, Finnish, capital-intensive, medium-low-technology goods manufactured by semiskilled, blue-collar workers were the most competitive on the international stage in 2010.

In today's economy, the specialization and intra-industry trade (IIT) become more and more important. From this point of view, it seems interesting to find out if it was possible to observe any changes in Finland on the intra-industry level in the years 1995–2010. Taking into consideration the goods groups that were subject to Finland's foreign trade according to their degree of technological advancement, one has to conclude that the medium-high-technology goods were characterized by the highest intensity of IIT in the total trade turnover of Finland in the analyzed period. Moreover, it should be noticed that, since 2005, the IIT has been continuously increasing (in 2005, the IIT index was 0.8859, while in 2010, it amounted to as much as 0.9942), especially in the domain of trade in research equipment (e.g., telescopes, microscopes, and accelerators), as well as electronic devices and chemicals.

Changes in the intensity of IIT within Finland's foreign trade in high-technology, medium-technology, and low-technology goods were generally not very significant, especially after 2005. The IIT index value for these goods groups ranged from 0.75 to 0.85, indicating a relatively strong IIT, but not as strong as in the case of medium-high-technology goods (Fig. 5.5)

As for the goods groups distinguished on the basis of the extent of use of tangible and intangible expenses in the production process, the intensity of IIT in Finland's foreign trade in the years 1995–2010 was the highest in the case of trade in labor-intensive goods and technology-related goods. In 2010, the values of IIT indexes for these goods groups were 0.9879 and 0.8452, respectively. It is in the domain of trade in labor-intensive goods that the intensity of intra-industry turnover increased the most in the 1995–2010 period (in 1995, the IIT index for this goods group was 0.6782, and in 2010 it was 0.9879) (Fig. 5.6).

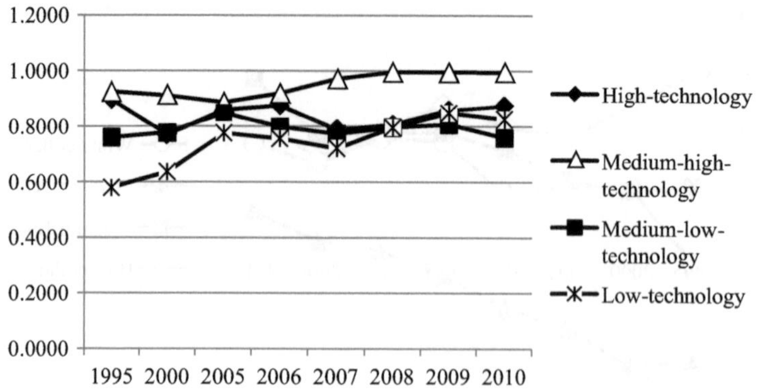

Fig. 5.5 Index of the intensity of intra-industry trade in Finland's foreign trade in chosen years of the 1995–2010 period according to the OECD classification of manufacturing industries based on the technology level. (Source: Own study on the basis of data from the United Nations Commodity Trade Statistics Database)

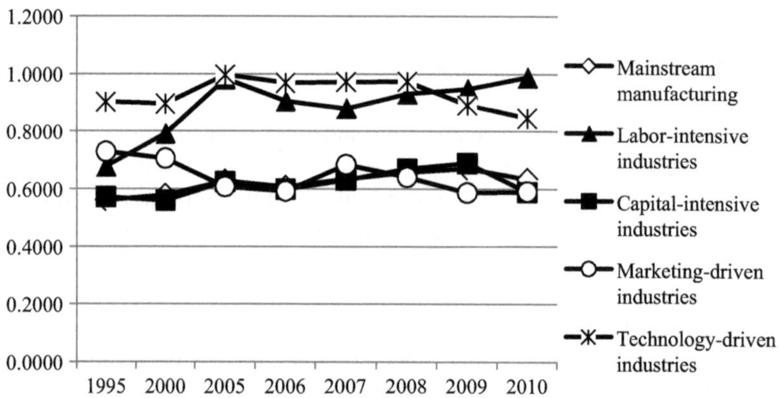

Fig. 5.6 Index of the intensity of intra-industry trade (IIT) according to the WIFO classification of manufacturing industries based on the extent of use of tangible and intangible expenses in the production process in the years 1995–2010. (Source: Own study on the basis of data from the United Nations Commodity Trade Statistics Database)

In the analysis of Finland's trade from the perspective of the skill level of the labor force employed to operate production processes, a large and increasing intensity of intra-industry turnover in the years 1995–2010 was observed in relation to goods manufactured by semiskilled, white-collar workers. However, the most dynamic increase in the intensity of IIT was noticed in the domain of goods produced by low-skilled, blue-collar workers (Fig. 5.7).

To sum up, over the analyzed years 1995–2010, an increasing trend in the intensity of Finnish IIT in labor-intensive, low-technology goods manufactured with the help of low-skilled, blue-collar workers could have been observed. Nevertheless, in

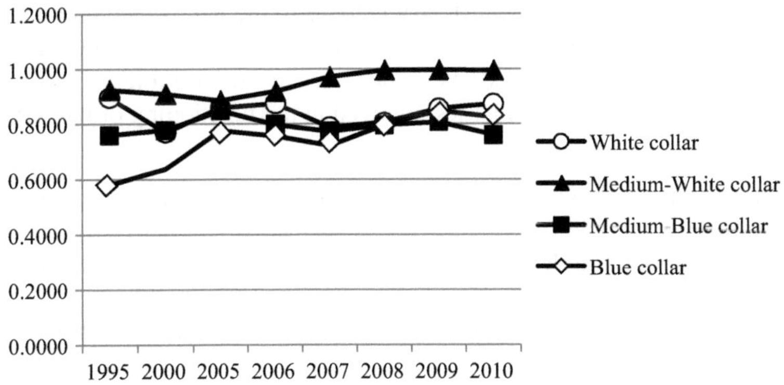

Fig. 5.7 Index of the intensity of Finland's intra-industry trade (IIT) according to the WIFO classification of manufacturing industries based on the skill level of labor force in the years 1995–2010. (Source: Own study on the basis of data from the United Nations Commodity Trade Statistics Database)

2010, the highest level of IIT in Finland's turnover was achieved by labor-intensive, medium-high-technology goods manufactured by semiskilled, white-collar workers.

5.1.2 NIS in Finland in the Years 1995–2010

Finland is one of the role examples of innovation-driven economy. The NIS in Finland is characterized by an exceptionally extensive structure of numerous institutions and their network of mutual relations. The main bodies of the NIS in Finland are:

- Research and Innovation Council,
- Ministry of Employment and the Economy,
- Ministry of Education and Culture,
- Ministry of Finance,
- Ministry of Social Affairs and Health,
- Finnish Funding Agency for Technology and Innovation,
- Finnish Industry Investment Ltd—A governmental company investing in the development of companies through venture capital and private equity,
- Strategic Centers for Science, Technology and Innovation,
- Academy of Finland,
- Public Research Organizations,
- Technical Research Center,
- Finnish Innovation Fund,
- an institution providing loans and guarantees for companies,
- innovative project fund independent from government control, and
- higher education facilities, companies, local governments, and associations.

The strategic position in the Finnish NIS belongs to the RIC, chaired by the prime minister. The council is a body responsible for overseeing and coordinating the innovation policy of the state. Its key objectives are to promote activities of the government and of various ministries involved in the functioning of the NIS in the following fields:[3]

- evaluation of condition and changes in the domain of research, technology, and innovation;
- issues concerning the development of science, technology, innovation, and human resources;
- issues concerning the development and allocation of public research and innovation funding; and
- coordination of activities of the government in the domain of science, technology, and innovation policy.

On the operational level, the most important role is attributed to the Ministry of Employment and the Economy and to the Ministry of Education. The Ministry of Education is responsible for the whole education system, with the exception of higher education and the Academy of Finland. What is more, the ministry has to oversee and coordinate the activity of national research institutions.

Tekes (Finnish Funding Agency for Technology and Innovation), which was founded in the 1980s and which is a part of the Ministry of Employment and Economy, is a unique organization in the Finnish innovation system. Financed from public resources, the institution is responsible for implementing projects from the R&D field and innovations. In 2010, Tekes funded 1,896 projects for the amount of EUR 633 million. Sixty percent of the funds were allocated to support projects of companies (with more than 61 % of that assigned to small and medium-sized enterprises (SMEs)), and the remaining 40 % went to research institutions, such as, among others, polytechnic schools and universities.[4]

Another element of the Finnish innovation system is the Sistra public fund created by the National Bank of Finland in 1967. In 1991, Sistra was transformed into an independent fund under the auspices of the Finnish Parliament. It is an unusual institution, as, despite the public nature, it is independent of government control.[5] It concentrates mainly on the study of trends in domestic and global economies, on the analysis of social trends, on activities aimed at sustainable economic development, and on the formation of innovative and competitive economy. Moreover, Sistra funds R&D projects and subsidizes companies.[6]

---

[3] http://www.minedu.fi/OPM/Tiede/tutkimus-_ja_innovaationeuvosto/tehtavat/?lang=en(accessed 25th July 2012).

[4] http://helsinki.trade.gov.pl/pl/finland/article/detail,1660,Otoczenie_instytucjonalne_systemu_innowacyjnosci.html (accessed 25th July 2012).

[5] http://www.sitra.fi/en/how-does-sitra-operate (accessed 25th July 2012).

[6] http://helsinki.trade.gov.pl/pl/finland/article/detail,1660,Otoczenie_instytucjonalne_systemu_innowacyjnosci.html (accessed 25th July 2012).

In Finland's innovation system, an important role is also played by public–private partnership, which forms the basis of the operation of, among others, special Strategic Centers for Science, Technology and Innovation. Due to the combination of the activity of various environments, i.e., universities, research centers, and companies, it is possible to increase the level of innovation of the whole economy. Currently, there are six centers:[7]

- Forestcluster Ltd (wood clusters),
- TIVIT Ltd (information and communication industry and services),
- FIMECC Ltd (metal products and mechanical engineering),
- CLEEN Ltd (energy and environment),
- RYM Ltd (development of environmental innovations), and
- SalWe Ltd (health and welfare).

In the 1990s, the OSKE program was created in Finland in order to develop a network of links between entrepreneurs, state institutions, and research centers. The program funds 13 so-called competence clusters.

While describing the NIS in Finland, it is important to mention the Otaniemi Science Park. It was created mainly, thanks to the actions of the Helsinki University of Technology, which initiated the need for a special system to support the commercialization of scientific research in Finland. In addition to R&D organizations and academic institutions, more than 600 companies are cooperating in the Otaniemi. The park has repeatedly been awarded in rankings for the most innovative region in Europe. According to its objectives, 15,000 new jobs are to be created in Finland, thanks to Otaniemi by the end of 2030.[8]

A special role in the Finnish innovation system is played by universities—there are 17 universities, and all of them are financed by the state.[9] Recent changes in legislation on higher education and R&D activities concern the intellectual property rights (IPR) to the results of research carried out in universities. According to the new provisions, academics are not entitled to them. They only have property rights in the case of open research conducted without external funding.[10] Thus, patenting has become an important element in the commercialization of university inventions, and the above-mentioned reform has increased the use of research and concepts developed at universities and, at the same time, has contributed to the emergence of a new source of funding for universities.

Another important institution in terms of budgetary outlays earmarked for the purposes of R&D is the Academy of Finland (AKA) with headquarters in Helsinki. In 2012, EUR 320.7 million has been earmarked in the state budget for the activities of this institution. These means are intended for scientific research, foreign scholar-

[7] http://helsinki.trade.gov.pl/pl/finland/article/detail,1660,Otoczenie_instytucjonalne_systemu_innowacyjnosci.html (accessed 25th July 2012).

[8] Ibid.

[9] http://www.tekes.fi/en/community/Universities_and_research_institutes/508/Universities_and_research_institutes/1382 (accessed 25th July 2012).

[10] Evaluation of the Finnish National Innovation System—Full Report, with quotes., p. 36.

ships and research trips abroad of Finnish researchers, and international cooperation between research centers and companies.

5.1.2.1 Innovative Capacity and Position of Finland

While evaluating the innovative capacity of Finland, it is important to pay particular attention to the funds allocated to the realm of R&D, which is primarily responsible for the creation of its own (endogenous) innovative solutions.

Finland, similarly to Sweden, has the highest index value indicating the level of funds allocated to R&D as the percentage of gross domestic product (GDP) in Europe. What is more, in the years 1995–2010, the share of total funds allocated to R&D as a percentage of GDP increased significantly. It was a result of the Finnish authorities' policy focused on the development of a knowledge-based economy. In 1995, the foregoing share amounted to 2.26 % of GDP, while in 2009 it reached a record value of 3.96 % of GDP.

The increase in the total expenditure on R&D in Finland did not result from an increase in the financial means from the state budget allocated to the field. The share of budget funds in relation to GDP did, in fact, remain almost unchanged. Interestingly, in the face of the effects of the global economic crisis, which did not omit Finland, budgetary financial outlays additionally increased. The share of budget expenditure in financing the R&D field in Finland in 2007 amounted to 3.47 % of GDP, while in 2008 it was 3.72 % of GDP, and in 2009—as much as 3.96 % of GDP. The increase in the expenditure on R&D in Finland resulted from a gradual increase in the business expenditure. While in 1995 the business accounted for 59.47 % of the total R&D funds, in 2008 the share amounted to as much as 70.29 %. In 2009, it slightly decreased to the level of 68.1 %. The share of foreign capital financing the R&D in Finland was relatively small in the analyzed period, but, since 2000, it had been gradually increasing to reach the level of 6.61 % of the total expenditure in 2009.

One of the indexes of innovative capacity is the percentage of households with Internet access. In the 2000–2009 period, the percentage increased systematically from the level of 30 % in 2000 to 77.8 % in 2009. It is one of the highest rates in Europe.

The result of outlays on the R&D, as well as of the degree of efficiency of use of human capital resources (especially those with specialized technical education), is the innovative position of a given economy. Its assessment includes, among others, the employment structure. From the point of view of innovation and, as a result, of competitiveness as well, the employment in the high- and medium–high-technology industries, as well as in knowledge-based services, is very important.

What draws attention in the case of Finland is the extraordinarily high, and increasing in the years 1995–2008, percentage of people employed in knowledge-intensive service sectors in relation to total employment in Finland. It increased from 36.77 % in 1995 to 41.06 % in 2008. A similar change, although on a much

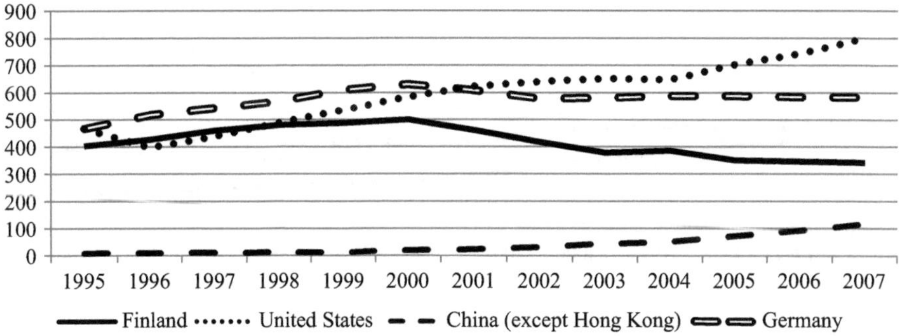

Fig. 5.8 Number of patent applications per million inhabitants in Finland in comparison with selected countries in the years 1995–2007. (Source: Own study on the basis of WIPO data)

smaller scale, was observed in the 1995–2008 period in Finland in the domain of employment in high- and medium-high-technology sectors (an increase from the level of 4.73 % in 1995 to the level of 6.95 % in 2008).

The innovative position of an economy is indicated by, *inter alia*, the number of patent applications. In the case of Finland, the number of patent applications per million inhabitants in 1995–2007 showed varied dynamics. From 1995 to 2000, this index visibly increased (while in 1995 it was 402.9, in 2000 it increased to 498.24). Such a trend was extremely beneficial in terms of competitiveness of the Finnish economy in international trade. However, in 2000, the upward trend reversed—the index began to decrease systematically to reach the level of 341.1 patent applications per 1 million inhabitants in 2007.

The comparison of the foregoing index in Finland and in selected countries of the world (US, Germany, and China) shows that, until 2000, Finland did not significantly differ in this respect from the US and Germany. However, since 2000, the differences between the countries have changed significantly (Fig. 5.8).

The comparison of the number of patent applications in Finland and of the value of funds allocated to R&D in this country shows a decreasing efficiency of the funding, from the viewpoint of the desired effect in the form of patent applications. After 2000, it was lower than the efficiency in Germany and the US, and especially in China (Fig. 5.9). Of course, these data do not determine the quality and the applicability of patent applications in Finland.

The increase in the number of patent applications in Finland in the years 1995–2000 proven above caused an increase in the share of export of high-technology goods in Finnish exports at that time. In 2000, the share reached a record value of approximately 23.5 %, i.e., almost one-fourth of total Finnish exports. After 2000, a downward trend was unfortunately observed—according to the data of Eurostat, the share decreased to 18 %.

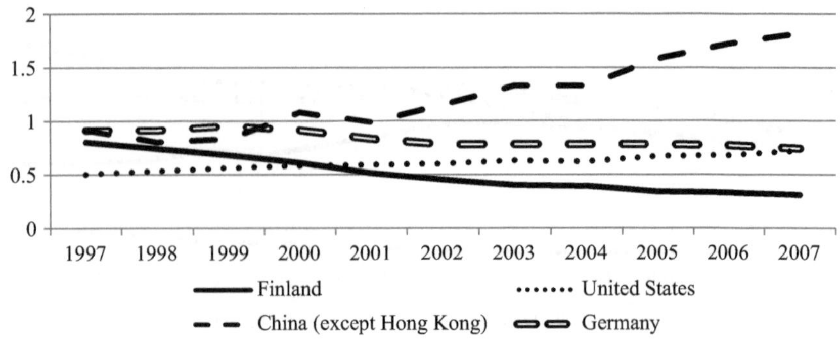

Fig. 5.9 Number of patent applications per US$ 1 billion of funds allocated to R&D in Finland in comparison with selected countries in the years 1997–2007. (Source: Own study on the basis of WIPO data)

5.1.2.2 Human Capital

While analyzing human resources as one of the determinants of long-term competitive advantages in Finland's foreign trade, it is important to begin by drawing attention to the size and the quality of these resources in the country.

In Finland, a permanent trend of population growth can be observed. While in 1950 Finland's population was 4 million, in 2010 it was already 5.36 million. According to OECD estimates, the upward trend will continue, although it will begin to somewhat weaken in 2030. The same estimates show that, in 2050, the population of Finland will be 5.75 million.

Nevertheless, a very unfavorable downward trend in relation to the working-age population has been observed recently. What is more, according to the OECD estimates, the situation will continue to worsen. Thus, unfavorable demographic changes—low birth rate and aging of the population (according to the UN, after about 20 years, up to 25% of the Finnish population will be over 65 years old)—increase the number of inactive population. As a result, Finland struggles with a growing problem of the so-called demographic burden. Over the past 50 years, the number of working-age population per one pensioner has drastically decreased. While in 1950 this index amounted to 9.56 and in 1990 to 5.03, in 2010 there were only 3.83 working-age persons per one pensioner. According to the OECD estimates, this trend is expected to continue in the future and, in 2050, the index will amount to just 2.06.

The analysis of Finland's human resources should include the broadly understood skill level of labor force in the country. The level can be deduced from, *inter alia*, the structure of labor force based on the level of education (Fig. 5.10).

From the perspective of the level of education, the quality of Finland's human capital in 1995–2007 deserves to be highly rated—the share of labor force with higher education systematically increased from the level of 20.7% of the total population in 1995 to 34.8% in 2007 (an increase by as much as 14.1 percentage points),

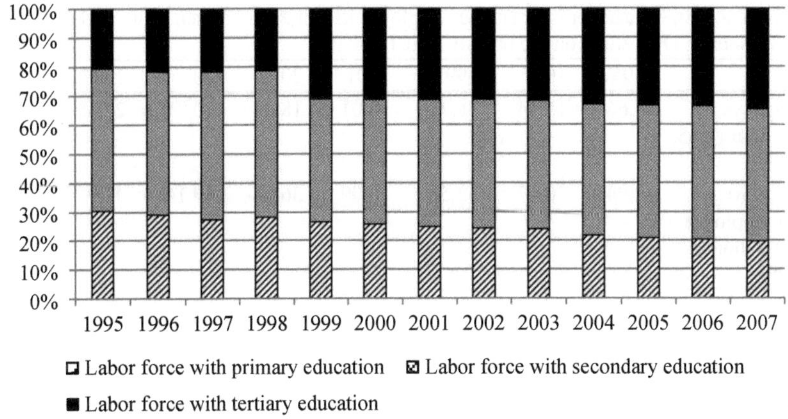

Fig. 5.10 Labor force structure in Finland based on the level of education in the years 1995–2007 (% of total). (Source: Own study based on data of the World Bank)

Table 5.7 Public expenditure on education (as % of GDP) and public expenditure on education (as % of total government expenditure) in Finland in the years 1999–2006. (Source: Own study based on data of the World Bank)

	1999	2000	2001	2002	2003	2004	2005	2006
Total public expenditure on education (as % of GDP)	6.09	5.89	6.06	6.22	6.43	6.42	6.30	6.18
Total public expenditure on education (as % of total government expenditure)	12.49	12.21	12.69	12.74	12.83	12.84	12.57	12.62

while a corresponding percentage of people with primary education decreased by 10.6 percentage points, from 30.3 % in 1995 to 19.7 % in 2007.

In the context of the gradual improvement in the education level of the Finnish population, attention should also be paid to the scale of public spending on education in the country. In the years 1999–2006, the public expenditure on education ranged between 12 and 13 % of total government expenditure, which was usually a little bit over 6 % of GDP (Table 5.7).

As for the extent of use of human resources (measured by the level of unemployment) in Finland, a gradual improvement since 2008 has been observed. The unemployment rate, which amounted to 15.3 % in Finland in 1995, decreased to 6.3 % in 2008. Unfortunately, the crisis caused an increase in the Finnish unemployment rate, which amounted to 8.2 % in 2009. The unemployment rate among young people (15–24-year-olds) in Finland considerably exceeded the total unemployment

Table 5.8 Average wage per hour and per employee in Finland in chosen years of the 1995–2009 period. (Source: Own study on the basis of OECD data)

	1995	1997	1999	2001	2003	2005	2007	2009
Average wage per hour (US$, PPP)	14.66	15.54	16.71	18.01	18.92	20.91	23.35	25.66
Average wage per employee (US$ thous., PPP)	28.94	30.67	32.56	36.09	36.92	40.14	42.97	45.00

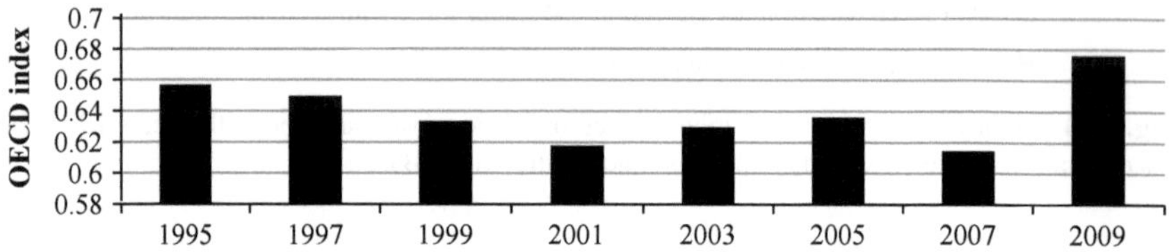

Fig. 5.11 Labor costs per production unit in Finland in chosen years of the 1995–2009 period. (Source: Own study on the basis of OECD data)

rate; nevertheless, it gradually declined (to 15.7 % in 2008 from 26.99 % in 1995). Unfortunately, as a result of the global crisis, this rate of unemployment increased in 2009 to the level of 20.47 %.

The assessment of Finland's human capital should also include labor costs, which gradually increased in the years 1995–2009. In the case of the average wage per hour, the increase amounted to as much as 75 % (from US$ 14.66 in 1995 to US$ 25.66 in 2009), while in the case of the average wage per employee, it was 55.5 % (from US$ 28.94 in 1995 to US$ 45 in 2009) (Table 5.8).

The labor costs per production unit in Finland in the years 1995–2009 decreased (except for the 2001–2005 period, in which a gradual increase was recorded), as a result of, *inter alia*, increasing innovativeness. As a consequence, the competitiveness of the whole economy was improving at the same time. Unfortunately, in 2009, in times of the global economic crisis, unit labor costs significantly increased (Fig. 5.11).

5.1.2.3 Innovation Policy

In recent decades, the economic growth in Finland was substantially based on the development of advanced technologies, on their implementation, and on the proceeds derived from exports. The innovation policy based on traditional knowledge and technology has significantly contributed to the improvement of international competitiveness of the Finnish economy.

Innovation policy reform was launched in 2008 with the implementation of the National Innovation Strategy. It implies the need to gradually abandon activities focused only on science and technology, in order to include the role of demand and the requirements of consumers and customers in innovative projects (40 % of public funds for R&D are to be allocated to projects based on demand[11]). The new National Innovation Strategy implies primarily:[12]

- renewal of the structure of the NIS by reducing its organizational, functional, and geographic fragmentation;
- enhancement of coordination and conduct of political activities at government level;
- wide promotion of the use of education, research, and innovative projects;
- research specialization of universities, polytechnic schools, and other research centers;
- enhancement of cooperation between educational and research centers and secondary and commercial sector through a clearer division of tasks;
- further development of the legal and institutional environment of companies and motivation systems;
- improvement in the functionality and efficiency of companies and innovation system services;
- consolidation of the position of SMEs in the NIS;
- elaboration of a long-term infrastructure policy and facilitation of the use of public data resources;
- diversification of the basis of financing the innovation and research activity;
- acceleration of internationalization of the NIS; and
- increase in the number of public export credits, venture capital funds, and capital investments.

As a result of the implementation of the new innovation strategy, the target direction of allocation of the funds from the Tekes fund will also change.[13] About one-third of the funds will be allocated to projects of already existing businesses with fewer than 500 employees. Less than one-third of the funds will be allocated to companies hiring more than 500 workers.

The special role of environmental and socially responsible projects in the new Finnish innovation policy should also be highlighted.[14]

The summarization of the assessment of the NIS in Finland should point out strengths and weaknesses of the system, as well as the opportunities and threats, which are presented in Table 5.9.

[11] http://www.proinno-europe.eu/sites/default/files/repository_files/12/03/Finland_TC_final.pdf, s.10 (accessed 25th July 2012).

[12] Ibidem, p. 7.

[13] Ibidem, p. 9.

[14] Ibidem, p. 10.

Table 5.9 SWOT analysis of the Finnish innovation system. (Source: Monitoring and analysis of policies and public financing instruments conducive to higher levels of R&D investments. The "POLICY MIX" Project. Country Review Finland, Universiteit Maastricht 2007, p. 7, http://ec.europa.eu/invest-in-research/pdf/download_en/finland.pdf (accessed 27th July 2012))

Strengths	Weaknesses
Science and technology policy has a long-term dimension and the investments in R&D projects are its key element	Dependence on global economic trends
Well-functioning system of education, research and development, highly rated in international rankings	Considerable distance from the centers of global market and geographic distance from the center of Europe
High education level of the society	Small domestic market, small labor force and consumer resources, language barrier, and quite harsh climatic conditions
Relatively small brain drain effect	Development funds targeted mainly at smaller projects of lesser importance
Relatively high percentage of scientists and researchers	Still small share of foreign faculty and students
Active international patent system	Large companies operate or invest a substantial part of their activity abroad
Good international reputation of Finland, which is considered a reliable and secure country	Small inflow of foreign direct investment
Large number of companies that base their activity on knowledge and innovation	
Good cooperation between the business sector and public R&D institutions	
High degree of internationalization of Finnish companies	
Opportunities	Threats
Increase in the internationalization of enterprises and organizations of the innovation system	Negative impact of the global economic crisis and the current recession in the eurozone
Improvement of the environment attracting new foreign investors and intellectual resources to the country, thus presenting Finland as a good place to operate an innovative business	Relatively low attractiveness of Finland to foreign investors, researchers, and students
Growing awareness of the importance of strategic modeling and evaluation of investment projects	Relatively low activity of Finland in the EU and in the domain of international R&D projects
Introduction of social innovations	Excessive emphasis on the national interest in international cooperation
Growing importance of international organization of cooperation and R&D projects	Aging population which weakens the performance of the national economy
Development of business and marketing concepts	Inadequacy of the education system and educational profiles to the market needs
Support for the creation and growth of innovative enterprises	Decreasing number of start-ups focused on innovation
	Declining business sector expenditure on R&D projects since 2008
	Relocation of companies' activity abroad
	Increase in the significance of the brain drain and outflow of highly qualified staff

5.1.3 Evaluation of the Relationship Between the Level of Innovation and Human Capital Accumulation, and the Advantages in Finland's Foreign Trade

Finland has achieved an undeniable economic success. In the early 1990s, Finland was a poor country confronted with huge economic problems. Its industry was based on low-technology (wood and cellulose) industries, which relied on their own forest resources. Today, however, it is one of the most open, and innovation- and knowledge-based, competitive economies in the world.

At the root of the economic success of Finland, there are, among others, developments in the sector of ICT, high level of education and highly effective system of funding R&D activity, and gradually increasing level of education of human resources. A specific exemplification of the scale of this success, and, on the other hand, its important stimulant, was the gradual growth in the significance of Finland in the international trade, which was mainly due to an increase in the export of high-technology goods. It is enough to say that in 2000 the share of the export of high-technology goods in Finnish exports reached a record value of approximately 23.5 %. After 2000, the share decreased. Nevertheless, it remained above 15 %.

In the context of Finland's high competitiveness on the international stage, it should be noted that the adopted strategy of modernization of the economy based on the construction of Finnish knowledge-based economy, in which the main emphasis was placed on innovation and improvement of the level of educating the society (human capital), proved to be extremely effective.

Thanks to this strategy, Finland achieved (Teichmann 2010, p. 209):

- concentration of productivity growth in selected industries (mainly in electrical and optical industries);
- increase in the efficiency of the use of capital allocated to innovative projects (especially in the domain of ICT and timber and paper industries, in which Finland had traditionally recorded lasting competitive advantages); and
- selective relocation of resources to more productive industries and companies, which had a positive impact on the improvement of labor productivity and, thus, their international competitiveness.

As a consequence, Finland, which in the 1990s was still competing on the basis of mainstream, capital-intensive, and low-technology goods produced by low-skilled, blue-collar workers, has become a country which, to a large extent, owes its high international competitive position mainly to capital-intensive, medium-low-technology goods produced by semiskilled, blue-collar workers.

In practice, it would not have been possible without a well-thought-out and consistently implemented policy promoting the innovativeness of the Finnish economy. Due to this policy, an extensive but well-functioning system of financing R&D, in which the main burden rests on the broadly understood business sector, was created. Another factor that contributed to Finland's success on the international stage was the creation of a perfectly well-functioning education system, which works in close cooperation with business.

All in all, Finland is a very good example of an economy with a very strong relationship between the level of innovation and human capital accumulation, and the advantages in foreign trade.

5.2 Case Study: Switzerland

Ireneusz Bil

Switzerland is one of the ten richest countries in the world (according to the International Monetary Fund (IMF)—eighth position, and according to the World Bank—ninth position), with the level of GDP per capita of approximately US$ 42,000. It is a relatively small country, with a population of 7.9 million. It is not a member of the EU, but it closely cooperates within the European Economic Area.

The following subchapter presents an evaluation of the relation between the accumulated human capital resources and the innovativeness of Switzerland, and the development of Switzerland's long-term competitive advantages in the international trade.

5.2.1 Switzerland's Foreign Trade Competitiveness

Switzerland's economy is characterized by a high turnover with foreign countries and a high share of exports in GDP. Since 2002, a slow but gradual increase in the balance of trade surplus can be observed. The surplus was maintained even during the 2008–2009 crisis. The appreciation of the Swiss franc against other major currencies of the world has an increasing impact on results in the foreign trade. Since the beginning of 2011, the euro has lost about 11 % in comparison with the franc, and the dollar has lost about 15 %.[15]

The EU is the largest recipient of Swiss goods. In 2010, the Union received almost one-third of total Swiss exports. As much as 20.6 % of the exports is sent to Asian countries and 14.3 % to the Americas. Other export destinations have a rather marginal role in Switzerland's exports. The domination of EU27 in Switzerland's turnover is even more visible in imports—in 2010, more than 77 % of the imported goods came from countries of the EU. The second position belonged to Asian countries (12.6 %) and the third to the Americas (7.2 %). The structure of Switzerland's trade with foreign countries shows close economic ties with the EU. For many years, Germany, the US, Italy, and France have remained Switzerland's most important trading partners in exports. China, which overtook Japan in 2010, has strongly reinforced its position.

[15] www.forex.com.

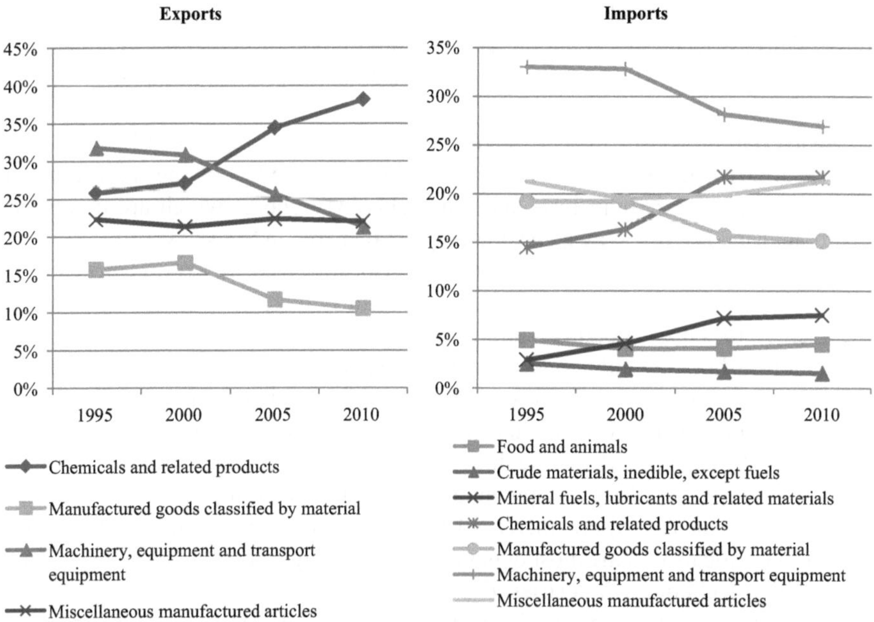

Fig. 5.12 Changes in Switzerland's exports and imports structure according to the SITC classification in the years 1995–2010. (Source: Own study, OECD data)

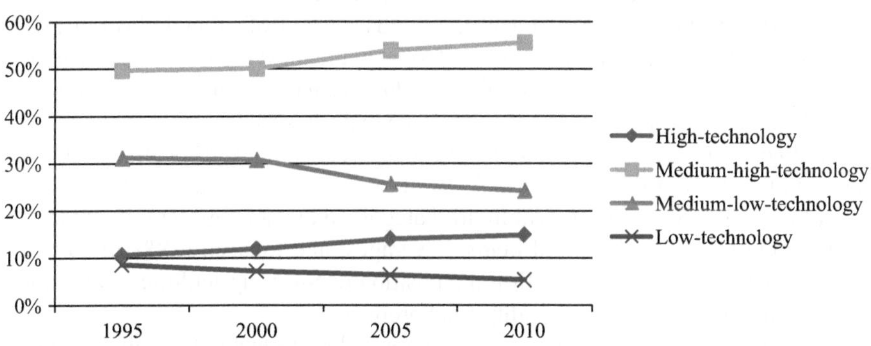

Fig. 5.13 Switzerland's exports structure according to the OECD classification based on the technology level. (Source: Own study, OECD data)

At the same time, Switzerland is one of the few countries that record trade surplus in trade with China. The achievement of trade surplus in trade with China coincided (2002–2003) with an emergence of surplus in the total trade.

In the structure of Switzerland's exports, the chemical industry has a particularly strong position (Fig. 5.13). Its share has been gradually increasing since 2000, while the share of other goods groups remained at a similar level (different industry products) or has decreased (e.g., machinery, equipment, and transport equipment). The share of individual sectors in imports is more stable and sustainable; an increase in the share of import of chemicals and related products stands out. It is, of course, possible to associate this situation with the importation of necessary intermediates (e.g., caprolactam), which undergo further treatment processes in the chemical industry (Fig. 5.12).

Switzerland exports mainly medium-high-technology goods (Fig. 5.13). In the 1995–2010 period, the group accounted for about 53 % of the total turnover of Swiss exports. Medium-low- and low-technology goods, which are traditional manufacturing sectors, play a considerably smaller role in Swiss exports.

Over the past 15 years, Switzerland has also improved its exports structure in terms of the share of high-technology goods, from 11 to 15 %. It is, however, below the average for OECD countries in this period, which amounts to about 22 %.

In high-technology industries, the manufacturing of pharmaceuticals has the biggest share in total exports. The last 15 years have seen a significant consolidation of this group in comparison with other industries in the sector—from 42 % in 1995 to 67 % in 2010. The second and the third place went to the export of telecommunications equipment and office equipment and computers. It is important to notice that the absolute values of exports increased also in these industries, although the export of pharmaceuticals was the most dynamic, hence the significant increase in its share in total high-technology industries' exports.

The increase in the export of high-technology goods means a decrease in the export of medium-low- and low-technology goods. The share of the first group in Swiss exports decreased to 24 % in 2010 (31 % in 1995), whereas the share of the second one decreased to 5 % (2010) from the level of 9 % in 1995.

In 2010, the production of metal goods (31 %) and of other industrial goods (24 %) as well as petroleum refining (18 %) were the most significant in the export of medium-low-technology goods. These industries have maintained their position over the past 15 years. Bigger changes in the exports structure took place in low-technology industries. One could observe a decrease in the significance of apparel and textiles (from 45 % in 1995 to 39 % in 2010) and paper and paper products manufacturing (18 vs. 14 %). Within this category, a dynamic growth was recorded by the industry producing food items, beverages, and tobacco (28 % in 1995 vs. 39 % in 2010). These ratios confirm the position of Swiss agriculture as an important international producer of quality food products.

Shifts in the goods structure of exports toward more technologically advanced and intensive goods are confirmed by the data analysis prepared according to the WIFO classification based on the extent of use of tangible and intangible expenses in the production process. In the years 1995–2010, the biggest increase in the share in exports was recorded by research-driven industries (from 26 to 42 %). However, it took place at the expense of the share in the value of exports of other industry categories (Fig. 5.14). The level of 42 % is higher than that noted by the countries of the economic triad (Radło 2011).

In the domain of imports, the largest growth in the share was recorded also by research-driven industries (from 23 to 31 %) and labor-intensive industries (from 16 to 18 %). A simultaneous large increase in the share of research-driven industries in both exports and imports (Fig. 5.14) may indicate an increase in intra-industry

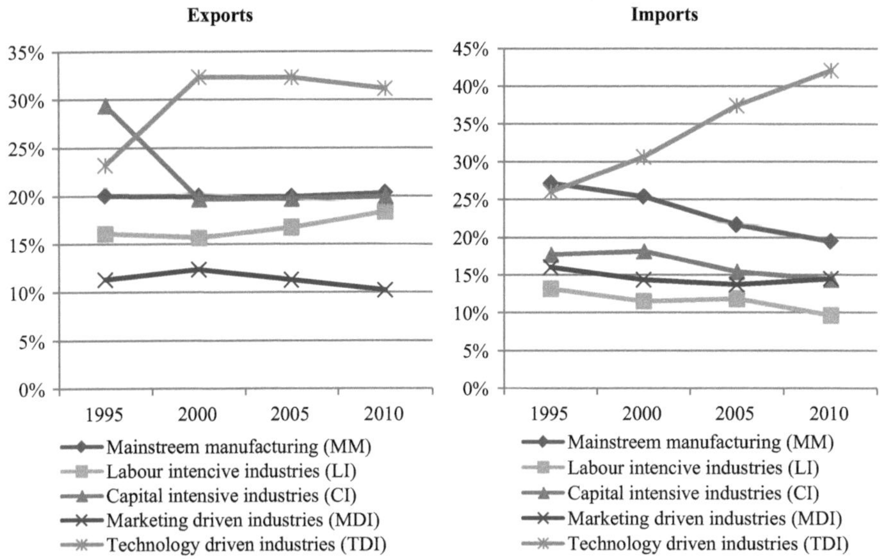

Fig. 5.14 Changes in the goods structure of Switzerland's exports and imports according to the WIFO classification based on the extent of use of tangible and intangible expenses in the production process in the years 1995–2010. (Source: Own calculations, WIFO data)

flows in the foreign trade and the development of intra-industry labor division, which, according to the theoretical considerations of Misala (2011), strengthens the international competitiveness of Switzerland.

A different section of the Swiss trade structure is shown by the data concerning the change in exports and imports structures according to the WIFO classification based on the skill level of the labor force (Fig. 5.15). The biggest share in the total foreign trade belongs to the production based on semiskilled, white-collar workers (in 2001, share in exports on the level of 56 %, and in imports 44 %). The second position is occupied by the production based on semiskilled, blue-collar workers, but these category figures for exports and imports vary considerably (shares of 24 and 33 %, respectively). Further positions belong to the production based on the most highly skilled and most low-skilled workers. Such a structure seems to be typical for most of the highly developed countries, in which employees of various levels and semi-highly skilled are essential.

However, the most important change of all seems to be that in the structure of the share of individual employee categories in exports. A gradual loss of significance of the goods manufactured by low-skilled workers in favor of those with higher education and high qualifications can be observed. In the domain of imports, this trend is also noticeable, although it is much weaker (Fig. 5.15). Thus, the foregoing data confirm the previously reported changes in the competitive position of Switzerland in the international division of labor.

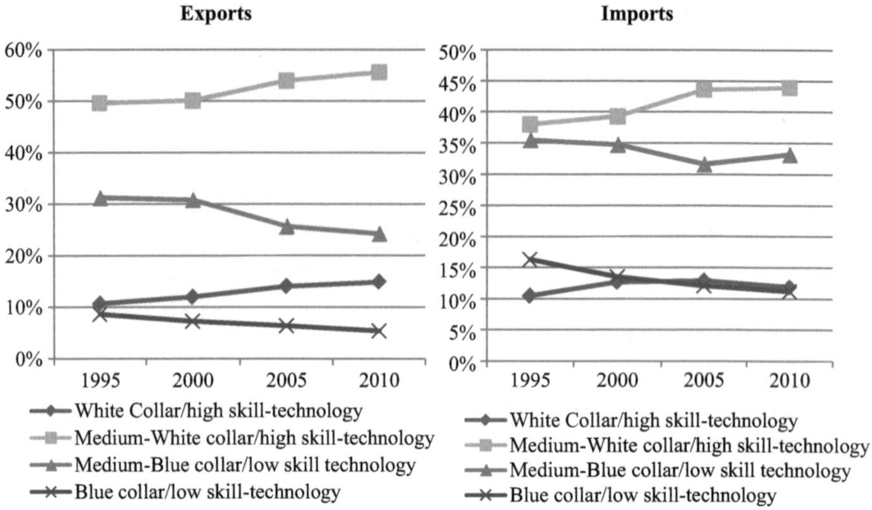

Fig. 5.15 Changes in goods structure of Switzerland's exports and imports according to the WIFO classification based on skill level of labor force in the years 1995–2010. (Source: Own calculations, WIFO data)

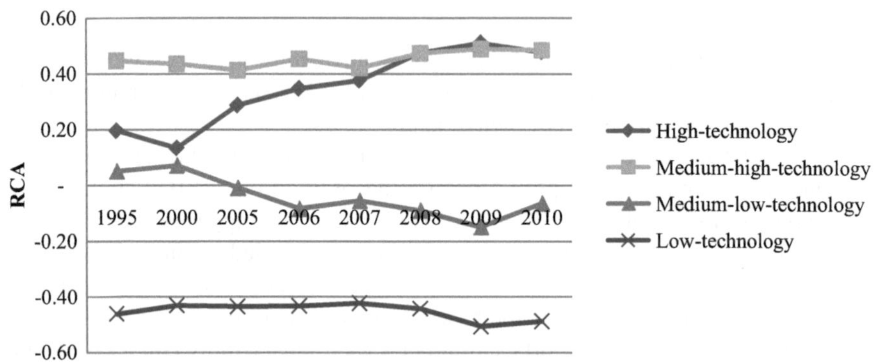

Fig. 5.16 Comparative advantages (RCA) of Switzerland according to the OECD industry classification by technology. (Source: SGH calculations, OECD data)

The analysis of the RCA of Switzerland shows, in turn, the trade specialization of the country (Fig. 5.16). Switzerland has the highest RCA in the production of medium-high- and high-technology goods, with no relative advantage in the production of medium-low- and (in particular) low-technology goods. While assessing the changes in the years 1995–2010 from the dynamic perspective, it is possible to notice positive shifts in the trade specialization of Switzerland, which strengthen the country's relative advantages in the trade in goods with the highest level of technical advancement. In 2000, a gradual increase in the RCA in the international trade in high-technology goods began. In 2009, the goods of this class of technology, and

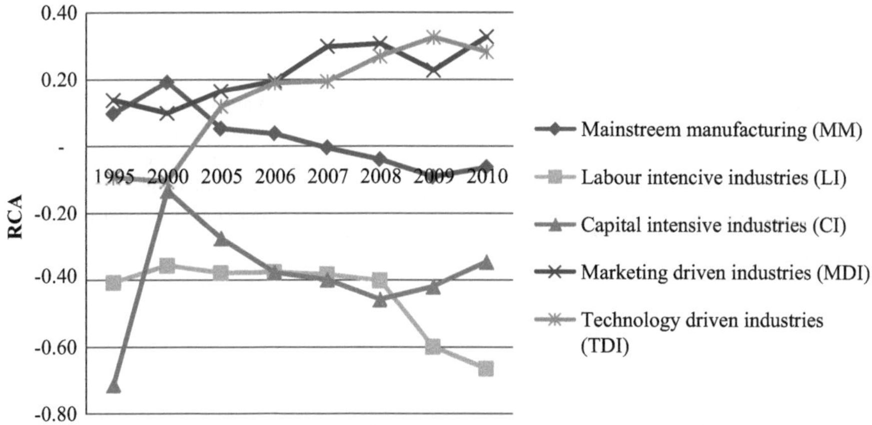

Fig. 5.17 Switzerland's comparative advantages (RCA) according to the WIFO classification (the extent of use of tangible and intangible expenses in the production process). (Source: SGH calculations, WIFO data)

especially pharmaceutical industry goods (RCA 0.78 in 2010) and telecommunications equipment (RCA 0.34 in 2010), recorded the highest rate of advantage in the Swiss trade. These results are consistent with the previous analysis of the dynamics of the main export sectors of industrial production in Switzerland.

Switzerland's advantages in the production of medium-low-technology goods have, in turn, considerably weakened—to such an extent that the RCA index for this class of production, since 2005, has been below zero, which means there are no RCAs. These shifts indicate Switzerland's increasing trade specialization in the knowledge-based production.

The above hypothesis can be verified by an analysis of indexes of comparative advantages by the WIFO classification based on the extent of use of tangible and intangible expenses in the production process (Fig. 5.17). In 1995, Switzerland still recorded the biggest advantages in the FT in the domain of mainstream and marketing-driven industries. Since 2000, however, an increase in comparative advantages related to research-driven industries has been observed. In the 2000–2010 period, the RCA index of these industries reached higher values than those of industry leaders from 1995 (Fig. 5.18). At the same time, the largest decrease in relative advantages was recorded by labor-intensive and mainstream industries.

The RCA index analysis confirms the strengthening of Switzerland' position in the trade in high-technology goods manufactured by research and—at least partially—marketing-driven industries. At this point, one should refer to another index of comparative advantages, which takes into account the WIFO classification based on the skill level of the labor force (Fig. 5.18). A closer look from this perspective also confirms the change in trade specialization of Switzerland. A strong increase in relative advantages related to industries using highly skilled, white-collar workers is particularly noticeable. Advantages in trade in goods produced by semiskilled workers remain at the same level.

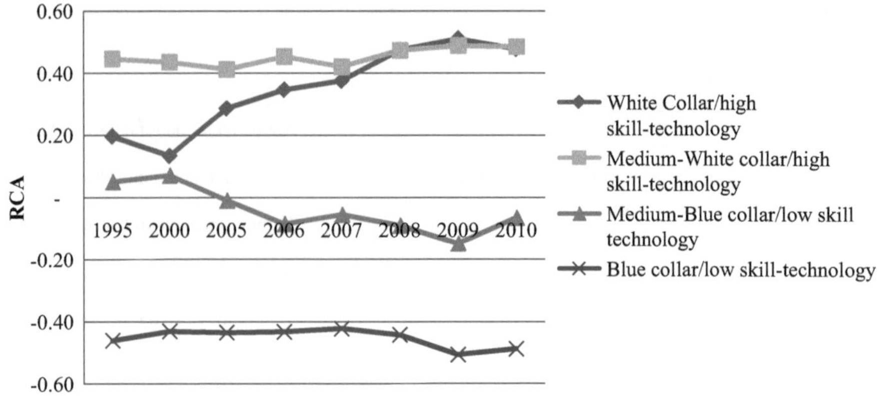

Fig. 5.18 Switzerland's comparative advantages (RCA) according to the WIFO classification (industries using various skill levels of labor force). (Source: SGH calculations, WIFO data)

In the case of industries hiring semiskilled, blue-collar workers, a significant change was also recorded. However, this time, the change was negative—the advantages had disappeared. In the years 1995–2000, the RCA index was still above zero. Nevertheless, in the subsequent years, it gradually decreased, until it reached the value of − 0.49 in 2010. It means that the goods produced by this type of workers in Switzerland had lost their international competitiveness for the goods manufactured in other countries.

The foregoing analysis of indexes showing the occurrence of the RCA indicated significant, positive shifts in the competitive position of Switzerland. Industries using more advanced technology, basing their production on research and highly skilled workers, to a greater extent than 15 years ago, gain importance both in trade and in creating added value.

In order to analyze changes in the IIT in goods which come under industries intensively using modern factors of production in comparison with the rest of manufacturing, one has to collate Switzerland's intra-industry trade (IIT) indexes with the total value of foreign trade of a given industry according to the classification of industries with regard to technology (Fig. 5.19) and the WIFO classification (Fig. 5.20). The analysis of the foregoing statistic data leads to surprising conclusions. The index of intra-industry exchange of medium-high- and high-technology industrial goods had, in the case of Switzerland, considerably decreased—from 0.80 to 0.71 and from 0.95 to 0.71, respectively, for both industries (2000–2010). A large decrease in the value of high-technology industries (which is visible in the diagram) is especially astonishing, as the IIT index of EU27 had simultaneously increased from 0.92 to 0.98.

The high level of IIT has continuously been recorded in medium-low technology industries, and the moderately high level in low-technology industries, in which the greatest increase in the intra-industry exchange in the past 15 years has been observed. The decrease in the IIT index for Switzerland in the domain of high

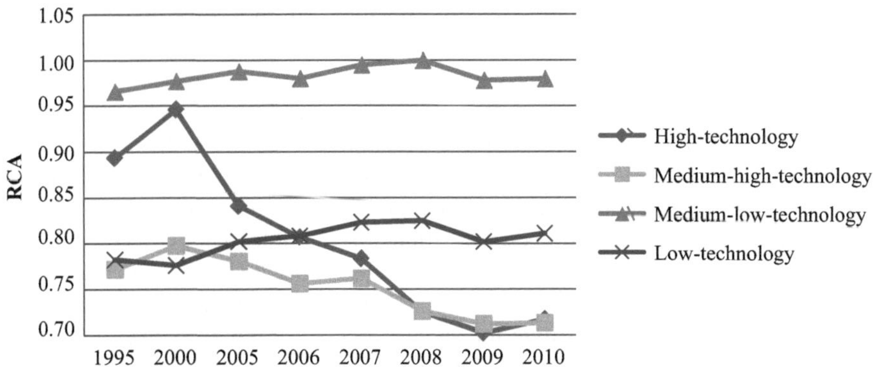

Fig. 5.19 Extent of Switzerland's intra-industry trade (IIT) according to the OECD technology classification. (Source: SGH calculations, OECD data)

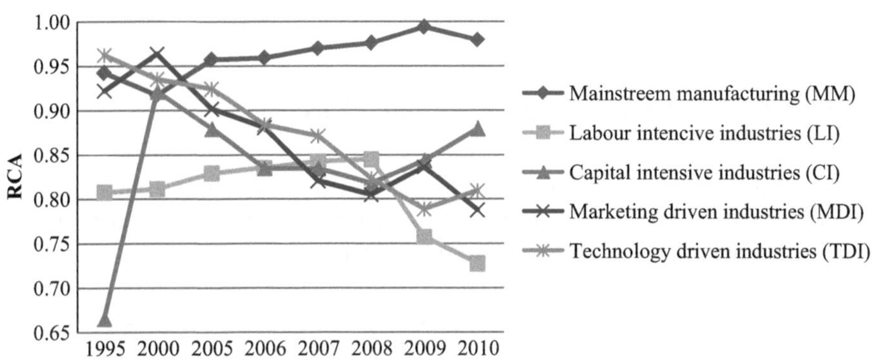

Fig. 5.20 Extent of Switzerland's intra-industry trade (IIT) according to the WIFO classification (the extent of use of tangible and intangible expenses in the production process in the years 1995–2010). (Source: SGH calculations, WIFO data)

technology may be a consequence of the high export specialization in certain sectors (pharmaceuticals, chemicals, watches, and precision instruments) and of a high demand for these goods observed at the same time as a sustainable demand for imported high-technology goods. This hypothesis is justified by the data on the growing share of these industries in Switzerland's total exports. Another explanation—complementary to the above-mentioned hypothesis—is a high degree of completion of added–value-creating processes, which does not involve a need to import semi-finished products of the high-technology industry sector. Such a situation may occur in the case of production of pharmaceuticals or precision instruments and watches.

The decrease in the intra-industry turnover is confirmed by the IIT index data for the WIFO classification according to the extent of use of tangible and intangible expenses in the production process. In the domain of research-driven industries, the decline in the value of IIT index from the level of 0.94 (2000) to the level of 0.81 (2010) indicates a decrease in the intra-industry exchange, but to a lesser extent than the OECD industry classification. A decrease in the intra-industry exchange is recorded also in the domain of labor-intensive and marketing-driven industries. A slight increase, on the other hand, is noted in capital-intensive and mainstream industries.

Both analyzed indexes of Switzerland's IIT imply a decline in the intra-industry exchange. The decrease in the IIT index for Switzerland according to the OECD classification is particularly different from the trends observed in the EU27. In the case of data according to WIFO classification, it is possible to notice a convergence with the EU27 trends, although it is not especially strong (EU27—decrease from 0.96 to 0.94, Switzerland—decrease from 0.94 to 0.98). From the theoretical point of view, this trend is not positive for Switzerland, since the exchange of IIT type generally stimulates innovativeness and facilitates the use of broadly understood economies of scale in production and sales in the field of production of the most advanced technologies. The explanation of these differences for Switzerland and the disclosure of the reasons is an interesting research task that can be undertaken during the next phase of the present study.

5.2.2 *Characteristics of Switzerland's NIS*

Switzerland is a unique country with a federal system. In fact, it is a union of states with far-reaching autonomy, which is also reflected in the system of science and the innovation policy. The legal basis governing the issues concerning the innovativeness is contained in federal legal acts and cantonal constitutions, in laws supporting economic development, or in regulations. A federative approach can be easily noticed here, e.g., in the anchoring of innovative thinking in the federal constitution as well as in individual constitutions of the cantons.

The federal government plays an important role in supporting innovativeness, although innovativeness should be understood in this context in a specific way. According to the current definition, innovativeness is the result of research, but it does not include technical or commercial exploitation of research results. That is why supporting innovativeness on the federal level includes, on the one hand, the Innovation Promotion Agency (KTI) and the support of research and education achievements on the other, without means of support such as the development of innovativeness of companies.

Two bodies are entitled to implement guidelines for innovativeness: the Federal Department of Economic Affairs (EVD) and Federal Department of Home Affairs (EDI). Two subdivisions of the Department of Economic Affairs support innovativeness. One of them is in charge of development and competition policy as well as of economic issues. It is also responsible for promoting location attractiveness—its field of action is the policy for SMEs, including, among others, export promotion, regional policy, and spatial planning policy. The second subdivision is the Federal Office for Professional Education and Technology (BTT). Its role consists in financing innovations with the funds provided by the Innovation Promotion Agency.

The activity of the Innovation Promotion Agency includes innovation management, transfer of knowledge and technology, and support for research projects as well as for business start-ups and entrepreneurship (Marxt 2009).

In Switzerland, the innovative activity on the level of cantons derives from the federal regional policy and is also a consequence of legal regulations of individual cantons with regard to economic development. The regional policy is based on three different, complementary directions: (1) support for the regional economy development, (2) coordination of regional policy performed by federal agencies; direction (3) know-how for the regional policy and its partners (SWTR 2007).

In Switzerland, an important role is played by both national and cantonal intermediate bodies. They are responsible for supporting the innovative activity of companies, in particular by facilitating the flow of information, knowledge, and technology between research institutions and industries. The main national agency funding the innovativeness and the transfer of knowledge and technology is the Innovation Promotion Agency (KTI), which receives a budget of approximately CHF 100 million from the federal government. Projects sponsored by the Innovation Promotion Agency are supporting market-oriented R&D projects, running companies in cooperation with academic institutions, and supporting the creation and development of knowledge-based companies. The Innovation Promotion Agency also offers the formation of platforms and networks of transfer of knowledge and technology.

Regional transfer operators are often supported by the Innovation Promotion Agency and serve academic institutions and companies in a given region. They are, *inter alia*, WTT CHost, WTT Nordwestschweiz, die Alliance, and W6 Mittelland. WTT CHost is a consortium of various institutions devoted to knowledge and technology transfer in the region of southern Switzerland. WTT CHost is committed to supporting the transfer of knowledge and technology (WTT) in machinery, metallurgical, and electrical industries (MEM) and textile industry, as these businesses are important to southern Switzerland's exports. From the scientific point of view, WTT CHost concentrates on supporting micro- and nanotechnology, ubiquitous computing, innovation technology, and transfer management.

A total of 23 science and technology parks ("Enabler") are located in Swiss cantons. They support innovativeness and entrepreneurship by helping young companies and start-ups and by offering them various specific services, which consist mainly in providing premises, advice, and infrastructure. What is more, the parks offer an opportunity to share knowledge among young entrepreneurs and partners. The parks' partners are vocational schools, federal universities, polytechnic schools, and companies. Transfer operators can be transregional or regional, but also thematic and associated with higher education (Marxt 2009).

In Switzerland, there are ten universities, two federal technical institutes, and numerous vocational schools, a large part of which enjoys very good reputation.

One of the most important instruments in the research carried out by universities is to develop research priorities (NFS/NCCR), which are funded by the Swiss Science Foundation (SNF). Currently, while cooperating or operating within their own university, academic institutions focus on 20 particularly innovative subjects. They include, among others, genetics, climate change, molecular oncology, and trade regulations (Marxt 2009).

Not only academic institutions but also the world's leading research institutes are located in Switzerland. In Geneva, there is the European Organization for Nuclear Research (CERN), which, with 3,000 employees and a number of facilities, examines the basis of elementary particles. The Swiss Federal Institute of Aquatic Science and Technology (EAWAG) is also a leading research institute, which conducts extensive research on water and aquatic resources (FUS 2011).

5.2.2.1 Innovative Capacity and Position

The most general index describing the innovative capacity of an economy is the share of R&D internal expenditure in GDP. In Switzerland, in the years 1995–2009, the value of this index ranged from 2.65 to 2.99% and was higher than those of the US, Germany, and the EU27. These economies recorded the following values: 2.79% of GDP in 2008 (US), 2.78% of GDP in 2009 (Germany), and 1.9% of GDP in 2009 (EU) (Table 5.10).

The data from Table 5.10 show that, in Switzerland, the share of public expenditure in the expenditure on R&D amounts to 22.84% of total R&D expenditure. It is a relatively small amount, as in the US or in Germany the share reached the value of 27.05 and 28.4% of total R&D expenditure, respectively, in 2008. As for the share of business expenditure on R&D in Switzerland, it remains at a stable level. In 2008, it amounted to 68.10%, which is a comparable value to those recorded in the US or in Germany. Such a structure of expenditure on R&D seems to characterize highly developed industrial countries with a strong innovative position. The values of other indexes describing the innovative capacity, such as the percentage of households with Internet access as a 7 percentage of all households or the importance of venture capital investments, indicate a strong position of Switzerland in the international comparison. As for the second case, the share of this type of investments is three times higher than in Germany and higher than in the US

In the evaluation of the innovative position, it should be emphasized that Switzerland's economy is characterized by a high index of number of patent applications per one million inhabitants, which amounted to 224.1 in 2007, while in the US or in Germany it was 800.17 and 581.67, respectively (Table 5.11). The latest data show, however, a relative weakening of its position in the international competition, as the value of the index did not increase in the years 2000–2007, while it soared in the US. Yet, in 1995, the number of patent applications in the three above-mentioned countries was similar. These changes may be partially—especially in the case of the US—related to the computer revolution as well as to software and hardware patents.

The number of patent applications per US$ 1 billion of funds allocated to R&D in Switzerland was considerably lower than those noted by the above-mentioned highly developed economies. It indicates a relatively smaller efficiency of use of expenditure on R&D in Switzerland than in the US or Germany.

Table 5.10 Chosen indexes of innovative capacity of Switzerland's economy. (Source: Own study on the basis of OECD, EUROSTAT, and WIPO data)

Index	Economy	1995	2000	2005	2006	2007	2008	2009
Share of total internal expenditure on Research and Development (R&D) (GERD) in GDP	US	2.50	2.71	2.57	2.61	2.67	2.79	…
	Germany	2.19	2.45	2.49	2.53	2.53	2.68	2.78
	Switzerland	2.65	2.53	2.90	…	…	2.99	…
	EU27	1.66	1.74	1.74	1.77	1.77	1.84	1.90
Share of government internal expenditure on R&D (GOVERD) in GDP	US	0.35	0.28	0.31	0.30	0.29	0.30	…
	Germany	0.34	0.33	0.35	0.35	0.35	0.38	0.41
	Switzerland	0.066	0.033	…	0.024	…	0.022	…
	EU27	0.28	0.25	0.25	0.24	0.23	0.24	0.26
Percentage of internal expenditure on R&D financed by the government	US	35.41	25.81	30.24	29.27	28.32	27.05	…
	Germany	37.88	31.40	28.38	27.53	27.51	28.40	…
	Switzerland	26.92	23.18	22.71	…	…	22.84	…
	EU27	39.55	35.40	35.10	34.10	33.95	34.56	…
Percentage of internal expenditure on R&D financed by other national sources	US	4.36	4.76	5.43	5.39	5.52	5.68	…
	Germany	0.26	0.41	0.29	0.36	0.35	0.31	…
	Switzerland	2.50	3.42	2.33	…	…	3.02	…
	EU27	1.82	2.13	2.36	2.60	2.49	2.64	…
Percentage of internal expenditure on R&D financed by foreign countries	US	…	…	…	…	…	…	…
	Germany	1.83	2.15	3.75	3.82	4.01	4.01	…
	Switzerland	3.10	4.31	5.23	…	…	5.95	…
	EU27	6.63	7.15	8.90	8.59	8.95	8.73	…
Percentage of internal expenditure on R&D financed by industry	US	60.23	69.42	64.33	65.34	66.17	67.27	…
	Germany	60.03	66.04	67.57	68.29	68.12	67.27	…
	Switzerland	67.46	69.08	69.73	…	…	68.19	…
	EU27	52.00	55.29	53.79	54.74	54.60	54.20	…

Table 5.10 (continued)

Index	Economy	1995	2000	2005	2006	2007	2008	2009
Share of business internal expenditure on R&D (BERD) in GDP	US	1.76	2.02	1.80	1.86	1.92	2.02	…
	Germany	1.45	1.73	1.72	1.77	1.77	1.86	1.88
	Switzerland	1.87	1.87	2.14	…	…	2.20	…
	EU27	1.02	1.11	1.08	1.11	1.12	1.15	1.16
Households with access to the Internet as a percentage of all households	US	…	41.50	…	…	61.70	…	68.70
	Germany	…	16.40	61.64	67.11	70.67	74.88	79.07
	Switzerland	…	…	…	70.5	73.9	77.0	…
	EU27	…	…	48.38	49.18	54.15	60.39	65.17
Investments in venture capital as % of GDP	US	0.04	0.27	0.04	0.04	0.05	0.05	0.05
	Germany	0.01	0.08	0.01	0.01	0.02	0.02	0.02
	Switzerland	0	0.02	0.02	0.02	0.06	0.06	0.06
	EU27	…	…	…	…	…	…	…

Remarks: (…)—no data
For Switzerland, 1995 column = data from 1996, 2005 column = data from 2004

Table 5.11 Selected indexes of innovative position of Switzerland's economy. (Source: Own study on the basis of OECD, EUROSTAT, and WIPO data)

Index	Economy	1995	2000	2005	2006	2007
Number of patent applications per million inhabitants	US	465.54	583.98	702.50	742.36	800.17
	Germany	466.71	629.32	586.48	582.84	581.67
	Switzerland	410.45	289.94	220.92	232.5	224.1
	EU27	…	…	…	…	…
Number of patent applications per US$ 1 billion of funds allocated to R&D	US	…	0.58	0.67	0.68	0.72
	Germany	…	0.91	0.78	0.77	0.74
	Switzerland	…	0.29	0.22	…	…
	EU27	…	…	…	…	…
Share of export of high-technology goods in total exports	US	25.87	29.95	26.15	26.13	…
	Germany	11.64	16.08	14.79	14.06	…
	Switzerland	16.20	19.91	21.77	21.28	…
	EU27	…	21.39	18.74	16.65	…

Remarks: (…)—no data

Advantages in the achieved level of innovation are indicated also by the index value of the share of high-technology goods' exports in total exports, which places Switzerland (21.28) between the US (26.13) and Germany (14.06), as well as the whole EU (16.65).[16]

5.2.3 *Human Capital*

Human resources constitute the crucial factor conditioning the innovativeness of a given economy, because the innovative activity—creating new ideas and solutions—is always a product of human thought (Weber 2001).

In terms of population, Switzerland is not a large country; it belongs rather to the group of small states. It currently has nearly 8 million inhabitants, which means an increase of nearly 1 million people in the population number in comparison with 1995. However, it is not a result of high birth rates, but, above all, of positive net migration. As an exceptionally prosperous, liberal, and welfare state, Switzerland is one of the most desired migration destinations in Europe. In order to keep the process under control, Switzerland has extremely restrictive and scrupulously observed immigration laws.

The positive balance of migration is, in turn, the reason why the population structure, which presents the proportion of people of working age in relation to the

[16] The foregoing index of the share of high-technology goods' exports in total exports cannot be compared with indexes presented in the previous subchapter, as it relates to the classification of goods by technology level. The methodology applied in the previous chapter referred to the technology level of entire industries.

Table 5.12 Data on skill level of labor force in Switzerland (in % in proportion to total labor force). (Source: Own study based on the World Bank Education Statistics data)

	1995	2000	2001	2002	2003	2004	2005	2006	2007
Share of labor force with primary education (%)	16.5	16.7	15.5	15.8	16.2	15.8	15.6	15.6	15.4
Share of labor force with secondary education (%)	60.2	59.7	59.8	59.7	58	57.2	56.6	55.9	54.8
Share of labor force with higher education (%)	20.7	23.2	24.2	24.2	25.6	27	27.7	28.5	29.7

total population, remains at a positive level. Immigrants in Europe are generally members of younger generations and have higher fertility rates. That is the reason why, thanks to immigrants and despite one of the highest life expectancies in the world, Switzerland has managed to maintain the percentage of working-age population at a favorable level.

Due to the country's attractiveness to immigrants, combined with the scrupulously observed immigration policy, Switzerland's demographic projections—for a European country—are really favorable. The pace of aging of its society will not significantly differ from the global average. It will allow the maintenance of control over the socioeconomic system stability process, as the proportion of working-age and retirement-age population is its essential component. The projections for Switzerland—with the assumption of continuation of current trends in the birth rate—are, in this case, much more favorable than those for Poland, as the estimated proportion for our country amounts to 1.90, while for Switzerland it is 4.2 in 2050.

An important factor in assessing the quality of human capital is the education level. (Weber 2008). In the examined period, Switzerland had been systematically consolidating its position, particularly in the case of the share of labor force with higher education, from the level of 20.7 % of the total labor force in 1995 to 29.7 % in 2007 (Table 5.12). A slight decrease in the share of people with primary education (16.5 % in 1995 to 15.4 % in 2007) and a strong one in the share of people with secondary education (from 60.2 to 54.8 %) are noticeable. The enrollment ratio for people over 25 years of age amounted to 10.3 in 2010, and it is similar to the average recorded in OECD countries.

Expenditures on higher education had risen in relation to Swiss GDP from 53.84 to 64.11 % in the years 1999–2003 to decrease to the level of 43.92 % in 2008. Although it is a significant decline in expenditures, the comparison of their value with those recorded in other OECD countries in 2008 shows that Switzerland remains one of the three countries that spend the most on higher education, following Denmark (52 %) and Norway (46 %) (Table 5.13).

Table 5.13 Skill level of labor force in Switzerland, 1999–2008. (Source: Own study based on the World Bank Education Statistics data)

	1999	2000	2001	2002	2003	2004	2005	2006	2007	2008
Expenditures on higher education as % of GDP per capita	53.84	53.13	55.43	59.65	64.11	62.31	55.21	53.05	46.73	43.92
Expenditures on secondary education as % of GDP per capita	27.32	26.85	27.64	29.09	28.89	27.65	27.83	26.26	25.21	31.11
Expenditures on primary education as % of GDP per capita	22.70	21.82	22.96	24.21	24.63	24.65	24.05	23.12	22.51	20.56

The observed positive trends in the level of education of human capital influence its use in a beneficial way, which is reflected in the unemployment level. The unemployment level in Switzerland is one of the lowest among OECD countries. The unemployed with higher education account for 17.9% (2007) of all unemployed, which constitutes a value below the average noted in OECD countries. In Switzerland, the extremely painful phenomenon of unemployment among the youth is also very rare, unlike in other highly developed countries. Its level is the lowest one (8.2% in 2007) among OECD countries. Professional activity of the youth is on a very high level in Switzerland.

Important information used to analyze the quality of human capital is provided by indexes of level and changes in labor productivity. The analysis of GDP per capita, based on the available statistic data, allows you to distinguish two principal factors which explain the development of labor productivity (per hour worked) and the effect of the use of labor. On the basis of the effect, it is possible to evaluate the intensity with which work is used in the production process. The average working time per employee, the effect of unemployment, the employment rate of people in the labor market, and the share of working-age population in the total population are presented in Table 5.14.

In comparison to the 1990s, Switzerland's labor market indexes, which comprise the effect of use of the existing labor force, have significantly improved. In Fig. 5.21, the correlation of GDP growth and labor productivity is noticeable. However, until 2004, the cumulative GDP growth was lower than the cumulative increase in labor productivity. Since 2003–2004, rapid growth of GDP and productivity, which could not keep up with it, could have been observed.

The summary of the assessment of Switzerland's NIS is the strengths, weaknesses, opportunities, and threats (SWOT) analysis, which, in the case of the Swiss innovation system, will allow the assessment of its strengths and weaknesses, as well as its opportunities and potential threats (Table 5.15).

Table 5.14 Labor productivity and effect of use of labor force in Switzerland (1995–2020). (Source: Own calculations: Bundesamt für Statistik, Neuchatel)

Year	GDP/inhabitant, prices, 2000, in proportion to the previous year (%)	Labor productivity by hours worked (%)	Effect of use of labor force (%)	Components of the "effect of use of labor force"			
				Average working time per employed person (%)	Unemployment effect (%)	Employment rate (people of working age) (%)	Share of people of working age in the total population (%)
	(1)=(2)+(3)	(2)	(3)=(4)+(5)+(6)+(7)	(4)	(5)	(6)	(7)
1995	−0.3	1.7	−1.9	−1.2	0.5	−1.0	−0.2
2000	3.0	2.9	0.1	−0.3	0.3	0.1	0.0
2005	2.0	2.2	−0.2	−0.3	0.0	−0.1	0.2
2010	1.7	1.0	0.7	1.0	−0.2	0.1	−0.2

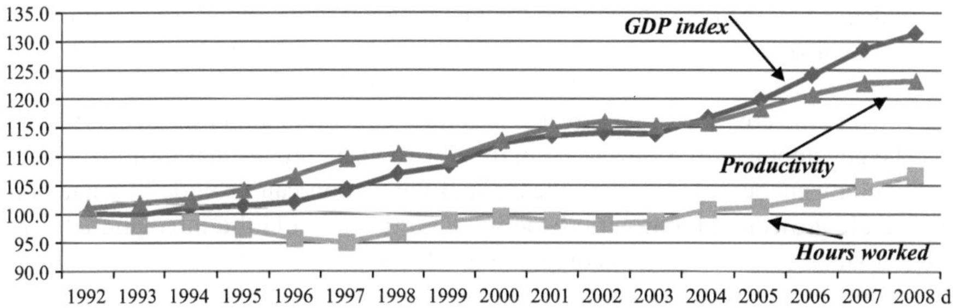

Fig. 5.21 Labor productivity by man-hours worked (1991 index = 100)

Table 5.15 SWOT analysis of Switzerland's innovation system. (Source: Study based on Marxt and Brunner (2009))

Strengths	Weaknesses
Leadership in research	Lack of a coherent innovation policy
Excellent system of education and training	Small average implementation of new products
Leadership in terms of overall innovation performance	Data quality
High number of people employed in knowledge-based services	Social climate of innovation
Very good infrastructure on an international scale	Framework conditions and financing business start-ups and spinoffs
	Number of projects supporting innovation
Opportunities	Threats
Balanced image of Switzerland (center for research, business, and services)	Strong changes in the macroeconomic environment
Growing international demand for knowledge-based services	Strengthening of regulations concerning innovations
Growing package of financial measures allocated to Switzerland's innovativeness	Reaching or exceeding the limits of the infrastructure
Growing importance of non-technical innovation	Deterioration in local activity
	Increase in administrative burdens for innovative organizations (particularly companies)

5.2.4 Evaluation of the Relationship Between the Level of Innovation and Human Capital Accumulation, and the Advantages in the Foreign Trade

The analysis of the case study of Switzerland makes it possible to positively verify the hypothesis concerning the relationship between human capital resources and innovativeness, and the development of long-term advantages in the FT. The data analyzed show that Switzerland is a country with high international competitiveness, as evidenced by the data concerning the international trade, the share of high-technology industries in exports, and the analysis of relative advantages in the FT. It is, to a large extent, an illustration of the increasing quality of human capital, the share of which in the creation of FT streams is getting larger and larger in Switzerland.

The development of Switzerland in the last 15 years proves that forming a competitive advantage based on knowledge and innovation can provide a sustainable and stable growth and long-term benefits of participation in the international division of labor. It is reflected in comparative advantages in the foreign trade in high-technology goods which use high-quality human capital.

The Swiss experience shows that the pro-innovation policy should place sustainable emphasis on infrastructure, research specialization, and creation of institutions leaders. It is important to focus the activity of research entities on market needs and collaboration with industry, and to promote the commercialization of the results of R&D.

Switzerland is an example of the way in which high expenditure—one of the highest in the world—on higher education translates into the creation of innovative advantages. The Swiss labor market records a high demand for well-educated labor force, especially in knowledge-intensive services. Switzerland strongly supports the process of lifelong learning, which facilitates a flexible response to changing economic conditions. Science also enjoys a high level of funding. The share of funds allocated to fundamental research is especially large.

Even though Switzerland is one of the leading countries in the world in terms of innovativeness and research, a number of studies indicate the need for further optimization of the NIS, particularly in the domain of implementation of new scientific achievements (BCG 2008). For example, the report "Education, research and innovation 2008–2011" contains objectives and support measures developed by the Federal Council, aimed at improving long-term security and quality of research and at increasing competitiveness and growth. They specify general measures of innovation policy. Support for the transfer of knowledge between academic institutions and the economy is planned in order to improve the future R&D projects which will be feasible in practice. It is supposed to be financed from a significantly increased budget of the Swiss Science Foundation (SNF) intended for fundamental research and from the funds provided by the Innovation Promotion Agency (KTI) (Marxt 2009).

5.3 Case Study: Ireland

Arkadiusz Michał Kowalski

This chapter discusses human capital and technical and organizational innovativeness of Ireland in the context of the international competitiveness of economy. Its aim is to determine whether there is a relationship between the accumulated human capital resources and the innovativeness of Ireland, and the development of Ireland's long-term competitive advantages in the international trade. The time scope of the study covers the years 1995–2010. It is a very interesting period from the point of view of the Irish economy, as it was the period of dynamic economic growth of this country. In the late 1980s/early 1990s, Ireland was one of the poorest countries of the EU, suffering from enormous structural problems connected with low exports, high unemployment, extreme budget deficit, and foreign debt. In the period studied, the macroeconomic indexes significantly improved. Ireland's experiences allow the identification of benefits connected with wide participation of a country in the global economy and indication of factors that are prerequisites of these benefits.

5.3.1 Evaluation of Ireland's Competitiveness in Foreign Trade

In 1995–2010, the participation of Ireland in global trade exchange went up; however, it was first of all the second half of the 1990s and the first two years of the new millennium that witnessed the growth. Ireland's exports increased from US$ 43.79 billion in 1995 to US$ 118.54 billion in 2010; however, the highest value, US$ 127.11 billion, was recorded in 2008. An increasing growth of Ireland's share in global export is observable; however, it was related mostly to the years 1995–2002, from 0.92 to 1.39%. Later on, the index decreased, reaching the value of 0.96% in 2010. A similar growth rate was shown by Irish imports, which went up from US$ 32.32 billion in 1995 to US$ 60.34 billion in 2010, reaching maximum values in 2007 (US$ 87.05 billion) and 2008 (US$ 84.95 billion). The share of Ireland in global import grew dynamically between 1995 (0.68%) and 2011 (0.84%); however, later it decreased, to 0.49% in 2010. The geographical structure of foreign trade of Ireland in the studied period is presented in Table 5.16 (exports) and Table 5.17 (imports).

The most important market for Irish commodities at present are the US. In 2010, Irish exports to the American market reached the amount of US$ 27,533.55 million, which constituted 23.23% of the total exports of Ireland. This signifies a dynamic growth of significance of the US as a trade partner, since in 1995 the share was only 8.32%. At that time, the most important country in terms of Ireland's exports was the United Kingdom, with a share totaling 25.44% However, in the last 15 years we have witnessed a drop in significance of this country, and in 2010 the index went down to 15.4%. The third export partner of Ireland is Belgium (14.34% in 2010), followed by Germany (8.07%) and France (5.03%).

In the entire period of 1995–2010, the biggest import partner of Ireland was the United Kingdom. The share of this country in Irish imports did not change significantly and totaled 35.58% in 1995, 31.3% in 2000, 31.21% in 2005, and 32.06% in 2010. The US ranked second for the entire studied period, and the share of this country in imports dropped from 17.69% in 1995 to 14.09% in 2010. The geographical structure of the foreign trade of Ireland reflects the most important directions of foreign policy of this country, which include the United States, a traditional place of economic emigration of the Irish, and the EU, whereby the priority

Table 5.16 Geographical structure of Ireland's foreign trade in chosen years of the period 1995–2010—main export partners. (Study based on the United Nations Commodity Trade Statistics Database)

	1995		2000		2005		2010	
	US$ million	%	US$ million	%	US$ million	%	US$ million	%
US	3,641.80	8.32	13,098.95	17.18	20,587.05	18.71	27,533.55	23.23
UK	11,140.12	25.44	16,653.07	21.84	18,856.49	17.14	18,257.65	15.40
Belgium	1,908.54[a]	4.36	3,660.26	4.80	16,639.90	15.13	16,998.60	14.34
Germany	6,328.61	14.45	8,629.70	11.32	8,234.30	7.49	9,566.99	8.07
France	4,143.41	9.46	5,823.03	7.64	7,040.36	6.40	5,957.39	5.03
Switzerland	795.37	1.82	1,875.70	2.46	3,988.99	3.63	4,717.78	3.98
Spain	1,073.62	2.45	1,881.91	2.47	3,669.57	3.34	4,457.60	3.76
The Netherlands	3,063.62	7.00	4,232.98	5.55	5,292.89	4.81	4,101.71	3.46
Italy	1,658.28	3.79	3,056.69	4.01	4,511.00	4.10	3,597.82	3.04
Japan	1,307.42	2.99	2,876.88	3.77	2,905.10	2.64	2,327.83	1.96
Total exports	43,789.34	100	76,261.88	100	110,003.09	100.00	118,538.57	100.00

[a] Together with Luxemburg

Table 5.17 Geographical structure of Ireland's foreign trade in chosen years of the period 1995–2010—main import partners. (Study based on the United Nations Commodity Trade Statistics Database)

	1995		2000		2005		2010	
	US$ million	%	US$ million	%	US$ million	%	US$ million	%
UK	11,498	35.58	15,854	31.30	21,935	31.21	19,345	32.06
US	5,717	17.69	8,424	16.63	9,941	14.14	8,503	14.09
Germany	2,256	6.98	2,945	5.81	5,359	7.63	4,634	7.68
China	347	1.07	825	1.63	4,618	6.57	3,336	5.53
The Netherlands	871	2.70	1,613	3.18	2,803	3.99	2,933	4.86
France	1,200	3.71	2,385	4.71	2,388	3.40	2,430	4.03
Norway	447	1.38	910	1.80	1,845	2.62	1,712	2.84
Belgium[a]	0	0	785	1.55	1,314	1.87	1,432	2.37
Switzerland	216	0.67	352	0.70	679	0.97	1,123	1.86
Japan	1,706	5.28	2,434	4.80	2,612	3.72	1,059	1.76
Italy	638	1.97	1,125	2.22	1,524	2.17	1,017	1.69
Singapore	1,293	4.00	1,898	3.75	1,078	1.53	673	1.12
Korea	269	0.83	1,178	2.33	1,140	1.62	323	0.54
Malaysia	530	1.64	604	1.19	699	0.99	272	0.45
Total imports	32,321	100	50,649	100	70,284	100	60,337	100

[a] Together with Luxemburg

is given to the bilateral relations with the United Kingdom (Kowalski 2008, p. 78). The next biggest partners of Ireland in imports are Germany (7.68 % in 2010) and China (5.53 %). A significant growth of this index is observable especially in the case of the latter (from 1.07 % in 1995), which reflects structural changes taking place in the global economy, and the promotion of Chinese economy to the position of the global export leader.

In 2010, chemicals and related products were a distinct majority in Ireland exports (58.88 %). The growing share of these commodities in Irish exports is noticeable, from 19.04 % in 1995. It is strongly connected with the inflow of direct foreign investments in the industry of chemicals and related products, first of all of American origin. In 1995, the prevailing export commodities of Ireland were machinery and transport equipment (34.44 %); however, their share shrank gradually, to 12.39 % in 2010. The production of electronic equipment, e.g., office machines and computers, had the biggest share in this commodity group. In the second half of the 1990s, Ireland became an important production center within this scope, supplying innovative products to many European countries. A key role was also played here by foreign direct investment (FDI), especially American and Japanese ones, for which Ireland constituted an attractive location, *inter alia* due to the unlimited access to a uniform internal market of the EU. These investments were very conducive to exports, as the modest Irish market was enabled to create a sufficient demand for the goods offered. Since 2005, however, a drop in Ireland's export in this group of products has been observed, first of all due to increased competition on the part of southern–eastern Asian countries, and relocation of production of transnational corporations operating in the electronic and computer industries to these countries (Table 5.18).

The most important group of goods imported by Ireland is constituted by machinery and transport equipment, whereby a downward trend is noticeable within this scope, from 42.04 % in 1995 to 26.55 % in 2010. It is worth noticing that it is a reverse trend when compared to the share of these goods in exports. It is a sign of gradual departure of the Irish economy from production of machinery and transport equipment, first of all in favor of chemicals and related products (Table 5.19).

Table 5.20 presents changes in the goods structure of foreign trade of Ireland in chosen years of the 1995–2010 period per OECD classification based on technology level, discerning high technology (HT), medium-high technology (MHT), medium-low technology (MLT) , and low technology (LT) industries.

In the entire period of 1995–2010, Ireland was characterized by a larger share of exports in high- and medium-high-technology industries, when compared to the whole global economy, which proves the technological advancement of this country. Particularly dynamic growth in export took place in the commodity group of medium-high-technology industries, from 40.35 in 1995 (36.96 % for the whole world, correspondingly) to 57.09 % in 2010 (39.73 % for the whole world). On the other hand, the biggest decrease concerned the commodity group of low-technology industries, from 18.5 % in 1995 (15.9 % for the whole world) to 7.13 % (12.14 % for the whole world). Changes in the goods structure of foreign trade of Ireland in

Table 5.18 Goods structure of Ireland's exports in chosen years of the 1995–2010 period by SITC groups. (Study based on the United Nations Commodity Trade Statistics Database)

SITC		1995		2000		2005		2010	
		US$ million	%	US$ million	%	US$ million	%	US$ million	%
0	Food and live animals	7,704.84	17.60	5,433.31	7.12	7,870.16	7.15	9,242.14	7.80
1	Beverages and tobacco	746.97	1.71	865.11	1.13	1,383.05	1.26	1,572.70	1.33
2	Crude materials, inedible, except fuels	677.68	1.55	548.50	0.72	1,000.19	0.91	1,459.66	1.23
3	Mineral fuels, lubricants, and related materials	193.97	0.44	263.90	0.35	771.12	0.70	1,362.58	1.15
4	Animal and vegetable oils, fats, and waxes	37.52	0.09	24.98	0.03	22.91	0.02	35.22	0.03
5	Chemicals and related products	8,335.94	19.04	25,341.61	33.23	50,497.12	45.91	69,793.01	58.88
6	Manufactured goods classified chiefly by material	2,110.37	4.82	1,809.25	2.37	2,200.11	2	1,926.58	1.63
7	Machinery and transport equipment	15,080.06	34.44	30,567.32	40.08	28,813.70	26.19	14,685.72	12.39
8	Miscellaneous manufactured articles	5,891.23	13.45	8,160.77	10.70	13,232.64	12.03	13,938.06	11.76
9	Commodities and transactions not classified elsewhere in the SITC	3,010.75	6.88	3,242.73	4.25	4,207.76	3.83	4,513.54	3.81
	Total exports	43,789.32	100	76,257.49	100	109,998.77	100	118,529.21	100

Table 5.19 Goods structure of Ireland's imports in chosen years of the 1995–2010 period by SITC groups. (Study based on the United Nations Commodity Trade Statistics Database)

SITC		1995		2000		2005		2010	
		US$ million	%	US$ million	%	US$ million	%	US$ million	%
0	Food and live animals	2,298.66	7.12	2,588.30	5.11	4,407.07	6.27	6,047.47	10.02
1	Beverages and tobacco	303.49	0.94	470.46	0.93	941.02	1.34	1,065.35	1.77
2	Crude materials, inedible, except fuels	635.86	1.97	743.05	1.47	1,162.73	1.65	855.76	1.42
3	Mineral fuels, lubricants, and related materials	1,068.64	3.31	2,065.73	4.08	4,830.18	6.87	7,359.33	12.20
4	Animal and vegetable oils, fats, and waxes	122.68	0.38	112.84	0.22	189.51	0.27	244.24	0.40
5	Chemicals and related products	4,118.69	12.75	5,559.56	10.98	8,976.65	12.77	11,532.88	19.12
6	Manufactured goods classified chiefly by material	3,648.31	11.29	3,942.92	7.79	6,102.83	8.68	4,672.28	7.74
7	Machinery and transport equipment	13,580.93	42.04	26,768.46	52.87	30,402.22	43.26	16,020.36	26.55
8	Miscellaneous manufactured articles	3,578.00	11.08	5,496.27	10.85	9,091.98	12.94	8,424.50	13.96
9	Commodities and transactions not classified elsewhere in the SITC	2,950.58	9.13	2,887.38	5.70	4,165.58	5.93	4,107.60	6.81
	Total imports	32,305.85	100	50,634.98	100	70,269.78	100	60,329.76	100

Table 5.20 Ireland's exports structure according to the OECD classification based on the technology level. (Study based on the United Nations Commodity Trade Statistics Database)

Industries	1995		2000		2005		2010	
	Ireland (%)	World (%)	Ireland (%)	World (%)	Ireland (%)	World (%)	Ireland (%)	World (%)
High technology (HT)	18.97	12.03	20.73	13.53	18.64	13.33	16.83	11.73
Medium-high technology (MHT)	40.35	36.96	44.62	40.80	49.84	39.14	57.09	39.73
Medium-low technology (MLT)	22.19	35.11	23.15	32.11	21.96	34.53	18.95	36.41
Low technology (LT)	18.50	15.90	11.50	13.56	9.56	13.00	7.13	12.14

Table 5.21 Changes in the goods structure of Ireland's exports and imports according to the WIFO classification based on the extent of use of tangible and intangible expenses in the production process. (Study based on the United Nations Commodity Trade Statistics Database)

Industries	1995		2000		2005		2010	
	Ireland (%)	World (%)	Ireland (%)	World (%)	Ireland (%)	World (%)	Ireland (%)	World (%)
Mainstream manufacturing (MM)	7.35	21.78	4.30	19.92	3.91	20.50	4.40	21.10
Labor-intensive (LI)	4.36	14.44	2.18	13.70	1.84	13.09	1.78	12.44
Capital-intensive (CI)	11.04	20.81	18.40	19.35	18.74	22.28	22.40	24.65
Technology-driven industries (TDI)	31.20	13.24	19.57	11.95	15.59	11.00	10.22	10.54
Marketing-driven industries (MDI)	46.06	29.74	55.56	35.08	59.93	33.13	61.21	31.27

1995–2010 according to WIFO classification based on the degree of use of tangible and intangible expenses in the manufacturing process, discerning mainstream manufacturing industries (MM), labor-intensive industries, capital-intensive industries (CI), technology-driven industries (TDI), and marketing-driven industries (MDI), are presented in Table 5.21.

Statistical data related to the goods structure of exports in line with WIFO classification, based on the degree of use of tangible and intangible expenses in the production process, confirm the high technological advancement of Ireland's economy. Exports of this country are above all based on marketing-driven industries, the share of which increased from 46.06% in 1995 (compared to 29.74% for the whole world) to 61.21% in 2010 (31.27% for the whole world, respectively). In addition, capital-intensive industries are of particular importance for Ireland's export of

Table 5.22 Changes in the goods structure of Ireland's exports and imports according to the WIFO classification based on the level of skills of the labor force. (Study based on the United Nations Commodity Trade Statistics Database)

Industries based on	1995		2000		2005		2010	
	Ireland (%)	World (%)	Ireland (%)	World (%)	Ireland (%)	World (%)	Ireland (%)	World (%)
Highly skilled workers (HS)	18.97	12.03	20.73	13.53	18.64	13.33	16.83	11.73
Semiskilled, white-collar workers (MW)	40.35	36.96	44.62	40.80	49.84	39.14	57.09	39.73
Semiskilled, blue-collar workers (WM)	22.19	35.11	23.15	32.11	21.96	34.53	18.95	36.41
Low-skilled workers (LS)	18.50	15.90	11.50	13.56	9.56	13.00	7.13	12.14

Table 5.23 Development and evolution of comparative advantages (RCA) in Ireland's foreign trade in the 1995–2010 period according to the OECD classification based on the technology level. (Study based on the United Nations Commodity Trade Statistics Database)

Industries	1995	2000	2005	2006	2007	2008	2009	2010
High technology	0.2012	0.1801	0.3443	0.3549	0.2891	0.4285	0.4059	0.5364
Medium-high technology	0.1349	0.2819	0.5770	0.5761	0.5942	0.7139	0.8381	0.8312
Medium-low technology	−0.1343	0.1706	0.1709	0.1962	0.0990	0.1337	0.1588	0.1052
Low technology	0.3338	0.1940	0.0615	0.1184	−0.0961	−0.0768	−0.2560	−0.2776

goods (22.4 % in 2010 compared to 11.04 % in 1995) and technology-driven industries (10.22 % in 2010 compared to 31.2 % in 1995). On the other hand, the lowest share is that of mainstream manufacturing industries (4.40 % for Ireland compared to 21.10 % for the whole world in 2010) and labor-intensive industries (1.78 and 12.44 %, respectively). Statistical data related to changes in the goods structure of foreign trade in chosen years of the 1995–2010 period according to WIFO classification based on the level of skills of the labor force (Peneder 1999), discerning industries using high skills (HS), medium/white-collar skills (MW), medium/blue-collar skills (MB), and low skills (LS) are presented in Table 5.22.

In the goods structure of Ireland, high-skills industries are of bigger importance when compared to the whole global economy (16.83 % in Ireland as compared to 11.73 % for the world 2010), as are industries using medium-skilled, white-collar workers (57.09 and 39.73 %, respectively). The biggest change in the years 1995–2010 was related to the drop in Irish exports in industries based on low-skilled workers, from 18.5 to 7.13 %, and the growth in exports in industries based on medium-skilled, white-collar workers, from 40.35 to 57.09 %. It confirms the gradual modernization of the structure of Ireland's economy in the studied period, manifesting itself in bigger (when compared to the whole world) growth of significance of the level of skills of the labor force in the goods structure of foreign trade, and it constitutes a proof of high usage of human capital.

Neoclassic theories of foreign trade anticipate that a given economy specialized in the production and export of goods in the case of which it has a comparative advantage. Table 5.23 presents the development of comparative advantages in foreign trade (RCA) in Ireland, in line with the OECD classification based on the technology level.

During the entire period of 1995–2010, Ireland had relative advantages in industries of high- and medium-high technology. Particularly fast growth of the index of RCAs can be observed in relation to the medium-high-technology industries, from 0.1349 in 1995 to 0.8310 in 2010. On the other hand, since 2007, Ireland has been characterized by a lack of relative advantage in low-technology industries, for which the RCA index dropped from the level of −0.0961 to −0.2776 in 2010. Table 5.24 presents the development of comparative advantages (RCA) in Ireland's foreign trade in line with WIFO classification based on the degree of use of tangible and intangible expenses in the production process.

Ireland has comparative advantages in technology-driven and capital-intensive industries. As for technology-driven industries, RCA index values were above 0 during the entire period under examination, and the index increased from 0.1641 in 1995 to 0.5907 in 2010. On the other hand, there were no comparative advantages in mainstream and labor-intensive industries, and from 2007 onwards, also in marketing-driven industries.

The assessment of advantages in FT in accordance with the classification of industries based on the skill level of labor force shows that in 2010 Ireland had the biggest relative advantages in industries based on semiskilled, white-collar workers, and industries based on highly skilled workers had the second biggest relative advantages within this scope. The RCA index had grown particularly dynamically in the case of industries based on semiskilled, white-collar workers, from 0.1349 in 1995 to 0.2819 in 2000, and then to 0.8312 in 2010. For industries based on low-skilled workers, the index had reached positive values until 2006, and since 2007, Ireland has not had a comparative advantage in this group of industries in the foreign trade. This confirms an even higher technological advancement of Irish economy, as well as the growing specialization in business activity characterized by the higher value added (Table 5.25).

In the contemporary global economy, comparative advantage is neither the only nor the primary prerequisite of the existence of international specialization and trade exchange. Changes within the scope of competitive advantages achieved by various economies may be indirectly connected, in the case of individual industries, with the IIT growth.

In 2010, the IIT in Ireland had the greatest intensity in low-technology (IIT=0.8037), medium-low-technology (IIT=0.6284), high-technology (IIT=0.4588), and medium-high-technology (IIT=0.3629) industries (Table 5.26). The IIT index

Table 5.24 Development and evolution of comparative advantages (RCA) in Ireland's foreign trade in the 1995–2010 period according to the WIFO classification based on the level of usage of tangible and intangible expenses in the production process. (Study based on the United Nations Commodity Trade Statistics Database)

Industries	1995	2000	2005	2006	2007	2008	2009	2010
Mainstream	−0.7626	−1.0989	−1.2953	−1.2708	−1.2399	−1.2734	−1.3812	−1.2333
Labor intensive	−0.6710	−1.3206	−1.6734	−1.6701	−1.7846	−1.8463	−2.0370	−1.7665
Capital intensive	−0.1619	0.6789	0.4362	0.3915	0.4321	0.2798	0.4038	0.2718
Marketing driven	0.8064	0.5013	0.1887	0.2771	−0.0762	−0.1899	−0.4535	−0.3982
Technology driven	0.1641	0.1781	0.3509	0.3717	0.3193	0.4764	0.5250	0.5907

Table 5.25 Development and evolution of comparative advantages (RCA) in Ireland's foreign trade in the 1995–2010 period according to the WIFO classification based on the skill level of labor force. (Study based on the United Nations Commodity Trade Statistics Database)

Industries based on	1995	2000	2005	2006	2007	2008	2009	2010
Highly skilled workers	0.2012	0.1801	0.3443	0.3549	0.2891	0.4285	0.4059	0.5364
Semiskilled, white-collar workers	0.1349	0.2819	0.5770	0.5761	0.5942	0.7139	0.8381	0.8312
Semiskilled, blue-collar workers	−0.1343	0.1706	0.1709	0.1962	0.0990	0.1337	0.1588	0.1052
Low-skilled workers	0.3338	0.1940	0.0615	0.1184	−0.0961	−0.0768	−0.2560	−0.2776

Table 5.26 Volume of intra-industry trade (IIT) and its changes in Ireland in the 1995–2010 period according to the OECD classification based on the technology level. (Study based on the United Nations Commodity Trade Statistics Database)

Industries	1995	2000	2005	2006	2007	2008	2009	2010
High technology	0.7528	0.7135	0.6234	0.6613	0.6957	0.6067	0.5258	0.4588
Medium-high technology	0.7841	0.6675	0.5281	0.5673	0.5643	0.4932	0.3760	0.3629
Medium-low technology	0.9155	0.7179	0.7001	0.7334	0.7842	0.7379	0.6270	0.6284
Low technology	0.6916	0.7072	0.7506	0.7699	0.8789	0.8383	0.8175	0.8037

increased only in the low-technology industry: from 0.6916 in 1995 to 0.7072 in 2000, and then to 0.8037 in 2010. At the same time, a constant decrease in the ITT ratio is noticeable for industries of high technology (from 0.7528 in 1995 to 0.7135 in 2000 and 0.4588 in 2010), medium-high technology (from 0.7841 in 1995 to 0.6675 in 2000 and 0.3629 in 2010), and medium-low technology (from 0.9155 in 1995 to 0.7179 in 2000, and then to 0.6284 in 2010), which confirms the deterioration of Ireland's international trade.

In 2010, the biggest volume of IIT in Ireland, according to WIFO classification based on the extent of use of tangible and intangible expenses in the production process, was recorded in marketing-driven industries. What is more, the most dynamic changes of the IIT index in the period under consideration may be observed for this group of goods. The index had grown from the level of 0.4957 in 1995 to 0.5738 in 2000, and to 0.9144 in 2009, whereby in 2010, it dropped slightly, settling at 0.8623. In terms of the IIT volume in Ireland in 2010, mainstream industries were a runner-up (IIT=0.7280). One should pay attention to the changes in IIT in capital-intensive industries, for which the IIT index had dropped from the level of 0.9292 in 1995 to 0.5039 in 2000, and then went up to 0.5589 in 2010, as well as in technology-driven industries, for which the IIT index had dropped from 0.7703 in

Table 5.27 Volume of intra-industry trade (IIT) and its changes in Ireland in the 1995–2010 period in goods trade according to the WIFO classification based on the extent of use of intangible and tangible expenses in production process. (Study based on the United Nations Commodity Trade Statistics Database)

Industries	1995	2000	2005	2006	2007	2008	2009	2010
Mainstream	0.7745	0.6682	0.6000	0.5697	0.5779	0.5903	0.6390	0.7280
Labor intensive	0.8184	0.5734	0.4540	0.4217	0.3815	0.3820	0.3918	0.5028
Capital intensive	0.9292	0.5039	0.5846	0.6452	0.6323	0.6713	0.5266	0.5589
Marketing driven	0.4957	0.5738	0.6920	0.6962	0.8691	0.8939	0.9144	0.8623
Technology driven	0.7703	0.7145	0.6205	0.6539	0.6820	0.5866	0.4810	0.4399

Table 5.28 Volume of intra-industry trade (IIT) and its changes in Ireland in 2000–2010 in goods trade according to the WIFO classification based on the skill level of labor force. (Study based on the United Nations Commodity Trade Statistics Database)

Industries based on	1995	2000	2005	2006	2007	2008	2009	2010
Highly skilled workers	0.7528	0.7135	0.6234	0.6613	0.6957	0.6067	0.5258	0.4588
Semiskilled, white-collar workers	0.7841	0.6675	0.5281	0.5673	0.5643	0.4932	0.3760	0.3629
Semiskilled, blue-collar workers	0.9155	0.7179	0.7001	0.7334	0.7842	0.7379	0.6270	0.6284
Low-skilled workers	0.6916	0.7072	0.7506	0.7699	0.8789	0.8383	0.8175	0.8037

1995 to 0.6205 in 2005, and then increased to 0.6820 in 2007, only to drop again in 2010, to 0.4399 (Table 5.27).

The next trade classification is based on the skill level of the labor force. In 2010, the IIT in Ireland was the most intensive in industries based on low-skilled workers. The IIT index for this group of goods increased from 0.6916 in 1995 to 0.8789 in 2007, to drop to 0.8037 in 2010. The IIT in industries based on semiskilled, blue-collar workers was of major importance, whereby the IIT value went down from 0.9155 in 1995 to 0.6284 in 2010. The biggest drop in volume of the IIT in Ireland was noted in the case of industries based on semiskilled, white-collar workers, from 0.7841 in 1995 to 0.6675 in 2000, and further to 0.3629 in 2011. The IIT index also decreased for industries based on highly skilled workers, from 0.7528 in 1995 to 0.7135 in 2000, and down to 0.4588 in 2010. However, in 2005–2007, a temporary growth was noted, from 0.6234 to 0.6957 (Table 5.28).

In general terms, a drop in the value of IIT of Ireland is noticeable, first of all in technologically advanced industries, which is a sign of low intra-industry specialization in foreign trade of technologically advanced goods. However, it should be noted that Ireland, because of its geographical location, when compared to such countries of western Europe as Germany, France, or the Netherlands, is one of the peripheral economies (e.g., in relation to the eurozone), which are characterized, as a rule, by a lower volume of IIT.

5.3.2 *Characteristics of Ireland's NIS*

Institutions that shape scientific, innovation, and educational policy in Ireland may be divided into two basic groups: entities responsible for preparation of strategic directions within the policy and organizations responsible for implementation of the policy with distribution of funds included (Pontikakis et al. 2006). Responsibility for preparation of strategic assumptions of the policy rests with a couple of organizations at the departmental level, which cooperate under the direction of the Cabinet Sub-Committee on Science and Technology. The main entity handling the implementation of innovation policy is the Department of Jobs, Enterprise and Innovation, which—through the Office of Science, Technology and Innovation (OSTI)—coordinates operations of the Inter-Departmental Committee (IDC) for Science, Technology and Innovation. OSTI is responsible for development, promotion, and coordination of Irish policy within the scope of innovations, science, and technology, and also the coordination of Ireland's influence on the development of the EU's policy in these areas, including the development of the European Research Area (ERA). The entity operates on the basis of the assumption that research, technological development, and innovation (RTDI) at the national and international level are of key importance for competitiveness and employment, and public investments in RTDI are a national priority.

Works of OSTI include all aspects of the NIS of Ireland and its participation in international R&D programs. The entity cooperates with Forfás, which is an advisory team for economic development, industry, innovativeness, science, and new technologies. In elaboration of strategic directions of innovation policy, also the following entities participate: the Office of the Chief Scientific Adviser and the Advisory Council for Science, Technology and Innovation. OSTI is also responsible for coordination and financing of a wide range of programs implemented against various executive agendas, of which the most important ones are the following:

- The Science Foundation Ireland (SFI) supports scholars and research teams, especially in the following fields:

1. biotechnology,
2. ICT, and
3. sustainable energy and energy-efficient technologies.

- The Enterprise Ireland, which is an agency of the Irish government, supports small and medium-sized Irish enterprises in five main fields of activity: research investments, innovativeness, productivity, support of development of companies, and exports. The organization constitutes a global network comprised of

the headquarters in Dublin, 13 regional offices in Ireland, and more than 30 commercial divisions operating at the embassies, including in Warsaw.
- The Industrial Development Agency Ireland (IDA-Ireland) promotes direct foreign investments in Ireland. The agency plays a key role in directing the activity of transnational corporations in the country toward innovativeness. The main tool within this scope is the R&D Capability Grant Scheme, the purpose of which has been to encourage transnational corporations to situate R&D centers at Irish branches.

In the field of the education policy of Ireland, the Department of Education and Skills is the key entity. One of its strategic aims is to ensure outstanding opportunities for the development of science, research, and innovation in higher education (Department of Education and Skills Statement of Strategy 2011–2014, p. 5). The department coordinates works of the IRCSET, whose members are experienced scientists and industry representatives. The council operates for the benefit of the development of human capital, which constitutes an important element of Ireland's NIS, *inter alia* by financing scientific research as part of master, doctoral, and postdoctoral dissertations at universities. When choosing the projects to be supported, the priority is given to exploratory research aimed at creation of new concepts, inventions, and innovations, which is conducted in cooperation between science and industry.

Of enormous importance in the Ireland's NIS are intermediating institutions, which aim at facilitating the flow of information, knowledge, and technology between research institutions and industry. In addition, the support for technological parks and incubators and innovative clusters is an important element of the economic growth policy of Ireland. Particular emphasis is laid on the development of *flagship parks* in locations called gateways, and on the creation of leading global scientific research clusters there. The concept originates from Ireland's National Spatial Strategy for 2002–2020 (Government of Ireland 2002, pp. 10–15), one of the main objectives being the development of local centers, the so-called hubs and gateways, located throughout the country and closely connected with one another. The basis for the activities undertaken is provided by the assumption that skilled integration of technology, science, and business at these centers is a key to competitiveness and innovativeness of the economy. The main *driving forces* of Ireland's economic growth are five highly developed municipal centers:

- Dublin, which has been considered an international gateway of Ireland, constituting an important science, technological development, and innovation center. There are various intermediating institutions, including Blanchardstown Business & Technology Park or Clonshaugh College Business & Technology Park.
- Cork, where within an area of 32 hectares the Cork Business & Technology Park is located, closely cooperating with, e.g., the University College Cork and the Cork Institute of Technology.
- Limerick/Shannon, which constitute an innovative and rapidly developing economic center, based *inter alia* on the Shannon Development-owned National Technology Park. The Shannon Development's operations are focused on supporting the development of the knowledge exchange network in the region, pro-

moting the Shannon Free Zone, ensuring optimum conditions for industry and supporting key innovative projects.

- Galway, where within an area of 38 hectares the Parkmore Business & Technology Park is located.
- Waterford, where in the area of 28 hectares, the Waterford Business & Technology Park is located.

To the above-mentioned highly developed municipal centers, four other centers have been added, at which, *inter alia* through the agency of IDA-Ireland, business and technology parks are being developed:

- Letterkenny, located in the northern-western part of the Border, near Derry, the second biggest town of the Northern Ireland, where the Letterkenny Business & Technology Park is located within an area of 44 hectares;
- Dundalk, situated in the eastern part of the Border, within the transport corridor between Dublin and Belfast, where within an area of 37 hectares the Dundalk Business & Technology Park was created;
- a triangle located in the Midlands, comprising the following centers: Athlone, where the 40-hectare Athlone Business & Technology Park is located; Mullingar with the 27-hectare Mullingar Business & Technology Park; and Tullamore, where the 31-hectare Tullamore Business & Technology Park may be found; and
- Sligo in the West, where the 42-hectare Sligo Business & Technology Park is located.

The implementation of the National Spatial Strategy takes place in the presence of strong financial support as part of the National Development Plan for 2007–2013 (Government of Ireland 2007), available in particular within the Gateways Innovation Fund used to build capacities in the R&D field and innovativeness in selected centers.

One of the key types of entities in Ireland's NIS are universities and research centers. The biggest scientific center in Ireland is Dublin where world-class universities operate, such as the Dublin Institute of Technology, the Trinity College Dublin, the University College Dublin, and the Dublin City University. National centers for science, engineering, and technology (CSET), financed with subsidies of the SFI, are also of importance. They bring together scientists and engineers from national and multinational enterprises.

Private entities in the NIS of Ireland, which deal with practical innovative activity, may be divided into two basic groups:

1. Transnational corporations implementing direct foreign investments in Ireland, mainly in high-technology sectors, with strong domination of two industries: ICT and biotechnology. The concentration of resources of multinational enterprises in these sectors has ensured a unique competitive advantage of Ireland on the global market. The majority of produced technologies are driven by demand from outside of Ireland. Among foreign producers of technologies and innovations on the Irish market, the dominating entities are American companies seeking particular location benefits, such as geographic closeness of and unlimited access to the EU single market, as well as common cultural and language her-

itage. These benefits, in combination with a more favorable tax system, skilled human resources, and effective institutions in Ireland, have ensured a comparative advantage to this country when compared to economies with similar resources (Pontikakis et al. 2006, p. 49).

2. Domestic enterprises, which are characterized by a lower level of innovativeness when compared to foreign companies. More than the internal reluctance to introduce innovations, the reason for this is the character of activity they pursue, as they operate first of all in labor-intensive sectors (e.g., foodstuffs, beverages, and tobacco products) (Barry 2005). For those domestic companies that will undertake the task of increasing the level of innovativeness of their activity, the main method consists in the recombination of existing products and processes, i.e., the implementation of imitative innovations (Pontikakis et al. 2006, p. 50).

The division into the two basic types of private entities in Ireland's NIS should be investigated against the background of dualism in this country's industry structure, which is characterized by a strong, modern sector of foreign enterprises and a less favorably developed, traditional, domestic sector, not engaged in production with the use of advanced technologies. While transnational corporations are of fundamental importance for the innovativeness of the Irish economy, their impact is partly limited by the fact that in most cases they are not strongly embedded in the national system of the country (Roper 2004). It is shown in a low level of integration of foreign companies with Irish business and scientific entities, and in weak connections with the economy of the country. While tax incentives are efficient in attracting multinational companies, they stimulate cooperation with domestic companies, *inter alia* within the scope of innovative activity, only to a limited extent. This mechanism applies in particular in the case of high-technology industries, where transnational corporations strive to protect their technological advantage (Louri et al. 2002).

An important element of the NIS in Ireland are clusters which on the one hand increase the geographic attractiveness for direct foreign investments, while on the other create new opportunities for export to foreign markets for local enterprises (Kowalski 2011). The development of Irish clusters takes place mostly in three areas:

1. ITC, with strong concentration around Dublin (where, e.g., Intel, IBM, and Microsoft operate), Limerick (Dell, Analog Devices) and Cork (Apple Computer, EMC), and to a lesser extent, Galway (Nortel, SAP) and Dundalk (Xerox);
2. the (bio)pharmaceutical sector which includes the following industries: pharmaceutical, biotechnological, and medical devices industry, with a strong concentration around Dublin (Johnson Diversey, Pfizer, and Wyeth), Cork (Pfizer, GlaxoSmithKline, and Johnson & Johnson), and Galway (Baxter Healthcare and Boston Scientific); and
3. internationally traded services centered in Dublin, where such corporations as Google, Yahoo, Ebay, Merrill Lynch, or Citibank have their headquarters.

Table 5.29 Innovative capacity of Ireland imports in chosen years of the 1995–2010 period. (Source: On the basis of: for indexes 1–6 and 9–10: data base of the Organization for Economic Co-operation and Development, OECD: OECD.Stat, for indexes 7–8: Eurostat Statistics Database (tsiir080))

	Index	1995	2000	2005	2009	2010
1	Share of total internal expenditure on research and development (R&D) (GERD) in GDP (%)	1.25	1.11	1.24	1.74	1.79
2	Share of government internal expenditure on R&D (GOVERD) in GDP (%)	0.11	0.09	0.09	0.10	0.08
3	Percentage of internal expenditure on R&D financed by the government (%)	22.55	23.45	32.00	31.35	–
4	Percentage of internal expenditure on R&D financed by other national sources (%)	1.95	1.90	1.93	1.82	–
5	Percentage of internal expenditure on R&D financed by foreign countries (%)	8.50	8.87	8.63	15.63	–
6	Households with access to the Internet (as % of all households)		20.40	47.18	66.67	–
7	Human resources for science and technology as a percentage of working population (%)	28.90	29.80	36.60	42.10	–
8	Investments in venture capital as % of GDP	0.00	0.11	0.02	0.02	–
9	Share of business internal expenditure on R&D (BERD) in GDP (%)	0.84	0.73	0.71	0.89	–
10	Percentage of internal expenditure on R&D financed by industry (%)	67.35	65.79	57.45	51.20	–

5.3.2.1 Innovative Capacity and Position

The technological development of Ireland in the period under consideration is evidenced by the share of total internal R&D expenses in GDP, in particular between 2000 (1.11 %) and 2010 (1.79 %). The favorable structure of these expenses should be pointed out, as the majority of them are financed by the industrial sector which makes them more effective. Despite the fact that more than a half of R&D expenses in GDP are financed by the industry, the constant decrease in this proportion, from 67.35 % in 1995 to 51.2 % in 2009, constitutes a negative trend. On the other hand, an increase in the percentage of internal R&D expenses financed by the government was noted, from 22.55 in 1995 to 31.35 % in 2009. It is a negative phenomenon in the context of the level of innovativeness of the Irish economy, as the results of R&D works financed from public sources are less usable in commercial terms and translate rather poorly into implementation of innovations (Table 5.29).

The increase in innovative capacities of Ireland is evidenced by the dynamic growth of the percentage of households with access to the Internet when compared to all households, from 20.4 % in 2000 to 66.67 % in 2009. The popularization of the Internet was one of the key factors influencing business and financial services on the Irish market. The dynamic increase in the share of human resources for science and technology in the working population, from 28.9 % in 1995 to 42.1 % in 2009, was also of significance for Ireland's innovative capacity in the period under consideration. On the other hand, insignificant investments in venture capital (0.02% of GDP in 2009), which in many developed economies constitute an efficient source for financing innovative activity, should be assessed negatively.

In Ireland, a drop in the share of employment in high- and medium-high-technology industrial sectors in the total employment was recorded, from 7.44% in 1997 to 5.24% in 2008, and a growth of the share of employment in knowledge-intensive service sectors in the total employment, from 29.38% in 1997 to 36.22 in 2008, was noted. Indicators related to intellectual property, which in the modern economy is becoming one of the key factors influencing the technical advancement of enterprises, are important in view of the analysis of the level of innovativeness of an economy (Kowalski 2010b). However, a constant decrease in the number of patent applications per US$ 1 billion of R&D expenses was noticeable in Ireland, from 0.74% in 1997 to 0.39% in 2010, as well as a decrease in the number of patent applications per one million citizens from 243.09 in 2000 to 163.57 in 2010, which may be a sign of decreasing effectiveness of R&D works.

When assessing the innovative position of Ireland, one should take into account the drop in the share of employment in high- and medium-high-technology industrial sectors in the total employment, from 7.44% in 1997 to 5.24% in 2008, as well as the growth of the share of employment in knowledge-intensive service sectors in the total employment, from 29.38% in 1997 to 36.22 in 2008. Moreover, a constant decrease in the number of patent applications per US$ 1 billion of R&D expenses was observed in Ireland, from 0.74% in 1997 to 0.39% in 2010, as well as a decrease in the number of patent applications per one million inhabitants from 243.09 in 2000 to 163.57 in 2010, which may be a sign of decreasing effectiveness of R&D works.

Ireland is characterized by a high share of export of high-technology products in the total exports. During the whole period under consideration, this index assumed the highest values among all member states of the EU, except for Luxemburg (since 2004), and Cyprus and Malta which acceded to EU in 2004. The above was related to the pro-export nature of direct foreign investments in high-technology sectors in the Irish market, first of all in computer and pharmaceutical industries. After 2000, however, we have observed a drop in the share of high-technology products export in the total exports of Ireland, which is connected *inter alia* with the relocation of a portion of the manufacturing activity of international corporations to emerging markets, mainly to China.

5.3.2.2 Human Capital

High-quality human capital has constituted one of the major causes of quick economic growth of Ireland. Availability of a skilled, English-speaking labor force played an important part in attracting direct foreign investments in sectors based on advanced technologies, characterized by a high added value, such as ICT, as well as pharmaceutical and biotechnological sectors. Nevertheless, Ireland is a small economy inhabited by more or less 4.5 million people. It should be noted, how-

Table 5.30 Innovative position of Ireland. (Source: On the basis of: for indexes 1, 2, and 5: Eurostat Statistics Database (tsc00011), (tsc00012), (tsiir160), for indexes 3–4: data bases of the World Intellectual Property Organization—WIPO)

	Index	1995	1997	2000	2005	2008	2009	2010
1	Share of employment in high- and medium-high-technology industries in the total employment (%)	…	7.44	6.94	6.02	5.24	…	…
2	Share of employment in knowledge-intensive service industries in the total employment (%)	…	29.38	31.76	34.02	36.22	…	…
3	Number of patent applications per US$ 1 billion of funds allocated to R&D	…	0.74	0.71	0.64	0.62	0.56	0.39
4	Number of patent applications per million inhabitants	232.76	216.65	243.09	189.67	210.36	203.64	163.57
5	Share of export of high-technology goods in total exports	32.882	37.524	40.544	29.538	…	…	…

ever, that a constant increase in population took place in the period under consideration. It was caused by one of the highest birth rates in Europe and by immigration. The inflow of people from abroad means a reversal of the phenomenon of material emigration, which had been occurring in Ireland since the 1980s, especially to the United Kingdom and the United States. However, since the 1990s, Ireland has been a place attracting labor force, and in particular highly skilled workers (Table 5.31).

The data shown in Table 5.30 confirm the growth of labor force resources in Ireland, from 1,466,000 in 1995 to 2,239,000 in 2009. They also indicate a growing quality of human capital in the period under consideration, which is expressed by the growth of the share of labor force with higher education, from 29.04 % in 1998 to 33.3 % in 2007, and also labor force with secondary education, from 31.8 % in 1995 to 34.2 % in 1998, and then to 38.4 % in 2007. The above took place at the expense of a drop in the share of labor force with primary education from 43.6 % in 1995 to 24.4 % in 2007. It means that more and more people continue education after graduating from primary school.

Due to the significance of people's education for competitiveness of the economy, Ireland attaches more and more importance to investments in human capital. It manifests itself in the growth of the share of expenses for education in GDP at each level of education: at the higher education level from 29.04 % of GDP in 1998 to 32.42 % of GDP in 2008, at the secondary school level from 17.25 % of GDP in 1998 to 27.24 % of GDP in 2008, and at the primary education level from 11.44 % of GDP in 1998 to 18.38 % of GDP in 2008. Favorable trends in terms of labor force

Table 5.31 Skill level of labor force in Ireland. (Source: World Bank data, EdStats, WB_HC_Extracts.xls file)

Index	1995	1998	2000	2001	2003	2005	2007	2008	2009	2010
Total labor force resources (in thousands)	1,466	1,626	1,759	1,802	1,895	2,055	2,197	2,213	2,239	–
Share of labor force with primary education (%)	43.6	40.3	–	30.9	29	26.8	24.4	–	–	–
Share of labor force with secondary education (%)	31.8	34.2	–	40.6	39.1	39	38.4	–	–	–
Share of labor force with higher education (%)	24.3	24.8	–	25.8	29.7	31	33.3	–	–	–
Expenditures on higher education as % of GDP	–	29.04	30.68	28.29	23.96	24.66	26.18	32.42	–	–
Expenditures on secondary education as % of GDP	–	17.25	16.42	17.29	19.27	21.68	23.19	27.24	–	–
Expenditures on primary education as % of GDP	–	11.44	11.16	11.88	13.36	14.63	15.72	18.38	–	–
Average number of years of education of persons aged 25+	10.90	–	11.19	–	–	11.40	–	–	–	11.62

Table 5.32 Human capital productivity in Ireland. (Source: OECD databases)

	1995	1998	2000	2005	2009
Remuneration per hour (in PPP), in US$	–	14.86	15.56	20.56	26.08
Remuneration per employee (in PPP), in US$	25,893.52	28,162.66	29,525.31	37,822.7	45,409.96
Labor costs per production unit	0.655317	0.595651	0.565965	0.576201	0.633266
Unit labor costs (ULC)	0.44301	0.46282	0.48344	0.5762	0.62294

education in Ireland are also confirmed by the index expressing the average number of years of learning of persons aged above 25, which had grown from 10.9 in 1995 to 11.62 in 2010. Table 5.22 presents data related to the level of usage of human resources in Ireland.

The usage of labor force resources in Ireland corresponds to changes in unemployment rate which had dropped rapidly from 12 % in 1995 to 4.3 % in 2000, and then slightly increased to 4.6 % in 2007 to quickly go up to the level of 11.7 % in 2007. A similar trend was recorded in the case of unemployment rate among youth. In 1995–2000, the rate went down from 19 to 6.5 %, to grow again to 9.1 % in 2007, and to 24.3 % in 2009. The vocational activity of people aged 24–25 years increased from 36.4 % in 1995 to 49.7 % in 2006, and then went down to 40.1 % in 2009.

Together with the fast economic development of Ireland, also remunerations grew at the end of 1990s/beginning of the first decade of the twenty-first century, from US$ 14.86 (in PPP) per hour in 1998 to US$ 26.08 per hour (in PPP) in 2009, and per employee, from US$ 25,893.52 (in PPP) in 1998 to US$ 45,409.96 (in PPP) in 2009. Initially, a decrease in labor costs per production unit was noted, from 0.655317 in 1995 to 0.565965 in 2000, and then they started to grow subsequently, up to the level of 0.633266 in 2009. Unit labor costs (ULC) went up from the level of 0.44301 in 1995 to 0.62294 in 2009. Particularly dynamic growth took place after 2000 (when the index value was 0.48344), which means diminishing of cost competitiveness of Ireland (Table 5.32).

5.3.2.3 Innovation Policy

Scientific and innovation policy in Ireland is connected with industrial policy. The strategic framework for these policies is defined in a document entitled "Strategy for Science, Technology, and Innovation" (Government of Ireland 2006). The strategy is based on a vision as a part of which Ireland would be one of the most important global centers of high-quality research and practical use thereof for the benefit of the economy and the society. Methods of achieving this aim included investments in human capital, education system, and modern industry.

Main instruments used as a part of the national innovation policy are tax incentives and direct financing of scientific and technological projects. There are a number of tax reliefs aimed at stimulation of innovative activity, including tax exemptions for expenses incurred for various scientific and technological research, preferential amortization rates for R&D, and relatively small tax on license receivables (Pontikakis et al. 2006, p. 45). Lower costs of innovations in Ireland, connected with more favorable solutions of the tax system, influenced the growth of total expenses for innovations (Hall et al. 2000). In addition, the overall liberalization of telecommunications market was conducive to the development of the ICT sector. The increased volume and quality of ICT infrastructure in Ireland played a special role in the context of overcoming the negative influence of peripheral geographic location.

Of high importance for the implementation of Ireland's scientific, innovation, and educational policy were regional policy instruments of the EU, which to a significant degree covered the development of modern ICT infrastructure and supported innovativeness. In the operating programs adopted within structural funds, significant emphasis was laid on priorities connected with RTDIs (Kowalski and Szlachta 2007). The usage of instruments offered within the innovation policy of the EU, and especially those related to research, technological development, and implementations, is also Ireland's strength.

A summary of the assessment of Ireland's NIS is the SWOT analysis.

Strengths:

- well-developed training infrastructure and scientific base, along with intermediating institutions, first of all business and technological parks and their strong connection (also in terms of location) with universities;
- well-developed modern sectors of high technology, including ICT and biotechnological sector;
- well-developed ICT infrastructure, leveling the negative impact of the peripheral geographic situation of the country;
- successful leveraging of high-quality FDIs characterized by a significant level of technological advancement;
- highly adaptive production base;
- well-educated labor force, equipped with high skills and professional abilities;
- determination of the Irish government to build a knowledge-based economy, and support for enterprises conducting innovative activities; and
- favorable legal and tax environment for business activity, including business connected with innovations.

Weaknesses:

- low level of technological advancement of national enterprises and insufficient level of R&D expenses in this sector;
- weak connections and low level of cooperation between transnational corporations and domestic business and scientific entities;
- low absorption capability of domestic enterprises within the scope of research and technology and weak knowledge commercialization structures; and
- low availability of seed capital.

Opportunities:

- adoption of activities aimed at the improvement of Ireland's status as a location of R&D activity;
- benefiting from programs and instruments offered as part of the scientific and innovation policy of the EU, e.g., framework programs;
- further utilization of investment attracting potential and development of ICT and biotechnological industry;
- high number of young, educated, and mobile people, skilled in the use of advanced technologies and production methods;
- using IPR policy to stimulate innovative activity;
- development of technological centers on demand basis (e.g., living labs);
- further development of innovative industrial clusters;
- using development potential of services based on knowledge (financial, ICT, medical services, including those related to the genome, clinical trials, and e-health);
- using development potential of biotechnological sector, including good environment for clinical trials, which is comprised, among other things, of high ethical standards, availability of patients, presence of active charity medical organizations, and the increasing number of world-class scientists; and
- development of new mechanisms in order to increase R&D activity in Irish industry.

Threats:

- competition on the part of developing economies (e.g., BRIC countries) and moving FDIs there;
- no coherent immigration policy for skilled workers and scientists;
- outflow of educated people with the related threat of shortage of skilled labor force;
- low international awareness related to the perception of Ireland as a center of world-class science and innovations; and
- threats resulting from the global financial downturn, and the related decrease in R&D expenditure, as well as limitation of FDI volume.

5.3.3 Evaluation of the Relationship Between the Level of Innovation and Human Capital Accumulation, and the Advantages in the Foreign Trade

The fast economic growth of Ireland, which had begun in the mid-1990s, was reflected in the growing share of this country in the global trade exchange. This growth was mainly related to innovative goods groups, which indicates a positive correlation between the innovation level of the Irish economy and the advantages in foreign trade.

When analyzing changes in Ireland's goods export structure according to the OECD classification, one may note a growing specialization of the country in medium-high- and high-technology industries, which is an evidence of Ireland's technological development. At the same time, industries using highly skilled workers and semiskilled, white-collar workers play a more significant role in the goods structure of Irish exports according to the WIFO classification when compared to the global economy, while a downward trend may be seen in industries based on low-skilled workers. This process confirms the development of human capital and gradual modernization of the structure of Irish economy, which influences competitive advantages in foreign trade.

FDIs, implemented first of all in technologically advanced sectors, with strong dominance of ICT and the (bio)pharmaceutical industry, constituted an important factor of innovativeness of the Irish economy. Concentration of the resources of multinational corporations in these sectors contributed to the creation of local clusters and ensured a unique competitive advantage for Ireland on the global market. What should be noted in this context is the duality of the Irish innovation system, in which, aside from an advanced sector of foreign enterprises, there is a less-developed sector of domestic enterprises which are not involved as much in production with the use of advanced technologies. A low level of cooperation between both groups of entities, especially in the field of innovative activity, poses an additional problem. The significant role of FDIs in the innovation system of Ireland is confirmed by the fact that in 2009, 15.63 % of the internal R&D expenditure in this country was financed from external resources, which constitutes one of the highest values of the index in the EU.

The competitive position of the Irish economy in foreign trade was strengthened by actions undertaken as part of the economic policy. Programs cofinanced from structural funds, under which projects connected with RTDI were given a priority, were of particular importance. An important role in the NIS was also played by the scientific sector, supported under the national scientific policy, e.g., by the SFI.

Competitive advantages in Irish foreign trade were also influenced by the dynamic development of human capital. It took place both in terms of quantity (increase in the number of labor force thanks to positive demographic trends and positive migration balance) as well as in terms of quantity (increase in the education level of the population, and increased expenses incurred for education at each level). The high-quality human capital constituted one of the factors attracting FDIs, which are one of the most important actors of the NIS of Ireland. Besides, it influenced relative advantages in Irish foreign trade, measured with the comparative advantage index (RCA), in industries based on semiskilled, white-collar workers, and highly skilled workers.

Chapter 6
International Competitiveness of Countries with Performing Innovation Systems

Andreas Bielig, Józef Olszyński, Grażyna Wojtkowska-Łodej, Oskar Kowalewski and Tomasz Napiórkowski

There are three types of performing national innovation systems (NIS). The first subsystem belonging to this group can be found in France, Germany, UK, Italy, South Korea, and Taiwan. The second subsystem can be observed in two countries: the USA and Japan. The third subsystem consists of: Spain, Austria, Norway, Canada, Australia, and New Zealand (Weresa 2012). In order to show the differences in using innovation and human capital for purposes of boosting competitiveness in international trade and to continue with the analysis of case studies, four examples of various performing NIS subsystems were selected: Germany, Austria, Spain, and the USA.

6.1 Case Study: Germany

Andreas Bielig, Józef Olszyński

Germany was selected as a case study that verifies the research hypotheses concerning the role of human capital and innovativeness in the development of competitive advantages in international trade for numerous reasons.

A. Bielig (✉) · O. Kowalewski (✉) · T. Napiórkowski (✉) · J. Olszyński · G. Wojtkowska-Łodej
World Economy Research Institute, Warsaw School of Economics,
Al. Niepodleglosci 162, 02-554 Warsaw, Poland
e-mail: post@andreasbielig.de

O. Kowalewski
e-mail: okowale@sgh.waw.pl

T. Napiórkowski
e-mail: tnapi002@gmail.com

J. Olszyński
e-mail: jolszy@sgh.waw.pl

G. Wojtkowska-Łodej
Foreign Trade and European Studies Institute, Warsaw School of Economics,
Al. Niepodleglosci 162, 02-554 Warsaw, Poland
e-mail: gwojtko@sgh.waw.pl

M. A. Weresa (ed.), *Innovation, Human Capital and Trade Competitiveness*, Innovation, Technology, and Knowledge Management, DOI 10.1007/978-3-319-02072-3_6,

Germany takes a significant position in the global economics and politics. In Europe, it is undoubtedly the economic and political leader. Many European countries ascribe special importance to German economy as they perceive it as the motor force of development processes. Germans are an important (basic) output market. For instance: Austria directs 22 % of its exports to Germany, Slovakia—17 %, the Netherlands—15 %, France—13 % (OECD Economic Survey 2012, p. 9).

The German economy is worth analyzing owing to the fact that the German government is very active in the area of the domestic and European economic policies and projects aimed at increasing competitiveness in international trade. The activities undertaken to support human capital and innovativeness deserve a closer look, especially since their positive effects are visible, e.g., in the process of revival after the 2008–2009 economic crisis.

6.1.1 The Competitiveness of Germany in International Trade

The German economy is definitely of proexport nature, and most certainly deserves to be referred to as an open economy. This is reflected both by the proportion of imports and exports to GDP (approximately 40 % and 35–45 %, respectively), and notable shares of German imports and exports in international trade: 9–10 % in exports and 7–9 % in imports (Deutschland in Zahlen 2012). In this context, the slogan referring to Germany winning the world championships in exports (Exportweltmeister) was coined. This title was formally won by Germany in the years 1986–1988, 1990, and from 2003 to 2008 (Export Weltmeister Deutschland 2011).

The traditional openness and proexport approach of the economy of the Federal Republic of Germany (FRG) was also visible in the years 1995–2010 (2011). These features of the German economy were best expressed by the condition of its exports and the changes that influenced this area. In the precrisis period, directly after 1995, FRG exports initially (until 2002) displayed small increases: from US$ 524 billion in 1995 to US$ 616 billion in 2002 (increase rate up to 4 % per annum with a slight drop in 1997 and 1999). The German economy picked up speed during the global trade boom in 2004, as Germany began to expand its exports. Undoubtedly, the 2004 European Union (EU) enlargement by 10 new output markets contributed to this rise. German exports started increasing by more than 10 %. The absolute export value reached its peak of US$ 1,466 billion (€ 984 billion, Deutschland in Zahlen 2012, p. 44). This golden age of German exports gave Germany the economic world championship in the years 2003–2008, yielded beneficial development effects, and resulted in the increased dependency of the economy on exports and the risk of further downturn in the phase of negative changes in the global economy (Płóciennik 2010).

Such risk became apparent in 2012, when owing to the global economic slowdown German exports fell by 23.7 % per annum, and their value amounted to US$ 1,128 billion (€ 8.03 billion, Deutschland in Zahlen 2012, p. 44).

German exports, quite unexpectedly, picked up very quickly, and annual increases of 18.5 % in 2010 and 11.4 % in 2011 allowed the surpassing of the exports value achieved in the record year 2008, both in US$ and euros (Olszyński 2012).

On the global market, the robust exports expansion of Germany encountered an even higher increase of competitiveness of the Chinese economy. Since 2009, China has enjoyed the position of largest global exporter, with Germany taking second place.

In service exports, Germany takes a more modest stand. Germany's share in global services amounts to approximately 6 % in exports, and 8–10 % in imports (Deutschland in Zahlen 2012, p. 131).

To a large extent, changes in Germany's imports took place parallel to exports development. In the years 1995–2010 (2011), absolute imports value increased from US$ 464 billion to US$ 1,204 billion in 2008, and US$ 1,067 billion in 2010. In 2011, German exports exceeded its 2008 level by approximately US$ 100 billion and reached the volume of € 902 billion (Deutschland in Zahlen 2012, p. 44). The rate of import changes was slightly, but visibly, faster than changes in exports, and in the recession year (2009), the drop in imports was slightly lower than in exports. This tendency resulted from the dependency of the German economy on imports of raw materials, especially in the area of energy. The increase in their prices in the first decade of the twenty-first century led to an overaverage increase in imports. Moreover, the constant strong expansion led to an increase in imports intensity, which was sometimes criticized as a "bazaar" economy (Sinn 2005).

The relative significance of the exports sector in the German economy in the years 1995–2010 was on the constant, though decreasing, rise. Although in 1995 the share of goods and services exports in GDP amounted to 24 % and was lower than the EU average (29.5 %), in 2008 it reached 47.3 % (in 2011—45 %), which corresponded to 6 % above EU average and 7 % above Polish average (Misala 2011, p. 155).

A characteristic feature of the geographical targets of German exports and imports that originates, in part, from historical events, and in part—from the generic trade structure, is the concentration of trading on European markets, especially—on the EU market. This is reflected by the following figures: approximately 60 % of German exports are destined to EU Member States' markets (40 % are dispatched to Eurozone countries), and 56.3 % of imports has its origin in this area. At the same time, one can note the tendency to decrease the share of the EU in German trade: in exports—from 64 % in 1995 to 59.3 % in 2011; in imports—from 62.7 % in 1995 to 56.3 % in 2011 (Deutschland in Zahlen 2011, p. 45). CEE countries, OPEC states, China, and the USA are becoming more and more important in German foreign trade, at the expense of EU markets.

In the years 1995–2010, the significance of main output and supply markets remained relatively stable for Germany. Since 1995, France, the USA, the Netherlands, UK, and Italy have remained Germany's major exports partners. In the analyzed period, the importance of Germany for the US and GB market fell, while the Chinese market notably gained significance. The share of China rose from 3.8 % in 2008 to 6 % in 2011.

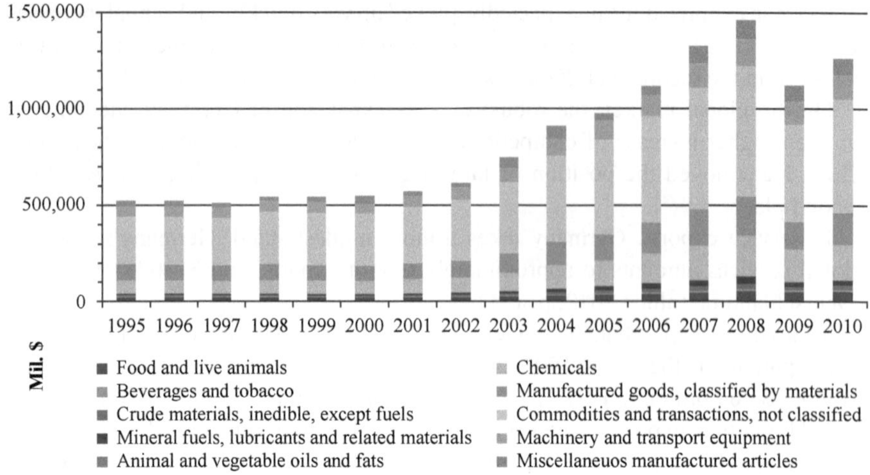

Fig. 6.1 Export structure of Germany 1995–2010 by SITC. (Source: OECD data)

In the area of imports, Germany's top five partners remained unaltered, only the changes in time were more notable and directed at different targets. The role of France (from 10 to 7.4 %) and the USA (from 7.5 to 5.5 %) diminished, the role of the Netherlands and China and CEE countries increased.

The concentration of German foreign trade on Europe catered for positive effects for many decades, but in the most recent period it turned out to hinder rapid switching to emerging markets, especially China and India (Misala 2011, p. 140).

In accordance with the regularities specified in theories on international trade, the goods structure of the German foreign trade and changes therein are typical for highly developed countries. In the most aggregated view, both in exports and in imports, processed industrial products are dominant, and the role of food products is constantly decreasing. And so, the share of food products in German imports fell from 10.3 % in 1995 to 7 % in 2010, and in exports from 5.1 to 5.0 %, respectively. Raw materials and industrial products constituted: 93 % in imports, including: readymade goods—64 %, and in exports—95 %, including readymade goods: 83 % (Deutschland in Zahlen 2012, p. 45).

More deaggregated data, according to the SITC specification, specify the evolution of the trade structure, including also the rate of change within this scope (Figs. 6.1 and 6.2).

In German exports, stimulants and food products had a stable position with a 5 % share. The share in the sales of clothing and textiles fell from almost 6 to a 3% low. In chemicals, Germany's share increased notably from 12.4 to 15.5 %. The share of machinery and electric engineering retained a fairly stable volume in the years 1995–2010 (approximately 33 %). Such changes indicate a structural evolution that is typical for modern developed economies.

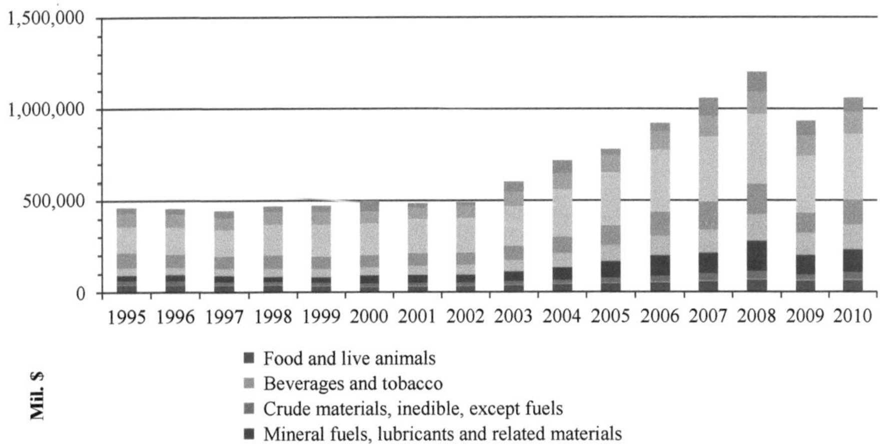

Fig. 6.2 Import structure of Germany 1995–2010 by SITC. (Source: OECD data)

In imports, goods structure was only slightly varied, but the changes were more dynamical (Fig. 6.2). The share of food products and stimulants in trading with Germany fell from 9 % in 1995 to 6.5 % in 2011. The share of energy carriers (oil, coal, gas, electricity) increased notably: from 8.3 % in 1991 to 13.3 % in 2011. The share of electrical engineering products and transport equipment was lower than in the case of exports, but remained stable (approximately 28 %). The annual goods structure of imports experienced greatest fluctuations in the area of fuels and raw materials. A radical example was the more than 50 % fall in the share in raw materials imports during the 2001 recession and the increase by more than 120–150 % during the 2003 and 2008 boom. This phenomenon was the effect of a series of factors: the increase in demand of the German exports sector and the robust increase in prices on global markets.

Changes in the international competitiveness of the German economy are closely connected with the changes in the qualitative structure of foreign trade. The qualitative dimension of German foreign trade, which corresponds to the achieved competitiveness level, becomes visible in the trade structure measured according to goods technological advancement, utilization of expenditure types, and the qualifications of the employed labor force. Revealed comparative advantages (RCA) indexes for individual export (import) areas indicate the level and changes in competitive advantages, and as a result delineate the current and future export specialization.

The generic structure of German foreign trade from the perspective of the technological advancement of goods that are the object of trade is an interesting case. German enterprises mainly export medium-high-technology (MHT) goods. This group of industrial goods (transport vehicles, machinery and equipment, chemicals) is dominant in German exports, and its share amounted to approximately 46 % in 1995, 54 % in 2008 (peak value), and 48 % in 2010 (Fig. 6.3.).

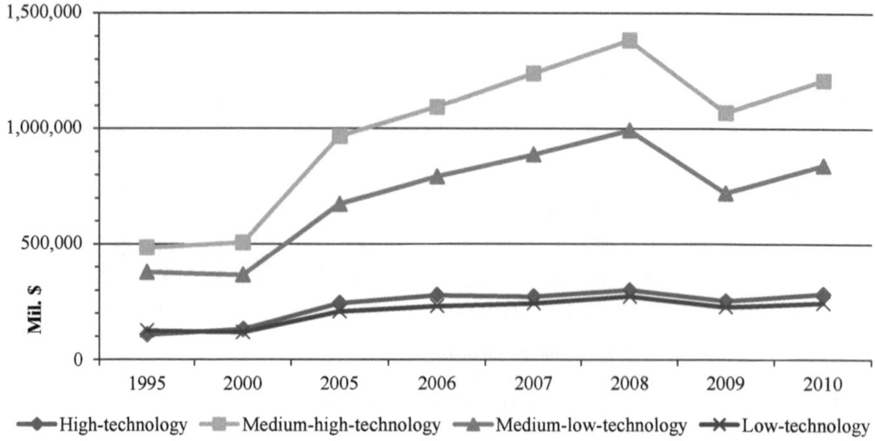

Fig. 6.3 Export structure according to technology level of Germany 1995–2010. (Source: OECD data)

The second largest share in German exports amounting to approximately 30–33 % belongs to medium-low technology (MLT) goods. The last and least significant group encompasses low-technology (LT) goods and holds a 10 % share in Germany's exports.

Currently, high-technology (HT) goods play the role of the key segment of the exports sector in the struggle for a competitive position in global trade. Within this scope, Germany slightly lags behind in comparison with the most developed OECD countries, but is constantly improving its position in the international ranking. As a result, the share of HT exports in total exports increased in the years 1995–2006 from 11.6 to 14.1 %. HT exports reached their all-time high in 2000 when their share in total exports amounted to 16.8 %. The most recent drop is connected with the strengthening of competitive advantages in the MHT sector (Gehrke, Krawczyk 2012). In the area of German imports (Fig. 6.4), these relations were shaped similarly as in the case of exports, owing to the level of technological advancement. However, the share of MLT and LT goods was higher.

The indexes of RCA (Fig. 6.5) confirm the trade specialization of Germany. They are positive and relatively high in the case of MHT goods. In the case of MLTs, they are several times lower and display a persisting downward trend. Germany completely lacks advantages in LT goods—here, RCAs are of negative value and are increasing.

In the HT segment, the RCAs are positive since 2007, but their value is very low and has been falling for the last 2 years.

Depicting the structure of trading according to input factors in the production process allows for enriching the data on the competitiveness of German industry (Figs. 6.6 and 6.7).

Germany, in line with the expectations following from theoretical considerations, mainly exports technology-driven industries (TDI) goods—their share amounts to 35 %. The next most significant exports group is the mainstream manufacturing

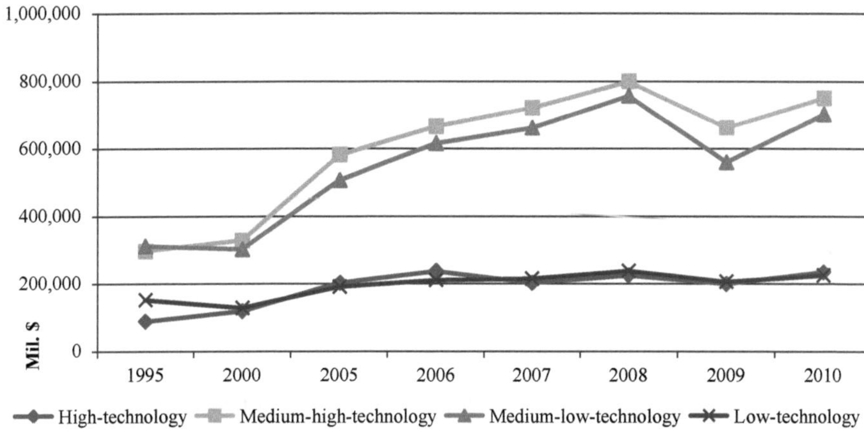

Fig. 6.4 Import structure according to technology level of Germany 1995–2010. (Source: OECD data)

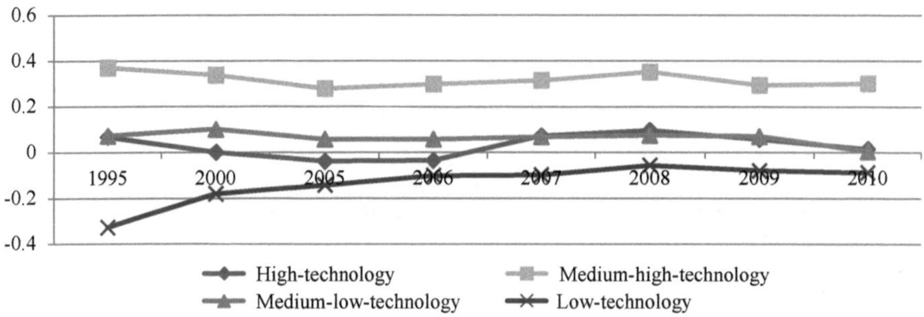

Fig. 6.5 RCA according to technology level of Germany 1995–2010. (Source: OECD data)

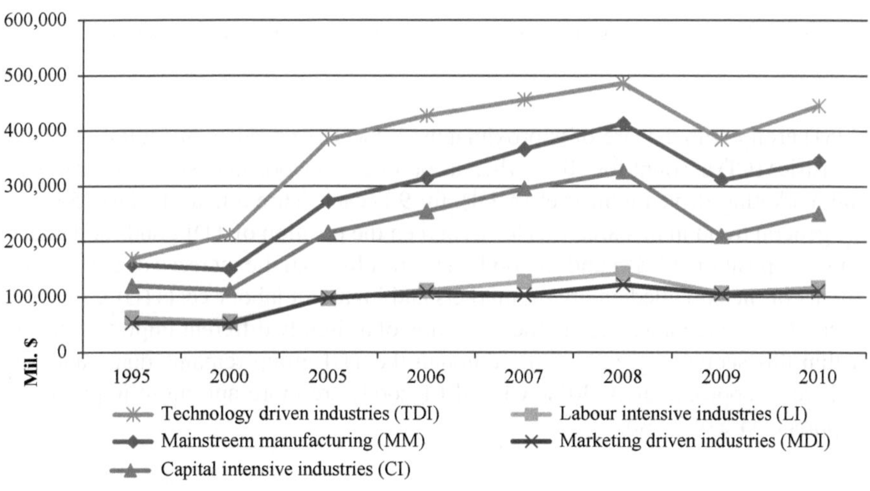

Fig. 6.6 Export structure according to input factors of Germany 1995–2010. (Source: OECD data)

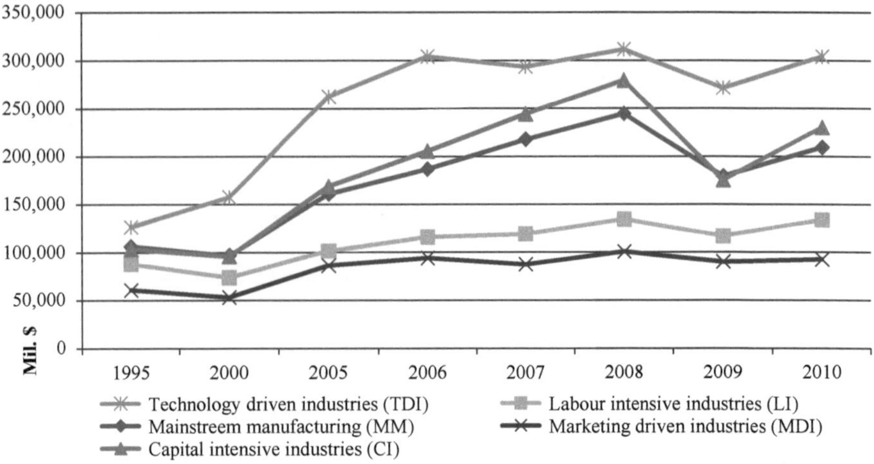

Fig. 6.7 Import structure according to input factor of Germany 1995–2010. (Source: OECD data)

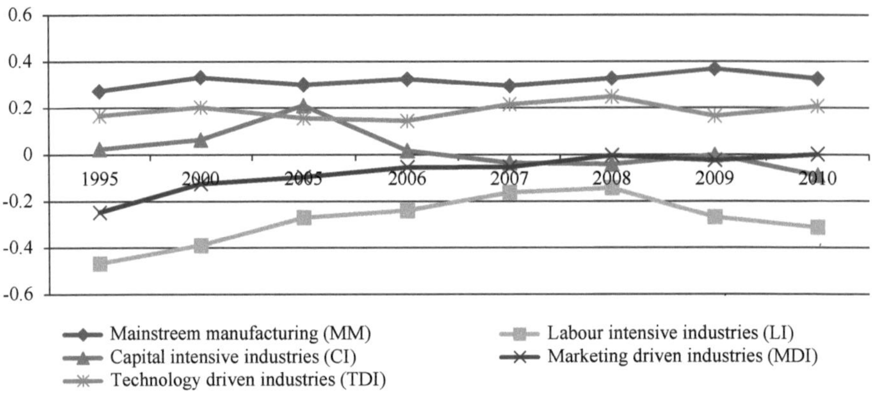

Fig. 6.8 RCA according to input factors of Germany 1995–2010. (Source: OECD data)

(MM) group with a share of approximately 27 %. The products of capital-intensive industries (CI) account for 20 %, whereas products of labor-intensive industries (LI) and marketing-driven industries (MDI), for 9 % each. Throughout the last 15 years, the general trend in this area was to strengthen the position of TDI goods and undermine the position of MM and LI goods. This results from the increase of competitive impact of the so-called emerging markets and high unit labour costs (ULC) in Germany. These phenomena led to the formation of a slightly different imports structure within this scope (Fig. 6.7). Still, although the TDI group remains dominant with a share of approximately 30 %, CI and LI goods are more and more important in imports (share of 22 %).

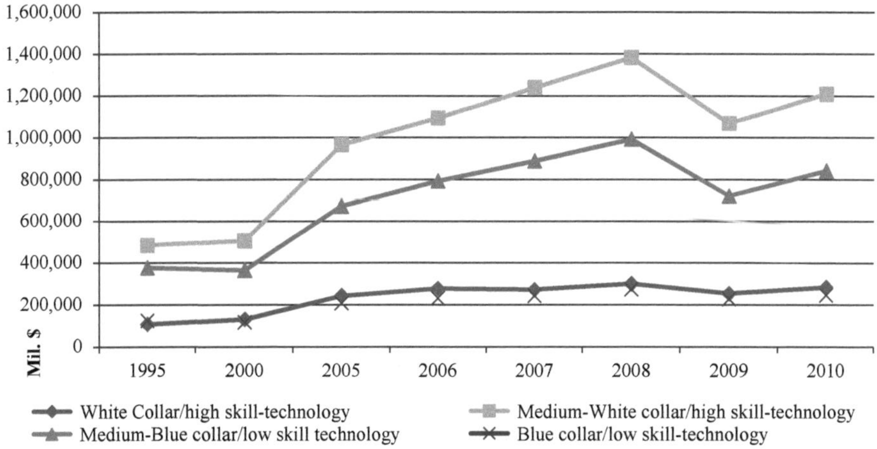

Fig. 6.9 Export structure according to labor input factors of Germany 1995–2010. (Source: OECD data)

RCA indexes (Fig. 6.8) indicate highest values of RCA for German MM products. However, their values have been decreasing in recent years. The second position is occupied by TDI products. This group exhibits an upward tendency. Competitive advantages are lacking in CI, MDI, and LI (negative RCA). RCA indexes for MDI and LI have increased since 1995, but LI dropped in 2009 and 2010 again. These data have been confirmed by the results of analyses performed by the Lower Saxony Institute for Economic Research (Niedersächsisches Institut für Wirtschaftsforschung). It follows from the studies that in the years 2008–2010, Germany maintained its position in global trade in the area of TDI products, whereby drastic drops in exports were noted in the area of MHT products (Gehrke, Krawczyk 2012, p. 2).

The specification of shares in total exports and their fluctuations presented according to labor force qualifications allows better comprehension of the quality of German foreign trade. Within this scope, significant changes in the FRG economy can be noted over time (Figs. 6.9 and 6.10).

The position of industries where production was based on medium-qualified white-collar employees was strengthened both in exports (48 %) and imports (35 %). In imports, industries where production was based on medium-qualified blue-collar employees take a position of similar significance (33 %), whereas in exports—their share is lower and amounts to approximately 30 %.

The share of industries where production is based on highly qualified employees amounts to 11 % in exports and imports. The share in trading of products manufactured by poorly qualified employees is decreasing. RCA indexes within this scope (Fig. 6.11) indicate that the revealed high and permanent comparative advantages occur in industries employing medium-qualified white-collar workers. Small, but positive, advantages also concern industries with medium-qualified blue-collar workers and the sector with highly qualified white-collar employees. However, advantages in the sector that employs poorly qualified workers are not visible (Budnikowski 2011).

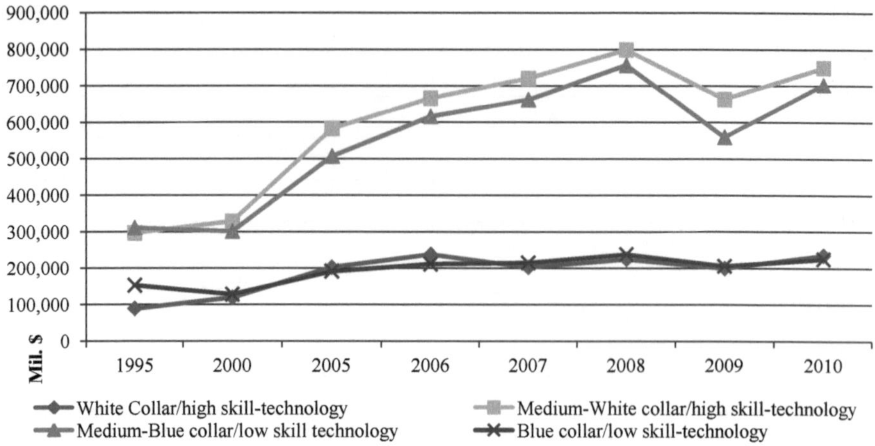

Fig. 6.10 Import structure according to labor input factor of Germany 1995–2010. (Source: OECD data)

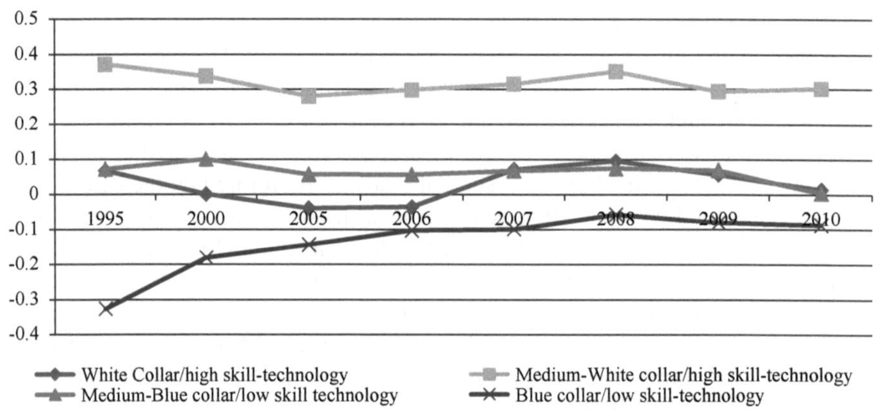

Fig. 6.11 RCA according to labor input factors of Germany 1995–2010. (Source: OECD data)

In German foreign trade, interindustry trade radically lost significance to the benefit of intra-industry trade (IIT). The occurring production factors and relevant competitive advantages influenced the volume, structure, and intensity of German IIT. Germany's IIT intensity indexes have been presented in Figs. 6.12, 6.13, and 6.14. These diagrams have been developed for various foreign trade profiles and provide information on the development of this trade over the last 15 years.

It is quite surprising that most intense IIT was noted in LT industries. More than 95 % of trading in this export segment is of intratrade nature. The case of HT goods is similar, though the intensity of this type of trade is a bit lower (88–91 %). MHT industries have the lowest score. However, although their result is low in comparison to other industries, their IIT intensity index is very high (76.6 %).

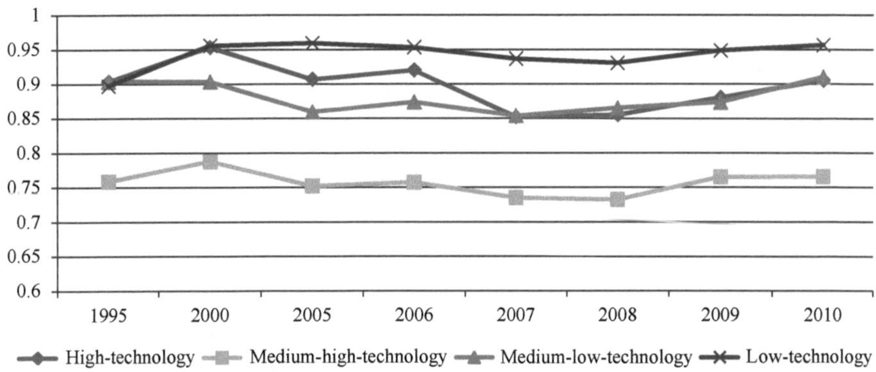

Fig. 6.12 IIT according to technology level of Germany 1995–2010. (Source: OECD data)

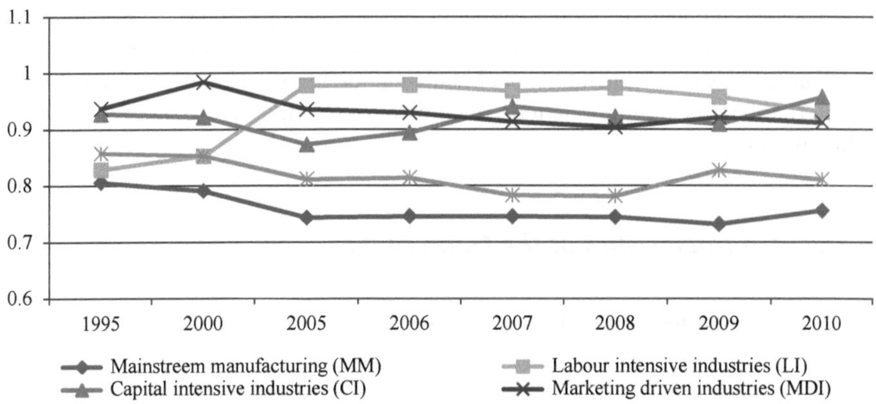

Fig. 6.13 IIT according to input factors of Germany 1995–2010. (Source: OECD data)

According to the WIFO classification that is based on the level of utilization of input factors in the production process, the case of Germany (Fig. 6.13) indicates that the highest intensity (0.957) occurs in CI, and slightly lower intensity is characteristic of LI (0.931). The last position in this ranking is taken by MM industries, with intra-industry intensity of approximately 0.755.

Similarly, as in the case of classification based on technology level, IIT intensity indexes are determined on the basis of labor force qualifications. The highest level of IIT is displayed by industries that employ poorly qualified workers (0.95). The lowest value of the index (0.76) is characteristic for industries that employ medium-qualified white-collar workers (Fig. 6.14).

Irrespective of the varied IIT intensity in particular foreign trade sectors, basically IIT has been gaining significance over the last 15 years (an exception to this rule was identified only once in the 2008–2009 crisis).

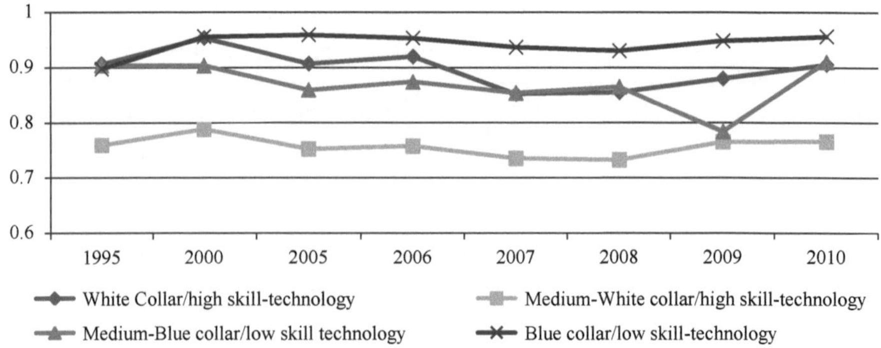

Fig. 6.14 IIT according to labor input factors of Germany 1995–2010. (Source: OECD data)

Germany's economy is assuming an international specialization, and the correlation between the state of global economy and international trade has become notably stronger. This resulted not only theoretically and empirically controversial in the "bazaar economy" phenomenon, but also in intense development of new forms of international economic relations, such as outsourcing and off-shoring. (Siebert 2005, p. 91).

6.1.2 Characteristics of the German NIS

6.1.2.1 Innovation Capacity and Position

Research in Germany is conducted at the governmental and nongovernmental level (BMBF 201, p. 38). The main player of the governmental research sector is the university sector, which includes universities (Universitäten) and universities of applied sciences (Fachhochschulen). Universities engage mainly in basic research with a wide spectrum of topics and research methods, while universities of applied sciences focus on applied research topics. Beside university research activities, there are four big research organizations in Germany:

1. Max-Planck-Gesellschaft (MPG) research institutes engage in basic research in new research fields with main emphasis on biological-medical, physical-chemical-technical, and human science areas.
2. Fraunhofer-Gesellschaft (FhG) is specialized in applied research and provides research facilities for clients from the business sector.
3. Helmholtz-Gemeinschaft Deutscher Forschungszentren (HGF), with its 16 research centers, is the largest scientific organization in Germany and contributes to research in six areas: earth and environment, health, key technologies, structure of materials, transport, and space. In addition, it provides high technology and infrastructure for other research groups.

4. Leibnitz-Gemeinschaft (WGL) conducts demand-oriented interdisciplinary research and provides research-based services for industry and public administration.

There are also scientific academies engaged in research at the level of the federal states. Since July 2008, they have been functioning in the new structure of the National Academy of Sciences in Germany.[1]

Another important factor in research activities and the main player in innovation activities on the nongovernmental level are private enterprises. About two-thirds of the invested R&D expenditures per year are raised from enterprises for internal or joint research projects with business or scientific partners (BMBF 2010, p. 39). In 2010, the overall R&D budget in Germany equalled € 56.2 billion (Stifterverband 2010, p. 1). With regard to its innovation capacity, the German economy showed a diverging development with respect to different indicators. The expenditures on R&D, measured as a share of GDP, were steadily increasing in the years 1995–2009 from 2.19 to 2.78 %. Thus, the relatively strong increase in the last 4 years, 2005–2009, by 0.3 % points (p.p.) is especially remarkable. However, the German economy could follow the international trend of increasing R&D expenditures and increasing R&D intensity only relatively late, from the year 2004, still it could not reach the growth rates of the leading Nordic countries or the USA (Schasse et al. 2011, pp. 82–91).

This development is accompanied by an increase in the share of governmental expenditures on R&D, which increased from 0.33 % in 1995 to 0.41 % in 2009. Despite the fact that the governmental share is on a relatively low level, a trend of expansion of the governmental engagement is observable, which indicates shared responsibility of economic policy for innovation in Germany. The annual average share was 0.36 %, fostered by a stronger increase of 0.06 p.p. in the years 2005–2009.

Since 1995, the share of domestic expenditures on R&D financed by the industry increased strongly from 60 to 67.27 % in 2008. The major growth of industrial investments, by 6 p.p., occurred in the years 1995–2000. From 2005 to 2008, even a slight decrease in industrial expenditure shares is visible. In accordance with the expansion of the industrial share of R&D expenditures, the complementary share financed by governmental institutions showed a declining trend from 1995 to 2008. According to the OECD data, it dropped from 37.88 to 28.4 %, showing especially a sharp decline by 6.5 p.p. in the years 1995–2000. From 2005 to 2008, a stabilization of the governmental share is observable. The average annual governmental R&D investment share in 1995–2008 was 31.51 %, which indicates a continued shifting trend of the R&D weights from governmental influences to an increased industrial responsibility. In addition, other national sources of financing R&D expenditures enlarged their shares in the overall relative R&D development, but on a very low level. In particular, the increase of R&D shares financed by abroad expanded in

[1] Union der Deutschen Akademien der Wissenschaften http://www.akademienunion.de/ and Nationale Akademie der Wissenschaften http://www.leopoldina-halle.de/cms/.

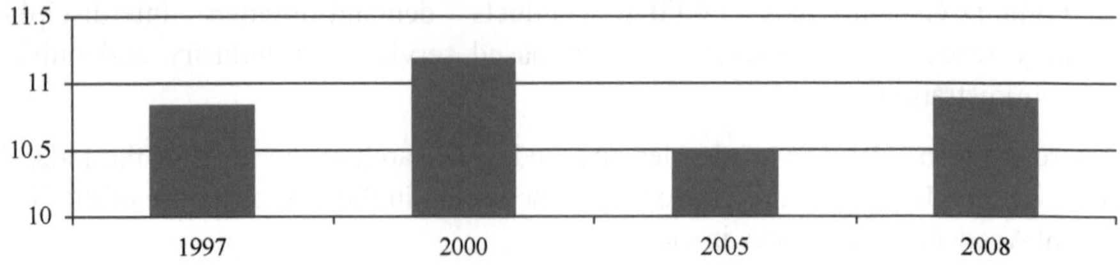

Fig. 6.15 Employment in high- and MHT manufacturing sectors in Germany, share in total employment (%). (Source: Data Eurostat)

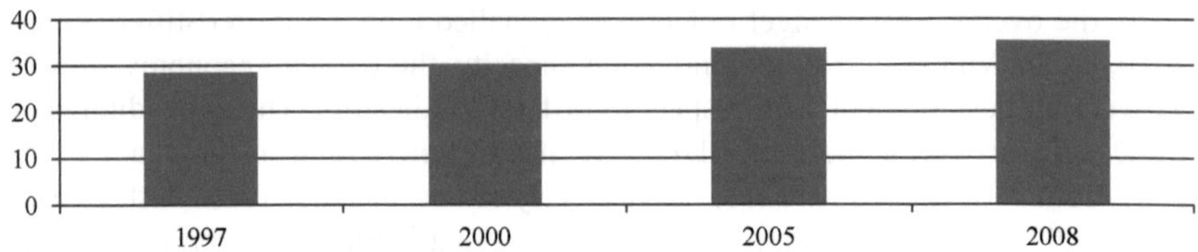

Fig. 6.16 Employment in knowledge-intensive service sectors in Germany as share of total employment (%). (Source: Data Eurostat)

1995–2008 from 1.83 to 4.01 % of GDP, which shows the growing importance of international sources for the innovation processes in Germany.

According to the Eurostat data, in 1995, in the German economy 36.7 % of the entire active population was employed in science and technology. This share increased by 2009 to a remarkable level of 42.4 %. Also, a very impressive development is visible in the share of households with Internet access in the total number of households. It increased from 16.4 % in 2000 to 79.1 % in 2009, which implies a very high level of usage of new media and information technologies.

For evaluating the innovation position of the German economy, we analyze five indictors. The share of employed persons in HT and MHT sectors showed a relatively volatile development from 1997 to 2008 (Fig. 6.15). Until 2000, the share increased from 10.84 to 11.19 % but declined in the following years to reach 10.5 % in 2005. In the following years until 2008, the share increased again to a comparable level of 1997, i.e., 10.89 %.

There was also an increase of employment in knowledge-intensive service sectors. Its share in total employment increased from 28.56 % in 1997 to 35.3 % in 2008 (Fig. 6.16), which means the growing relevance of knowledge-intensive services for innovation activities in Germany. One of three people works in this area, mirroring the ongoing transformation of the German industrial structures. A study by DIW Berlin states that R&D-intensive industries have driven the economic development of the German economy during the economic crisis since 2008 as the most important factor with growth rates above the industrial average (Belitz et al. 2011, p. 1 f.).

The overall structural change in German R&D is relatively slow. Traditional strong industrial sectors, such as chemicals, electronics, and machinery, have lost R&D dynamics in the last decade.

As an output indicator of innovative activities in R&D, we analyze the number of resident patent filings per million US$ of R&D expenditures. This indicator declined from 0.91 in 1997 to 0.74 in 2007 (World Intellectual Property Organization; WIPO 2012). This trend shows the problems connected with increased R&D intensities in the German economy, leading to a smaller overall input–output ratio of patent filings and R&D spending. A different trend can be observed with regard to resident patent filings per million population. The number of patent filings increased from a very high level of 466.71 filings in 1995 to 581.67 filings in 2007. In 2000, the filings reached their maximum with 629.32 filings per million population, since 2005, the filing development seems to have stabilized on a high level.

As a result of innovation activities in the German economy, the share of exports of HT products in the whole exports increased in 1995–2006 from 11.64 to 14.06%. The export share reached its maximum in 2000 with 16.08%, then slightly decreasing until 2006. The relatively low level of the HT export share reflects the structural weaknesses of the German economy with comparative advantages mainly in high-quality products in automotive, machinery, or chemical sectors. But the increasing trend shows that the German industry is coping widely with the challenges in the international competition in innovation. German foreign trade was shifting between 2002 and 2008 slightly to more non-research-intensive products on the export and the import side (Cordes, Gehrke 2011, pp. 1–4). Nevertheless, the German economy could hold its worldwide leading position in the exports of R&D-intensive products, successfully competing with China and the USA.

Seen from a synoptic perspective, the innovation system of the German economy could improve its competitiveness concerning its innovation capacities for all indicators outlined here, with the exception of the share of R&D expenditures related to GDP financed by the government.[2] With regard to innovation position, all indicators show a positive development with the exception of stable development in the employment share in HT and MHT sectors and a slightly negative trend in patent filings per million US$ of R&D expenditures. Therefore, since 1995, the overall innovation development in the German economy could, therefore, in general contribute to enhance the competitiveness of the German industry in the international perspective.

6.1.2.2 Evaluation of Germany's Human Capital and its Utilization

Germany is the largest European economy and the fourth largest economy in the world. Currently, Germany has a population of 82.8 million and ranks second in Europe (after Russia), fourth among OECD states (after the USA, Japan, and Mexico)

[2] Analogous Indications also show the European Union Innovation Scoreboard (2011, p. 33) and the European Competitiveness Report (2011, pp. 7–12).

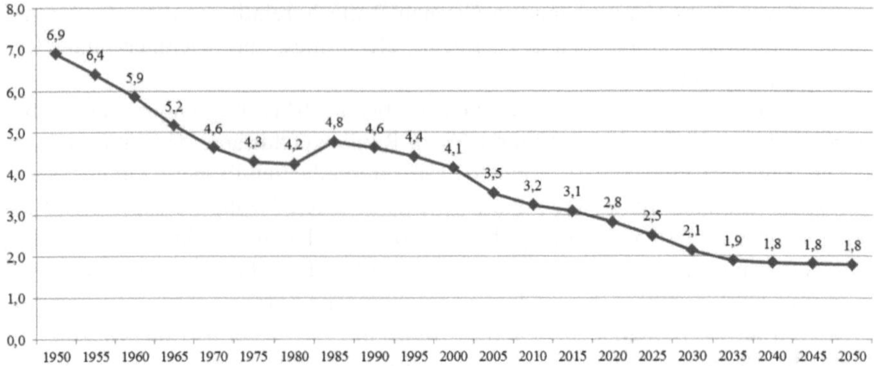

Fig. 6.17 Number of persons in productive age (15-64 years) per pensioner (above 64) / ratio working population to pensioners. (Source: World Bank data, Education Statistics (EDStats) database, file: WB_HC_Extracts.xls (http://databank.worldbank.org/ddp/home.do))

and 14th in the world in terms of human resources potential. The German society is a relatively old society, the average age of its residents is 43.4 (Deutschland in Zahlen 2011, p. 7), whereby the population growth rate, including the migration balance, makes it obvious that the current demographic situation can neither be enhanced nor maintained.

According to current OECD data, the FRG population will reach its peak in 2015. After that, the downward trend will set in (81 million in 2030, 74 million in 2050). As a result of the irreversible aging of the society, the age median will rise to 46.8 in 2050 (a higher median can be identified only in Japan, Slovenia, Italy, and the Czech Republic; Dickmann 2005, p. 28). The population will also become overburdened toward persons in postproductive age. According to OECD data, the demographic dependency ratio is constantly decreasing from 6.9 in 1950 to 3.2 in 2010, 2.1 in 2030, and 1.8 in the years 2040–2050 (Fig. 6.17).

In recent decades, FRG enjoyed a favorable situation in terms of labor force (active participants of the labor market, according to ILO).[3] The major part of this force was constantly increasing, and in 2009 it amounted to 42.3 million persons. According to the forecasts of research institutes specializing in labor market studies, this situation is undergoing significant changes. A significant drop in the number of active participants of the labor market will become visible in 2015: the number of the professionally active will fall to 35–40 million in 2020, to 32–35 million in 2040,

[3] Total labor force encompasses persons of 15 or more years of age who meet the criteria specified in the definition of professionally active persons drafted by the International Labor Organization (ILO). This means that all the persons in the given country that contribute to the manufacturing of products and rendering of services in the given period constitute that country's labor force. It embraces both the employed and the unemployed. States adopt different practices with regard to including such groups as employees of armed forces and seasonal employees or part-time workers in this category. Generally, this category does not cover the performance of household duties and other unpaid services provided by guardians and employees in the nonformal sector.

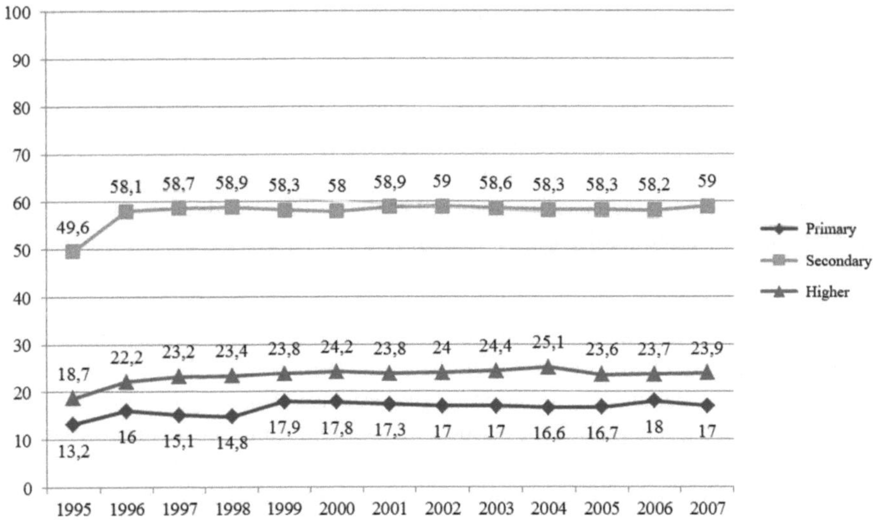

Fig. 6.18 Education of the labor force (%). (Source: World Bank data, Education Statistics (EDStats) database, file: WB_HC_Extracts.xls (http://databank.worldbank.org/ddp/home.do))

and to 25–35 million in 2050. At the same time, the coefficient of the professionally active will drop from approximately 51 to 42.5% (Schäfer and Seyda 2005, p. 118).

Gaps in the amount of human capital can be compensated by increasing its quality. In this regard, the situation of Germany has experienced visible changes. For a few decades, Germany has been implementing the education expansion project (Bildungsexpansion), which has had particular influence on the level of education and qualifications of women (Plünnecke and Seyda 2005). In the last 15 years, the pace of changes in education has slowed down notably, but the general picture is still appealing: almost 60% of the society has secondary education, 24% higher education, and 17% primary education. At the same time, the average number of years of education has been extended from 9.4 in 1995 to 12.2 in 2009. Although this decreases the labor force in the short term, it should increase its quality in the long term (cf.: Fig. 6.18).

Forecasts in this field indicate that in the future the percentage of persons with academic education will become stable, and minor increases will occur only in the group of younger women (Plünnecke and Seyda 2005).

As regards education and qualifications of the human capital in Germany, it is necessary to point out two specific phenomena. First, the so-called dual vocational education system has been the traditional education system in Germany (legally adopted in 1938). This system provides for the integration of education offered by vocational schools and employers. In this system, it is the entrepreneurs and their organizations (Industry and Trade Chambers and Chambers of Crafts) that participate in the process of developing and managing qualifications, as well as adapting them to their own needs and shaping the ethos of the given profession. What counts here is not just knowledge and skills, but also the "calling" (the German word “Beruf” denotes “profession” and is derived from the verb “berufen,” which means “to call”; Miller-Idriss 2002).

Second, in the most recent decade, in the opinion of experts and the OECD, Germany has achieved notable progress in the area of “enhancing the education system within the scope of quality and providing equal access” (OECD 2012, p. 14). The efforts of the federal government and the governments of individual federal states are expected to increase the quality of research, the effectiveness of university courses, and promote studies in exact sciences. This objective follows from the fact that these areas are of key importance in innovation development and maintenance/enhancement of competitiveness of the economy (Gorynia-Pfeffer 2012).

The results of assessments of human capital utilization in Germany spark criticism. The relative evaluation of 26 OECD countries performed by the IdW placed Germany in the penultimate position (above Poland). This negative opinion was the resultant of: a relatively long period of education, short period of professional activity, low number of working hours per annum, and low retirement age (Anger 2007, p. 13). Currently, basic indexes of human capital utilization are more favorable, and reforms of the social insurance system and raising retirement age to 67 should cater for further improvement. The activity (employment) rate is notably higher than the average for OECD and in 2010 amounted to 76.1 % in the case of men (OECD—72.7 %) and 66.1 % in the case of women (OECD—56.7 %; OECD 2012, p. 12). The unemployment rate, which is a direct indicator of waste of human capital, ranged from 7.5 to 11.1 %.

In recent years, as countries all over the globe struggled with the effects of the global economic crisis, Germany experienced a “small employment miracle” (OECD 2012, p. 3). The unemployment rate has been falling since 2010 and currently amounts to 6.5 %—a record low in the last 20 years. At the same time, the percentage of the unemployed with higher education in the total unemployment rate is notably falling, and the value of this index indicates that academic education is profitable, as it provides profit-yielding employment opportunities. The “employment miracle” especially concerns the youth. While the youth (15–24 years of age)[4] unemployment rate amounts to 25 % in France, 30 % in Poland, and 40 % in Spain, in Germany it is only slightly higher than the general unemployment rate. The effective system of vocational education and the state support policy yield a high percentage of professionally active youth (ranging from 41.4 to 48.7 % in the years 1995–2009) (Fig. 6.19).

The relatively short period of employment in Germany has a notably negative influence on human capital utilization. On average, men work 40.1 h a week (OECD average: 41.2 h), and women—30.5 h (OECD: 34.8; OECD 2012). Annually, Germans work for 1,655 h, which makes Germany one of the top shortest working states (in comparison to other OECD countries—Germany ranks fourth after France, Denmark, and Greece; Deutschland in Zahlen 2011, p. 145).

[4] In demography, it is assumed that persons between 15 and 24 years of age belong to the population of youth.

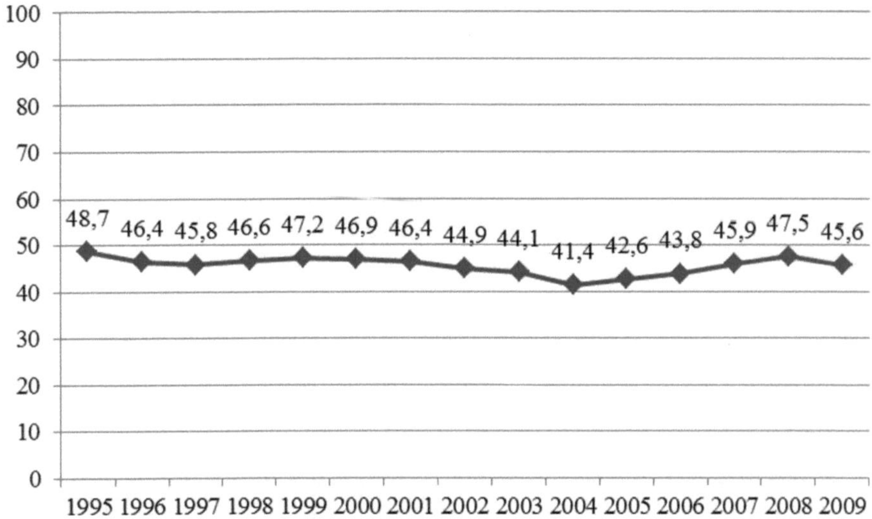

Fig. 6.19 Activity of persons aged 15–24 (proportion of the number of employed in the given age group to the total population belonging to the given age group). (Source: World Bank data, World Development Indicators (WDI) and Global Development Finance database, file: WorldBank2.xls (http://databank.worldbank.org/ddp/home.do))

Germany, as a highly developed country that is prone to stabilization and relatively low GDP growth rates (exception: GDP fluctuations in 2009–2010), is also characterized by a slow pace of changes in labor productivity. In the period from 1995 to 2010, the average annual increase in productivity, measured as GDP per one employed person, amounted to 0.8 % (average for OECD—1.5 %, and measured as GDP per one earned hour—1.3 % (OECD—1.5 %; OECD 2012, p. 17). In the same period, remuneration per hour and per employee, measured in US$ and according to purchasing power parity (PPP), increased at a very steady pace (Fig. 6.20 and 6.21). As a result, the relatively high-quality human capital that accounted for constant GDP increases was available on the market for a reasonable price that was increasing only slightly.

Thanks to the moderate increase in remuneration (approximately 1.5 % per annum; to compare: in France—4.3 %; Fabre 2012) and reforms of the labor market and social insurance, in the years 2000–2008, total labor costs in Germany increased by 17 % (in France—an increase by 56 % in the same period). Figure 6.22[5] indicates an almost constant decrease in labor costs per production unit. This tendency was briefly interrupted by the crisis, which led to a drop in GDP in 2009.

[5] The index presented in Fig. 6.22 is calculated on the basis of this database and corresponds to total labor costs divided by production volume. This index is used to determine total labor costs. It refers to employee remuneration adapted to self-employed persons, so essentially it refers to income from labor. This index is also referred to as the real unit labor cost.

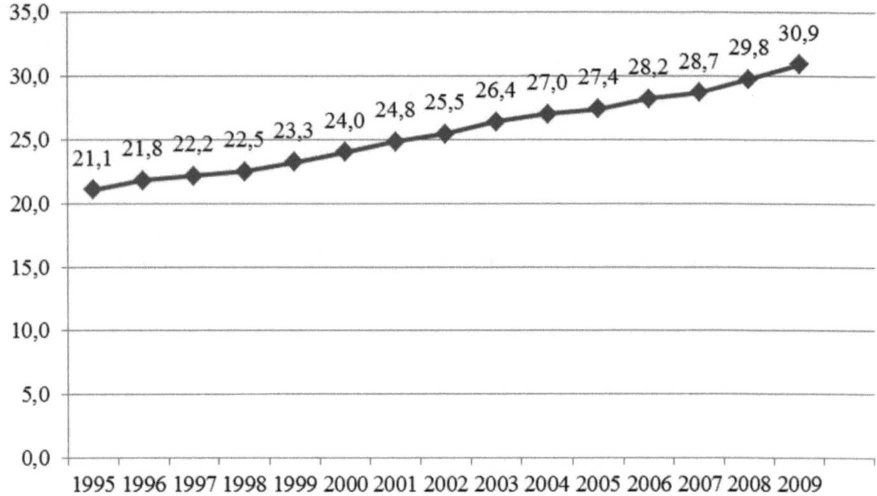

Fig. 6.20 Remuneration per hour (US$ according to PPP). (Source: OECD data (http://stats.oecd.org/OECDStat_Metadata/ShowMetadata.ashx?Dataset=ULC_ANN&ShowOnWeb=true&Lang=en))

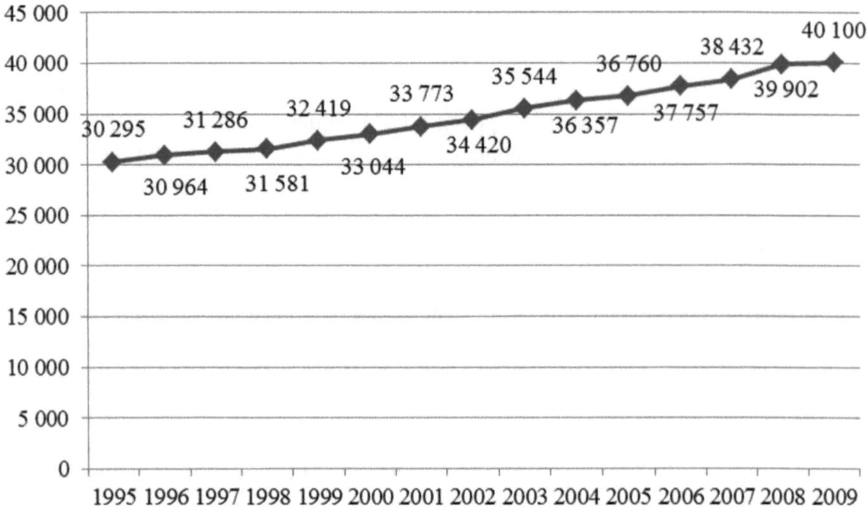

Fig. 6.21 Annual renumeration per employee (US$, according to PPP). (Source: OECD data (http://stats.oecd.org/OECDStat_Metadata/ShowMetadata.ashx?Dataset=ULC_ANN&ShowOnWeb=true&Lang=en))

Germany managed to elicit a drop in ULC, the index that reflects cost competitiveness, in the years 2003–2007, when the economic situation was beneficial. However, after the rapid deterioration of Germany's economic standing in 2009, the downward tendency returned. The data presented in Fig. 6.22 confirm the effectiveness of the reforms implemented in the first decade of the twenty-first century

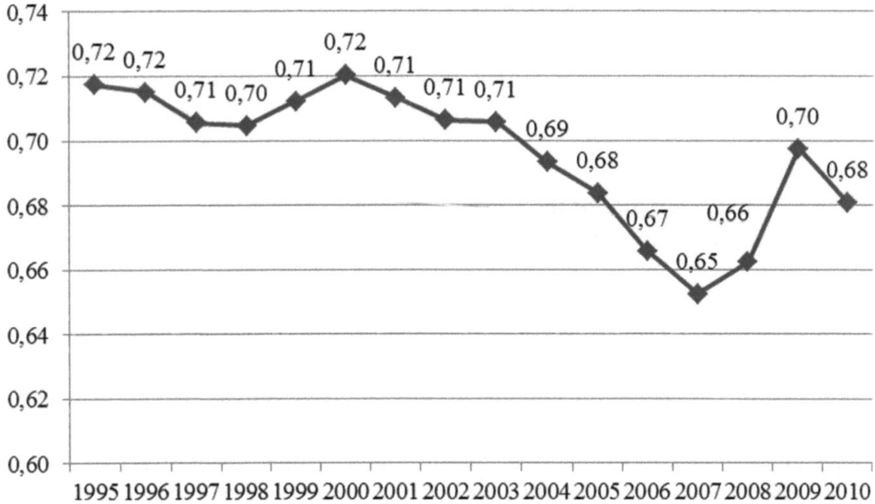

Fig. 6.22 Labor costs per production unit: the OECD index. (Source: OECD data (http://stats.oecd.org/OECDStat_Metadata/ShowMetadata.ashx?Dataset=ULC_ANN&ShowOnWeb=true&Lang=en))

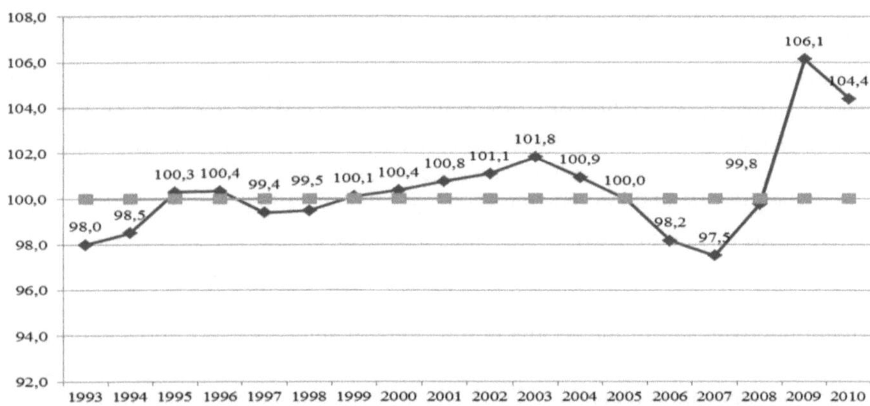

Fig. 6.23 Unit labor costs (average labor cost per product unit; year 2005=100). (Source: OECD data (http://stats.oecd.org/OECDStat_Metadata/ShowMetadata.ashx?Dataset=ULC_ANN&ShowOnWeb=true&Lang=en). Base year: 2005)

and the significance of increasing the competitiveness of German economy in the postcrisis period.

In conclusion, we can state that human capital has been a powerful trump card that helped Germany ensure the competitiveness of its economy. In terms of quantity (human capital as resource), the favorable current situation will notably deteriorate owing to demographic factors. The population and labor force supply will decrease in Germany. Signals indicating a decrease in human capital resources have led to—it seems—effective activities aimed at increasing its quality. New rules of creating and managing competences allow their enhancement and very effective utilization, which is tantamount to an increase in productivity. It is probable that, if combined with innovation system enhancement, this will allow the maintaining of development processes in the German economy in the first half of the twenty-first century.

6.1.2.3 Characteristics of the German Science, Innovation, and Education Policy

The efficiency of the German innovation system is based on the simultaneous improvement of the interrelated policies in the education, science, and innovation sectors. Professional education activities on the secondary level are bound to the structures of the successful traditional German "dual system" model with close interactions of school and enterprise education. This pattern has been adapted by other countries and the EU for their education policy targets for 2020 (BMBF 2012a, p. 2). The structures of the dual system contributed not only to the stabilization on the labor market during the global economic crisis but also to the good international position of the German economy in youth unemployment, professional qualification level, and the access to the labor market after finishing secondary education. In the field of tertiary education, especially the implementation of the Bologna Reform Program at universities for increased international mobility of students is now reaching a high level. Positive effects observed are the increased numbers of students, the implementation of an internationally recognized quality assessment, and the increased level of internationalization of German universities (BMBF 2012b, p. 1). With the public program "Hochschulpakt 2020," the federal and local governments supported the academic sector with additional financial funds for the enlargement of university capacities. In addition, the "Qualitätspakt Lehre" is implemented until 2020 to improve the quality of teaching and curricula development. For high-level research, the successful concept of the "Exzellenzinitiative" at universities was enhanced and continued in 2010 for its third edition (GWK 2010). A remaining challenge is the future improvement of the relatively low level of students in the so-called MINT-Subjects, which are decisive for the further development of technological development and innovation potentials. The political innovation strategy in Germany was refined by the "High-tech Strategie 2020," which is setting new accents and is now focused on five strategic fields of demand for innovation activities (EFI 2011, p. 13).

6.1.3 Synoptic Analysis of Strengths, Weaknesses, Opportunities, and Threats of the German Innovation System

Based on the findings in this case study for the German innovation system, we summarize the key conclusions underlying the innovation activities and institutional structures in the German economy from 1995 to 2010. They are presented as a SWOT analysis.

The German innovation system is based on a strong system of institutional structures in the academic and business sectors, which foster the innovation activities and provide continued input for the industrial development. Threats for the innovation development derive from finance restrictions for the academic sector and insufficient engagement of business R&D. Also, the R&D dynamics of the industrial core sectors in Germany is to slow for providing sufficient knowledge input. The performance in the HT sector, measured by the exports or the employment share, may be regarded as a weakness. But also the education of students in the MINT-subjects is causing future restrictions if there is no progress observed. In general, the unsuitable financing situation of the academic sector, especially for universities, endangers the innovative potentials for the future. The demographic development shows thereby a negative trend, deteriorating strongly the balance of working and senior population in the next decades (Table 6.1).

6.1.4 Interrelationship Between the Level of Innovativeness and the Accumulation of Human Capital with Respect to the Foreign Trade Competitiveness

The analyses conducted earlier proved that Germany belongs to the group of countries, which derive their main advantages in economic competitiveness from close interactions between the unique structure of the NIS, in particular traditionally developed strengths of the system of education and human capital accumulation. The innovation system is characterized by a functional task division between the governmental and the private sector, with strong emphasis on the role of public institutions, such as universities or research centers, for the promotion of innovation processes. However, the private sector with enterprises and privately financed institutions are playing a strong role in the innovation process, which in turn leads to an improvement of the country's position in the global competition. These innovation processes are supported by a strongly diversified national education and human capital accumulation system developed in the recent years. One of the main strengths of the human capital system seems to derive from the broad variety of educational structures, leading to a steady input stream of highly qualified employees for the processes of innovation. Another main factor was increasing efforts for higher efficiency in the utilization of human capital in the economy. Many indicators show that the German economy was able to profit from global trends of increasing labor productivity and the reduction of labor cost shares for products. This led in the recent years to increased competitiveness of the German economy on the global markets regardless of the surrounding higher pressure of competition from other traditional industrial countries but also from emerging economies. The main challenge for the future consists in handling the structural problems resulting from demographical pressure. Furthermore, further improvement of the economic competitiveness in the foreign trade structure is a challenge as many indicators show that the trade pattern development is significantly slower in Germany than in comparable other industrial countries.

Table 6.1 SWOT analysis of the German innovation system. (Source: Own analysis)

Strenghts	Weaknesses
Structure of academic education	Too low efficiency of technology transfer
Structure of excellent nonuniversity research organizations	Small numbers of students in MINT-subjects
System of academies of sciences	Low share of foreign sources in industrial R&D
Networks of universities and business research	Small share of venture capital investments in R&D
Institutionalized technology transfer from universities to businesses	Decreasing share of HT products in German exports
Dual system of professional education (theory and practice)	
Increasing share of employment in R&D in total employment	
Increasing share of employment in knowledge-intensive service sectors in total employment	
Increasing patent numbers per million of population	
Large size of population	
Reducing unemployment rate of highly qualified people	
Shrinking unemployment rate of young persons	
Reducing labor costs per working hour	
Increasing numbers of households with Internet access	
Opportunities	Threats
Comprehensive, but economic cycle-sensitive structures of business R&D	Increasing government expenditures on R&D as percentage of GDP
Time-limited financing of enlargement of universities capacities with Hochschulpakt 2020	Overstrain of university capacities
Time-limited financing of teaching and curricula quality with Qualitätspakt Lehre	Underfunding of universities
Time-limited financing of academic excellence with Exzellenzinitiative	Declining share of industrial R&D financed by governmental institutions
Stagnating share of R&D expenditures as percentage of GDP financed by industry	Volatile but stagnating employment share in HT and MHT manufacturing sectors in Germany of total employment
Increasing but too low business enterprise expenditures on R&D as percentage of GDP	Slow structural change in sectoral R&D
Slow dynamics of structural change in R&D in key industries	

6.2 Case Study: Austria

Grażyna Wojtkowska-Łodej

Austria is a country located in Central Europe and its territory covers 83,858 km^2. It is a country with a relatively small population that amounted to 8.4 million in 2011. In Austria, similarly as in other European countries, one can observe a decreasing birth rate. The forecast for the years 2010–2015 expects this rate to amount to 0.2 %. Low birth rate translates into low increase in population until 2030, which is estimated to reach 8.6 million. At the same time, the life expectancy, which in 2011 equaled 80 years of age, is on the rise.[6] As a result, Austria belongs to a group of countries with an aging society.

Currently, the Austrian economy is one of the wealthiest and most stable EU economies. In the analyzed period of 1995–2010, the pace at which gross GDP was growing—despite the economic crisis—was higher in 2010 than the analogical EU27 index, as well as the index for the Eurozone states (EU-EA 17). The following factors are also indicative of the good standing of the Austrian economy: low inflation amounting to 1.8 %, twice as low as in other countries, unemployment rate of 4.4 %, and relatively lower public debt.

As a result of these processes, the wealth of the Austrian society kept increasing. In the period from 1995 to 2010, GDP per capita increased by three-fourth.

6.2.1 Competitiveness of Austrian Foreign Trade

In the analyzed period from 1995 to 2010, which was convergent with Austria's accession to the EU, imports volume increased more than twofold and reached US$ 150 billion in 2010, in comparison with US$ 66.4 in 1995, whereas exports volume in the same period increased 2.5 times and reached the level of US$ 144.9 billion in comparison with US$ 57.6 billion in 1995 (Table 6.2).

The observed growth rates of the volume of Austrian imports and exports were also higher than the growth rate of external EU imports and exports, and were comparable with the growth rate of international trade. As a result, Austrian foreign trade per capita doubled and reached the level of US$ 18,000 in imports and US$ 17,200 in exports in 2010.

Austria's main trade partners are EU Member States, including Germany, Hungary, Italy, the Netherlands, France, UK, the Czech Republic, Slovakia, and, moreover, Switzerland, the USA, Russia, as well as Poland (Table 6.3). In the analyzed period, trading with China went up. This is reflected by the increase of the share of the Chinese market in total imports by 1.8 p.p. in 2010 as compared with 1995, and in total exports by 1.98 p.p. Trading between Austria and Russia remains at a stable level. In the most recent decade, trading with Russia has been on the rise. This increase concerned exports to a larger extent than imports.

[6] Cf. *Human Development Report 2011—Sustainability and Equity: A Better Future for All.*

Table 6.2 Austrian foreign trade in the years 1995–2010. (Source: Own calculations, OECD database)

Item	Unit	1995	2000	2005	2006	2007	2008	2009	2010
Imports volume	Billions of US$	66.4	68.4	119.9	134.4	156.1	175.0	136.4	150.6
Imports growth rate[a]	%	–	102.9	175.4	112.0	116.1	112.1	77.9	110.4
Exports volume	Billions of US$	57.6	63.7	117.7	134.2	156.6	172.2	131.4	144.9
Exports growth rate[a]	%	–	110.6	184.8	114.0	116.7	110.0	76.3	110.3

[a] Growth rate indexes concern the years: 2000/1995, 2005/2000, 2006/2005, 2007/2006, 2008/2007, 2009/2008, 2010/2009 and 2010/1995

It is worth mentioning that in the period from 1995 to 2010, the share of other countries that have not been mentioned in the table is increasing, and amounts to almost one-fifth of total imports and exports.

When analyzing the generic structure of Austrian trade according to the SITC classification, one generally observes a very similar generic structure of trade, as well as rate of change (these assumed their highest value in the period from 2000 to 2005).

In 2010, farming products, mineral raw materials, and processed products (SITC 0–5 goods group) accounted for approximately 30 % of Austrian imports (Table 6.4). The following raw materials are dominant in Austrian imports: oil, coal, and natural gas. This is the result of relatively scarce natural resources, their depletion, and the imposition of restrictions on their utilization aimed at environmental protection. A significant share in imports is held by machinery, equipment, and transport equipment (SITC-7), industrial goods classified according to raw material (SITC-6), and various industrial goods (SITC-8). In the analyzed period, the growth rate of imports of oil, animal and vegetable oils and fats (SITC-4), nonedible raw materials (SITC-2), chemicals and related products (SITC-5), and imports of various industrial goods was higher than the growth rate of total imports.

Export is dominated mainly by machinery, transport tools, and equipment, as well as industrial goods, products, and chemicals. In the analyzed period, the exports of oil (SITC-3), beverages and tobacco (SITC-1), chemical products (SITC-5), machinery, transport equipment (SITC-7) demonstrated higher growth rate than total exports.

In order to analyze the changes in the structure of Austrian trade, the generic foreign trade structure was studied according to the OECD classification based on the technological level, and then according to the WIFO classification based on the utilization level of input factors in production and changes in the generic structure of trade based on the qualification of the labor force.

In the structure of imports and exports determined according to the former criterion, MHT and MLT goods are dominant, followed by LT industries, and the last position is taken by HT goods (Table 6.5).

Table 6.3 Geographic structure of Austrian imports and exports—main partners in the years 1995–2010 (%). (Source: Own calculations, OECD database)

No.	Main imports partners	Share in imports				No.	Main exports partners	Share in exports			
		1995	2000	2005	2010			1995	2000	2005	2010
1.	Belgium-Luxemburg	2.6	Data not available	Data not available	Data not available	1.	China	0.7	0.7	1.3	2.6
2.	China	1.9	2.6	3.3	3.7	2.	Czech Republic	2.7	2.9	3.1	3.8
3.	Czech Republic	4.9	4.4	4.0	2.8	3.	France	4.2	4.4	4.2	4.2
4.	France	1.9	3.5	2.6	2.8	4.	Germany	37.4	33.2	31.8	31.6
5.	Germany	43.1	40.6	42.2	39.5	5.	Hungary	3.6	5.0	3.4	3.1
6.	Hungary	8.7	7.1	6.6	6.8	6.	Italy	8.6	8.7	8.6	7.8
7.	Italy	3.4	3.0	2.8	2.8	7.	The Netherlands	2.8	2.3	1.8	1.6
8.	Japan	2.5	2.7	2.0	1.6	8.	Poland	1.3	1.6	2.0	2.5
9.	The Netherlands	0.7	1.7	2.3	2.0	9.	Russia	1.4	0.9	1.78	2.3
10.	Russia	0.8	1.0	0.9	1.1	10.	Slovenia	1.6	1.8	1.9	2.0
11.	Slovakia	0.8	1.4	1.8	2.3	11.	Spain	2.0	2.7	2.8	1.8
12.	Switzerland	3.9	3.2	3.5	5.4	12.	Switzerland	5.7	6.8	4.9	5.1
13.	UK	2.9	2.9	1.9	1.5	13.	UK	3.1	4.4	4.1	3.0
14.	USA	4.2	5.5	3.3	2.9	14.	USA	2.8	5.0	5.6	4.5
	Total	82.2	79.5	77.2	75.2		Total	77.8	80.4	77.3	76.0
15.	Other countries	17.8	20.5	22.8	24.8	15.	Other countries	22.2	19.6	22.7	24.0
	Total	100.0	100.0	100.0	100.0		Total	100.0	100.0	100.0	100.0

Table 6.4 Structure of Austrian foreign goods trade in selected years of the period of 1995–2010 according to SITC (%). (Source: Own calculations, database received from OECD)

SITC	Item	Imports								Exports							
		Goods structure of imports				Rate of change in imports				Goods structure of exports				Rate of change in exports			
		1995	2000	2005	2010	2000/ 1995	2005/ 2000	2010/ 2005	2010/ 1995	1995	2000	2005	2010	2000/ 1995	2005/ 2000	2010/ 2005	2010/ 1995
0	Food and live animals	5.1	4.7	5.2	6.0	95.8	195.8	140.7	264.0	3.0	3.5	4.2	5.2	128.7	220.4	151.7	430.0
1	Beverages and tobacco	0.4	0.5	0.6	0.8	128.8	201.9	165.5	0.4	0.7	1.1	1.9	1.5	180.0	308.8	97.7	544.0
2	Crude materials, inedible, except fuels	4.6	3.3	3.7	5.0	74.4	194.4	167.8	242.6	4.1	3.0	2.6	2.8	79.9	158.5	133.6	169.1
3	Mineral fuels, lubricants, and related materials	4.5	5.5	12.3	11.0	125.1	395.6	110.5	232.7	1.0	1.3	4.6	3.2	136.0	677.4	84.8	781.1
4	Animal and vegetable oils, fats and waxes	0.2	0.2	0.2	0.4	85.6	188.6	270.4	436.6	0.1	0.1	0.1	0.2	97.5	188.6	255.3	469.6
5	Chemicals and related products	10.6	9.9	10.5	12.5	97.0	185.4	146.8	264.2	9.0	6.7	8.5	11.4	80.9	237.2	164.8	316.3
6	Manufactured goods, classified chiefly by material	19.6	16.9	15.7	16.3	88.9	163.1	128.5	186.2	29.0	21.4	21.0	22.0	81.1	182.3	127.8	188.9
7	Machinery and transport equipment	36.5	40.7	36.3	32.8	114.9	156.6	111.9	201.5	39.2	41.1	41.3	38.4	115.2	186.8	113.7	245.6
8	Miscellaneous manufactured articles	17.8	16.2	14.7	14.9	93.9	158.5	126.4	188.1	13.3	12.7	12.0	11.5	105.4	175.4	117.1	216.4
9	Commodities and transactions not classified elsewhere in the SITC	0.8	2.2	0.8	0.5	281.1	66.2	70.8	131.9	0.6	9.1	3.9	3.8	1425.0[a]	80.7	117.9	1375.0[b]
	Total	100.0	100.0	100.0	100.0	102.9	175.4	123.9	224.1	100.0	100.0	100.0	100.0	110.6	184.8	122.0	249.2

[a] This is indicative of an increase from US$ 0.36 billion in 1995 to US$ 5.74 billion in 2010

[b] This is indicative of an increase from US$ 0.36 billion in 1995 to US$ 4.4 billion in 2010

Table 6.5 Goods structure of Austrian imports and exports according to OECD classification based on the technological level (1995–2010, %). (Source: Own calculations, OECD database)

No.	Item	ISIC Rev. 2	Imports				Exports			
			1995	2000	2005	2010	1995	2000	2005	2010
			Share within group	Share within group	Share within group	Share within group	Share within group	Share within group	Share within group	Share within group
	HT industry		*100.0*	*100.0*	*100.0*	*100.0*	*100.0*	*100.0*	*100.0*	*100.0*
1	Airplanes and spaceships	3845	5.3	10.1	2.9	2.5	1.7	8.6	2.1	2.9
2	Computers and office equipment	3825	27.0	21.4	23.2	20.6	20.1	15.2	15.7	14.1
3	Electronic and telecommunications equipment	3832	50.5	54.1	56.1	50.3	60.0	60.7	60.1	50.4
4	Pharmaceuticals	3522	17.2	14.5	17.7	26.5	18.1	15.5	22.1	32.6
	Share of group in total imports/exports		*8.8*	*11.1*	*9.7*	*8.2*	*7.0*	*9.7*	*8.0*	*8.6*
	MHT industries		*100.0*	*100.0*	*100.0*	*100.0*	*100.0*	*100.0*	*100.0*	*100.0*
5	Fine goods	385	18.3	18.9	18.9	19.5	18.0	16.9	17.4	18.6
6	Electric machinery and apparatus	3843	19.8	19.9	21.1	18.2	18.1	20.4	20.6	16.2
7	Electronic equipment	383–3832	11.3	13.0	11.7	11.1	13.4	16.2	13.7	13.2
8	Chemicals	351+351+3522	24.2	23.3	23.6	25.9	20.3	19.5	21.1	23.9
9	Other transport vehicles and equipment	3842+3844+3849	1.6	1.7	1.8	2.2	1.9	2.0	2.2	2.8

Table 6.5 (continued)

No.	Item	ISIC Rev. 2	Imports				Exports			
			1995	2000	2005	2010	1995	2000	2005	2010
			Share within group	Share within group	Share within group	Share within group	Share within group	Share within group	Share within group	Share within group
10	Machinery, excluding electric	382–3825	24.7	23.1	23.0	23.2	28.2	24.9	25.0	25.3
	Share of group in total imports/exports		*37.7*	*38.2*	*38.6*	*38.1*	*36.5*	*37.9*	*40.4*	*41.0*
	MLT industries		*100.0*	*100.0*	*100.0*	*100.0*	*100.0*	*100.0*	*100.0*	*100.0*
11	Rubber and plastic products	355+356	18.7	21.0	20.8	17.2	16.8	19.1	17.8	15.0
12	Shipbuilding	3841	1.2	1.4	1.3	1.2	2.2	2.6	2.5	2.4
13	Other industrial products	39	16.1	14.7	13.0	13.1	14.8	13.2	12.2	11.9
14	Nonferrous metals	372	3.8	3.4	3.5	5.8	3.3	3.9	3.1	5.9
15	Nonmetal mineral products	36	8.9	8.4	8.0	9.6	8.8	8.6	8.1	9.1
16	Metal products	381	30.7	29.9	19.3	30.0	34.0	33.6	34.8	36.0
17	Refinery products	351+354	16.5	17.6	19.5	18.3	12.2	12.2	13.1	11.8
18	Iron products	371	4.1	3.7	4.7	4.7	8.0	6.8	8.5	8.8
	Share of group in total imports/exports		*36.7*	*35.4*	*37.1*	*39.4*	*38.8*	*35.7*	*35.2*	*36.8*
	LT industries		*100.0*	*100.0*	*100.0*	*100.0*	*100.0*	*100.0*	*100.0*	*100.0*
19	Paper products	34	19.2	18.3	17.2	15.6	27.3	25.3	22.7	19.7

Table 6.5 (continued)

No.	Item	ISIC Rev. 2	Imports				Exports			
			1995	2000	2005	2010	1995	2000	2005	2010
			Share within group	Share within group	Share within group	Share within group	Share within group	Share within group	Share within group	Share within group
20	Textiles and clothing	32	45.0	42.6	41.7	43.2	40.5	34.0	32.7	35.5
21	Farming and food products, stimulants, and tobacco	31	18.2	21.6	25.0	24.6	13.3	21.0	25.2	24.8
22	Wood and related products	33	17.6	17.5	16.1	16.6	18.9	19.7	19.4	20.0
	Share of group in total imports/exports		*16.8*	*15.3*	*14.6*	*14.3*	*17.7*	*16.7*	*15.4*	*13.6*

The share of MLT goods imports in total imports amounted to 37.7 % in 1995 and increased to 38 % in 2010, while their share in total exports of these goods equaled 36.5 % and 41 %, respectively. MLT goods imports are dominated by machinery, excluding electrical machinery, chemicals, precise goods, electrical machinery, and apparatus. The highest share in the MLT group is held by metal goods, both in imports (approximately 30 % in the analyzed period) and exports (approximately 35 %).

The most important products manufactured in Austria in the analyzed period were machinery, transport vehicles, electrical, electric engineering, and metal goods. The following products can be listed here: engines, trucks and off-road vehicles, rolling stock, machine tools, optic equipment and scientific research apparatus, products of oil processing, nitrogen fertilizers, and soda.

The group of products manufactured by LT industries, accounting for approximately 15 % of imports and exports, is dominated by textiles and clothing.

The group encompassing HT goods constitutes approximately 8 % of total imports and exports in the analyzed period. This group is dominated by electronic products, radio and television equipment, pharmaceuticals, computers, and office equipment.

The goods structure of Austrian imports and exports presented according to the OECD classification, clearly dominated by MHT and MLT goods, which account for more than two- third of total imports and exports, is very similar to the structure of EU trading. The difference definitely concerns a larger share in total EU imports and exports of HT goods, which in 2010 amounted to 14.3 % in imports, in comparison with 8.2 % in Austria, and in exports: 12.7 % in the EU and 8.6 % in Austria.[7]

However, in the analyzed period one can note certain changes in the imports and exports structure. It is worth emphasizing that in imports, and to a larger extent in exports, the share of MHT goods (including chemicals) is on the rise. Moreover, a minor increase of HT goods has been noted, especially pharmaceuticals by 14.5 p.p., in the period from 1995 to 2010. Furthermore, exports of electronic and telecommunications products, computers, and office equipment fell by approximately 6 p.p. These changes were accompanied by a drop in trading in LT products.

Another classification that is interesting from the point of view of analyzing Austrian foreign trade structure and its fluctuations is the WIFO classification. This classification takes into account the level of utilization of material and nonmaterial input factors in the production process (Table 6.6). The analysis of the data allows the drawing of a conclusion that the structure of Austrian imports and exports is dominated by MM, CI, and TDI goods. Imported and exported LI and MDI goods hold a smaller share here.

The changes occurring in the trade structure are reflected by an almost threefold increase in Austrian TDI goods imports, 2.5-fold increase in MI imports, and twofold increase in imports in other groups. Similarly, as in imports, in exports in the analyzed period, MM, CI and TDI goods are dominant. The rate of change in this structure of exports is slightly lower than in the case of exports. In the period from

[7] Cf. Own calculation performed on the basis of the database.

Table 6.6 Changes in the structure of Austrian imports and exports in selected years of the period of 1995–2010 according to the WIFO classification based on input factor utilization level (%). (Source: Own calculations, OECD database)

Item	Imports					Exports				
	1995	2000	2005	2010	2010/1995	1995	2000	2005	2010	2010/1995
MM industries										
Index reflecting share in total imports/exports	33.8	28.4	28.9	30.9	–	26.6	24.6	24.4	26.3	–
Rate of change[a]	–	89.4	190.6	125.7	234.1	–	93.8	166.1	131.6	205.1
LI industries										
Index reflecting share in total imports/exports	15.1	13.6	12.9	13.3	–	17.8	16.3	14.8	14.7	–
Rate of change[a]	–	96.0	176.3	134.0	205.9	–	82.4	152.9	121.1	171.0
CI industries										
Index reflecting share in total imports/exports										
Rate of change[a]	22.8	19.3	19.9	22.7	–	18.9	19.4	22.9	24.2	–
TDI industries	–	153.7	181.0	101.2	232.2	–	98.6	198.6	129.0	252.7
Index reflecting share in total imports/exports										
Rate of change[a]										
MDI industries	18.4	26.6	25.7	22.1	–	24.3	28.8	25.9	23.3	–
Index reflecting share in total imports/exports	–	–	–	–	281.β	–	120.2	150.9	109.6	198.9
Rate of change[a]										
	9.9	12.1	12.6	11.0	–	12.4	10.9	12.0	11.5	–
	–	130.3	194.2	102.4	259.1	–	97.6	181.7	117.4	208.3

[a] Rates of change calculated for the period: 2000/1995, 2005/2000, 2010/2005, and 2010/1995

Table 6.7 Trade according to WIFO classification based on material and nonmaterial input in the production process in Austrian and EU imports and exports (%). (Source: Own compilation based on database)

Item	Imports				Exports			
	2000		2010		2000		2010	
	Austria	EU	Austria	EU	Austria	EU	Austria	EU
MM	28.4	17.3	30.9	19.1	24.6	22.9	26.3	26.0
LI	13.6	16.7	13.3	18.2	16.3	12.1	14.7	10.2
CI	19.3	15.7	22.7	20.3	19.4	17.10	24.2	21.5
TDI	26.6	38.1	22.1	32.5	28.8	36.5	23.3	33.1
MDI	12.1	12.2	11.0	9.9	10.9	11.4	11.5	9.1
Total	100.0	100.0	100.0	100.0	100.0	100.0	100.0	100.0

1995 to 2010, the share of CI goods exports in total exports increased. The share of MM and MI goods exports did not change relatively, whereas the share of goods manufactured by other industries fell. One can observe an increase in the share of CI goods by approximately 5 p.p., as well as a decrease in the share of LI goods (by 1.8 p.p. in imports and by 3.1 p.p. in exports in 2010 in comparison with 1995).

These changes in the structure of Austrian trade are to a certain extent compliant with the trends observed in the EU (Table 6.8). However, the rate of change was different. Although in the period of 2000–2010, imports and exports increased twofold both in Austrian and in EU Member States, the growth rate of Austrian imports was higher than the growth rate of EU27 imports (Table 6.7).

The structure of Austrian trade can be approached from a different perspective, mainly according to the WIFO classification based on labor force qualifications (Table 6.8).

The classification indicates that Austrian trade is dominated by products manufactured by medium-qualified white-collar and blue-collar workers. In 2010, in total they constituted 77.5 % of imports (in comparison with 74.4 % in 1995) and 77.8 % of exports (in comparison with 75.3 % in 1995). Although this trend is similar to the general EU tendency, the share of products manufactured by industries employing highly qualified workers is higher in the EU than in Austria (by 3 p.p. in imports and by 5 p.p. in exports in 2000, in comparison with an increase by 6 p.p. in 2010, respectively). However, it is worth noting the rapidly changing growth rate of exports of goods manufactured in Austria by means of highly qualified employees. In the period from 1995 to 2010, the value of these exports rose threefold.

The results of the analysis of Austrian trade and the changes in its structure evoke the question how this situation influences and determines comparative advantages in foreign trade, especially in reference to industries that intensely employ modern production factors. This analysis was performed similarly as the previous analyses concerning industry groups according to three types of industry processing sectors classifications. Table 6.9 presents the indexes of revealed comparative advantages (RCA) in Austria in the period from 2000 to 2010 according to the OECD classification based on technological level.

Table 6.8 Changes in the structure of Austrian foreign trade in selected years in the period from 1995 to 2010 according to WIFO classification based on labor force qualification. (Source: Own calculations, OECD database)

Industries employing	Imports					Exports				
	1995	2000	2005	2010	2010/1995	1995	2000	2005	2010	2010/1995
Highly qualified employees										
Index reflecting share in total imports/exports	8.7	11.1	9.8	8.2	n/a	7.0	9.6	9.2	6.6	–
Rate of change[a]	n/a	130.5	147.8	103.0	197.5	–	148.2	174.0	117.6	305.9
Medium-qualified white-collar workers										
Index reflecting share in total imports/exports										
Rate of change[a]	37.7	38.3	38.6	38.1	–	36.5	37.9	40.4	41.0	–
Medium-qualified blue-collar workers	–	104.6	168.8	121.1	213.8	–	106.3	199.0	125.4	281.2
Index reflecting share in total imports/exports										
Rate of change[a]										
Poorly qualified employees	36.7	35.5	37.0	39.4	–	38.8	35.7	35.2	36.8	–
Index reflecting share in total imports/exports	–	99.8	174.5	130.4	227.1	–	99.6	183.6	128.9	236.6
Rate of change[a]										
	16.9	15.5	14.6	14.3	–	17.7	16.8	15.2	15.6	–
	–	94.4	159.2	120.5	180.9	–	102.3	172.1	109.3	192.5

[a] Indexes reflecting the rates of change calculated for the period: 2000/1995, 2005/2000, 2010/2005, 2010/1995

Table 6.9 RCA indexes in Austria according to the OECD classification based on the technological level in the years 2000–2010. (Source: OECD database)

Item	2000	2005	2006	2007	2008	2009	2010
HT industries	−0.1539	−0.0322	−0.015	−0.0018	0.0261	0.0847	0.1231
MHT industries	−0.0194	0.0940	0.1186	0.1108	0.1375	0.1495	0.1474
MLT industries	−0.0038	−0.0012	−0.0059	−0.0122	0.0077	−0.0081	0.0063
LT industries	0.0748	0.1015	0.0917	0.0742	0.0603	0.0216	0.0230

Table 6.10 RCA index in Austrian foreign trade according to WIFO classification based on the utilization of expenditure on the production process in the years 2000–2010. (Source: OECD database)

Item	2000	2005	2006	2007	2008	2009	2010
MM industries	0.1681	0.2546	0.2591	0.2479	0.2978	0.2713	0.2547
LI industries	−0.1529	−0.0590	−0.0185	−0.0089	0.0041	−0.0353	−0.0380
CI industries	−0.0241	−0.0582	−0.0912	−0.0517	−0.0113	0.0082	−0.0004
TDI industries	−0.0511	0.0739	0.0689	0.0357	−0.0038	−0.0491	0.0170
MDI industries	0.1086	0.1337	0.1188	0.0371	0.0457	0.0107	0.0178

It follows from the index analysis that, in 2000, the values of all of them, except for LTIs, were lower than zero, which indicates lack of competitive advantages. However, after 10 years, the economic situation and foreign trade standing in Austria experienced notable changes, which are reflected by the positive value of these indexes. Austria developed a relative advantage in TDI goods. The goods manufactured in these industries also require relatively high volumes of human capital.[8] These industries include, among others, machinery construction, transport vehicles, and mechanic equipment.

Moreover, Austria is also gaining advantages in foreign trade in RI industries, which is reflected by the indexes. However, it incurs losses in CI (Table 6.10).

Indexes of Austrian comparative advantages, both those referring to TDI goods and goods requiring relatively high expenditure on human capital and research, indicate the increase in competitive advantages of these industries, contrary to traditional branches and sectors. However, it ought to be emphasized that the gap between Austria's economy and major players in the global economy within the scope of technological and TDI goods is significant.

In conclusion, one can state that it follows from the analysis of comparative advantage indexes that at the beginning of the twenty-first century, Austria is a country that participates in international trade and develops comparative advantages in HT and TDI.

When analyzing the volume of IIT in Austria and the IIT indexes that reflect the intensity of intra-industry trade (IIT) in individual sectors, it is also possible to

[8] Cf.: J. Misala, *Zarys teoretycznych podstaw rozwoju długookresowych przewag konkurencyjnych w handlu międzynarodowym ze szczególnym uwzględnieniem ich istoty, czynników determinujących i metod analizy* (*Outline of theoretical basis for development of long-term competitive advantages in international trade, with particular attention paid to their essence, determinants, and analysis methods*), pp. 86–87.

Table 6.11 Changes in IIT in Austria from 2000 to 2010 in goods trade according to the OECD classification based on technological progress. (Source: Own calculations, OECD database)

Item	2000	2005	2006	2007	2008	2009	2010
HI industries	0.8879	0.9745	0.9986	0.9992	0.9950	0.9764	0.9578
MHT industries	0.9547	0.9624	0.9415	0.9430	0.9394	0.9441	0.9457
MLT industries	0.9625	0.9900	0.9963	0.9956	0.9958	0.9772	0.9838
LT industries	0.9982	0.9586	0.9549	0.9612	0.9779	0.9920	0.9922

Table 6.12 Changes in Austrian IIT from 2000 to 2010 in goods according to the WIFO classification based on the utilization level of material and nonmaterial expenditure on the production process. (Source: Own calculations, OECD database)

Item	Years						
	2000	2005	2006	2007	2008	2009	2010
MM industries	0.9516	0.8826	0.8719	0.8750	0.8601	0.8837	0.8924
LI industries	0.8884	0.9611	0.9901	0.9972	0.9940	0.9636	0.9617
CI industries	0.9764	0.9615	0.9538	0.9758	0.9863	0.9853	0.9805
TDI industries	0.9813	0.9426	0.9414	0.9798	0.9852	0.9866	0.9896
MDI industries	0.9389	0.9724	0.9663	0.9804	0.9901	0.9567	0.9892

observe changes. The results of the IIT index analysis lead to the conclusion that at the beginning of the twenty-first century, Austria belongs to the group of countries with high shares of IIT both in HT and RI industries, as well as industries based on high- and medium-qualified employees (Table 6.11).

In regard to changes in Austrian intra-industry trading in products based on technological progress, in 2000 the highest indexes in this area were achieved by LT products. The LT product index still remains at the high level of 0.9922. This means that the international trade of the given sector is constituted almost in its entirety by Austrian IIT. It is worth emphasizing that in the decade from 2000 to 2010, it was most apparent that the intra-industry cooperation within the scope of HT products increased notably (increase in IIT index from 0.8879 in 2000 to 0.9578 in 2010). These values are comparable with EU27 indexes.

However, taking into account the classification of trading based on the level of utilization of input factors in the production process (Table 6.13), one can observe the change in the level of IIT in industries based on marketing research. IIT is also on the rise in LI and, to a lesser extent, in CI (Table 6.12).

The values of the discussed IIT indexes observed in Austria are comparable with their average values for the EU27. However, in Austria, in the analyzed decade, one could observe notable changes in HT industries based on research and industries employing highly and medium-qualified workers.

6.2.2 *Description of the Austrian NIS*

The Austrian R&D and innovation policy *Forschungs-Technologie und Innovationspolitik* (FTI) is formed by entities operating at the national level and at the level

Table 6.13 Austria's innovation indexes in selected years of the period from 1995 to 2000. (Source: OECD and Eurostat (index 7) data, Eurostat (tsiir080) index 8, 11, and 12). (tsiir160) index 15); WIPO (index 13 and 14))

No.	Item	Unit	1995	2000	2005	2010
1	2	3	4	5	6	7
1.	Share of gross domestic expenditure on R&D (GERD) in GDP	%				
	Austria		1.55	1.94	2.48	2.75
	EU27		1.66	1.74	1.73	1.91
2.	Share of government expenditure on R&D (GOVERD) in GDP	%				
	Austria		n/a	n/a	0.13	n/a
	EU27		n/a	0.25	0.25	n/a
3.	Percentage of domestic expenditure on R&D financed by the government	%				
	Austria		46.91	38.00	35.91	41.20
	EU27		39.55	35.40	35.10	n/a
4.	Percentage of domestic expenditure on R&D financed by other national entities	%				
	Austria		0.40	0.33	0.43	0.43
	EU27		1.82	0.65	8.22	n/a
5.	Percentage of domestic expenditure on R&D financed by foreign entities	%				
	Austria		7.04	19.86	18.04	15.05
	EU27		6.63	7.15	8.90	n/a
6.	Households with Internet access (percentage of all households)	%				
	Austria		n/a	19.00	46.7	n/a
	EU27		n/a	n/a	48.4	n/a
7.	Human resources for science and technology as percentage of professionally active population	%				
	Austria		23.3	29.1	35.7	36.5
	EU27		n/a	31.7	35.6	38.1
8.	Investments in *venture capital* as % of GDP	%				
	Austria		n/a	0.029	0.012	0.07
	EU27		n/a	n/a	n/a	n/a
9.	Share of business expenditure on R&D (BERD) in GDP	%				
	Austria		n/a	n/a	1.73	n/a
	EU27		1.02	1.10	1.08	n/a
10.	Percentage of domestic expenditure on R&D financed by industry	%				
	Austria		45.66	41.81	45.62	43.32
	EU27		52.00	55.29	41.51	n/a
11.	Share of employment in HT and MHT industries in total employment	%				
	Austria		6.70	6.77	6.29	n/a
	EU27		n/a	n/a	n/a	n/a
12.	Share of employment in KI service sectors in total employment	%				
	Austria		25.63	28.17	31.09	n/a
	EU27		n/a	n/a	n/a	n/a

Table 6.13 (continued)

No.	Item	Unit	1995	2000	2005	2010
1	2	3	4	5	6	7
13.	Number of patent applications expressed as US$ per 1 billion of expenditure on R&D	Number				
	Austria		n/a	0.41	0.37	n/a
	EU27		n/a	n/a	n/a	n/a
14.	Number of patent applications per one million inhabitants	Number				
	Austria		217.27	244.77	275.71	n/a
	EU27		n/a	n/a	n/a	n/a
15.	Share of HT product exports in total exports	%				
	Austria		7.82	14.05	12.81	n/a
	EU27		n/a	21.39	18.74	n/a

of states and communes (*Gemeinde*). Also, the instruments for supporting bottom-up initiatives within this scope are varied. Three levels of management, impact, and support that constitute part of R&D and innovation policies may be discerned, namely: ministries and institutions responsible for this policy, supporting institutions, and entities targeted by this system, which is aimed at stimulating activities as well as knowledge, technology, and innovation development processes. Creating the development vision and putting legislative initiatives forward to the parliament is the task of the federal government. The Ministries and Departments responsible for broadly understood economic development policy are tasked with drafting visions of development strategies in cooperation with each other. Besides specifying policy goals, these policies determine priorities for further actions, boundary conditions, appropriate regulations, tools, and procedures. Financial frameworks for the effective accomplishment of the set goals are specified by the Ministry of Finance. Before such programs are created, they are subjected to a vast discussion between the industry groups concerned, associations, agencies, groups of scientists and researchers representing research institutes and universities, and experts in the field of research, development, and innovation.

The Ministry of Transport, Innovation, and Technology (*Das Bundesministerium für Verkehr, Innovation und Technologie*; BMVIT), the Ministry of Economy, Family, and Youth (*Das Bundesministerium für Wirtschaft. Familie und Jugend*; BMWFJ), the Ministry of Research and Development (*Das Bundesministerium für Wissenschaft und Forschung*; BMWF), and the Ministry of Finance (*Das Bundesministerium für Finanzen*; BMF)[9] play a special role in the process of establishing the innovation policy and innovation system in Austria.

The major task of the first ministry, i.e., MBVIT, is enhancing the competitiveness of the economy and industry through innovation and technology development,

[9] Cf.: A.J. Maj, *Die Innovationsförderung in der EU am Beispiel Österreichs und Polens. Eine landeskundliche Studie*, http://othes.univie.ac.at/6960/1/2009-10-06_0107672.pdf—praca, (25 July 2012).

as well as creating workplaces for highly qualified employees, developing science at the highest level, and supporting stable growth. Moreover, the Ministry also caters for basic research and its applications, including patent development.

The activity of the Ministry of Economy, Family, and Youth (BMWFJ) within the scope of supporting technology transfer and innovation management especially concerns crafts. The Ministry pays special attention to small- and medium-sized enterprises that actively participate in adapting to new scientific, technical, and technological solutions in the economic practice.

The Ministry of Science and Research, in its competence acts, foresees supporting and catering for the development of science, scientific research, and education.

The activity of the Ministry of Finance is of great significance for the process of supporting the entire system of R&D and innovation. Besides managing the federal budget, it is responsible for funds that are necessary for the performance of R&D projects.

Apart from the ministries, federal agencies (*Förderungagenturen*) also play an important role in the structure of supporting innovation, e.g., agencies such as the Austrian Economic Services Company (*Austria Wirtschaftsservice GmbH;* AWS) and the Research Support Association (*Forschungsförderungsgesellschaft;* FFG), conducting numerous research programs. The function of the owner of the first two agencies is held jointly by the ministries BMVIT and BMWFJ.

These few central institutions are governed by the federal government and the National Council with its standing Committee for Research, Innovation, and Technology, which creates conditions conducive to the development of scientific research, supporting innovation, and education in Austria. The above institutions indirectly manage and support bottom-up initiatives of enterprises, research institutes, and academic groups.

In the case of direct activities aimed at supporting key and priority research projects, an important role is played by specialist institutions that have been established many years ago in Austria, namely:

- Fund for Supporting Scientific Research (Fonds zur Förderung der Wissenschaftlichen Forschung; FWF).
- Fund for Supporting Research for Crafts (Forschungsförderungsfonds für die Gewerbliche Wirtschaft; FFF).
- AWS.
- FFG.[10]

In practice, an important role in Austria's economic activity in the area of R&D is also played by research institutes belonging to so-called cooperation areas (*kooperativen Bereich*) of the enterprise sectors, including:

[10] Cf.: *Österreichischer Forschungs- und Technologiebericht 2012*, BMWF, BMVIT, Vienna, 2012, pp. 50–56. Since 1998, *Christian Doppler Forschungsgesellschaft* (CDG) has also been in operation. It manages single support instruments and has therefore been assigned a special position.

- Austrian Institute of Technology (AIT).[11]
- Joanneum Research Company (*Joanneum Forschung* GmbH).
- Competence Centers for Excellent Technologies (COMET).

All these institutions provide a wide range of opportunities and impact tools, thanks to which not only enterprises can fulfill their ideas and introduce them to the market, but also new technologies can be developed, cooperation between knowledge and economy can be established, and the process of obtaining highly qualified new researchers can be facilitated.

6.2.2.1 The Innovation Capacity and Position of Austria

The index determining the share of external expenditure on R&D in GDP is one of the basic indexes describing the innovation capacity of the economy. Upon its accession to the EU, Austria allocated 1.55 % of GDP to R&D (Table 6.13). During 15 years, this index went up and reached the level of 2.75 % in 2010. It is worth underlining that in the period from 1995 to 2010, the same index increased for EU27 by 0.25 p.p., and in the case of Austria—by 1.2 p.p.

In Austria, the rate of growth of the share of R&D expenditure in GDP was higher when the share of public funds (41 % in 2010) and of funds of enterprises (43 % in 2010) was high in that expenditure and when the share of human resources employed in the R&D system in the total working population was on the rise. The observed changes in these indexes were also the manifestation of economic reforms, including restructuring, privatization, and liberalization, which were initiated in Austria at the beginning of the 1990s and which Austria has continued as the EU Member State. Joining the EU and adopting the *aquis communautaire* entailed many changes for Austria. It was a challenge and, at the same time, a stimulus to undertake further reforms. This may be perceived as one of the causes of the trade creation and shift effects obtained by Austria, stronger integration with the internal market, as well as introduction of the single currency.[12]

On 8 March 2011, the Austrian government adopted a long-term framework program concerning R&D: the so-called Research, Technology and Innovation Strategy.[13] This program defines the strategic and operational goals, centers of gravity, and tools for their implementation. Austria's main goal is to achieve the position of the most innovative Member State of the EU by 2020. The expected increase in the percentage share of R&D expenditure in GDP up to 3.76 % should facilitate this process. However, owing to the situation on the international financial

---

[11] Cf.: www.ait.ac.at.

[12] More in, cf.: A. Belke, F. Schneider. Privatization in Austria: Response to International and External Pressures, CESifo DICE Report, no.1/2005.

[13] More in: *Strategie der Bundesregierung für Forschung, Technologie und Innovation, Potenziale ausschöpfen, Dynamik steigern, Zukunft schaffen: Der Weg zum Innovation Leader*, www.bmvit.gv.at (25 July 2012).

market and the economic crisis, it is assumed that in the medium term it will not be possible to maintain the current upward trend within the scope of expenditure from the national budget. Nevertheless, it is considered that in the face of growing international competition, as well as social, economic, and ecological factors, there is no alternative but to further support research, development, and innovation.[14]

It is worth noting that in the analyzed period, Austrian research and innovation development were not only financed from own sources, i.e., from public funds and funds of private enterprises, but were also cofinanced by foreign entities, including especially the EU. The following entities participated in the R&D process: enterprises, universities, research institutes, public–private, and private institutions (*Gemeinnütziger Sektor*).

The structure of Austria's R&D expenditure by type, besides basic research, is dominated by expenditure on experimental research and applied research. The last group of research is mainly conducted by enterprises. In turn, universities lead the way in basic research. Moreover, in the enterprise sector, it is visible that expenditure varies depending on economic field and technological input. In the case of universities, expenditure differs depending on fields of knowledge. Furthermore, in the period from 2002 to 2009, correlation between the technological advancement of manufactured goods or rendered services and R&D expenditure intensity was additionally observed.[15] Moreover, as has been mentioned earlier, R&D activities are focused on enterprises. Seventy-one percent of total R&D expenditure was attributed to large enterprises, employing more than 250 workers, which constituted 14 % of the total number of research enterprises in 2009.

As regards small enterprises, employing up to 50 workers and constituting 59 % of the total number of research enterprises, the situation was notably different. These enterprises received 11 % of R&D expenditure.[16] A concentration of R&D enterprises in the relevant sector is therefore observed, as is the fact that it is large enterprises that are mostly active in this field. In Austria, the following enterprises are considered to be most innovative: enterprises operating in the electronic sector (data processing), design offices, industry, including the automotive industry, machinery construction, IT and ICT product manufacturing, chemicals, pharmaceuticals, the gas, and chemical industry.[17] Moreover, the number of patent applications

---

[14] This program is coordinated with the EU strategy *Europe 2020* and the proposal of the European Commission of 30 November 2011 to establish a new Research and Innovation Framework Program "Horizon 2020." Cf.: Europäische Kommission, *Europa 2020: Eine Strategie für intelligentes, nachhaltiges und integratives Wachstum,* Mitteilung der Kommission an das Europäische Parlament, den Rat, den Europäischen Wirtschafts- und Sozialausschuss und den Ausschuss der Regionen, KOM(2010)2020, Brüssel, 3 March 2010; Europäische Kommission, *Horizont 2020–das Rahmenprogramm für Forschung und Innovation,* Mitteilung der Kommission an das Europäische Parlament, den Rat, den Europäischen Wirtschafts- und Sozialausschuss und den Ausschuss der Regionen, KOM(2011) 808 endgültig, Brüssel, 30 November 2011.

[15] Cf.: *Österreichischer Forschungs- und Technologiebericht 2012,* …, op. cit., pp. 39–40.

[16] Ibid., p. 41.

[17] Cf.: *Österreichischer Forschungs- und Technologiebericht 2012,* …, op. cit., …, p. 94.

is increasing in Austria, as is the previously discussed share of HT product exports in total exports.

As a result of the analysis of indexes determining the innovation capacity of Austria's economy (Table 6.13), it may be stated that important factors determining this capacity are the volume and increasing growth rate of the share of R&D expenditure in Austrian GDP, high participation of enterprises in the financing of cooperation with research centers, and implementation of new technological, scientific, and organizational solutions, as well as the comparable with EU27—but growing—percentage of highly qualified employees in the total of working population. These phenomena are additionally supported by the transparent and R&D-oriented institutional system.

Important indexes describing the innovation status of Austria also embrace indexes related to human capital, including indexes regarding the percentage of persons employed in the HT and MHT sectors in total employment, or the growing rate of employment in the KI services sector.

6.2.2.2 Human Resources

As regards the population of Austria, and—in particular—the changes in it, in accordance with the forecasts, Austria's population will go up from 8.4 million in 2010 by 3 % to 8.7 million residents by 2020, and by a further 7 % in 2050 to just under 9 million. During the period of 10 decades, the percentage of persons aged 15–64 in total population has been gradually changing. The downward trend that can be observed here is indicative of the aging of Austria's population. This is confirmed by the decreasing number of working-age population per pensioner. The relevant index rose almost twofold in 2010, in comparison with 1950, and equaled 3.8, whereby forecasts expect it to amount to 2.2 in 2020. As a result, fewer and fewer persons will be professionally active in the economy. This might have a major impact on the further economic development and competitiveness of Austria.

Nowadays, in well-developed economies, knowledge is becoming more and more important. In practice, this translates into an increase in demand for highly qualified labor force on the labor market. The number of university graduates with relevant high qualifications, skills, and knowledge will be the key factor determining competitive advantages and innovation performance in the nearest future, both at the level of enterprises and the entire economy. It is assumed that a well-educated employee with high qualifications is the basic condition for R&D and innovation, including the transfer of scientific theory to economic practice.[18] It is expected that demand for, especially, higher university education in the area of natural and technical sciences will increase. Among the working population in Austria, an increase in the group of employees who have graduated from universities may be observed (Table 6.14). The largest amount of funding is directed at this group.

[18] Cf. *Österreichischer Forschungs- und Technologiebericht 2012*, …, op. cit., p. 133.

Table 6.14 Indexes reflecting the level of qualifications held by the Austrian population. (Source: World Bank data; own compilation)

No.	Qualification level—indexes	Unit	1995	2000	2005	2010
1	Labor force	%	3,862,262	3,878,645	4,081,455	4,297,150[a]
2	% of labor force with higher education	%	8.6[b]	9.6	18.1	17.7[c]
3	% of labor force with secondary education	%	66.8[b]	69.1	64.1	62.8[c]
4	% of labor force with primary education	%	24.7[b]	21.3	17.8	18.9[c]
5	Expenditure on university education as % of GDP	%	…	44.1	50.3	43.5[d]
6	Expenditure on primary education as % of GDP	%	…	27.9	26.5	27.3[d]
7	Expenditure on secondary education as % of GDP	%	…	23.1	23.6	24.1[d]
8	Average number of years of education per person aged 25+	years	8.7	9.1	9.5	9.8

[a] Data for 2009
[b] Data for 1996
[c] Data for 2007
[d] Data for 2008

In the analyzed period, it may also be noted that the average number of years of education per person aged 25+ is low, although it is increasing. In 2009, total R&D expenditure in the higher education sector amounted to € 1.9 billion, € 1.5 billion of which was allocated to universities. From 2002 to 2009, a 44 % increase in employment in the R&D area at universities could also be observed, from 9.147 in 2002 to 13.134 in 2009. At the same time, the number of scientific staff went up by, on average, 51 %. However, growth in other personnel categories is not uniform.[19] The highest growth—by 70 %—was observed in the category of assistants and scientific personnel. The number of employed professors and assistant professors went up by 7 %. The percentage of the latter group in the employment structure amounted to 21 % of scientific personnel in 2002, and in 2009 it dropped to 16 %. The highest employment level is characteristic of such fields as natural sciences and medicine, then in technical and social sciences.

The level of utilization of human capital can be evaluated by means of a few measurement methods, namely: the unemployment rate, its structure, especially with regard to young, educated people who are unemployed.

As follows from the data presented in the table below, in the years 1995–2010, Austria displayed a relatively low unemployment rate of 4.4 % in August 2010. However, the indexes reflecting unemployment among persons with higher education and among youth, which were around 10 % in 2007, are unfavorable (Table 6.15).

Another important determinant of the level of human capital utilization is its efficiency. In the analyzed period, an increase in remuneration per working hour

[19] Cf.: *Österreichischer Forschungs- und Technologiebericht 2012*, …, op. cit., p. 135 and next.

Table 6.15 Indexes reflecting the level of human capital utilization in Austria. (Source: World Bank data, own compilation; Österreichisches Institut für Wirtschaftsforschung. Institut für Höhere Studien. Assessing the Lisbon Strategy 2005–2010 and Estimating Expected Effects from Reaching the EU 2020 Goals, S. Ederer, J. Janger. S. Kaniovski. D. Kletzan-Slamanig. J. Berger. I. Fortin. H. Hofer. I. Paterson. E. Skriner. K. Schönpflung. U. Schuh. W. Schwarzbauer. Vienna. December 2010, p. 2, cf.: www.WIFO_IHS_Lissabon_31032011.pdf)

No.	Indexes reflecting the level of human capital utilization in Austria	Unit	1995	2000	2005	2009
1.	Unemployment rate (unemployment as % of total employment)	%	3.7	3.5	5.2	4.8
2.	Unemployed with higher education as % of total unemployed	%	4.5	4.6	9.6	10.0[a]
3.	Unemployment among youth (% of unemployed aged 15–24)	%	5.2	5.1	10.3	9.9
4.	Professional activity of persons aged 15–24	%	58.3	53.4	53.0	53.1

[a] Data for 2007

[b] Data as at August 2010

Table 6.16 Indexes depicting the efficiency of human capital in Austria. (Source: OECD data)

No.	Indexes depicting the efficiency of human capital in Austria	Unit	1995	2000	2005	2009
1.	Remuneration per hour (in PPP)	US$ (according to PPP)	19.02	21.75	24.49	29.12[a]
2.	Annual remuneration per employee (in PPP)	US$ (according to PPP)	31,592.15	36,166.89	40,325.08	6,031.83[a]
3.	Labor cost per production unit	Average labor cost per production unit)	0.73	0.71	0.67	0.67
4.	Unit labor costs (ULC, OECD index)		0.69	0.67	0.70	0.75

[a] Data for 2009

and remuneration per employee may be observed (Table 6.16). This tendency is also present in other European states. In Austria, remuneration per working hour increased by 53 % in 2009 (in comparison to 1955), in Sweden—by 84 %, in Finland—by 75 %, in Germany—by 47 %, and in Denmark by 61 %. This means that the trends in the Austrian economy resemble those present in the economies of the states considered to be the leaders in innovation.

Labor costs and their changes are interesting from the point of view of development of competitive advantages. In 2009, labor cost per production unit fell from 0.73 to 0.67. It is currently comparable with the countries mentioned earlier. Still, it is the lowest in Finland, where it equals 0.63. Another means of measurement, i.e., unit labor costs (ULC), is displaying an upward trend. This is the outcome of the drop of this index in processing and industry, and its increase in construction, trade, finance, business, and market services. Growing ULC lead to a drop in Austria's price competitiveness.

6.2.2.3 Austria's Scientific, Innovation, and Education Policy

As mentioned earlier, significant changes have been taking place in the Austrian economy since the beginning of the 1990s. The accession to the EU facilitated the previously launched privatization and restructuring processes, and contributed to the increase in Austrian trade growth. In 2002, Austria—as the EU Member State—decided to implement a long-term economic program: the so-called Lisbon Strategy. The objective of this program was to make the development of EU Member States' economies more dynamic by means of employing common modern technological solutions, such as computers, mobile telephones, ensuring that households have access to the Internet, and developing scientific research, knowledge, and new fields of research, in particular nanoscience and nanotechnology. Significant changes also occurred on the Austrian labor market and had positive influence on the employment of the elderly, youth, and women.

In order to effectively implement the objectives of the Lisbon Strategy, the Austrian authorities had to implement a stable policy that would include and support activities aimed at increasing the level of education, as well as boosting expenditure on research and new technologies.[20] In the last 15 years, Austria has actively participated in the integration process and introduced the desired changes to its economy.

The R&D program for the next few years (*FTI-Strategie*),[21] adopted for implementation in 2011, is the expression of particular importance attached to further R&D, and the need for reforms in the education system. The goals of this program are correlated with the objectives of the EU economic strategy *Europe 2020* that concerns changes in the labor market, intensification of activities aimed at furthering research, development, and innovation, counteracting of climate change, and energy, and intensification of education and social inclusion.

All factors characterizing the previous system of support for innovation development, which are presented in the table below, are already sufficient to determine Austria's competitive advantage over other EU Member States (Table 6.17).

The advantages, if used wisely, will be conducive to further development of innovation in Austria. However, this is also dependent on the restriction or elimination of all factors classed as the disadvantages of this system.

The factors listed as "opportunities" are conditions that, if used correctly, might influence the growth rate of R&D and innovation growth.

---

[20] Cf. A. Szymański, *Austria. Prawie pełna kontynuacja* (Austria. Almost Complete Continuation) (in:) *Zmiana i kontynuacja. Polityka europejska wybranych państw Unii Europejskiej (Change and Continuation. European Policy of Selected EU Member States)*, L. Jesień (scientific ed.), PISM, Warsaw 2008, p. 41–47; Österreichisches Institut für Wirtschaftsforschung, Institut für Höhere Studien, *Assessing the Lisbon Strategy 2005–2010 and Estimating Expected Effects from Reaching the EU 2020 Goals,* S. Ederer, J. Janger, S. Kaniovski, D. Kletzan-Slamanig, J. Berger, I. Fortin, H. Hofer, I. Paterson, E. Skriner, K. Schönpflung, U. Schuh, W. Schwarzbauer Vienna, December 2010, Cf.: www.WIFO_IHS_Lissabon_31032011.pdf.

[21] Cf.: *Strategie der Bundesregierung für Forschung, Technologie und Innovation, Potenziale ausschöpfen, Dynamik steigern, Zukunft schaffen: Der Weg zum Innovation Leader*, …, op. cit.

Table 6.17 SWOT analysis of Austria's innovation system. (Source: Own elaboration)

Strengths	Weaknesses
R&D strategy as part of Europe 2020 program	(Too) high concentration of R&D expenditure in enterprise sector[a]
Institutional structure that allows for creating an environment conducive to the development of new knowledge, techniques, and entrepreneurship	Relatively low participation of SMEs in innovation development
Policy supporting research, development, and innovation and other policies, among others, macroeconomic, structural, and others, as well as their mutual interactions	Low participation of women in scientific research
Experience in the role and significance of R&D in economy	Relatively low rate of establishing innovative enterprises
High percentage of R&D expenditure in GDP	Low quality of university research
R&D expenditure structure	Internationalization in the R&D area
Enterprises have a large share in financing innovation	Poor participation in European studies
Cooperation between all participants of the innovation process	Education structure
Positive trends in human capital utilization	Low percentage of persons with higher education
Opportunities	Threats
Need for university reform aimed at introducing basic research and using the opportunities provided by new research-supporting structures (IST-A)	Economic crisis
Need for further changes in education policy	
Vast institutional opportunities for developing innovation that can be used, thanks to the mobility of R&D participants	
Need for further utilization of human capital for innovation purposes	
Need for further adaptation of innovation policy to future challenges	
Need for drawing the attention of enterprises, including SMEs, to innovation	
Increasing the percentage of youth, women, and immigrants in the labor market through education, enhancing qualifications, and the requirement system, lower taxes, and benefits	

[a] In the business sector, 50 % of total R&D expenditure is held by 33 companies—-when utilized, they will support growth

The phenomena connected with the threats to the innovation system do not necessarily have to occur in reality. Their actual occurrence is indirectly dependent on the effectiveness of anticrisis activities and determination with which the current research, technology, and innovation strategy will be implemented in Austria until 2020.

These activities also encompass objectives connected with the following five areas: supporting the education system, strengthening the scientific environment, including scientific and research cooperation with foreign universities, strengthening cooperation between economic practice and science in the area of supporting research, innovation, and entrepreneurship development. These goals can be indirectly dependent on the effectiveness of management of the R&D system, and the management of funds for R&D, which are planned at 3.76 % of GDP by 2020.

6.2.3 Interdependencies Between the Level of Innovation, Accumulation of Human Capital, and Advantages in Foreign Trade

Austria is a relatively small European country, and international trade plays an important role in its economy. The open development policy that has been implemented by the Austrian government is, among other things, expressed in the share of foreign trade in GDP. In the analyzed period, an increase of the share of exports in GDP from 34.9 % in 1995 to 51.9 % in 2011 may be observed. Apart from the economic policy, also the trade policy and large-scale promotion campaigns carried out by the government in cooperation with various institutions (e.g., the Austrian Economic Chamber) were conducive to boosting Austrian exports. One of the special goals was to overcome export barriers for Austrian SMEs.[22]

In the period from 1995 to 2010, the structure of international trade changed. This change led to, among others, an increase in the percentage of highly processed products in total international trade, as well as in exports. In order to manufacture such products, highly qualified labor force is required. This is a practical expression of activities aimed at supporting R&D, innovation, and the implementation of new solutions in the economic system. During the economic crisis, the growth rate of export decreased, especially in the machine industry and in vehicle exports. It seems that Austria has yet to discover its vast potential within the scope of rendering and exporting office services and developing transit services.

6.3 Case Study: Spain

Oskar Kowalewski

This subchapter contains an evaluation of significance of human capital and innovativeness in the process of development of comparative advantages in Spain's foreign trade and of their transformation into dynamic competitive advantages.

6.3.1 Spain's Competitiveness in Foreign Trade

In the 1995–2010 period, a slight increase in Spain's share in the international foreign exchange was recorded. The biggest increase of the Spanish share in world's

[22] Cf.: www.go-international.at; www.aba.gov.at; www.fiw.at; www.ibwe.at.

Table 6.18 Spain's foreign trade in the 1995–2010 period (value and dynamics of exports and imports). (Source: Own work on the basis of WTO data)

Year	Exports (US$ billion)	% share in world's exports	Imports (US$ billion)	% share in world's imports	Balance (US$ billion)
1995	89.62	1.91	113.40	2.39	−23.78
2000	113.34	1.86	152.90	2.43	−39.55
2005	192.80	1.94	289.61	2.82	−96.81
2006	214.06	1.84	329.98	2.77	−115.91
2007	253.75	1.92	391.24	2.87	−137.48
2008	279.23	1.83	418.73	2.66	−139.50
2009	223.13	1.88	287.50	2.39	−64.37
2010	246.27	2.04	315.55	2.55	−69.28

exports took place in 2003, when it reached the value of 2.15 %. In the subsequent years, the index decreased even to the level of 1.83 % in 2008, which means it was lower than in 1995, when it amounted to 1.91 %. In 2010, however, it reached the value of more than 2.04 %. Nevertheless, it has to be stressed that, in absolute values, Spain's exports increased over the analyzed period from US$ 89.62 billion in 1995 to US$ 246.27 billion in 2010.

Spanish imports also showed high dynamics in the years 1995–2010. Their value increased from US$ 113.40 billion in 1995 to US$ 315.55 billion in 2010, and the maximum value of US$ 418.73 billion was reached in 2008. The share of Spain in global imports dynamically increased from 2.39 % in 1995 to 2.55 % in 2010, and its maximum value (2.87 %) was reached in 2007.

Trade deficit could be observed throughout the examined period. It increased from US$ 23.78 billion in 1995 to US$ 69.28 billion in 2010 (Table 6.18).

In the 1995–2010 period, the most important output market for Spanish goods was France. In 2010, Spain's exports to the French market reached the value of US$ 24,008 million (M), which constituted 18.28 % of the country's total exports. It means that France was gradually losing its role as the main trade partner of Spain, since in 1995 the share had amounted to 20.53 %. In the analyzed period, the most important trade partners, except for France, were Germany, Portugal, and Italy, although their share in total exports of the country had also decreased since 1995 and was equal to 10.47 %, 8.92 %, and 8.77 %, respectively, in 2010. At the same time, since 1995, it has been possible to observe growing export of Spanish goods to developing countries. Nevertheless, the share of individual countries from Europe in Spain's total exports has been relatively high (Table 6.19).

In 2010, Spain's biggest partner in imports was Germany, with a share of 11.74 %. However, the share had decreased from 15.24 % in 1995. Yet, the decline in the import of German goods to Spain was smaller than that recorded by French goods. Since 1995, the share of French goods imported to Spain had decreased from 17.22 to 10.73 % in 2010. As a result, France had moved from the first to the second place in the ranking of Spain's most important trade partners. Spain's other biggest import partners were China and Italy with the share of 7.92 and 7.01 %, respectively, in 2010. A significant increase in the share of import of goods is visible especially in the case of China, the share of which was only 1.98 % in 1995 (Table 6.20).

Table 6.19 Geographical structure of Spain's foreign trade—main partners in exports in the years 1995–2010. (Source: Based on WTO data)

Year	1995		2000		2005		2010	
Country	US$M	%	US$M	%	US$M	%	US$M	%
UK	7,161	7.99	9,425	8.32	16,202	8.40	15,238	6.19
Italy	8,153	9.10	9,938	8.77	16,128	8.37	21,600	8.77
Portugal	7,486	8.35	10,619	9.37	18,456	9.57	21,973	8.92
Germany	13,810	15.41	14,023	12.37	21,901	11.36	25,785	10.47
France	18,401	20.53	22,034	19.44	36,991	19.19	45,008	18.28
Total exports	89,616	100.00	113,343	100.00	192,798	100.00	246,265	100.00

Table 6.20 Geographical structure of Spain's foreign trade—main partners in imports in the years 1995–2010. (Source: Based on WTO data)

Year	1995		2000		2005		2010	
Country	US$M	%	US$M	%	US$M	%	US$M	%
Portugal	3,309	2.92	4,108	2.69	9,340	3.22	11,319	3.59
USA	7,527	6.64	8,034	5.25	9,306	3.21	12,572	3.98
The Netherlands	4,890	4.31	5,721	3.74	11,681	4.03	14,179	4.49
UK	8,812	7.77	10,737	7.02	16,455	5.68	14,298	4.53
Italy	10,321	9.10	13,432	8.78	24,747	8.54	22,127	7.01
China	2,241	1.98	4,304	2.82	14,557	5.03	25,006	7.92
France	19,524	17.22	26,145	17.10	40,841	14.10	33,872	10.73
Germany	17,286	15.24	22,732	14.87	42,713	14.75	37,030	11.74
Total imports	113,399	100.00	152,898	100.00	289,611	100.00	315,547	100.00

[a] Together with Luxembourg

In 2010, Spain exported primarily machinery, equipment and transport equipment (33.8 %), manufactured goods (17.1 %), chemicals and related products (14.4 %), as well as food and live animals (12.2 %). In the years 1995–2010, a significant change in the exports structure could be observed. The share of export of machinery, equipment, and transport equipment decreased by 8 p.p. At the same time, a decrease of almost 7 p.p. in the share of chemicals and of almost 4 p.p. in the share of related products and of mineral fuels, lubricants, and related materials in Spain's exports was recorded (Table 6.21).

In 2010, Spain imported primarily machinery, equipment and transport equipment (28.8 %), mineral fuels, lubricants and related materials (18.4 %), and chemicals and related products (14.2 %). In the case of the imports structure, significant changes could have also been observed in the years 1995–2010. For example, over this period, the share of mineral fuels, lubricants, and related materials in imports had increased by 10 p.p. At the same time, in the analyzed period, a decrease of more than 6 p.p. in the share of machinery, equipment, and transport equipment in imports was recorded. It should be noted that the observed changes in the imports structure are very similar to those in the exports structure (Table 6.21).

The analysis of changes in the goods structure of Spain's foreign trade in the years 1995–2010 according to the OECD classification based on the technology level shows that, throughout the examined period, Spain was characterized by a

Table 6.21 Goods structure of Spain's exports and imports by SITC groups. (Source: Based on WTO data)

SITC goods groups	Exports								Imports							
	1995		2000		2005		2010		1995		2000		2005		2010	
	US$M	%	US$M	%	US$M	%	US$M	%	US$M	%	US$M	%	US$M	%	US$M	%
0 Food and live animals	11,486	12.84	12,611	11.14	21,807	11.32	29,931	12.15	11,739	10.38	10,974	7.20	20,645	7.14	25,620	8.12
1 Beverages and tobacco	1,353	1.51	1,640	1.45	2,917	1.51	3,726	1.51	1,871	1.65	1,911	1.25	4,011	1.39	4,060	1.29
2 Crude materials, inedible, except fuels	2,054	2.30	2,496	2.20	3,873	2.01	6,104	2.48	6,842	6.05	6,330	4.15	10,526	3.64	13,992	4.43
3 Mineral fuels, lubricants, and related materials	1,455	1.63	4,037	3.56	8,043	4.17	12,508	5.08	9,446	8.36	18,557	12.18	40,691	14.07	58,232	18.45
4 Animal and vegetable oils, fats, and waxes	967	1.08	1,068	0.94	2,409	1.25	3,277	1.33	808	0.71	321	0.21	950	0.33	1,559	0.49
5 Chemicals and related products	7,718	8.63	10,732	9.48	23,399	12.14	35,511	14.42	13,704	12.12	16,250	10.66	33,229	11.49	44,742	14.18
6 Manufactured goods classified chiefly by material	17,594	19.67	19,595	17.30	32,487	16.86	42,083	17.09	17,553	15.53	20,436	13.41	36,404	12.59	35,193	11.15
7 Machinery, equipment, and transport equipment	37,912	42.38	47,910	42.31	77,239	40.08	83,111	33.75	39,785	35.19	61,522	40.37	107,941	37.34	91,017	28.84
8 Miscellaneous manufactured articles	8,224	9.19	11,348	10.02	17,534	9.10	21,791	8.85	11,170	9.88	15,414	10.11	33,252	11.50	39,559	12.54
9 Commodities and transactions not classified elsewhere	683	0.76	1,808	1.60	3,009	1.56	8,224	3.34	143	0.13	677	0.44	1,458	0.50	1,575	0.50
Total exports	89,447	100	113,245	100	192,717	100	246,265	100	113,061	100	152,392	100	289,107	100	315,547	100

Table 6.22 Changes in the goods structure of Spain's exports in comparison with the rest of the world according to the OECD classification based on the technology level. (Source: Own work on the basis of WTO data)

Industries	1995		2000		2005		2010[a]	
	Spain (%)	World (%)	Spain (%)	World (%)	Spain (%)	World (%)	Spain (%)	World (%)
HT	6.66	12.03	7.38	13.53	7.87	13.33	7.85	11.73
MHT	38.98	36.96	40.09	40.80	39.86	39.14	40.34	39.73
MLT	36.31	35.11	35.78	32.11	36.91	34.53	35.51	36.41
LT	18.05	15.90	16.75	13.56	15.35	13.00	16.30	12.14

[a] Data for Spain for 2009

Table 6.23 Changes in the goods structure of Spain's exports and imports in comparison with the rest of the world according to the WIFO classification based on the extent of use of tangible and intangible expenses in the production process. (Source: Based on WTO data)

Industries	1995		2000		2005		2010[a]	
	Spain (%)	World (%)	Spain (%)	World (%)	Spain (%)	World (%)	Spain (%)	World (%)
Mainstream	31.67	21.78	35.45	19.92	48.65	20.50	53.71	21.10
Labor-intensive	15.67	14.44	14.81	13.70	11.78	13.09	10.62	12.44
Capital-intensive	23.44	20.81	22.14	19.35	17.61	22.28	15.88	24.65
Technology-driven	10.28	13.24	9.71	11.95	7.72	11.00	6.96	10.54
Marketing-driven	18.94	29.74	17.90	35.08	14.24	33.13	12.83	31.27

[a] Data for Spain for 2009

relatively high share of the export of MHT and MLT goods (Table 6.22). The 1 p.p. decrease in the share of the export of HT goods has to be assessed positively, but it still remains on a considerably lower level than the analogical share of the global economy. The share of the export of LT goods is, in turn, high, although it had decreased by 3 p.p. in the years 1995–2010. The above-mentioned data concerning the exports structure illustrate the fact that, despite Spain's high expenditures on R&D in the last two decades, changes in the Spanish economy take place relatively slowly.

The goods structure of Spain's foreign trade in the years 1995–2010 according to the WIFO classification based on the extent of use of tangible and intangible expenses in the production process (Table 6.23) confirms the low technological advancement of the Spanish economy. Exports of the country are based principally on mainstream industries, the share of which increased from 31.67% in 1995 (in comparison with 21.78% for the entire world) to 53.71% in 2010 (21.10% for the entire world). The decrease from 23.44% in 1995 to 11.04% in 2010 in the share of the capital-intensive industry and from 10.28% in 1995 to 6.96% in 2010 in the share of the TDI in the goods structure of Spain's exports also has to be assessed negatively. On the other hand, the 5 p.p. decrease in the share of the labor-intensive industry in the analyzed period has to be given a positive assessment.

The analysis of changes in the goods structure of foreign trade in the years 1995–2010 according to the WIFO classification based on the skill level of labor force (Peneder 1999) confirms the previous findings on the exports structure of Spain. Of the largest significance in exports are industries hiring low-skilled workers, the

Table 6.24 Changes in the goods structure of Spain's exports in comparison with the rest of the world in the 1995–2010 period according to the WIFO classification based on the skill level of labor force. (Source: Based on WTO data)

Industries based on	1995		2000		2005		2010[a]	
	Spain (%)	World (%)	Spain (%)	World (%)	Spain (%)	World (%)	Spain (%)	World (%)
High-skilled workers	6.66	12.03	7.38	13.53	7.87	13.33	7.85	11.73
Semiskilled white-collar workers	38.98	36.96	40.09	40.80	39.86	39.14	40.34	39.73
Semiskilled blue-collar workers	36.31	35.11	35.78	32.11	36.91	34.53	35.51	36.41
Low-skilled workers	18.05	15.90	16.75	13.56	15.35	13.00	16.30	12.14

[a] Data for Spain for 2009

Table 6.25 Development of comparative advantages (RCA) in Spain's foreign trade and their evolution in the years 1995–2010 according to the OECD classification based on the technology level. (Source: Based on WTO data)

Industries	1995	2000	2005	2006	2007	2008	2009
HT	−0.3711	−0.3848	−0.2617	−0.2314	−0.2205	−0.3335	−0.2744
MHT	−0.0466	−0.0106	0.0324	0.0598	0.0628	0.0660	0.0002
MLT	−0.0147	0.0289	0.0573	0.0750	0.0493	0.0674	0.0396
LT	0.1497	0.2173	0.1469	0.1410	0.1461	0.1491	0.0596

share of which amounted to 16.30 and 12.14% correspondingly in 2010. However, it should be highlighted that the share of industries hiring low-skilled workers in Spain's export structure had decreased by 2 p.p. in comparison with 1995. At the same time, in the years 1995–2010, the share of industries hiring high-skilled workers in Spain's exports increased by 1 percentage point. In 2010, this share was equal to 7.95% for Spain and was considerably lower than the analogous share for the world, which amounted to 11.73%. Nevertheless, in the years 1995–2010, an increase from 38.98 to 40.34% in the share of the export of goods produced by semiskilled white-collar workers was recorded. Simultaneously, a decrease from 36.31 to 35.51% in export of goods produced by semiskilled blue-collar workers was observed. The foregoing data confirm slow changes in the structure of the Spain's economy in the analyzed period. These changes, however, were still based on human capital only to a small extent (Table 6.24).

The use of the index of comparative advantages in foreign trade RCA of Spain according to the OECD classification based on the technology level allows the analysis of advantages in production and export of Spanish products, in which, in theory, this economy specializes. Throughout the examined period of 1995–2010, Spain recorded relative advantages in relation to low, medium-low, and MHT industries.

The continuously low level of technological advancement of the Spanish economy was confirmed by the lack of comparative advantages in high-level industries, even though the RCA index increased from the level of −0.3711 in 1995 to −0.2744 in 2010. This increase reflects the changes in the Spanish economy, which, however, had been taking place quite slowly in the last years of the period (Table 6.25).

Table 6.26 Development of revealed comparative advantages (RCA) in Spain's foreign trade and their evolution in the years 1995–2009 according to the WIFO classification based on the extent of use of tangible and intangible expenses in the production process. (Source: Based on WTO data)

Industries	1995	2000	2005	2006	2007	2008	2009
Mainstream	0.0056	0.0167	0.0209	0.0622	0.0785	0.1793	0.1088
Labor-intensive	0.1407	0.2089	0.0599	0.0941	0.0465	0.0196	0.0223
Capital-intensive	0.0002	0.0227	0.1177	0.1114	0.1050	0.2168	0.1297
Marketing-driven	0.1047	0.2546	0.2518	0.2402	0.2489	0.2680	0.1700
Technology-driven	0.1540	0.0286	0.0353	0.0174	0.0349	0.0312	0.0722

Table 6.27 Development of revealed comparative advantages (RCA) in Spain's foreign trade and their evolution in the years 1995–2010 according to the WIFO classification based on the skill level of labor force. (Source: Based on WTO data)

Industries based on	1995	2000	2005	2006	2007	2008	2009	2010
Highly skilled workers	−0.3711	−0.3848	−0.2617	−0.2314	−0.2205	−0.3335	−0.2744	−0.3711
Semiskilled white-collar workers	−0.0466	−0.0106	0.0324	0.0598	0.0628	0.0660	0.0002	−0.0466
Semiskilled blue-collar workers	−0.0147	0.0289	0.0573	0.0750	0.0493	0.0674	0.0396	−0.0147
Low-skilled workers	0.1497	0.2173	0.1469	0.1410	0.1461	0.1491	0.0596	0.1497

The RCA index calculated according to the WIFO trade classification based on the extent of use of tangible and intangible expenses in the production process shows that Spain had comparative advantages in MDI and CI. These industries recorded the highest values of the index of comparative advantages in 2009. In the same year, the lowest RCA values were noted in relation to technology-driven and mainstream industries. The decrease from 0.154 in 1995 to 0.0722 in 2009 in the value of RCA index for TDI has to be assessed very negatively (Table 6.26).

The analysis of the evolution of the RCA index in foreign trade according to the WIFO classification based on the skill level of labor force in the years 1995–2010 shows that in 2010 Spain had only relative advantages in industries hiring low-skilled workers (Table 6.27). In the years 2005–2009, the country also recorded an advantage in industries hiring semiskilled white-collar workers. In 2010, however, the advantage was lost. What is more, over the analyzed period, Spain noted no comparative advantage in relation to foreign trade in goods produced by highly skilled workers and low-skilled workers. These data confirm a slow evolution of the Spanish economy and a continued lack of comparative advantage in HT sectors.

While analyzing Spain's foreign trade, it is also important to take into consideration the indexes describing the intensity of IIT generated by the processes of internationalization and production relocation. The IIT index reflects whether a country exports and imports products manufactured by the same industries. The intra-industry exchange makes it possible to augment the economic benefits associated

Table 6.28 Extent of Spain's IIT and its changes in the years 1995–2009 in the goods trade according to the OECD classification based on the technology level. (Source: Based on WTO data)

Industries	1995	2000	2005	2006	2007	2008	2009
HT	0.7057	0.6707	0.6776	0.6796	0.6844	0.6466	0.7420
MHT	0.8600	0.8463	0.8149	0.8156	0.8170	0.8320	0.8740
MLT	0.8756	0.8656	0.8270	0.8230	0.8105	0.8327	0.8935
LT	0.9572	0.9590	0.8708	0.8552	0.8575	0.8726	0.9034

Table 6.29 Extent of IIT of Spain and its changes in the years 1995–2009 according to the WIFO classification based on the extent of use of tangible and intangible expenses in the production process. (Source: Based on WTO data)

Industries	1995	2000	2005	2006	2007	2008	2009	2010
Mainstream	0.8856	0.8596	0.8094	0.8168	0.8246	0.8875	0.9278	0.8856
Labor-intensive	0.9527	0.9548	0.8282	0.8323	0.8091	0.8096	0.8849	0.9527
Capital-intensive	0.8830	0.8625	0.8564	0.8407	0.8375	0.9061	0.9382	0.8830
Marketing-driven	0.9348	0.9776	0.9226	0.9040	0.9082	0.9315	0.9583	0.9348
Technology-driven	0.9594	0.8654	0.8163	0.7953	0.8036	0.8152	0.9096	0.9594

with the increasing specialization in a limited number of goods produced by particular industries. Owing to these reasons, the increase in the value of the intra-industry turnover should be assessed positively.

According to the data contained in Table 6.28, in 2010, the largest extent of IIT in Spain was observed in relation to: LT (IIT = 0.9034), MLT (IIT = 0.8935), MHT (IIT = 0.8740), and HT (IIT = 0.7420) industries. The growth from 0.7057 in 1995 to 0.7420 in 2009 has to be assessed positively, although in the years 2000–2008 the index remained on a significantly lower level than in the last year. Nevertheless, an improvement in competitiveness can be noticed also, thanks to an increase of the IIT index in MHT industries, from 0.8600 in 1995 to 0.8740 in 2010, and in MLT industries, form 0.8756 in 1995 to 0.9034 in 2010. At the same time, a continuous decrease of the IIT index is noticeable in relation to LT industries, from 0.9572 in 1995 to 0.9034 in 2010, which can confirm a lasting nature of the increase in competitiveness of Spain in the international trade.

Changes in the extent of IIT were also recorded for the Spanish economy in relation to goods classified according to the WIFO classification based on the extent of use of tangible and intangible expenses in the production process.

In 2010, the biggest intensity of IIT in Spain occurred in relation to TDI and LI. Moreover, in the case of the TDI, the IIT index had decreased from the level of 0.9594 in 1995 to 0.7953 in 2006. Afterward, a dynamic increase to 0.9594 was observed in 2010. A similar situation also occurred in the case of capital and MDI. As for MDI, the highest value of the IIT index was noted in 2000 and was equal to 0.9776, while in 2010 it amounted to 0.9348, which means it was similar to the value recorded in 1995. Among the analyzed industry groups, mainstream industries were characterized by the most stable IIT index, the value of which in 2010 was equal to its value observed in 1995 (Table 6.29).

Table 6.30. Extent of Spain's IIT and its changes in the years 2000–2009 in the goods trade according to the WIFO classification based on the skill level of labor force. (Source: Based on WTO data)

Industries based on	1995	2000	2005	2006	2007	2008	2009
High-skilled workers	0.7057	0.6707	0.6776	0.6796	0.6844	0.6466	0.7420
Semiskilled white-collar workers	0.8600	0.8463	0.8149	0.8156	0.8170	0.8320	0.8740
Semiskilled blue-collar workers	0.8756	0.8656	0.8270	0.8230	0.8105	0.8327	0.8935
Low-skilled workers	0.9572	0.9590	0.8708	0.8552	0.8575	0.8726	0.9034

The development of Spain's IIT in goods according to the WIFO classification based on the skill level of labor force in the years 1995–2010 indicates that it was the most intensive in industries hiring low-skilled workers (Table 6.30). The IIT index for this goods group, however, decreased from 0.9572 in 1995 to 0.9034 in 2007, which illustrates the changes in the economic structure that were taking place in the country. These changes are also reflected in the increase in the IIT related to industries hiring high- and semiskilled white-collar workers (from 0.7057 and 0.8600 in 1995 to 0.7420 and 0.8740 in 2009, respectively). In turn, the decline in the IIT related to industries hiring semi- and low-skilled workers can be considered as a proof of a lasting nature of changes taking place in Spain's economic structure.

6.3.2 Characteristics of Spain's NIS

In the 1980s, the Spanish government, following international solutions, took a number of actions aimed at restoring the country's scientific potential. Augmenting the value of public expenditures allocated to science was not only supposed to allow Spanish researchers to compete on a global scale, but also to increase the innovativeness and competitiveness of the economy. In 1986, in order to accomplish the foregoing goals, the Spanish parliament passed a law on the development of science and technology. The law was embodied in operational and functional way in the national plan for the development of science and technology (Muñoz, Espinosa Díaz 2000a).

In the last decade, the most important changes in the policy related to research, development, and innovations were introduced in 2006. Their purpose was to implement new instruments supposed to remove major barriers in the innovation system and to improve the financing and the transfer of knowledge from science to industry. New instruments, including, among other things, the approval of the Spanish 2010–2015 innovation strategy as part of the ERA 2020 Vision by regional and national governments in 2009, were applied. The strategy introduced a significant financial support to the national research plan for R&D&I, which provided a financial support for the cooperation between research centers and companies through, *inter alia*, a greater use of venture capital (VC) funds. The new innovation strategy also concentrates on the diffusion of information and communication technologies (ICT) by companies and consumers. These actions are to contribute to the economic recovery of Spain through an intensive and general use of ICT. As a consequence, the government continues to encourage activities supposed to increase scientific potential and innovativeness of the economy, but with a smaller financial support owing to the economic crisis (OECD 2009).

In Spain, institutions shaping science, innovation, and education policies can be divided into two basic groups, which is owing to the political division of the country. The first group includes institutions operating on the central level, which form the science, innovation, and education policies for the entire state. Among all central institutions, the Ministry of Science and Innovation (MICINN) plays the most important role. It was established in April 2008 and is politically responsible for almost all actions related to R&D in Spain. In 2011, owing to its central role, the Ministry received for its activity as much as 68 % of funds allocated by the government to science in the central state budget (GBAORD). The Spanish Ministry of Science and Innovation is formally responsible for creating, planning, and coordinating the crucial instrument of science and innovation policy, which is the 4-year National Plan for Scientific Research, Development, and Innovation. Nevertheless, the final version of the plan must first be approved by the Interministerial Commission for Science and Technology (CICYT), chaired by the prime minister of Spain. The Commission also includes representatives from the ministries of the highest level, which are also involved in R&D. Eventually, the National Plan is approved by the Spanish parliament, although, during the legislative process, the parliament consults the parliamentary committee responsible for the research and the parliamentary committee on industry (Heijs 2011).

Another important central institution in Spain responsible for accomplishing goals set in the National Plan is the Ministry of Industry, Tourism, and Trade (MICYT). However, it is primarily responsible for shaping the innovation policy concerning the ICT. What is more, the Ministry proposes and puts in practice the R&D policy in the domain of industry, trade, energy, small- and medium-sized enterprises, tourism, telecommunications, and the information society. In 2011, the Ministry had spent 26.8 % of funds from the GBAORD budget for its activities in this area. A significant role in the Spanish innovation system is also played by the Ministry of Defense, which is in charge of the science and innovation policy in the domain of national security. The Ministry of Public Works is, in turn, responsible for R&D of innovation policy in matters of technology, engineering, construction, and environmental protection (Heijs 2011).

Since 2008, as a result of reforms in the Spanish innovation system, the Ministry of Education has again been responsible for the higher education, however, it has little influence on research. For these reasons, the Ministry received only 2.5 % of the GBAORD budget in 2011. As for the higher education, it is under the authority of individual regions in Spain, which are responsible for its financing.

The second group in Spain's innovation system is constituted by institutions operating on the regional level, which shape, independently of each other, the science, innovation, and education policies in 17 autonomous communities of Spain. The dualism of the Spanish innovation system is a result of a very strong position of regional authorities of the autonomous communities, which have been trying to become dependent on the central government in recent years. Although central institutions still remain the main decision makers in the innovative policy of the country, the fact that regional authorities are responsible for the science, innovation, and education policies on the regional level, as well as for the funding of actions taken in these domains, makes the Spanish innovation system very complex (Muñoz and Espinosa Díaz 2000b).

Competencies of regional authorities in the areas of science, innovation, and education are the same in all regions of Spain. Most of the autonomous communities of Spain have their own Regional Plan of Research, Development, and Innovation, which is, to a large extent, patterned after the National Plan. The actions of the autonomous and national authorities are coordinated by the General Council. Unlike governments of other states, the Spanish government is authorized to form only general framework for the research, development, and innovation policy. In turn, the autonomous authorities can formulate and finance their own science policies, research projects, infrastructure, and other kinds of R&D activities (Sanz-Menéndez and Cruz-Castro 2005). As a result, different areas overlap to a very large extent, because the high degree of decentralization of education in a region is not accompanied by any coordination. Owing to these reasons, since 2008, the Spanish government is trying to achieve better integration of central and regional policies in the area of science, innovation, and education (Uyarra 2010).

In Spain's innovation system, an important role is played by intermediate bodies, which are responsible for supporting innovative activity of companies, in particular by facilitating the flow of information, knowledge and technology between research institutions and industry.

In Spain, there are three main entities that are responsible for the management and implementation of the research, development, and innovation policy at the central level:

1. CICYT, which is in charge of elaborating, coordinating, and evaluating the research, development, and innovation policy, in particular in relation to the National Plan. Moreover, the interministerial commission coordinates the work of the ministries in the domain and oversees the Spanish institutions involved in international projects. The actions taken by the CICYT are supported by the General Council of Science and Technology, which in turn coordinates relations between the autonomous regions, as well as between the regions' authorities and the central administration.
2. The Spanish Foundation for Science and Technology (FECYT) is the primary instrument used by central authorities to influence the research activity of subordinate entities. The priority objective of the FECYT is to financially support academic institutions, research centers, and companies in research, which will make the Spanish model more knowledge- and innovation-based. Another important goal of the FECYT is to play the role of the leader in large scientific networks, technological parks, or even in centers such as technological museums.

3. The Center for the Development of Industrial Technology (CDTI) is supposed to support the innovativeness and technological development of Spanish companies. Its mission is "to help Spanish companies to increase their technological profile." Since 2009, the CDTI is directly subordinated to the Ministry of Science and Innovation, which deals with the funding and support of R&D projects of Spanish companies. At the national level, the Center assesses and manages funds allocated to companies engaged in research, development, or innovation projects. The CDTI also offers a support program for the creation and consolidation of technology companies. At the international level, the Center organizes and promotes the participation of Spanish companies in technological cooperation programs, e.g., European Framework Programmes such as EUREKA or IBEROEKA. What is more, the CDTI manages and supervises Spanish participation in large international research institutions such as the European Space Agency (ESA).

The above-mentioned institutions are complemented by public or private funds or foundations, which have been established by banks or companies to finance research in specific areas. Some of these foundations are national, e.g., Instituto Salud Carlos III Health Research, the role of which is to finance research in the field of biomedicine and other health sciences. The institution is also responsible for promoting research and for commenting on regulatory issues such as biobanks.

Some of the nonprofit foundations created with public or private funds concentrate their activity on issues related to the science, innovation, and education policy. Among these entities, the FECYT has a special place at the national level, while the Catalan Foundation for Research (Fundació Catalana per la Recerca) constitutes a good example of such enterprise at the regional level.

Since 1980, science and technology parks have become key players in the Spanish system of innovation. In 1988, the Association of Science and Technology Parks of Spain (APTE), which today includes 53 members (19 partners and 34 affiliates), was established. Only 18 parks operate in Spain. The rest of them is still at the planning or early development stage. In Spanish science and technology parks, the dominant role is played by companies in the ICT sector, which hire almost 25 % of all researchers.

In recent years, the activity and the development of science and technology parks was supported primarily by local authorities. The actions of Catalan government aiming at the creation of a park specialized in biomedicine in Barcelona in 2005 may serve as an example of cooperation in the domain. Owing to the cooperation of Catalan government, the City Council of Barcelona and the Pompeu Fabra University, a new technology park was built next to a hospital. Today, however, it is financed mainly by the private sector. The involvement of the private sector in the construction of the park is especially owing to the fact that the majority of pharmaceutical companies and their laboratories are located in the region of Catalonia. What is more, three large international companies—Novartis, Bayer, and Sanofi—have their headquarters in Barcelona.

The cooperation between research institutions and companies is supported by communications offices (ILO). These offices try to determine technological needs of socioeconomic sectors and, afterward, stimulate the technology transfer between public and private sectors. In this way, they contribute to the use and commercialization of the results of R&D of universities and public research centers. As for the European Economic Innovation Centers, they offer entrepreneurs integrated consulting and services related to the preparation of actions, which are necessary to transfer technology and knowledge.

In the Spanish innovation system, besides ILOs, there are also bodies and organizational units subordinated to the regional or local administration, the objective of which is to support innovations and facilitate the performance of activities in this area. These are public entities of different legal forms, e.g., the Spanish Patent and Trademark Office (OEPM) or regional and local development agencies.

In the Spanish innovation system, important entities in the national and regional innovation systems are universities, research centers, and institutions, which promote and transfer knowledge and technology to companies. Currently, there are about 60 public universities across the country, which are controlled by regional authorities. The authorities finance the educational and research activity of the universities, which are, however, autonomous organizations operating on the basis of their own charter. Nevertheless, all Spanish public academic institutions play a social role by representing the local community and the private sector. The Council approves academic institutions' budgets and, as a consequence, indirectly determines the direction of their development in the field of education and research. As for the contact with the government, academic institutions are represented by the Association of Rectors (Conferencia de Rectores de Universidades Españolas), which includes all rectors of Spanish universities.

The activity of universities in the domain of cooperation and knowledge transfer to companies is supported by University–Enterprise foundations (FUE) established by universities and chambers of commerce in regions. Their task is to create strong links with companies, in order to make the exchange and transfer of knowledge possible. In 2011, there were 33 FUEs.

Besides universities, an important role in the field of research is also played by public research centers, which are supervised by individual ministries. National research institutions and universities constitute the core of the Spanish innovation system and are responsible for the implementation of most of the activities set in the National Plan for Scientific Research, Technological Development, and Innovation. The largest public center in the Spanish innovation system is the CSIC, which associates over 100 research centers related to all fields of science. Unlike universities, these centers do not have strong links with industry. However, in recent years, some of the research centers, in particular those related to biology, agriculture, or oceanography, have started to informally cooperate with companies from their surroundings. As a result, industry is informally influencing their research programs, which has recently been supported by central and regional authorities (Heijs 2011).

Table 6.31 Spain's innovative capacity in the selected years of the 1995–2009 period. (Source: Own work on the basis of OECD data and Eurostat Statistics Database)

Index		1995	2000	2005	2007	2009
Share of total internal expenditure on R&D in GDP (%)	Spain	0.79	0.91	1.12	1.27	1.38
	EU27	1.66	1.74	1.74	1.77	1.90
Share of government internal expenditure on R&D in GDP (%)	Spain	0.15	0.14	0.19	0.22	0.28
	EU27	0.28	0.25	0.25	0.23	0.26
Share of business internal expenditure on R&D in GDP (%)	Spain	0.38	0.49	0.60	0.71	0.72
	EU27	1.02	1.11	1.08	1.12	1.16
Percentage of internal expenditure on R&D financed by the government (%)	Spain	43.56	38.64	42.99	43.66	NDA
	EU27	39.55	35.40	35.10	33.95	NDA
Percentage of internal expenditure on R&D financed by other national sources (%)	Spain	5.22	6.75	4.97	3.88	NDA
	EU27	1.82	2.13	2.36	2.49	NDA
Percentage of internal expenditure on R&D financed by foreign countries (%)	Spain	6.69	4.88	5.74	7.01	NDA
	EU27	6.63	7.15	8.90	8.95	NDA
Percentage of internal expenditure on R&D financed by industry	Spain	44.53	49.73	46.29	45.46	NDA
	EU27	52.00	55.29	53.79	54.60	NDA
Households with access to the internet (as % of all households)	Spain	NDA	NDA	35.51	44.63	53.96
	EU27	NDA	NDA	48.4	54.1	65.2
Human resources for science and technology as a percentage of working population (%)	Spain	24.30	31.30	36.50	37.70	37.20
	EU27	NDA	31.7	35.6	37	38.1
Investments in VC as % of GDP	Spain	0.00	0.03	0.01	0.01	0.00
	US	0.04	0.27	0.04	0.05	0.05

6.3.2.1 Assessment of Innovative Capacity and Position of Spain

In order to establish to what extent the competitiveness of Spain's economy was conditioned by the technological advancement, an empirical analysis of innovative capacity and position will be conducted in this subchapter. The essential data are contained in Table 6.31.

The fact that the government undertakes significant actions for the technological development of Spain is proven by the fact that the share of the total internal expenditure or R&D in GDP almost doubled in the analyzed period. In 1995, the share of the total internal expenditure or R&D in GDP was 0.79 % of GDP. It increased to the level of 1.38 % of GDP in 2009. However, in 2009, despite the efforts undertaken by the government, the total internal expenditure on R&D in the GDP in Spain was lower than the average recorded in the EU Member States, which amounted to 1.90 % of GDP.

An increase in expenditure on R&D is not always accompanied by an increase in innovativeness of the economy. The extent of the involvement of companies in the commercialization of scientific research can be illustrated by, *inter alia*, indexes describing the share of different sources in the R&D funding, including government expenditure, of other national sources, international programs (mainly EU) and companies. In 2007, the share of the government expenditure in expenditures on R&D in Spain amounted to 43.66 %. At the same time, the analogical share recorded in the EU Member States was equal to 33.95 % of GDP. The level of

Table 6.32 Spain's innovative position. (Source: Based on Eurostat Statistics Database and WIPO data)

Index	1995	2000	2005	2007	2008
Share of employment in HT and MHT industries in the total employment (%)	5.25	5.37	4.67	4.47	4.78
Share of employment in knowledge-intensive service sectors in relation to the total employment (in %)	22.18	24.55	26.86	28.19	28.89
Number of patent applications per US$ 1 billion of funds allocated to R&D	n/a	0.33	0.25	0.22	n/a
Number of patent applications per million inhabitants	51.97	67.31	70.05	72.80	n/a
Share of export of HT goods in total exports	5.81	6.37	5.65	4.92	n/a

financing expenditures on R&D from public resources in Spain, higher than the average noted in other EU Member States, has to be assessed negatively. As for the share of business expenditure on R&D (BERD) in Spain, it increased in the analyzed period and amounted to 45.46% at the end of 2007. Nevertheless, the share was lower than the average share of BERD in the EU27, which was almost 55% in the same period. In 2007, the percentage of internal expenditure on R&D financed by foreign countries was also 1 p.p. lower than the average noted in the EU27. The comparison of these data with the results recorded in such Member States as Finland or Germany shows a relatively weak position of Spain. It confirms that the cooperation of research centers and companies in Spain is weak, which has a negative influence on the level of the economy's competitiveness. The weakness of the link between science and economy is also proven by the low prevalence rate of VC investments, which is an effective source of funding innovative activity in the case of many developed economies. In 2009, the share of such investments in the GDP in the USA reached the value of 0.05%, while in Spain it was virtually zero.

Another index illustrating the innovative capacity of an economy can be the one indicating the spread of use of information technology. It can be measured by the percentage of households with Internet access. This is the domain in which a significant improvement can be observed in Spain, as the share of households with Internet access increased from 35.51% in 2000 to 53.96% in 2009. Despite the increase, the index of Spain remained on a lower level than the EU average, which was equal to 65.2%. Moreover, in the past, Spain noted a higher percentage of people employed in science in relation to the working population than the average recorded in the EU. Unfortunately, in 2009, the index had deteriorated in Spain by decreasing to 37.2%, while the average of its value in the EU had increased over the same period to 38.1%.

Table 6.32 presents indexes concerning Spain's innovative position.

In the 1995–2008 period, a decrease from 5.25 to 4.78% in the share of employment in HT and MHT industries in the total employment was recorded. The decline was, to some extent, compensated by an increase from 22.18% in 1995 to 28.89% in 2008 in the share of employment in knowledge-intensive service sectors in the total employment.

While assessing the level of innovativeness of an economy, it is also important to take into consideration the indexes concerning the intellectual property protection, which indicates, to a large extent, the technological advancement of companies. In Spain, a continuous decline in the number of patent applications per US$ 1 billion of funds allocated to R&D form 0.33 in 2000 to 0.22 in 2007 was noted. At the same time, however, another index had been improving, i.e., the number of patent applications per million inhabitants increased from 51.97 in 1995 to 72.87 in 2007, which may indicate a positive trend in the effectiveness of R&D. Nevertheless, a relatively low competitiveness of Spanish companies was confirmed by the relatively low share of export of HT goods in the total exports. In the 1995–2008 period, the index had decreased from 5.81 to 4.92 % in 2007.

6.3.2.2 Assessment of Human Capital

The demographic structure of a country and the quality of human capital has a crucial meaning for the long-term competitive advantage of the country. In terms of the population number, Spain, which is home to more than 46 million inhabitants, is one of the biggest countries in Europe. Unfortunately, just like in the case of other European developed economies, it is possible to observe negative long-term demographic trends there, such as the negative birth rate, which leads to a decrease in population. The persistence of low rates of population growth in Spain may have negative consequences for the economy and the sustainability of public finance. Both these problems are illustrated in the projection of the proportion of working- and retirement-age population until 2050 in comparison with the data from the years 1995–2010. According to the OECD data, in comparison with 2010, the proportion will decrease from 4.01 to 2.94 in 2025. If the current trends remain unchanged, in 2050, the number will decrease to 1.48. It means that the percentage of working-age population in relation to the total population will drastically sink, influencing the economic growth and innovativeness of the economy. Furthermore, the aging population will become a huge burden for the public finance. As a result, it may be very difficult to maintain the current social security system in Spain in the future.

The second factor forming the innovativeness of the economy is the quality of human resources. In Spain, it is possible to observe a positive trend in the recent years in the level and structure of education of the labor force, as opposed to the demographic changes, as it is shown in Table 6.33.

The World Bank's data confirm the increase in labor force resources in Spain from 16.4 million people in 1995 to 22.9 million people in 2009. At the same time, it is possible to observe an increase from 16 % in 1995 to 24 % in 2007 in the share of labor force with secondary education, as well as of labor force with higher education, form 20.3 % in 1995 to 31.8 % in 2007. These changes took place at the expense of a decrease in the share of labor force with primary education, from 56.1 % in 1995 to 43.8 % in 2007. It means that more and more people decided to continue their education after completing primary school. As a result, the quality of human capital in Spain had been increasing throughout the examined period.

Table 6.33 Data on the skill level of labor force in Spain. (Source: Based on the World Bank Education Statistics data)

Index	1995	1998	2000	2001	2003	2005	2007	2008	2009	2010
Total labor force resources (in thousands)	16.389	17.123	18.049	18.046	19.545	20.929	21.601	22.765	22.911	n/a
Share of labor force with primary education (%)	56.1	52.8	53.3	52	n/a	45.3	43.8	n/a	n/a	n/a
Share of labor force with secondary education (%)	16	17.6	19.2	19.6	n/a	23	24	n/a	n/a	n/a
Share of labor force with higher education (%)	20.3	27.8	26.9	27.9	n/a	31.3	31.8	n/a	n/a	n/a
Expenditures on higher education as % of GDP	n/a	n/a	20.54	21.44	22.66	22.69	25.09	27.25	n/a	n/a
Expenditures on secondary education as % of GDP	n/a	n/a	23.24	22.65	23.39	23.30	24.15	25.80	n/a	n/a
Expenditures on primary education as % of GDP	n/a	n/a	17.53	17.96	18.57	19.04	19.40	20.33	n/a	n/a
Average number of years of education of people aged 25+	7.73	n/a	9.09	n/a	n/a	9.76	n/a	n/a	n/a	10.35

Owing to the growing significance of education of population for the economy's competitiveness, the Spanish government attaches more and more importance to investing in human capital. It is reflected in a visible increase in the share of expenditure on education in the GDP on every education level: on the higher education level from 20.3 % of GDP in 1995 to 31.8 % of GDP in 2008, on the secondary education level from 23.24 % of GDP in 2000 to 25.8 % of GDP in 2008, and on the primary education level from 17.53 % of GDP in 2000 to 20.33 % of GDP in 2008. In Spain, the positive trend concerning the education of labor force is also confirmed by the index related to the average number of years of education of people aged 25 years and over, which had increased from 7.73 in 1995 to 10.35 in 2010.

The positive trends observed in the domain of the level of education of the human capital do not, unfortunately, directly translate into its use. For many years, the biggest problem of the Spanish economy has been the high unemployment rate, especially among the young. Although during the economic boom the unemploy-

Table 6.34 Productivity of human capital in Spain. (Source: OECD data)

Item	1995	1998	2000	2005	2009
Wage per hour (in PPP), in US$	15.73	16.81	17.62	20.24	24.45
Annual wage per employee (in PPP), in US$	26,237	2,8274	29,725	33,083	39,503
Labor costs per production unit	0.68	0.67	0.67	0.63	0.62

ment rate had declined very fast from 22.7 % in 1995 to 8.3 % in 2007, afterward it increased again to 18 % in 2009 due to the crisis. A similar trend could have also been observed in the case of the unemployment rate related to the young in Spain. In the years 1995–2007, it decreased from 40.45 to 18.19 %. Then, it increased to 37.85 % in 2009. Taking into consideration the current demographic trends in Spain, it may result in a significant decline in competitiveness of the country.

During the last decade, a significant increase in average wages could have been observed in Spain, from the level of US$ 15.73 (in PPP) per hour to US$ 24.45 (in PPP) per hour in 2009. A similar trend was recorded in relation to the wage per employee, which increased from US$ 26,237 (in PPP) in 1995 to US$ 39,503 (in PPP) in 2009. Over the analyzed period, a decrease from 0.68 in 1995 to 0.62 in 2009 in the labor costs per production unit was noted at the same time, which means that the competitiveness of Spain had increased despite the increasing wages (Table 6.34).

6.3.2.3 Characteristics of Spain's Science, Innovation, and Education Policy

In 2011, the Spanish government implemented a number of measures set in the National Reform Program (NRP) and the new National Plan for Scientific Research, Technological Development, and Innovation.

In Spain, the main instruments used within the national and regional innovation policy are tax incentives and direct funding of scientific and technological projects. In recent years, the use of instruments of the EU's regional policy, which, to a large extent, included the development of modern R&D infrastructure and the support of innovativeness, was very significant for the science, innovation, and education policy in Spain. In the adopted operational plans cofinanced by the structural funds, emphasis was placed on priorities related to research, technological development, and innovations.

The summary of the assessment of Spain's innovativeness is the SWOT analysis of the NIS (Table 6.35).

6.3.3 The Relationship Between the Level of Innovation, Human Capital Accumulation, and the Advantages in the Foreign Trade

Spain is an example of a country in which, for years, great importance has been attached to the improvement of innovativeness of the economy. As a result, large expenditures have been allocated to R&D. Unfortunately, despite the actions

Table 6.35 The assessment of Spain's NIS. (Source: Own work)

Opportunities	Threats
Increase in expenditure on R&D, including the EU funds	Serious economic problems, which may be accompanied by a decline in expenditure on R&D
Benefits of the increase in competition and of the opportunity to participate in the single European market and of the inflow of foreign investment	High unemployment rate, in particular among the young
Greater use of the opportunities arising from a strong position of large Spanish companies	Deteriorating demographic situation, which may affect the economic growth in the future
Opportunity to develop international links and international cooperation within the EU and with Latin America	Outflow of high-skilled labor force
Growing number of patents per capita	Continuously low level of internationalization of the whole society
Regions with strong research centers, which are associated with companies operating in their area	Persistent stratification in the economic and scientific potential between the regions
Further development of innovative industrial clusters	
Strengths	Weaknesses
Government's determination to develop a knowledge-based economy and support for companies engaged in innovative activities	High government expenditure on R&D
High total expenditure on R&D	Weak links and cooperation between national and regional research centers
Relatively open society and high level of internationalization of universities and research centers	Weak links between companies and universities and research centers
High rate of international publications	Low efficiency of universities and public R&D centers in the commercialization of research
Universities and research centers, which are among the best in the world	Lack of fully developed R&D system and its integration on the national, regional, and industrial level
Improvement in research infrastructure	Low share of companies in R&D funding, including VC
Improvement in structure of education of the population	
Favorable legal and tax environment for business, including the one related to innovations	

undertaken, the transfer of knowledge from universities and research centers to the economy remains at a relatively low level.

As for the turnover of Spanish goods and services, the exchange of goods of relatively high or medium level of technological advancement has been crucial. The export of high and medium technology, as well as the import of this group of goods, constitutes almost 50 % of Spain's foreign trade today. While analyzing the commercial results in the context of the National Innovation Plan, it is important to point out the fact that the increase in the value of Spanish exports over the last two decades, including the change of the exports structure, was possible thanks to the expenditures on R&D and to the increasing internationalization of Spanish companies. In the past, an important factor influencing the growth in exports was a very high inflow of foreign direct investment to Spain. The inflow was related to an increasing availability of skilled labor force in Spain and slightly lower labor costs in comparison with, for example, Germany. It indirectly indicates a positive relationship between the expenditures on the development of human capital and exports from Spain.

Until the middle of the 1990s, an important characteristic of Spain was the development of economy compliant with a strategy, which is typical for so-called catch-up countries. It was reflected in a negative balance of disembodied technical knowledge, including the patent and license turnover (Misala 1992, 1993). The strategy is more typical for developing countries than for highly industrialized countries, which rely on domestic innovations. At the time, the problem was in the commercialization of knowledge and in the weakness of its transfer from academic institutions and research centers to the industry. What is more, as noticed by Laaser (1997), Spain had, to a large extent, lost the ability to economically catch up with the most developed countries of the world in terms of the development of international division of work of interindustry nature, already at the beginning of the 1990s, which has been proven by the current economic crisis.

As a result, it is possible to conclude that the efficient functioning of the NIS had contributed to the improvement in the competitiveness of Spanish exports in the first period, i.e., in the second half of the 1990s. However, the weakness of the system and, in particular, insufficient links between companies and universities and research centers constitute the factor, which reduces the rate of growth of the Spanish economy. On the one hand, Spanish companies play, to a large extent, a significant role in foreign markets and have a competitive advantage, in particular in the financial and construction sector. In Spain, the emergence of competitive international companies was possible thanks to the increase in the quality of human capital, as well as to the high openness of the society.

On the other hand, in this case also the majority of Spanish companies operated in sectors characterized by a low level of innovativeness. In the face of the present crisis and the decline in the competitiveness of the Spanish economy, the Spanish government is again trying to lead to an increase in the significance of the NIS in the creation of competitive advantages of companies. With this end in view, a reinforcement of links between universities and research centers and companies has now become the priority of the innovation policy.

6.4 Case Study: The USA

Tomasz Napiórkowski

The idea behind this work is to look at the connection between the human capital and the trade RCA in the US economy.

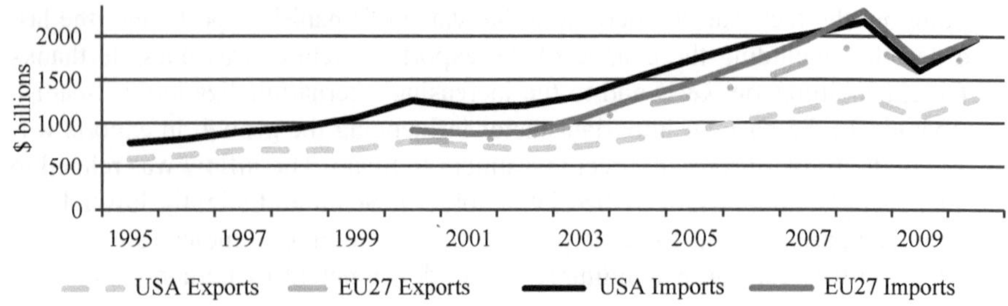

Fig. 6.24. US and EU exports and imports and trade deficit (USA only) from 1995 to 2010 (in US$). (Source: Author's own graph based on data from UN Comtrade)

6.4.1 Evaluation of Trade Competitiveness

To put it in perspective, the US trade, in some of the following sections, USA will be compared with the EU.

US exports (Fig. 6.24) increased in the last decade from $ 582,964 million (M, 1995) to $ 1,277,109 M (2010). Owing to the dotcom bubble, exports saw their first decline. The financial recession of 2007 caused exports to plummet in 2009 to a level only $ 19,682 M higher than the one seen in 2006. Exports of the EU from 2000 followed nearly the exact pattern of the US exports.

US imports (Fig. 6.24) moved from 1995 ($ 770,821 M) to $ 1,966,496 M in 2010 with dips in 2001 and 2009. Comparing the US imports with those of the EU (Fig. 6.24), it is apparent that the US imports have been significantly higher from 2000 to 2007 and that both values are nearly identical in 2010 with EU imports standing at a level of $ 1,975,721 M. The US trade deficit has increased from $ 187,856 M in 1995 to $ 689,387 M in 2010 with a small dip in 2001 and a sharp decline in 2009 (down by $ 319,751 M).

In 1995, US exports to its top 10 destinations (Fig. 6.25) amounted to $ 378,620 M with the top three spots being taken, respectively, by Canada ($ 126,021 M), Japan ($ 64,259 M), and Mexico ($ 46,309 M). In 2000, the list of top 10 destinations of US products has stayed the same but the value of exports to those countries increased in size to $ 536,672 M with Mexico moving to the number two slot ($ 111,710 M) beating Japan ($ 65,251 M) while Canada remained the number one recipient of US exports ($ 176,408 M). In 2005, China moved from the number 11 (1995, 2000) to the number 4 spot as a recipient of US exports. In 2010, the total value of US exports to its top 10 trading partners reached $ 798,730 M with Canada ($ 248,186 M) and Mexico ($ 163,320 M) at the top two spots and China importing $ 91,878 M.

Examining top countries that export their goods to the USA (Fig. 6.26) in 1995, the top three spots are taken by Canada ($ 148,277 M), Japan ($ 127,195 M), and Mexico ($ 62,745 M). In 2000, the top three stays without change but with significant increases in the amount of goods imported by the USA. Just like in the case of exports, the biggest change in 2005 comes from the movement of China ($ 259,829 M) as now the second biggest exporter to the USA behind Canada

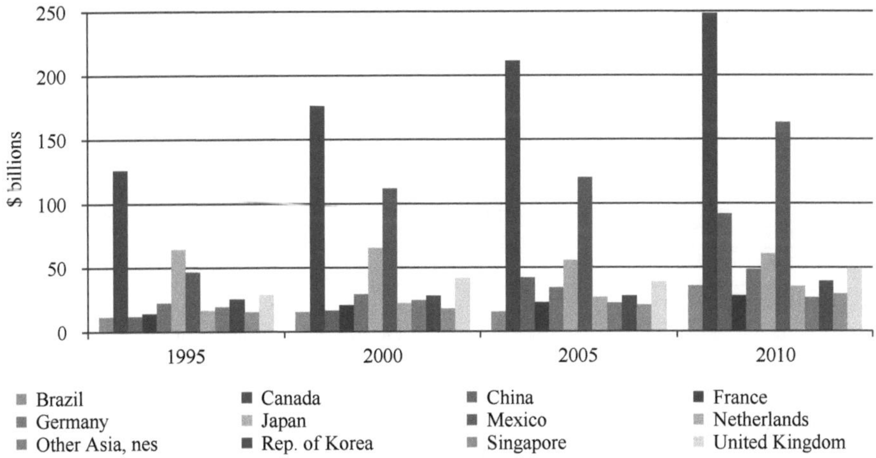

Fig. 6.25 Top recipients of US exports in 1995, 2000, 2005, and 2010 (in US$). (Source: Author's own graph based on data from UN Comtrade)

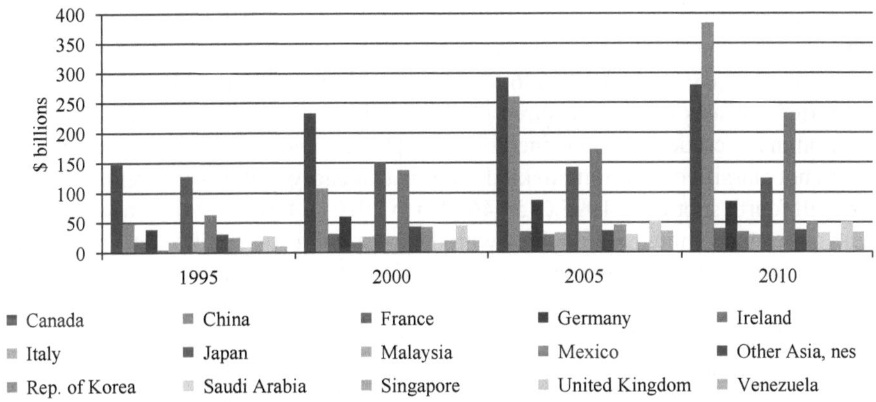

Fig. 6.26 Top exporters to USA in 1995, 2000, 2005, and 2010 (in US$). (Source: Author's own graph based on data from UN Comtrade)

($ 291,865 M) and ahead of Mexico ($ 172,480 M). In 2010, China is the biggest exporter to the USA ($ 382,953 M). Canada ($ 279,275 M) moved to the second and Mexico ($ 231,919 M) stayed at the third place.

In summary, the biggest change came in the form of China highly increasing its position as a very significant trading partner to the USA. Also, with few exceptions, both lists of top 10 partners stayed unchanged, which suggests little change in the nearby future.

While examining the US exports as classified according to the SITC (Fig. 6.27), it is very apparent that all categories can be divided into the three groups. The first group is the one in which US exports are the greatest in value: machinery and

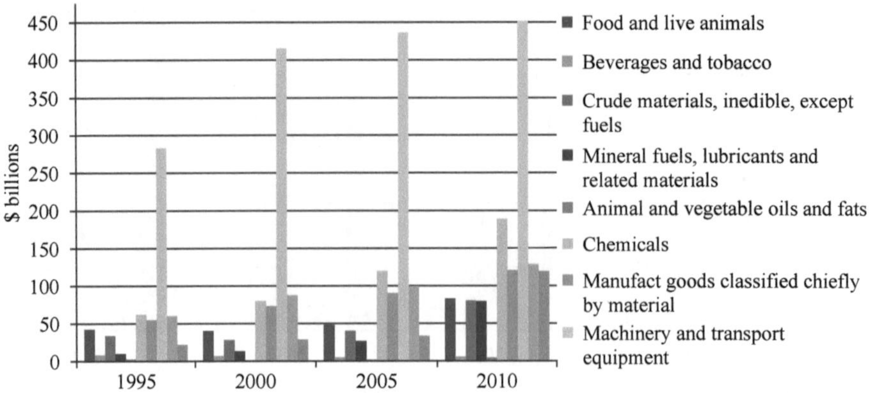

Fig. 6.27 US exports according to the SITC classification in 1995, 2000, 2005, and 2010 (in US$). (Source: Author's own graph based on data from UN Comtrade)

transport equipment, chemicals, and miscellaneous manufacturing articles. Machinery and transport exports constantly stay at the overall top spot with their value increasing from $ 282,736 M in 1995 to $ 451,752 M in 2010. Chemicals stand as the second highest category with the value of US exports in that category increasing from $ 62,089 M to $ 188,836 M. The second group in the order of the amount exported consists of manufactured goods classified chiefly by material, food and live animals, crude materials, inedible, except fuels and commodities and transactions not classified according to kind. The first category in this group remains at the overall fourth spot from 1995 ($ 54,468 M) to 2010 ($ 120,304 M). Similarly, other categories have more than doubled with the most interesting shift in this period being made by commodities and transaction-oriented exports that increased in importance from the overall seventh to fifth place and from $ 21,694 M to $ 118,686 M in value. The last group that comprises of mineral fuels, lubricants and related materials, beverages and tobacco, and animal and vegetable oils and fats is admittedly the most interesting owing to its dynamic. Mineral fuels, lubricants, and related materials remained number one it this group ($ 10,101 M in 1995, $ 79,394 M in 2010). Beverage and tobacco exports have been decreasing from 1995 ($ 8,086 M) to 2005 ($ 4,458 M) then slightly increasing in 2010 ($ 5,457 M). Animal and vegetable oils and fats remained at their last spot registering a small dip in 2000 with their final value in 2010 being equal to $ 4,230 M.

The main conclusion is that US exports from 1995 to 2010, owing to being able to be grouped into three above categories, have been highly structured and mostly focused on machinery and transport equipment and chemicals paying the least attention to animal and vegetable oils and fats.

US imports classified according to SITC (Fig. 6.28) can similarly be divided into groups representing high, medium, and low imported goods. Again, machinery and transport equipment are the goods with the highest value imported ($ 347,848 M in 1995, $ 720,428 M in 2010) followed by manufactured goods classified chiefly

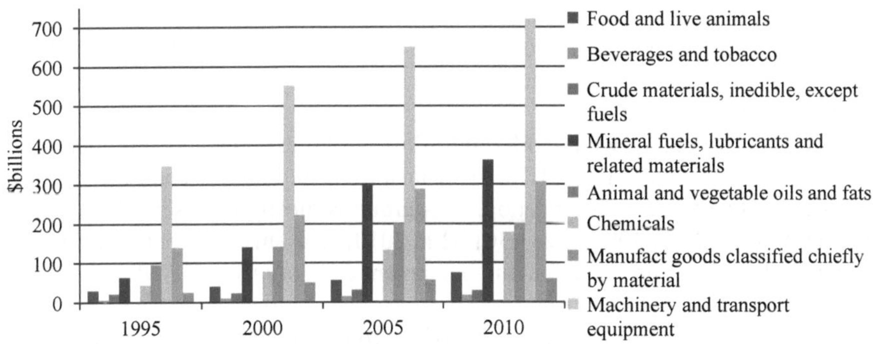

Fig. 6.28 US imports according to the SITC classification in 1995, 2000, 2005, and 2010 (left-hand axis in US$). (Source: Author's own graph based on data from UN Comtrade)

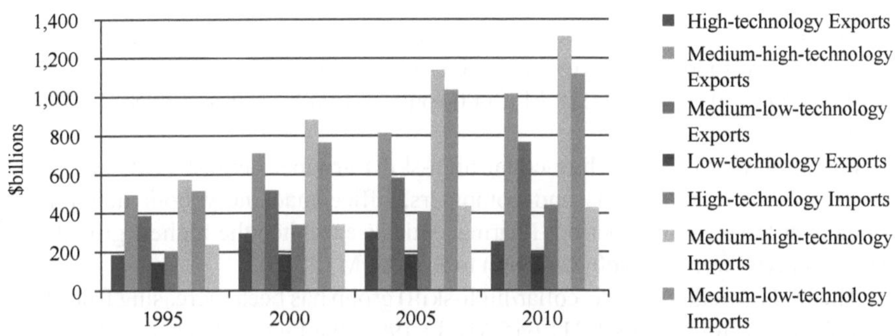

Fig. 6.29 US trade classified according to the technological level and collar/skill assignment (left-hand axis in US$). (Source: Author's own graph based on data from UN Comtrade)

by material ($ 94,359 M in 1995, $ 201,503 M in 2010) and miscellaneous manufactured articles ($ 138,686 M in 1995, $ 306,369 M in 2010). The biggest shift in importance was made by Mineral fuels, lubricants, and related goods that were number four in 1995 ($ 62,952 M) and number two in 2010 ($ 361,993 M). The middle of the group consists of food and live animals with the exception being the chemicals ($ 42,85 M in 1995, $ 178,293 M in 2010). Crude minerals, inedible, except fuels stayed at the eighth spot while only increasing from $ 21,410 M to $ 29,282 M. Beverages and tobacco and animal and vegetable oils and fats played the least significant role increasing from $ 5,673 M and $ 1,511 M to $ 17,474 M and $ 4,525 M, respectively.

This section will look at US exports and imports (Fig. 6.29) as classified according to technological intensity and according to collar/skill assignment (Hatzichronoglou, 1997).

US exports in HT (white-collar/high-skill) increased from 1995 ($ 185,039 M) to 2005 ($ 300,140 M) registering a drop in 2010 ($ 250,709 M). The two biggest components of this group (electronics–communications and computers, office machinery) have only seen growth in value from 1995 to 2000. In addition, pharmaceuticals (a group with the lowest value in 1995) has changed places with the aerospace category value of which dropped significantly from $ 22,031 M in 1995 to just $ 4,781 M in 2010.

The MHT (medium white-collar/high-skill) group exhibits a steady increasing trend reaching a level of $ 1,012,984 M making it the biggest contributor to exports according to this classification. This group is mostly represented by chemicals ($ 123,334 M in 1995, $ 281,301 M in 2010). The smallest contributor to this group is the other transportation and equipment category ($ 3,457 M in 1995 and $ 7,056 M in 2010).

The MLT (medium blue-collar/low-skill) group's biggest categories are petroleum refining and fabricated metal products that have increased from $ 104,065 M and $ 98,336 M in 1995 to $ 226,152 M and $ 222,853 M in 2010, respectively. Ferrous metals have increased the most-fold from $ 6,666 M (1995) to $ 19,537 M (2010).

The LT (blue-collar/low-skill) group contributes the least to exports and is dominated by food, beverages, and tobacco category that moved in value from $ 74,341 M (1995) to $ 93,228 M (2000) only to finally decrease to $ 83,13 M in 2010.

For imports, the HT (white-collar/high-skill) group's biggest contributors are electronics–communications and computers, office machinery both doubling in value over the examined period. Pharmaceuticals exhibited the highest growth rate as they moved from $ 8,860 M (1995) to $ 75,84 M (2010).

The MHT (medium white-collar/high-skill) group has been increasing from 1995 ($ 572,311 M) to 2010 ($ 1,311,645 M). In 1995, the biggest backer to this group were imports in the motor vehicles category ($ 128,249 M), which placed at the number four spot in 2010 ($ 241,473 M) behind scientific instruments ($ 321,110 M), chemicals ($ 312,429 M), and nonelectrical machinery ($ 275,599 M).

The MLT group (medium blue-collar/low-skill) is the second biggest group ($ 1,118,508 M in 2010). Year 1995 has been led by rubber and plastic products ($ 118,518 M) and fabricated metal products ($ 117,465 M). Since 2000, the latter category continued its dominance till 2010 when it reached $ 315,917 M. Shipbuilding has been the smallest contributor throughout the period.

LT (blue-collar/low-skill) imports have been dominated by the textile and clothing category ($ 200,955 M). This set is the only one that exhibited a decrease in value in 2010 from 2005.

This part examines changes in the US trade by dividing it into groups according to the type of inputs required for production (Peneder 1999; Fig. 6.30). The industry group that contributed the most to exports and imports from 1995 to 2010 has been the technology-driven one with LI and MDI playing the smallest role in exports and imports, respectively.

The MM group has been increasing steadily reaching a value of $ 263,531 M in 2010. The group is led by other special purpose machinery and machinery for production, use of mechanical power both more than doubling in value from 1995 to 2010.

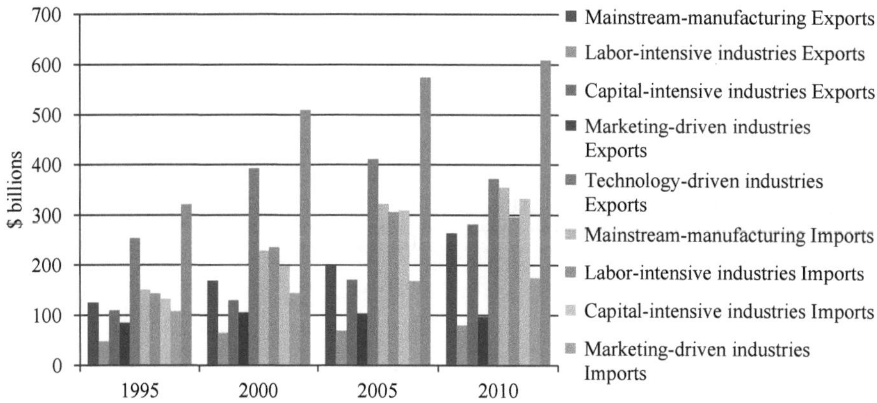

Fig. 6.30 US trade classified according to inputs (in US$). (Source: Author's own graph based on data from UN Comtrade)

The LI group has also enjoyed a continuous growth. In 1995, it was championed by electrical equipment n.e.c. category ($ 7,447 M) that has increased in value to $ 10,905 M in 2010. The most significant growth has been seen in the jewelry and related articles category as it moved to the dominant spot in 2010 ($ 21,441 M; $ 3,263 M in 1995). Leather clothes and other wearing apparel and accessories categories both decreased by over a third in value.

CI ($ 281,416 M in 2010), led by basic chemicals, have more than doubled in value. Refined petroleum products category has exhibited the most drastic growth, growing in value from just $ 5,736 M in 1995 to the second spot from the top in 2010 with $ 60,972 M.

The MDI group has experienced a decline in 2010 ($ 96,235 M) after increasing in 2000 to $ 104,925 M. The group's biggest contributor is the reproduction of recorded media category while the biggest decline from 1995 ($ 5,263 M) to 2010 ($ 494 M) has been seen in the tobacco products category.

TDI have moved from $ 253,810 M in 1995 to $ 372,092 M in 2010. Pharmaceuticals have exhibited the highest growth rate ($ 9,077 M in 1995, $ 51,679 M in 2010) while the leader of the group in 1995 (electronic valves and tubes, other electronic components) has increased from $ 42,607 M to only $ 51,715 M in 2010.

From the import point of view, the MM group constantly grew from 1995 ($ 149,978 M) to 2010 ($ 345,937 M). The main driver of this group is the machinery for production, use of mechanical power category ($ 17,373 M in 1995, $ 48,258 M in 2010). The biggest growth was seen in the weapons and ammunition category that increased from $ 708 M (1995) to $ 4,047 M (2010).

The cumulative value of LI decreased around $ 10 M between 2005 and 2010 to $ 295,349 M. Leather clothes and other wearing apparel and accessories led the group in each period by a significant margin also registering biggest gains throughout the timeframe.

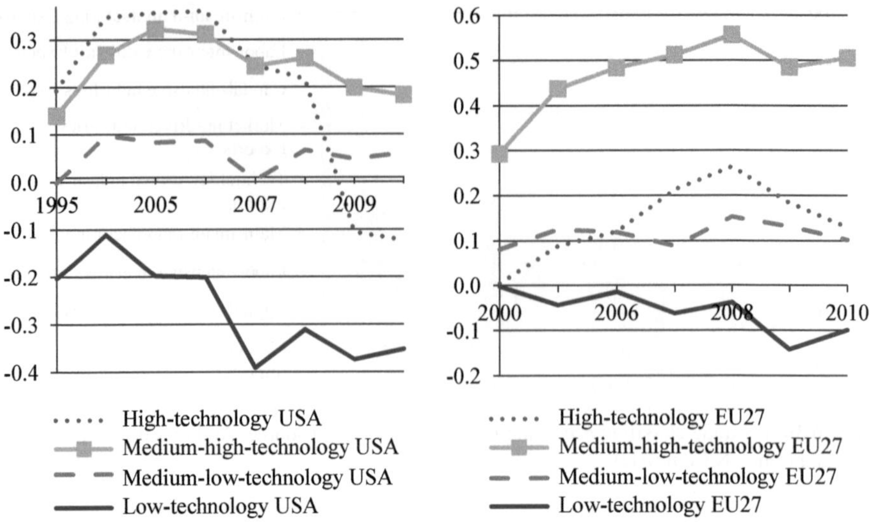

Fig. 6.31 RCA for the USA (*left* graph) and the EU (*right* graph) trade classified according to the technological level and collar/skill assignment. (Source: Author's own graph based on calculation of data from UN Comtrade)

CI ($ 132,766 M in 1995, $ 332,954 M in 2010) saw greatest gains in refined petroleum products ($ 12,372 M in 1995 to $ 74,504 M in 2010) and basic chemicals ($ 25,578 M in 1995 to $ 75,643 M in 2010). The category that saw a decline in value ($ 707 M in 1995, $ 685 M in 2010) is cement, lime, and plaster.

Reproduction of recorded media has contributed the most to the MDI group ($ 173,651 M, 2010) despite actually decreasing $ 15,236 M from 2000 to 2005. It is worth noticing that as exports of tobacco products decreased their imports have increased from $ 189 M in 1995 to $ 733 M in 2010.

In the TDI group, pharmaceuticals exhibit the biggest change in the period increasing sharply from $ 8,860 M (1995) to $ 74,898 M (2010). Motor vehicles have the highest value ($ 139,497 M) in 2010 with the lowest value ($ 772 M) being assigned to pesticides, other agrochemical products.

Revealed comparative advantage (RCA) (Fig. 6.31) has been calculated using the formula presented in Chap. 1. The USA has the greatest RCA value in HT (white-collar/high-skill) till 2008 when the MHT (medium white-collar/high-skill) category, which is the definite continuous leader for the EU, takes the number one spot.

The USA has enjoyed advantages in the HT (white-collar/high-skill) group till 2006. The only category that kept its advantage in all years is pharmaceuticals. Comparatively, the EU has kept an RCA through the entire period with the RCA value increasing till 2008 (0.2643) and declining thereafter till 0.1290 in 2010. By far, the category that has the biggest advantage in this group is again the pharmaceuticals in which EU has a significantly greater advantage than the USA. In regard to the leading category for the USA, aerospace, the EU not only kept the advantage throughout the entire time but was also able to keep that advantage at a higher level.

However, the EU kept a constant disadvantage in the computers and machinery category with the lowest RCA equaling to −0.4474 (2000).

The USA has enjoyed advantages in the HT (white-collar/high-skill) group till 2006. The only category that kept its advantage in all years is pharmaceuticals. Comparatively, the EU has kept an RCA through the entire period with the RCA value increasing till 2008 (0.2643) and declining thereafter till 0.1290 in 2010. By far, the category that has the biggest advantage in this group is again the pharmaceuticals in which EU has a significantly greater advantage than the USA. In regard to the leading category for the USA, aerospace, the EU not only kept the advantage throughout the entire time but was also able to keep that advantage at a higher level. However, the EU kept a constant disadvantage in the computers and machinery category with the lowest RCA equaling to −0.4474 (2000).

The MHT (medium white-collar/high-skill) set of categories for the USA enjoyed a continuous advantage with a maximum value of 0.3116 (2005), which was the beginning of a nearly steady decline to 0.1733 (2010). In 2000, the biggest RCA (0.4801) has been achieved in chemicals while in 2010 the RCA of this category has decreased to 0.3267, still remaining the highest in the group. The EU has also exhibited a constant RCA with a dip from its maximum value (0.5567 in 2008) in 2009 (0.4851) bouncing back to 0.5057 (2010). Similarly, the EU has also had a revealed comparative disadvantage in other transportation equipment in years 2000 to 2010 with motor vehicles being the leader in each period moving from 0.6051 in 2000 to 0.9984 in 2010 with a dip in 2009.

The shipbuilding category had the biggest RCA value in the MLT (medium blue-collar/low-skill) for the USA. The RCA for the group has been volatile, decreasing from 2000 (0.0876) to 2005 (0.0733) to then increase in 2006 (0.0763), decrease in 2008 (0.0566) and 2009 (0.0381), and finally increasing in 2010 (0.0507). Opposite to the shipbuilding in the USA, in the EU this category has seen its RCA value decrease significantly to 0.3690 (2010) from 0.6505 (2000). The entire group for the EU has also been very volatile in its RCA value. Other manufacturing is also at a revealed comparative disadvantage in the EU, although not to the same extent as in the USA. The biggest disadvantage is seen here in nonferrous metals (−0.6940 in 2000, −0.3857 in 2010).

The LT (blue-collar/low-skill) group has been at an increasing revealed comparative disadvantage (−0.1208 in 2000, −0.3624 in 2010). The only category that had an RCA in this group is the food, beverages, and tobacco that enjoyed the biggest RCA value in 2005 (0.6679) before it decreased to 0.4250 in 2010. By far, the biggest disadvantage is seen in the wood and furniture category that stood at −1.0489 in 2000 and is at −1.2770 in 2010. EU has also seen a continuous disadvantage in this set but at a lower level than the USA. The group as a whole is more evened out with two categories (paper printing and food, beverages, and tobacco) having an RCA (although the latter not as big as in the case of the USA) with the remaining two categories that are at a revealed comparative disadvantage having negative RCA values bigger than those in the USA.

The USA has the highest RCA value in TDI till 2009 in which that spot is taken by CI. For the USA, it is also important to note that the labor-intensive group has, by far, the highest revealed comparative disadvantage. This is translated into the analysis of the EU, but at a smaller rate. The MM group is, in the case of the EU, the absolute leader with the highest RCA.

The MM group has been enjoying RCA that, after reaching its maximum of 0.1792 (2008), fell to 0.1339 (2010). The biggest advantage has been exhibited by the paint, coating, printing ink category that is one of the few categories that have shown an RCA from 2005 to 2010; in 2000 the highest RCA was in weapons and ammunition. In 2000, other transportation equipment n.e.c. had the biggest disadvantage (−2.0274). The biggest disadvantage in 2010 was in the knitted and crocheted articles (−1.8500). The EU had enjoyed a significantly higher RCA in this group (0.4204 in 2000, 0.5204 in 2010) with a higher RCA in the US leading category in almost every examined year. In 2010, the highest RCA is seen in the other special purpose machinery that grew from 1.1334 (2000) to 1.5657 (2010). Parallel to the USA, other transportation equipment n.e.c. had the lowest RCA value in 2000 (−1.5287) and also enjoyed a very significant improvement in 2005 (0.0864) obtaining an RCA that lasted till 2008 with the final RCA value of −0.4640.

LI (USA) have exhibited a constant disadvantage. Year 2000 was dominated by negative RCA with the lowest (−3.0216) being assigned to cutting, shaping, finishing of stone category with bodies for motor vehicles (0.4243) being the bright spot. In 2010, the category with the highest RCA value was the railway locomotives and rolling stock one that increased from 0.2082 (2000) to 1.1976 (2010). The set of categories that are labor-intensive (EU) is at a constant, though not as severe, disadvantage with its RCA values ranging from −0.1779 (2000) to −0.4152 (2009). The category with the lowest RCA value (−1) in 2000 was the leather clothes eventually giving way to made-up textile articles that had the lowest RCA value of −1.5026 in 2010.

CI stayed at a slight advantage till 2008 (0.1336) and eventually grew to 0.2635 (2010). Few categories enjoyed an RCA in 2000 but they were all vastly overshadowed by ceramic tiles and flags (−2.7416) and cement, lime, and plaster (−2.1045). In 2010, the category with the highest advantage was the basic chemicals (0.6296) followed by pulp, paper, and paperboard (0.5506). In 2000, the EU was a region of more extremes with the highest RCA (2.5661) in ceramic tiles and flags and the highest disadvantage (−1.8778) in coke oven products. Only two categories were at a revealed comparative disadvantage in 2010; namely, man-made fibers (−0.4202) and basic precious and nonferrous metals (−0.3734).

The category that has the highest RCA value in MDI in 2000 is the tobacco products (2.4270). Meat products (1.4719) and prepared animal feeds (1.3092) lead in terms of RCA in 2010. Footwear (−2.5) and luggage, handbags, saddler, and harness (−1.6) categories have relatively the highest comparative disadvantages. Tobacco products have also been the source of the greatest RCA for this group for the EU, but unlike in the case for the USA, their RCA increased throughout the timeframe. Biggest disadvantages are seen in fish and fish products and games and toys for the entre period. For the group, the EU had a continuous RCA but of a smaller than US magnitude.

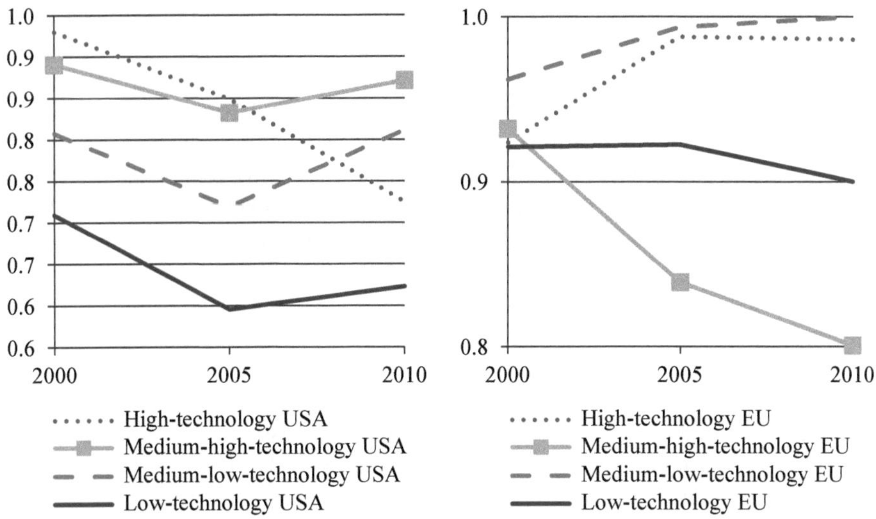

Fig. 6.32 IIT for the USA (*left* graph) and the EU (*right* graph) trade classified according to the technological level and collar/skill assignment. (Source: Author's own graph based on calculation of data from UN Comtrade)

Last, the TDI group turned from a positive to a negative RCA. The pesticides, other agrochemical products category has had the continuous positive RCA value. This was paralleled by the aircraft and spacecraft category till 2009 when the RCA value became negative (−0.6891) and in 2010 became the category with the second highest disadvantage. Motor vehicles have had the highest revealed comparative disadvantage from the group (−0.9469) in 2000 with TV, radio, and recording apparatus being the highest in 2010. For the EU, the group had an increasing RCA value that reached its maximum (0.2966) in 2008 ending the decade with the RCA value of 0.2212. Opposite to the USA, motor vehicles had the highest RCA value through almost the entire period from 1.0201 (2000) to 1.3643 (2010). Similarly, to some degree to the USA, in addition to office machinery and computers, TV, and radio and recording apparatus had the biggest disadvantage in this group.

The intra-industry trade (IIT) (Fig. 6.32) coefficient has been obtained following the formula presented by Misala (2011). For the USA, the highest value of IIT in 2000 is seen in the HT group (white-collar/high-skill, 0.9299), which was substituted as a leader in 2008 by the MHT (medium white-collar/high-skill) group. The lowest values are, by far, present in the LT (blue-collar/low-skill) group. In the EU, the highest IIT value can be seen in the MLT (medium blue-collar/low-skill) group that reached a level of 0.9998 in 2010.

The HT (white-collar/high-skill) group's IIT value has been exhibiting volatility in examined years eventually decreasing from 0.9299 in 2000 to 0.7249 in 2010. Pharmaceuticals have had the highest IIT in 2000 but, just like all other categories in this set, have seen it decrease overall with some small increases in between. Unlike, this group of categories in the EU has seen an increase in its IIT value from

0.9239 (2000) to 0.9863 (2010). The categories within this group are much more volatile. The biggest difference between the two regions is in the value of the IIT for the pharmaceuticals, it having the lowest IIT values in the group for the EU. Aerospace was the leading category in the years 2000 and 2010 with electronics–communications taking over the number one spot from 2005 to 2009.

The second group, MHT (medium white-collar/high-skill), is more balanced than the previous one with its IIT value oscillating between 0.8906 (2000) and 0.8324 (2005) with the final value reached in 2010 of 0.8715. Chemicals have the highest IIT value with the exception of 2009. The Other transport equipment category has the lowest IIT. The EU started with a higher IIT value (0.9321) but exhibited a steep decline thereafter. Overall, categories within this group are more evened out with scientific instruments (0.9635, the highest value in 2000) and electronic machinery (0.9699, the highest value in 2010) leading the group.

The MLT (medium blue-collar/low-skill) group has also experienced a decline in 2005 (0.7194) from 2000 (0.8074) further increasing to 0.8118 in 2010. Shipbuilding and petroleum refining categories have been at an almost even level in 2000 (0.9633 and 0.9502, respectively) but the latter registered a sharp increase in its IIT value from 2008 ending as the number one category in 2010 (0.9871). Other manufacturing has the lowest IIT value (around 0.5) throughout the entire timeframe. As its entirety, this set had significantly higher IIT values for the EU with the lowest one (0.9621) being registered in 2000 and the highest (0.999) in 2010. Rubber and plastic products and petroleum refining have the highest IIT value with the exception of 2008 in which ferrous metals category's IIT value reached a level of 0.9968. Nonferrous metals on the other hand had the lowest IIT value in 2000 but exhibited a significant growth in it in 2008.

The LT (blue-collar/low-skill) group for the USA is led by food, beverages, and tobacco (0.9967 in 2010). IIT values assigned to the wood and furniture category are seen the lowest. The group as a whole for the EU has been declining from 2005 (0.9223) till 2010 (0.9). Unlike in the case for the USA, the food, beverages, and tobacco category has seen a rather decreasing trend with the paper printing category having the highest IIT value from 2007 on.

The IIT (Fig. 6.33) is the highest for the USA in 2000 for TDI (0.8706) that continues to remain at the number one spot until 2009 where CI see their IIT value jump significantly to 0.9078 and 0.9161 a year later. The lowest IIT values can be seen assigned to the labor-intensive group; similar to what is seen in the EU. There, the technology-driven group also has the highest IIT value in 2000. From 2005 to 2006 (and then from 2009 to 2010), the highest value is seen in MDI followed by the capital-intensive group (2007 to 2008).

MM sees a big drop in 2005 (0.7661) from 5 years earlier (0.8482). Year 2000 sees three categories (plastic products, 0.9736, isolated wire and cable, 0.9645, and machinery for production, use of mechanical power, 0.9941) having IIT values very close to one. At the end of the explored timeframe, the other general purpose machinery is at the top (0.9981) after enjoying a nearly constant growth. The IIT value that is closes to zero (0.1510) is present in 2000 for the other transportation equipment category that makes a significant recovery in terms of the IIT value to 0.5127

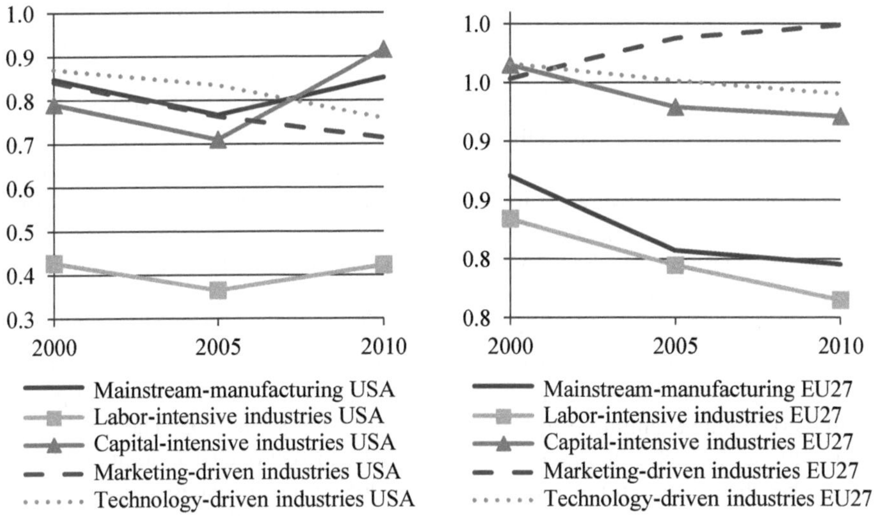

Fig. 6.33 IIT for the USA (*left* graph) and the EU (*right* graph) trade classified according to inputs. (Source: Author's own graph based on calculation of data from UN Comtrade)

in 2005. This is very similar to the EU where this category's 2000 value (0.3128) is the lowest one in the set and then triples in 2005 (0.9872). As a collective, the values for the EU are higher than those seen in the case of the USA till 2009 in which the USA saw its IIT value increase but the EU saw a drop to 0.7703.

LI have the lowest IIT (0.3644 in 2005, 0.4267 in 2000) values as a group. On one side, there are categories like the bodies for motor vehicles, trailers that enjoy values of above 0.9 in 2000, 2005, and 2006 with values still above 0.8 thereafter. On the other hand, there are categories like leather clothes and cutting, shaping, and finishing of stone where the latter reaches 0.03 in 2006. The EU has significantly higher IIT values in this group with the same category having the lowest (0.1649) value in the collective. Another similarity is in high IIT values of the textile waving category. The cutting, shaping, and finishing of stone differs in that it has a higher IIT value in 2000 reaching a maximum of 0.9957 in 2008.

The group that collects CI shows a constant increase in its IIT values from 2005 (0.7095) to 2010 (0.9161). Alike the previously described group, there is a wide range between high-IIT-values categories (textile fibers, basic chemical and man-made fibers that have the highest IIT value in 2010, 0.9977) and low-IIT-values categories (ceramic tiles and flags 0.1319 in 2010 and cement, lime, and plaster that has later increased to a moderate 0.5625). Worth mentioning is the parts and accessories for motor vehicles category because of its high IIT value (0.9779) in 2000 and then a fall to 0.8351 five years after. By comparison, IIT values for this group for the EU are around 0.9 falling to 0.8714 in 2009; the only year where it was lower than the corresponding value for the USA with the overall lowest IIT values being assigned to the ceramic tiles and flags category. The two groups that constantly enjoy high IIT values are the textile fibers and refined petroleum products.

The MDI group reaches its lowest value (0.6436) in 2007. The footwear category has the lowest IIT values that are above 0.1 only in 2000. The highest value in 2010 (0.9885) is assigned to the tanks, reservoirs, and central heating radiators. The EU sees two dips in this group: one in 2000 (0.9536) and the second in 2008 (0.9549) showing that, overall, it has higher IIT values than the USA. The collective has three leading categories: meat products, tanning, and dressing of leather and cutlery, tools, and general hardware. At the other side of this spectrum are categories like dairy products, ice cream, and tobacco products.

Last, the TDI group in the USA is somewhat volatile in its IIT values falling from 0.8769 (2008) to 0.7587 (2010). There are a lot of reversals between the years 2000 and 2005 where in the former year a category (pharmaceuticals, for example) has a relatively high (0.9088) value but sees it decreasing (to 0.8121) and staying around that level till 2010. Opposite can be said about categories like medical equipment with its IIT value changing from 0.7943 in 2000 to 0.9481 in 2005. The EU has IIT values higher in each examined year with even the lowest one (0.9366 in 2008) being higher than any seen in the case of USA, 2005 sees some categories (industrial process control equipment, for example) with lower values than those in 2000. Interestingly in 2010, aircraft and spacecraft, a category with an IIT value below 0.5 in 2010 in USA, has the third (after medical equipment and optical instruments and photographic equipment) highest IIT value in the EU.

6.4.2 The Characteristics of the US NIS

The US NIS has evolved moving through three distinctive stages that are designed around Joseph Schumpeter and described by Daneke (1998). The first stage, prior to World War 2, is called the Pre-Schumpeterian Period and is defined best by Daneke (1998) as a period in which "small entrepreneurial firms struggle for capital while large conglomerates struggle with bureaucratization" (p. 102) with modest support for university research and their ties with the business sector as well as high focus on solely in-house research. The Schumpeterian Period (1945 to 1985) saw high defense and space program R&D and small start-up companies coming into existence with the help of VC. The government also introduced two important legislations in that period. The Bayh-Dole Act (1980) "allowed universities and other nonprofit institutions automatically to retain title to patents derived from federal funded R&D" (Jaffe 1999). Second, the 1984 National Cooperative Research Act, "which reduced the antitrust penalties for collaboration among firms in precommercial research" (Mowery 1996). In 1986, a year after the start of the Post-Schumpeterian Period, which is where the USA is today, the Federal Technology Transfer Act (amended in 1989) allowed federal laborateries to enter Cooperative R&D Agreements starting joint research ventures and allowing for extensive technology transfers and encouraged partnerships between the private and the public sectors.

The US NIS can be divided into three levels: the government level, the connecting level, and the R&D level.

The House of the Representatives and the US Senate that both comprise the US Congress and the US president stand at the top of the government level. Arne Duncan is the current (confirmed by the US Senate on January 20, 2009 to the President's Cabinet) US Secretary of Education. As the head of the Department of Education (DE), he is responsible for the US Education policy. DE is the Cabinet-level department of the US government and its mission is to "promote student achievement and preparation for global competitiveness by fostering educational excellence and ensuring equal access."[23] Its roles are to form policies regarding the establishment and the distribution of the federal education funding, gather data, and conduct research on US schools, identify and bring attention to present issues, as well as fight discrimination in programs with federal funding. The government level also consists of the Federal Reserve System that was enacted by the US Congers in 1913 with the Federal Reserve Act with the mission to "provide the nation with a safer, more flexible, and more stable monetary and financial system."[24] The Federal Reserve Board, being independent from the government, is led by its chairman and other board members chosen by the president and confirmed by the Senate, currently that position is held by Ben Bernanke. In addition, the system also comprises the Federal Open Market Committee (FOMC; charged with buying and selling of government bonds on the open market) as well as the 12 Federal Reserve Banks that implement the monetary policy, aimed at achieving the maximum level of employment, nonfluctuating prices, and practical long-term interest rates. The next important actor at the government level is the US Department of Commerce, led by John Bryson since October 21, 2011, that serves in order to promote "job creation, economic growth, sustainable development and improved standards of living for all Americans by working in partnership with businesses, universities, communities and [US] workers."[25] Last, the described level also consists of the budget of the US government that is proposed by the president to the US Congress for the coming fiscal year that begins on October 1. Borrowing a buzzword from the literature on US economics, this group of actors enacts decisions that trickle down to lower levels, as presented in this view, to affect the level and the potential for the innovation in the USA.

The next two levels, the intermediate and the R&D, are harder to divide and can be seen as one general level but this research distinguishes between the two for the purpose of presentation.

US banks, based on the current Federal Reserve policy and the present economic conditions, provide funds to private investors and the VC[26] sector (with the two

[23] US Department of Education; http://www2.ed.gov/about/what-we-do.html

[24] Board of Governors of the Federal Reserve System: http://www.federalreserve.gov/aboutthefed/mission.htm.

[25] US Department of Commerce: http://www.commerce.gov/about-department-commerce.

[26] That is described as "a useful indicator of the degree of change" (as will be seen in further analysis conducted in this section) by David Hart in his article "Accounting for change in national systems of innovation: A friendly critique based on the US case" when talking about the Internet boom and bust (p. 649).

interacting on their own) that are later distributed to Research Parks (some of which are highly linked with local universities like the University of Arizona Science and Tech Park) as well as other research investment destinations in the private business and educational sectors. Thanks to legislations like the ones mentioned in the Introduction, the latter two sectors work together with the second of the two being the result of actions taken by the US DE. A staggeringly important role in this level and in the entire US NIS is played by the US Patent and Trademark Office (USPTO) that is a part of the US Department of Commerce with David Kappos, the Under Secretary of Commerce for Intellectual Property, as its director.

Last, the R&D level consists of the private and public universities and the military R&D sectors. All of which are interconnected with each other working together, again mostly owing to described changes resulting from legislations and the current innovation policy of the first level, and are connected with the other two levels through various links presented in the diagram (the link between the US educational system and the university R&D, for example) as well as other possible links not shown in this general representation of the US NIS (e.g., the US Federal Budget impacting the R&D level not only directly as shown, but also through spending allocated to other sectors in the NIS).

6.4.2.1 Innovation Capacity and Position

The capacity and the performance of the US innovation are assessed in this work through their comparison to the EU27 during 1995, 2000, 2005, and 2008 (or other depending on the years for which the data are available). Data, unless other specified, have been obtained from OECD.

Gross domestic expenditures on R&D (GERD; as a percentage of GDP, are significantly higher in the USA than in the EU reaching their maximum value of 2.78 in 2008 (1.84 for the EU) after falling to 2.57 (almost to the level seen in 1995) in 2005.

When analyzing government expenditures on R&D (GOVERD) and BERD, both as a percentage of GDP, it is evident that, for the USA and the EU, the business sector is responsible for greater spending on R&D than the government. At the same time, R&D financing provided by the government in both regions is less volatile than the one coming from the business sector on which variations in the economic well-being, like recessions or extreme growth, have a much higher impact.

For the USA and the EU, the highest value of GOVERD (0.35 and 0.28, respectively) was reached in the first analyzed year (1995). While GOVERD for the USA saw an increase in 2005, EU's GOVERD has been steadily declining over the compared period.

BERD exhibit a wave-like trend in the USA and the EU increasing from 1995 (1.76 and 1.02, respectively) to 2000 (2.02 and 1.11) and 2008 (2.02 and 1.15, highest values for both series) with a dip in 2005 (1.8 and 1.08). It is worth pointing out that with the increase in 2008, the US BERD has, more or less, returned to the level from 2000 but BERD for the EU have surpassed their 2000 value.

Examining GERD financed by the government, other national sources and by the industry (all expressed by a percentage of GERD) the first, and not a surprising conclusion is that, other national sources contribute the least to GERD in the USA and the EU. Another parallel between the examined regions regarding this source of financing is that its relative value is increasing across the examined years reaching the maximum in 2008 (5.68 for the USA and 2.64 for the EU, this also pointing to the extent of the difference between the two). The government is the second biggest from the set of looked-at financiers of GERD. Similar to GOVERD, the highest value is seen in 1995 (35.4 for the USA and 39.55 for the EU) with this being the one category where the EU spends relatively more than the USA. In 2000, the percentage of GERD financed by industry (in the USA and in the EU) has reached the highest relative level (69.42 and 55.28, respectively).

It is very interesting to see that when the percentage of GERD financed by the industry in the USA increases (years 2000 and 2008), its government counterpart decreases putting forward the hypothesis that the US government has a desired level that GERD must maintain and sees government funding of it as rather supplementary, in addition to complementary, to industry spending. Similar effect can be seen, though to a much lower level, in the EU.

The VC investment as a percentage of GDP in the USA (data have been obtained from the Eurostat) in 1995 stood at a level of just 0.04. Five years later it reached a level of 0.27 being massively driven by the dotcom boom. In 2005, after the recession caused by the dotcom bust, VC investment fell to 0.04 as a percentage of GDP moving to 0.05 in 2008.

Number of patents filed by residents (per million US$ spent on R&D and per population in million) can be used to express the innovative position of the USA. Both variables follow the same increasing trend with a break toward a greater rate of increase in 2006. It is interesting to see that despite a decrease in spending on R&D seen in 2005, the number of patents per money invested has not decreased but only slightly leveled out while still maintaining a positive change from 2005 (0.67) to 2006 (0.68). The number of patents per population (in million) has almost doubled from 1995 (465.54 patents) to 2007 (800.17). Noteworthy is the fact that in 2005 that number was 702.5 from 583.98 5 years earlier further adding to the hypotheses that the US position as an innovative power is strong and increasing and that, as much as it is impacted by the fluctuations in the home economy, it is rather resistant to them exhibiting a continuous growth.

Last, in this section the research looks at exports of high technology as a share of total exports for both the USA and the EU (Fig. 6.34). In 1995, the share of exports of high tech stood at 25.87 % in the USA; a level not reached by the EU in the examined timeframe. At the turn of the century, both examined regions reached their highest values in the 4 examined years: 29.95 % for the USA and 21.39 % for the EU. After the year 2000, the measured value began decreasing. The difference being that for the USA the change between 2005 and 2006 has been very small (from 26.15 to 26.13) but for the EU it has been equal to almost 2.1 %.

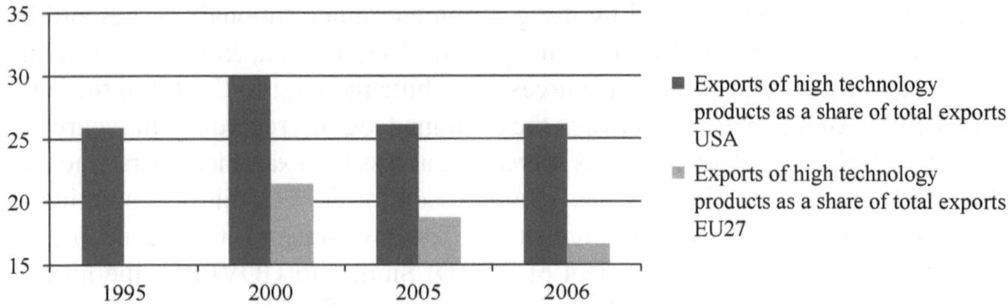

Fig. 6.34 Exports of HT products as a share of total exports in the USA and the EU. (Source: Author's own graph based on data from Eurostat)

In conclusion to this section, it is possible to say that not only does the USA have a greater innovation potential that is achieved by high level of inputs like funding but also has a stronger and a better innovative position as compared with the EU.

6.4.2.2 Human Capital in the USA

In this section, this work will look at the human capital present in the USA in four parts; namely, education and spending on it, population and employment, costs of labor, and productivity.

At the turn of the century, the vast majority of the labor force (51.5 %, as measured by the percentage of the total; Fig. 6.35) was classified as having secondary education followed by the labor force with tertiary (34.8 %) and primary (13.7 %) education. In 2005, the la tter remained in the last place and saw its share later declining from 2005 (10 %) to 2007 (9.5 %). In the meantime, the percentage of the labor force with tertiary education (59.9 %, 2000) became larger than the one with secondary education (30.1 %, 2000). The distance between the two values increased as more of the labor force attained the level of tertiary education (61.1 % in 2007). It is worth to point out that it is 2002 that was the year in which the share of the labor force with tertiary education increased the most (from 35.7 % in 2001 to 58.9 %) and the share with the secondary education saw the biggest drop (from 50.8 to 30.8 %).

The first reason for such a switch over the examined years is that the US labor is not seen as a low-cost one and therefore it is required of it to present a certain level of skills (education). The second reason is that as more and more people entered the labor market with tertiary education, in addition to those already present in the market that had or decided to upgrade their skills, its value became diluted setting it to be the new standard. This changed the position of the tertiary education from giving a competitive edge to being a requirement; therefore, forcing it on a bigger share of the labor force. By comparison, the share of labor force with secondary education in the EU is the prevailing one reaching 49.07 % in 2005. Similar to the USA, the share with the primary education has been decreasing, but is still significantly higher than in the USA; 30.02 % in 2000 and 26.57 % in 2005. The most interesting, and significantly contributing to the reason why the USA is such a strong innovative

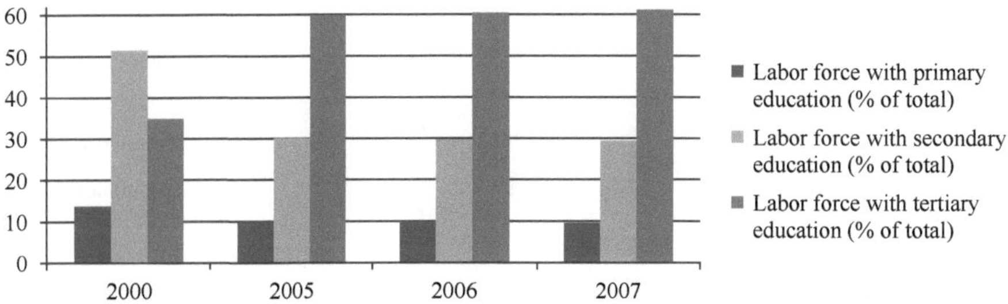

Fig. 6.35 US labor force with primary, secondary, or tertiary education (in % of total labor force). (Source: Author's own graph based on data from the World Bank)

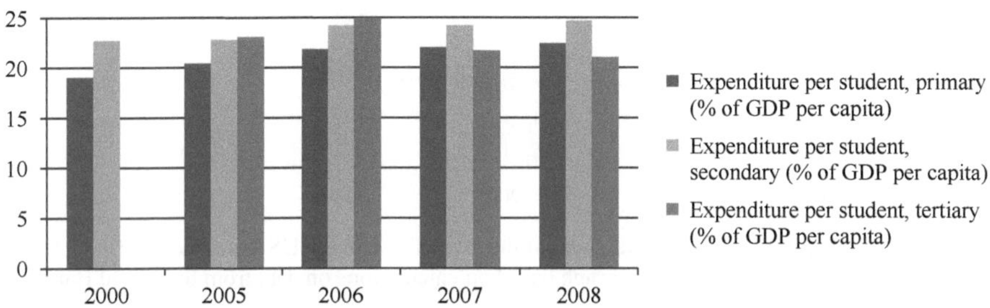

Fig. 6.36. US expenditure per student in primary, secondary, and tertiary education (in % of GDP per capita). (Source: Author's own graph based on data from the World Bank)

power and is considered to have a high-level human capital, is the gap in the share of labor force with tertiary education. At the turn of the century, 20.61 % of the labor force present in the EU was classified as one with tertiary education when in the USA that share was nearly twice as large, 34.8 %. Five years later this gap increased in size significantly with the value for the EU reaching only 24.17 % (almost 60 % for the USA).

In 2000, expenditures per student in the primary education (19.06 %, shown as a percentage of GDP per capita; Fig. 6.36) in the USA were lower than in the secondary (22.71 %) and tertiary (26.76 % in 1999) education. Spending on the latter reached its maximum (31 %) in 2001 and then, almost uninterruptedly, declined to just 21.05 % in 2008. This is surprising as 2002 was the year in which the share of the labor force with the tertiary education increased the most. Spending per student in the secondary education continued increasing, as a percentage of GDP per capita, till 2003 (25.41 %) after which there were 2 years of declining expenditures. In 2006, this value was at 24.19 % and increased thereafter to 24.68 % in 2008. Last, the expenditure per student at the primary level was less volatile throughout the period falling in 2005 from 21.61 % a year earlier to 20.43 % only to make it up a year later and eventually reaching 22.44 % in 2008.

The EU devotes less of its GDP per capita to spending per student in the primary education in 2000 and in 2005 as compared with the USA at the same time placing a

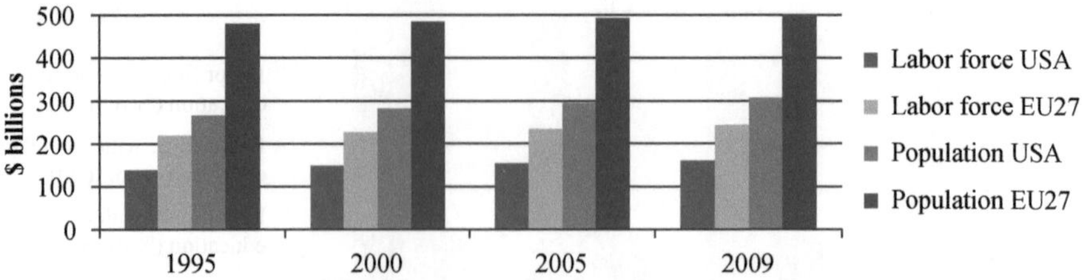

Fig. 6.37 Labor force and population for the USA and the EU (*left-hand axis*). (Source: Author's own graph based on data from the World Bank)

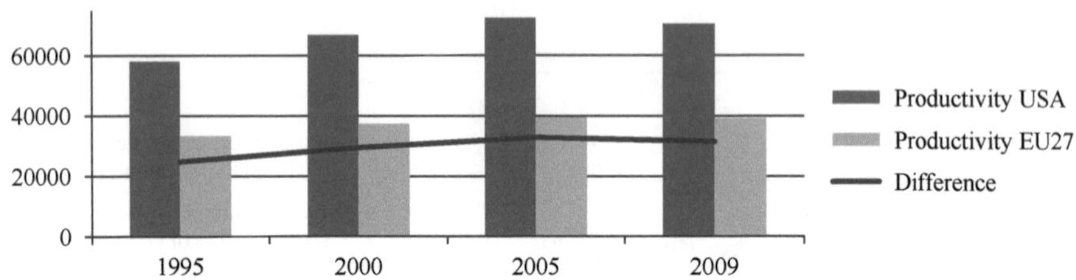

Fig. 6.38 Productivity, and the difference, of the labor force in the USA and the EU (in GDP per labor force). (Source: Author's own graph based on calculations on data from the World Bank)

slightly higher focus on spending on the secondary education (23.59 % in 2000 and 24.15 % in 2005) with both of those values increasing. In 2000, the expenditures per student in the tertiary education in the EU reached a level of 30.22 % and 25.26 % of GDP per capita 5 years later, more than in the USA.

The EU has a much higher population and the labor force (Fig. 6.37) than the USA; however, per 100 of its population, there are around 52 workers in the USA whereas there are only between 46 (1995) and 48 (2009) in the EU.

The possible advantage for the EU may be found in the fact that the ratio is slightly decreasing for the USA but is exhibiting some growth for the EU. This suggests that human capital in the USA is not only better educated (higher share of the labor force with tertiary education) but is also more highly utilized than the one present in the EU.

The ULC (as measured by the OECD index 2005 = 100) for the USA has been steadily increasing from 1995 (83.84) to 2009 (108.34). The biggest growth has been in the Construction sector where the ULC more than doubled in the examined period reaching the highest level from all the looked-at sectors of 134.61. A hypothesis can be stated here that such a dramatic increase in this sector is a result of the housing bubble during which costs of properties increased substantially transferring those increases back to the sector in the form of higher pay.

Finally, this work looks at the productivity (GDP, constant 2000 US$, divided by labor force) in the USA and the EU (Fig. 6.38). The reason the measure of GDP per capita is not appropriate is because, for example, a newborn baby does not contribute to the economy through purchases or production but is included in the GDP per capita calculation.

The productivity of the US labor force has increased from \$ 58,070.86 per member of the labor force (MLF) in 1995 (when the EU value in that year equaled just \$ 33,241.88 per MLF). The difference increased reaching its maximum in 2005 when the productivity of both regions was at the highest value from the presented years.

In summary, the USA has a better potential for innovation owing to its investments in R&D, has a more educated, more utilized, and vastly more productive labor force than the EU; hence, justifying the overall increase of ULC in the US economy and explaining why the USA has the advantage over the EU in many areas including innovation.

6.4.2.3 The US Scientific, Innovation, and Educational Policy

On September 2009, president Obama introduced "A Strategy for American Innovation: Driving Towards Sustainable Growth and Quality Jobs" initiative that was later updated on February 4, 2011. The idea of this initiative is to "improve America's economic growth and competitiveness on many critical dimensions."[27] There are three dimensions that president Obama choose to focus on.

First, investment in the building blocks of American innovation that focuses on educating students with skills needed in the twenty-first century that will in turn create a world-class workforce. This goal is to be achieved by various tools at different levels of education (early childhood, elementary and secondary, and college). The focus is on introducing competition (Head Start Program), public–private partnerships (in science, technology, engineering, and mathematics; STEM, education) allowing for training that matches skills required in the job market (Task Force on Skills for America's Future) and investment in community colleges and the public workforce system. Additionally, the first aim is aided by strengthening and broadening of American leadership in fundamental research through a large increase in federally funded research focusing on three key research agencies: the National Science Foundation, the Department of Energy's Office of Science, and the National Institute of Standards and Technology laboratories. Last, this first dimension is to be reached by building the leading infrastructure in terms of roads, rails, and runways and in terms of a virtual infrastructure to advance the US IT ecosystem.

Second, promotion of market-based innovation as "American businesses are … engines of innovation."[28] The first part of this dimension is to increase the speed of business innovation with a simplified and a permanent R&E tax credit that will allow businesses to better plan their expanses, thanks to the predictability assured by the Research and Experimentation Tax Credit. Entrepreneurs of all sizes will be assured stable capital markets (in addition to two \$ 1 billion initiatives for early-stage seed financing) as well as (through internal initiatives and three-track patent prioritization based on its value) an accelerated patent issuance (decreasing the delay in the processing time from 35 to 20 months) in addition to the overall promo-

[27] White House: http://www.whitehouse.gov/innovation/strategy/executive-summary.

[28] Ibid.

tion of entrepreneurship (Startup America) and to the Affordable Care Act that will make it easier to start and join new businesses without the loss of health coverage. To further aid the entrepreneurs, the initiative wants to catalyze innovation hubs that would allow for contacts between those that are established and those that are new to the entrepreneurial ecosystem. Last, the aim of this section is to promote innovative, open (the National Export Initiative), and competitive markets (Horizontal Merger Guidelines).

Third, acceleration of breakthroughs for national priorities that has the following four subsections: unleashing a clean energy revolution (with a goal of doubling the supply of renewable energy by the end of 2012 through investments in research via the Department of Energy's Office of Science and new Energy Innovation Hubs), acceleration via investments in biotechnology (National Institute of Health), nanotechnology (the National Nanotechnology Initiative), and advanced manufacturing, developing advanced space capabilities (guided by the National Space Policy), and application (global communication, navigation and other use in the commerce world), driving breakthroughs in health care technology (adoption and promotion of the IT aspects like transfer of information; the Recovery Act and the Affordable Care Act), and by creating a massive growth in innovation and usage of educational technologies (Advanced Research Project Agency—Education).

Education has enjoyed increasing funding in three of its levels (primary, secondary, and tertiary) reaching levels of 22.44, 24.68, and 21.05 % (each of GDP per capita) in 2008. The "Strategic Plan For Fiscal Years 2007–2012"[29] of the US DE has overall 16 objectives that will be satisfied as means of fulfilling the three main goals of the plan. These goals are: Goal 1: Improve student achievement, with focus on brining all students to grade level in reading and mathematics by 2014, Goal 2: Increase the academic achievement of all high school students, and Goal 3: Ensure the accessibility, affordability, and accountability of higher education, and better prepare students and adults for employment and future learning. The plan for the years from 2011 to 2014 takes some of those goals to a new level and introduces new areas that need the attention. The new plan, which has six goals, focuses on increasing the overall access, quality, completion, and readiness for college (Goal 1 and Goal 2), improving the health of the youngest students (from birth through third grade, Goal 3), promote equality and fight any discrimination in all aspects of the educational system (Goal 4), continuous improvement of the system as a whole through the use of data, research and evaluation, transparency, innovation and technology (Goal 5), and improving the Department's capacity to implement the set goals (Goal 6). The education sector was harmed by the recent recession that was experienced by the USA starting in 2007. As a response to decreases in employment in the education segment, part of the Recovery Act has awarded $ 97.4 billion in a grant that created "approximately 275,000 education jobs, such as teachers, principals, librarians, and counselors"[30] showing the understanding of the US government that without a solid foundation, that is the education, of its citizens, it cannot

[29] U.S Department of Education: http://www2.ed.gov/about/reports/strat/index.html?src=ln.

[30] US Department of Education: http://www.ed.gov/recovery.

maintain its current levels of advancement, innovation, or competitiveness on the international scale.

The list of strengths of the US NIS starts with its streamline level that translates into operational effectiveness further exhibited in the quality of the US human capital. Thanks to the current structure and the workings of the US NIS, the USA has a very high potential for innovation that is a result of not only a strong base (education) of a large labor force but also a result of high spending on R&D and the government's recognition of the importance of such investments. All this translates into a very productive labor force and high innovation (as measured by the number of patents by the US residents).

The USA should be commended for the high percentage of its labor force that comprises workers with tertiary education (61.1 % in 2007). Where the NIS fails is in not furthering that accomplishment by higher spending that has actually been decreasing, falling to just 21.05 % of GDP per capita. Additionally, the patent system needs updating to decrease, and eventually eliminate, the bottlenecks. It is important to point out at this point that the initiative presented by president Obama, as described earlier, addresses those weaknesses.

The USA, owing to the size and the quality of human capital and the possibility for high investment, can be said to have a self-sustainable NIS. The opportunity that exists for the USA is to import talent from abroad to supplement the possible lacks in its own human capital. This would not only limit the required investment in human capital by domestic actors (e.g., the US government) but additionally assure that the USA has the best in the world talent and that its competitiveness on the international level will be sustained owing to its procurement of this scarce resource.

With the exception of a severe economic slowdown, domestic and worldwide (like the one seen at the turn of the last decade), there are no direct threats that can harm the workings of the US NIS. With the new initiative in place, there is little threat that domestic innovators will look for funding or patent registration abroad. If the opportunity of importing foreign talent will be taken advantage of, which would be easier if the immigration laws would allow for a simpler and a more expedient way for foreigners to enter the US job market, there will be no need for the US-based companies, or the foreign companies that choose to join the US NIS, to go elsewhere to make use of resources such as labor or patent laws.

6.4.3 The Interdependencies Between the Level of Innovation, Human Capital Accumulation, and RCA in International Trade

When comparing the USA with the EU, the USA continuously outspends the EU countries. In regard to investments in education, the USA has been continuously increasing its expenditures per student in the primary and secondary education with spending on tertiary education falling from the peak reached in 2006. Counterintuitive to spending information, the percentage of labor force with tertiary education in the USA increased from 34.8 to 61.1 % between years 2000 and 2007. In the same timeframe, the share of the labor force classified as having only primary education fell from 13.7 to 9.5 %. As compared with the USA, the EU's labor force mostly (49.07 % in 2005) comprises workers with secondary education; only 24.17 % are classified as ones with tertiary education in 2005 versus 59.9 % in the USA. Additionally, the innovation-related policy of the USA has translated into increasing number of patents filed by its residents. Last, when comparing the two economic regions, it is important to note that the population and the labor force in the USA are smaller than in the EU. Yet, in the USA the labor force constitutes a greater share of general population, it is better educated and significantly more productive.

The USA has exhibited the highest RCA in trade related to HT (white-collar/high-skill, years 2000 to 2007) and MHT (medium white-collar/high-skill, years 2008 to 2010) when trade is classified according to the technology level and collar assignment. According to the same division, the EU has enjoyed the highest RCA in the MHT group. When classifying trade according to inputs, Technology-driven (2000, 2005 to 2008) and Capital-intensive (2009, 2010) industries in the USA have exhibited the highest levels of RCA. Within the same classification, the MM group has towered in regard to displayed RCA values over other clusters. It is also important to note that for the USA the HT (white-collar/high-skill) group and the TDI set have not only lost their top spot as designated above but their RCA values turned negative showing that the revealed advantage that the USA had turned into a revealed disadvantage; something that has not been present when looking at the leading categories for the EU. This can be hypothesized to be the outcome of the severe recession that the USA found itself in during those years.

By examining the two earlier paragraphs, it is evident that with the exception of the recession years, higher investments in the development of human capital as exhibited by its innovation and its composition according to the level of education or, simply put, its quality rather than the sheer size of the labor force will result in RCA in trade that is related to high levels of technology and of required skills.

The comparison between the labor force in the USA and the one in the EU provides a great example of the above statement. In the USA, the high portion of the tertiary education allows it to achieve advantages in the categories of trade related to TDI that require a certain high level of education and training. In the EU, where the majority of the labor force is said to have secondary education, the advantages are present in groups of products concerned with a lower level of technology and with industries focused on MM. Examining this relationship from the other end, both of the examined regions have high disadvantage when looking at LT and labor-intensive sets of products that would be associated with countries in which the labor force is of low quality (low level of education); therefore, putting the importance on the size of the labor force. Here again, the USA with its higher-quality labor has higher disadvantage in those groups than the EU.

Chapter 7
International Competitiveness of Countries with Catching-up Innovation Systems

Mariusz-Jan Radło and Marcin Gomułka

A *catching-up* national innovation system (NIS) can be divided into three different subtypes. Greece, Portugal, Hungary, Poland, and Slovenia form one subgroup. The second, slightly different, subtype is represented by Malaysia and Malta. The third subgroup comprises Estonia, Latvia, Lithuania, Slovakia, and Ukraine (c.f.: Chap. 3). What these three subtypes of *catching-up* innovation systems have in common is their average innovative capacity and move into higher positions in innovativeness rankings—all are close to the global average. In the last decade, all the ratios used to describe these systems underwent notable changes. This means that these countries are gaining momentum and *catching up* with global averages (Weresa 2012). This *catching-up* process lends its name to this type of innovation systems, and the sharp improvement in the value of indexes induces to seek the causes of this process and point out to what extent it is reflected in the improvement of competitive advantages of national trade.

The analyses of the two case studies discussed in this chapter—Poland and Hungary—are driven by the search for an answer to this question.

7.1 Case Study: Poland

Mariusz-Jan Radło

Most of the studies devoted to human capital developments in Poland indicate that it is markedly diversified in terms of both space and demography. For example, Roszkowska (2006) indicates that average human capital level in Eastern Poland

M.-J. Radło (✉) · M. Gomułka
World Economy Research Institute, Warsaw School of Economics,
Al. Niepodleglosci 162, 02-554 Warsaw, Poland
e-mail: mjradlo@sgh.waw.pl

M. Gomułka
e-mail: mrgomulka@gmail.com

M. A. Weresa (ed.), *Innovation, Human Capital and Trade Competitiveness*, Innovation, Technology, and Knowledge Management, DOI 10.1007/978-3-319-02072-3_7,

is lower than in the western part. Kotowska and Więzak (2007), in turn, show that human capital is also significantly influenced by demographic factors, including among others age and employment in the nonagricultural sector. Similar results have been presented by Liberda (2005), who argues that the demographic structure of households and their income to a large extent determine investments in education. Many researchers underline that the condition of human capital in Poland has been steadily improving and has had positive influence on economic growth (Janc 2009). The economic transformation was the main factor strengthening these changes. Furthermore, many empirical studies show that lower human capital is one of the key factors limiting innovativeness of small and medium-sized companies (Gagacka 2007; Baczko et al. 2010).

In light of the above remarks, the goal of this section is to search for the answer to the question whether the recent improvement in human capital in Poland influences innovativeness of the economy—and if yes, then to what extent. This section also tries to tackle the issue of the extent of influence that accumulated human capital resources and the resulting innovations have on long-term competitive advantages in foreign trade.

7.1.1 Competitiveness of Poland's Foreign Trade

An analysis of Polish exports provides insight into the changes in the structure of competitive advantages of the Polish economy. Especially, the following question arises: is Poland shifting toward more technologically advanced and unique activities in which production processes require more knowledge and technology? The more competitive advantages are based on unique resources, including knowledge and technologies, which are more stable and more resistant to changes in prices or exchange rates (Porter 1990, 1998). This means that pro-competitive structural changes result in shifting production and exports toward industries and products that base their development on very unique advantages resulting from the applied technologies, labor force qualifications, or utilized resources.

As follows from the data presented in Fig. 7.1, the value of Polish exports climbed continuously from EUR 23 billion to EUR 172 billion in the years 1995–2008. However, it is worth noting that in the years 1995–2002, the increase in the value of exports was relatively slow and it was not until 2003 that growth dynamics stepped up. Furthermore, in 2009 the most recent crisis in the global economy yielded a temporary decrease in the value of Polish exports from EUR 172 billion in 2008 to EUR 137 billion. In 2010, Polish exports picked up again and reached the volume of EUR 157 billion.

Polish imports in the years 1995–2010 exceeded the value of exports in the entire analyzed period, and the fast pace of changes in imports was very similar to export dynamics. This means that until 2003, Polish imports were relatively stable and growing, but this growth was minor. A period of rapid growth did not occur until 2003. From 1995 to 2008, Polish imports rose from EUR 29 billion to EUR

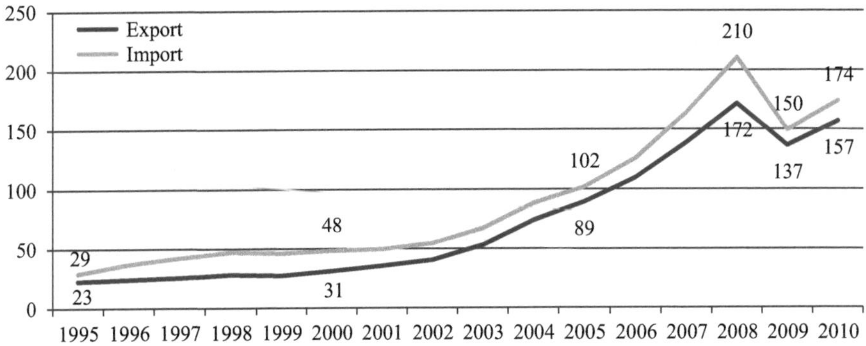

Fig. 7.1 Polish exports and imports in the years 1995–2010 (billions of EUR). (Source: Own study based on the Comtrade database)

210 billion. The global economic crisis resulted in a temporary decline in the value of imports from EUR 210 billion in 2008 to EUR 150 billion in 2009. Afterward, imports went up to EUR 174 billion. As follows from the data presented in Table 7.1, in the years 1995–2010, Germany, France, Great Britain, Italy, Czech Republic, Russia, Sweden, Hungary, and Spain were Poland's top ten export partners. At the same time, it is worth noting that from 1995 to 2010, the structure of Poland's export partner was quite volatile. A significant drop was visible in the case of Germany, from 38.4 % in 1995 to 26 % in 2010, while countries such as France, Great Britain, or Czech Republic gained in importance. The following countries were Poland's top ten import partners in 2010: Germany, Russia, China, Italy, France, Czech Republic, the Netherlands, South Korea, Great Britain, and the USA. The shares of individual economies also fluctuated throughout the period from 1995 to 2010. One significant development here is the shrinking of Germany's share by 5 % points (p.p.) and the enlargement of China's and Russia's shareby 7.9 and 3.7 p.p., respectively.

More information about Polish foreign trade in the years 1995–2010 can be derived from its commodity structure. It follows from Table 7.2 that in 2010 Poland exported mainly machinery and transport equipment (41.5 %), manufactured goods (20.1 %), manufactured articles (12.7 %), food (9.2 %), and chemicals (8.6 %). At the same time, in the years 1995–2010, a significant change in the structure of exports took place, as the share of machinery and transport equipment rose notably (by 20.5 p.p.), and Poland witnessed a decline in the share of manufactured goods and articles (by 7.9 and 7.5 p.p., respectively) and mineral fuels (by 3.8 p.p.). At the same time, Poland imported mainly machinery and transport equipment (34.4 %), manufactured goods and articles (17.6 and 10.1 %, respectively), chemicals (14.2 %), and fuels (10.9 %). Here, it is worth noting that the changes in the structure of Polish imports in the years 1995–2010 were notably smaller than the changes in exports and they concerned mainly the growth of imports of machinery and transport equipment (by 4.9 p.p.) and the drop in manufactured goods (by 4 p.p.).

Table 7.1 Poland's key trading partners in the years 1995–2010 (% of total exports or imports). (Source: Own study based on the Comtrade database)

Exports of goods					
Country	1995	2000	2005	2010	2010–1995[a]
Germany	38.4	35.6	28.2	26.0	−12.4
France	3.6	5.2	6.2	6.8	3.2
Great Britain	4.0	4.5	5.6	6.3	2.3
Italy	4.9	6.4	6.1	6.0	1.1
Czech Republic	3.1	3.8	4.6	5.9	2.8
The Netherlands	5.6	5.1	4.2	4.4	−1.2
Russia	5.6	2.7	4.4	4.2	−1.4
Sweden	2.5	2.7	3.1	3.0	0.5
Hungary	1.2	2.0	2.8	2.8	1.6
Spain	1.1	1.6	2.5	2.7	1.6
Ukraine	3.2	2.5	2.9	2.5	−0.7
Belgium[b]	2.5	3.0	3.0	2.4	−0.1
Denmark	3.0	2.7	2.1	1.8	−1.2
USA	2.7	3.2	2.1	1.8	−0.9
Listed countries—in total	81.4	80.9	77.8	76.7	−4.7
Imports of goods					
Germany	26.7	24.2	24.7	21.7	−5.0
Russia	6.8	9.6	8.8	10.5	3.7
China	1.6	2.8	5.4	9.5	7.9
Italy	8.6	8.4	7.1	5.6	−3
France	4.9	6.5	6.0	4.3	−0.6
Czech Republic	3.1	3.2	3.6	3.7	0.6
The Netherlands	4.5	3.5	3.4	3.7	−0.8
South Korea	0.9	1.5	1.6	2.8	1.9
Great Britain	5.2	4.5	3.1	2.7	−2.5
USA	3.9	4.4	2.4	2.5	−1.4
Belgium[b]	2.6	2.6	2.6	2.4	−0.2
Spain	1.6	2.4	2.2	2.0	0.4
Sweden	3.1	2.9	2.3	1.9	−1.2
Listed countries—in total	73.5	76.5	73.1	73.2	−0.3

[a] Difference between 2010 and 1995 values
[b] 1995 data for Belgium also include Luxembourg

The most positive changes in export structure in terms of the share of goods produced by industries differentiated by level of technological advancement were (Table 7.3):

- the drop, in the years 1995–2009, of the share of goods produced by low-technology (LT) and medium-low-technology (MLT) industries from 22.1 to 16.4% and from 51.3 to 40.7%, respectively, and
- the increase, in the years 1995–2009, of the share of goods produced by high-technology (HT) and medium-high-technology (MHT) industries from 21.6 to 33.8% and from 5.1 to 9%, respectively.

Table 7.2 Commodity structure of Polish foreign trade in selected years of the period from 1995 to 2010 (%). (Source: Own study based on the Comtrade database)

Exports of goods	1995	2000	2005	2006	2007	2008	2009	2010	2010–1995[a]
Food and live animals	9.1	7.4	8.6	8.3	8.3	8.3	9.3	9.2	0.1
Beverages and tobacco	0.7	0.4	0.6	0.7	0.8	0.9	1.4	1.3	0.6
Crude materials, inedible, except fuels	4.5	2.8	2.1	2.2	2.2	2.2	1.8	2.3	−2.2
Mineral fuels, lubricants, and related materials	8.0	5.1	5.0	4.4	3.7	4.2	3.0	4.2	−3.8
Animal and vegetable oils, fats, and waxes	0.1	0.1	0.1	0.2	0.2	0.2	0.2	0.2	0.1
Chemicals and related products	8.0	6.8	6.6	7.0	7.1	7.6	7.5	8.6	0.6
Manufactured goods classified by material	28.0	25.0	22.3	22.9	22.9	21.5	19.2	20.1	−7.9
Machinery and transport equipment	21.0	34.3	38.5	40.0	40.8	41.1	42.9	41.5	20.5
Miscellaneous manufactured articles	20.2	18.0	13.8	13.0	12.7	12.4	12.9	12.7	−7.5
Commodities and transactions not classified elsewhere in the SITC	0.4	0.1	2.2	1.3	1.3	1.8	1.8	0.1	−0.3
Total exports = 100	100.0	100.0	100.0	100.0	100.0	100.0	100.0	100.0	6.9
Imports of goods									
Food and live animals	8.0	5.2	5.2	4.9	5.2	5.6	6.8	6.5	−1.5
Beverages and tobacco	0.7	0.4	0.5	0.5	0.6	0.5	0.6	0.6	−0.1
Crude materials, inedible, except fuels	5.3	3.3	2.9	2.9	2.9	2.9	2.6	3.0	−2.3
Mineral fuels, lubricants, and related materials	9.2	11.0	11.4	10.4	10.0	11.4	9.5	10.9	1.7
Animal and vegetable oils, fats, and waxes	0.6	0.3	0.3	0.3	0.3	0.4	0.4	0.3	−0.3
Chemicals and related products	14.9	14.1	13.8	13.1	12.6	12.6	13.5	14.2	−0.7
Manufactured goods classified by material	21.6	19.9	20.2	20.5	20.9	18.4	17.1	17.6	−4
Machinery and transport equipment	29.5	36.9	34.6	35.1	34.9	34.6	34.7	34.4	4.9
Miscellaneous manufactured articles	9.6	8.7	8.6	8.2	8.8	9.2	10.8	10.1	0.5
Commodities and transactions not classified elsewhere in the SITC	0.5	0.2	2.4	3.9	3.9	4.4	4.1	2.2	1.7
Total imports = 100	100.0	100.0	100.0	100.0	100.0	100.0	100.0	100.0	6.0

[a] Relation of 2010 value to 1995 value

Table 7.3 Polish export structure by technology level in the years 1995–2009 (%). (Source: Own study based on the Comtrade database)

	1995	2000	2005	2006	2007	2008	2009
High technology	5.1	6.4	6.6	6.9	6.9	7.5	9.0
Medium-high technology	21.6	27.4	32.0	32.5	33.1	33.8	33.8
Medium-low technology	51.3	45.7	44.0	44.0	43.7	43.0	40.7
Low technology	22.1	20.6	17.4	16.6	16.3	15.8	16.4

Table 7.4 Polish export structure by input factors in the years 1995–2010 (%). (Source: Own study based on the Comtrade database)

	1995	2000	2005	2006	2007	2008	2009
Mainstream manufacturing (MM)	24.4	23.8	26.1	26.6	27.9	26.6	25.1
Labor intensive (LI)	35.2	29.3	22.0	19.8	18.9	17.9	18.3
Capital intensive (CI)	25.2	20.2	20.4	21.2	21.8	22.4	18.2
Marketing driven (MD)	10.8	10.6	12.6	12.5	11.9	12.4	13.8
Technology driven (TD)	7.5	16.2	19.0	20.0	19.4	20.7	24.5

Despite these positive structural changes that took place in the years 1995–2009, in 2009 more than half of the Polish exports value was generated by exports of LT and MLT commodities.

The most positive changes in the export structure in the years 1995–2009 in terms of the share of goods produced by industries differentiated by input factors (Table 7.4) were:

- the drop in the share of goods produced by labor-intensive (LI) and capital-intensive (CI) industries from 35.2 to 18.3 % and from 25.2 to 18.2 %, respectively, and
- the increase in the share of goods produced by RDI, marketing-driven industry (MDI), and mainstream-manufacturing (MM) industry; the share of RDIs increased from 7.5 to 24.5 %, MDIs from 10.8 to 13.8 %, and MM industries from 21.4 to 25.2 %.

One should also positively assess the fact that the total share of RDI and MDI amounted to 38.3 % and was higher than the total share of LI and CI industries (36.5 %). The scale of this change is reflected by the fact that in the mid-1990s, the share of LI and CI industries in exports amounted to more than 60 %, while the share of RDI and MDI totaled to a mere 18.3 %.

The key development aspects in the export structure in the years 1995–2009 as regards the share of goods produced by industries differentiated by labor force qualifications (Table 7.5) were:

- the drop in the share of goods produced by industries based on medium-qualified, blue-collar workers and industries based on semiqualified employees, from 51.3 to 40.7 % and from 22.1 to 16.4 %, respectively, and
- the increase in the share of goods produced by industries based on highly qualified and medium-qualified employees, from 5.1 to 9 % and from 21.6 to 33.8 %, respectively.

Table 7.5 Polish export structure by labor force qualifications in the years 1995–2009 (%). (Source: Own study based on the Comtrade database)

Industries	1995	2000	2005	2006	2007	2008	2009
Based on highly qualified employees	5.1	6.4	6.6	6.9	6.9	7.5	9.0
Based on medium-qualified, white-collar employees	21.6	27.4	32.0	32.5	33.1	33.8	33.8
Based on medium-qualified, blue-collar employees	51.3	45.7	44.0	44.0	43.7	43.0	40.7
Based on semiqualified employees	22.1	20.6	17.4	16.6	16.3	15.8	16.4

Table 7.6 Comparative advantages in foreign trade by technology level. (Source: Own study based on the Comtrade database)

	1995	2000	2005	2006	2007	2008	2009
High technology	−0.50	−0.47	−0.36	−0.32	−0.15	−0.14	−0.10
Medium-high technology	−0.53	−0.31	−0.16	−0.11	−0.08	−0.08	−0.11
Medium-low technology	0.48	0.35	0.17	0.17	0.17	0.18	0.14
Low technology	0.32	0.45	0.31	0.31	0.35	0.32	0.24

Despite these positive structural changes which took place in the years 1995–2009, in 2009 still more than half of the Polish exports value was generated by exports of commodities manufactured by industries based on medium-qualified, blue-collar employees and semiskilled workers.

In the years 1995–2009, the revealed comparative advantage (RCA) in foreign trade also changed significantly (Table 7.6.). The main positive trends witnessed in the analyzed period involved the increase in the revealed advantage rate of goods produced by HT and MHT industries and the drop in the values of these rates for commodities produced by LT and MLT industries. Despite these positive changes, in 2009 the value of the revealed advantage rate still showed only a minor comparative advantage in trading in LT and MLT commodities and a slight lack of such an advantage in trading in HT and MHT commodities.

The data concerning the process of developing comparative advantages in the trading of goods produced by industries with varied input factors are presented in Table 7.7. The analysis of this table leads to the conclusion that the major positive changes in the revealed advantage development process in the years from 1995 to 2009 in foreign trade included a drop in the revealed comparative advantage for LI and CI goods, as well as an increase in the value of revealed advantage rates in RDI, MDI, and MM industry. At the same time, it is worth noting that in 2009 Poland developed a revealed advantage in the trading of goods manufactured by MDI and LI industry. In reference to RDI and MM industry, the RCA value indicates a slight lack of revealed advantages. However, in reference to CI industries, the lowest values were noted for the revealed advantage rate.

Table 7.7 Comparative advantages in foreign trade according to input factor classification. (Source: Own study based on the Comtrade database)

	1995	2000	2005	2006	2007	2008	2009
Mainstream manufacturing (MM)	−0.24	−0.10	−0.08	−0.04	−0.01	0.02	−0.04
Labor intensive (LI)	1.19	1.08	0.70	0.66	0.62	0.55	0.46
Capital intensive (CI)	0.12	−0.07	−0.29	−0.27	−0.25	−0.14	−0.25
Marketing driven (MD)	0.02	0.18	0.28	0.33	0.38	0.41	0.33
Technology driven (TD)	−0.86	−0.41	−0.19	−0.20	−0.16	−0.14	−0.05

Table 7.8 Comparative advantages in foreign trade according to labor force qualification level. (Source: Own study based on the Comtrade database)

Industries	1995	2000	2005	2006	2007	2008	2009
Based on highly qualified employees	−0.50	−0.47	−0.36	−0.32	−0.15	−0.14	−0.10
Based on medium-qualified, white-collar employees	−0.53	−0.31	−0.16	−0.11	−0.08	−0.08	−0.11
Based on medium-qualified, blue-collar employees	0.48	0.35	0.17	0.17	0.17	0.18	0.14
Based on semiqualified employees	0.32	0.45	0.31	0.31	0.35	0.32	0.24

The changes in revealed advantages in trading classed by labor force qualifications of various industries are presented in Table 7.8. The analysis of this table allows for concluding that, from 1995 to 2009, the RCA value of industries based on highly qualified employees and medium-qualified, white-collar employees increased. This should be considered as a positive development. Similarly, also the drop in the value of comparative advantage rates in the trading of commodities produced by industries based on medium-qualified, blue-collar employees and semiqualified workers should be positively evaluated. Despite these positive trends, in 2009 Poland still witnessed modest revealed advantages in the trading of goods produced by industries based on medium-qualified, blue-collar workers and semiqualified employees, as well as a slight lack of such advantages in trading goods produced by industries based on highly qualified employees and semiqualified, white-collar employees.

In order to evaluate the nature of Polish foreign trade, it is also important to take into account the indexes of intra-industry trade (IIT) intensity. They show whether a country exports or imports products manufactured in the same sectors. An increase in IIT is usually the result of growth in the diversification of goods that are the object of trading or adapting to the preferences of consumers and prices of production factors. Generally, the increase in the value of IIT should be evaluated as a positive phenomenon.

The analysis of changes and values of the IIT index in trade, analyzed from the perspective of classifying industries according to technology level, shows that in the period from 1995 to 2009, a notable increase in the intensity of IIT in HT and MHT

Table 7.9 IIT index by technology level. (Source: Own study based on the Comtrade database)

	1995	2000	2005	2006	2007	2008	2009
High technology	0.65	0.57	0.76	0.78	0.84	0.83	0.91
Medium-high technology	0.63	0.64	0.86	0.88	0.88	0.86	0.90
Medium-low technology	0.88	0.95	0.98	0.98	1.00	0.99	0.97
Low technology	0.96	1.00	0.91	0.91	0.91	0.94	0.92

Table 7.10 IIT index according to input factors. (Source: Own study based on the Comtrade database)

Industries	1995	2000	2005	2006	2007	2008	2009
Mainstream manufacturing (MM)	0.76	0.74	0.90	0.91	0.91	0.91	0.93
Labor intensive (LI)	0.56	0.69	0.72	0.75	0.78	0.83	0.82
Capital intensive (CI)	0.94	0.75	0.79	0.80	0.79	0.83	0.83
Marketing driven (MD)	0.89	0.87	0.92	0.91	0.89	0.90	0.88
Technology driven (TD)	0.50	0.60	0.84	0.83	0.84	0.83	0.93

Table 7.11 IIT index according to labor force qualification level. (Source: Own study based on the Comtrade database)

Industries	1995	2000	2005	2006	2007	2008	2009
Based on highly qualified employees	0.65	0.57	0.76	0.78	0.84	0.83	0.91
Based on medium-qualified, white-collar employees	0.63	0.64	0.86	0.88	0.88	0.86	0.90
Based on medium-qualified, blue-collar employees	0.88	0.95	0.98	0.98	1.00	0.99	0.97
Based on semiqualified employees	0.96	1.00	0.91	0.91	0.91	0.94	0.92

industries occurred. The IIT intensity also increased in MLT industries. However, it dropped in LT industries. It is also worth noting that in 2009, in all types of industries, IIT indexes were high and close to 1 (Table 7.9).

The changes in the intensity of IIT in commodities produced by industries with various input factors are presented in Table 7.10. Its analysis allows for drawing the conclusion that, from 1995 to 2009, the highest IIT growth was observed in the trading of commodities produced by RDI and LI industry. Such growth also took place in MM industries. In MDIs, a clear upward or downward trend was not discerned, though IIT intensity in CI industries dropped. In 2009, however, very high IIT ratios were noted in all the listed types of industries.

IIT intensity in various sectors based on diversely qualified labor force is presented in Table 7.11. Its analysis leads to the conclusion that from 1995 to 2009, the highest IIT growth was observed in the trading of commodities produced by industries based on highly qualified employees and industries based on medium-qualified, white-collar employees. The minor increase in intensity concerned trading in goods produced by industries based on medium-qualified, blue-collar employees, and a slight drop took place in the trading of commodities produced by industries based on semiqualified workers. In 2009, however, very high IIT ratios were noted in all the listed types of industries.

7.1.2 Characteristic Features of the NIS

The NIS embraces all market and nonmarket institutions that influence the directions and speed of innovation and technology diffusion in the economy. According to the Organisation for Economic Co-operation and Development (OECD 1999), national innovation system should be considered in the wider context composed of market and nonmarket institutions. Therefore the NIS and its role in the process of developing innovative processes in Poland should be considered in the broader context embracing the development of market and nonmarket institutions in Poland and in the context of the general level of the economy's development.

In the above perspective, the NIS encompasses research institutions, the education system, supporting institutions and enterprises, as well as industrial clusters operating on a regional level and regional innovation systems. All these institutions operate in a wider environment that encompasses global innovation networks, macroeconomic and regulatory environments, transport infrastructure, production factors, the product market, and the education and training system. Interaction between all these elements shapes the national capacity to innovate and the country's innovative position.

According to the Central Statistical Office (Główny Urząd Statystyczny, GUS), the main entities pursuing research and development (R&D) activity in Poland include all institutions and persons concerned with creative works, works performed in order to expand knowledge resources, and works carried out to identify new applications for that knowledge. This means that R&D in Poland embraces:

- Entities whose core activity has been classed in section 72 of the 2007 Polish Classification of Activities (PKD 2007) "Scientific research and development," including:
  - state R&D units,
  - scientific units of the Polish Academy of Science,
  - enterprises,
  - associations, and
  - foundations and units whose operation is supported by other legal forms.
- Universities: public and nonpublic, pursuing R&D activity.
- Entities that, apart from their core activity, systematically or sporadically pursue research activity and development.

The above entities, according to the Frascati Manual (OECD 2002), can be divided into the following types by institutional sectors: enterprises, government institutions, universities, and private nonprofit institutions. This division is consistent with the triple helix system and is therefore used to analyze the dependencies and the R&D activities of various institutions belonging to the Polish NIS.

In order to better comprehend the role of individual Polish NIS institutions and their interrelations—instead of an analysis of their competences and mutual dependencies—it is worth approaching this task by taking a closer look at the structure of research financing sources and differentiating them by the origin of these resources and the entity that is responsible for their disbursement. According to the data provided by the Central Statistical Office (GUS 2011), in 2009 Polish R&D activities were mainly financed from budget funds which constituted 60.4 % (PLN 5.48 billion) of total R&D expenditure. In 2009, enterprise budgets were the second most important source of R&D financing (27.1 % of total R&D expenditure in Poland, PLN 2.46 billion), and the third position was held by universities (6.7 % of R&D expenditure), with foreign funds amounting to 5.5 % of R&D outlays.

However, sources of financing are only one dimension of innovative activities. The dependencies between individual elements of the innovation system are equally significant. This also includes cash flows between the government sector, universities, and enterprises. The data concerning this issue have also been presented by GUS (2011). Their analysis leads to the following conclusions:

- Government funds allocated to R&D are mainly allocated to government research centers and universities. Almost half of the government funds for R&D (49.8 %) have been assigned to government institutions, 44.4 % to universities, while the remainder, just under 6 %, have been assigned to private institutions (mainly enterprises). At the same time, government funds are the key source of research financing at government institutions (87.7 %) and universities (72.4 %). The share of government funds in the financing of research conducted by enterprises and nonprofit private institutions was low and amounted to 12.3 and 2.4 %, respectively.
- Of enterprise funds, 87.9 % was allocated to financing research conducted at enterprises. Another 8 % of these funds was earmarked for financing research in the government sector, while the remaining few percent was allocated to financing research at universities.
- The funds obtained from universities amounted to 6.7 % of R&D outlays in 2009 (PLN 0.61 billion) and almost all of them were allocated to financing research at universities.
- Foreign funds helped finance 5.5 % of R&D expenditure in Poland in 2009 (PLN 0.50 billion). They were used most often by government and local government units and public universities, less often by enterprises. However, the highest share of foreign funds in total R&D resources was noted at institutions that cooperate with enterprises (26.2 %), foundations and associations (16.3 %), and nonpublic universities (8.5 %). In the financing structure of R&D centers, foreign funds had a 6.1 % stake. In the case of offices of the Polish Academy of Sciences, it was 9.2 %. Moreover, in 2009 funds used at enterprises with a majority of foreign capital constituted 13.6 % of funds earmarked for R&D.

The situation described above indicates that there are very strong dependencies and interrelations between the government and government R&D centers, and universities. Nevertheless, dependencies between the government sector and enterprises are very weak. The interrelations between the private sector and state research centers are even feebler, and the weakest ties bind the enterprise sector and universities. The facts described above mean that the dependency structure of the Polish innovation system is far from the model that is described by Etzkowitz and Leydesdorff (2000). The greatest disadvantage of this structure is the fact that enterprises and universities are not close partners, and that they are very weakly bound with state research centers. In addition, the sector of universities and state research centers are financially dependent on government subsidies that finance the overwhelming majority of expenses incurred by the said institutions. However, enterprises finance mainly their own research and, to a minor extent, they seek support by cooperating with state research centers and, to an even lesser extent, by cooperating with universities.

To sum up the above analysis, it is necessary to point out that in Poland public R&D funds were channeled mainly to sectors that are part of the public sector or are closely related to the public sector, including mainly the government sector and universities. To a very small extent, they were distributed to the private sector, including enterprises and private nonprofit organizations. At the same time, the private sector financed its R&D activity mainly from its own funds. On the other hand, the government sector and universities financed their R&D needs mainly from public funds. Moreover, cash flows between the private sector and universities and the sector of government research centers were very low. This situation indicates very weak ties between enterprises and the two institutional sectors mentioned above.

7.1.2.1 Evaluation of Poland's Innovative Position and Capacity from 1995 to 2010

Despite the positive structural changes connected with the increase in the significance of advantages in foreign trade resulting from technology or employment of more advanced input factors, the innovative position of the Polish economy remains relatively weak. As follows from the data presented in Table 7.12, the Polish economy is characterized by a very low patent application rate per 1 million inhabitants. In 2007, this rate amounted to 62.75, while in the USA or in Germany, it equaled 800.17 and 581.67, respectively. Furthermore, the value of this rate in the years 1995–2007 did not improve in Poland. However, it is worth noting that the number of patent applications per USD 1 billion of R&D expenditure in Poland was similar to that characteristic of the above-mentioned highly developed economies. This indicates a relatively similar effectiveness of R&D expenditure utilization in Poland and in the USA, or in Germany. Nevertheless, this expenditure is much lower in Poland. This innovation gap between Poland and highly developed countries is also

Table 7.12 Selected innovative position rates of the polish economy. (Source: Own study based on OECD, EUROSTAT, and World Intellectual Property Organization (WIPO) data)

Rate	Economy	1995	2000	2005	2006	2007
Number of patent application per 1 million inhabitants	USA	465.54	583.98	702.50	742.36	800.17
	Germany	466.71	629.32	586.48	582.84	581.67
	Poland	67.25	62.52	53.14	56.55	62.75
	EU27	..	..	..	..	..
Number of patent application per USD 1 billion of R&D expenditure	USA	..	0.58	0.67	0.68	0.72
	Germany	..	0.91	0.78	0.77	0.74
	Poland	..	0.82	0.73	0.73	0.78
	EU27	..	..	..	..	..
Share of HT commodities in total exports	USA	25.87	29.95	26.15	26.13	..
	Germany	11.64	16.08	14.79	14.06	..
	Poland	2.05	2.84	3.20	3.11	..
	EU27	..	21.39	18.74	16.65	..
Share of employment in HT and MHT industry sectors in total employment (%)	USA	..	..	..	..	..
	Germany	..	11.19	10.50	10.72	10.72
	Poland	..	:	4.90	5.13	5.50
	EU27	..	7.40	6.58	6.59	6.69
Share of employment in RDI sectors in total employment (%)	USA	..	..	..	..	..
	Germany	..	30.37	33.75	34.13	34.79
	Poland	..	:	24.49	24.66	24.81
	EU27	..	30.36	32.47	32.80	32.96

(..) data not available

indicated by the share of HT commodities in total exports, as this share was markedly lower in Poland than in the USA, Germany, or EU27.[1]

Moreover, detailed employment rates indirectly indicate differences in the achieved innovation level. As follows from the data presented in the above-mentioned table, the share of employment in HT and MHT industry sectors in total employment in Poland was 50 % lower than in German economy and amounted to approximately 5 %. It was also slightly lower than that in the entire European Union (EU), where it oscillated between 6.58 and 7.4 %. In addition, the share of employment in RDI sectors in total employments was slightly lower than that in the EU27 or Germany.

The innovative position of the given economy is connected with many factors influencing its innovative capacity. Among others, they encompass various indexes describing R&D expenditure, its structure, access to the Internet, or venture capital development. The selected indexes describing the innovative capacity of the Polish economy against the backdrop of the USA, Germany, and EU27 are described in Table 7.13.

[1] The rate reflecting the share of HT commodities in total exports is not comparable with the rates presented in the previous section, as it refers to the classification of commodities by technological level. In the previous section, the method referring to the technology level characteristic for entire sectors was used.

Table 7.13 Selected innovative capacity rates of the Polish economy. (Source: Own study based on OECD, EUROSTAT, and WIPO data)

Rate	Economy	1995	2000	2005	2006	2007	2008	2009
Share of Gross Domestic Expenditures on R&D (GERD) in GDP	USA	2.50	2.71	2.57	2.61	2.67	2.79	..
	Germany	2.19	2.45	2.49	2.53	2.53	2.68	2.78
	Poland	0.63	0.64	0.57	0.56	0.57	0.60	0.68
	EU27	1.66	1.74	1.74	1.77	1.77	1.84	1.90
Share of Government Intramural Expenditure on R&D (GOVERD) in GDP	USA	0.35	0.28	0.31	0.30	0.29	0.30	..
	Germany	0.34	0.33	0.35	0.35	0.35	0.38	0.41
	Poland	0.22	0.21	0.21	0.21	0.20	0.21	0.23
	EU27	0.28	0.25	0.25	0.24	0.23	0.24	0.26
Percentage of internal expenditure on R&D financed by the government	USA	35.41	25.81	30.24	29.27	28.32	27.05	..
	Germany	37.88	31.40	28.38	27.53	27.51	28.40	..
	Poland	60.24	66.55	57.71	57.45	58.61	59.79	60.44
	EU27	39.55	35.40	35.10	34.10	33.95	34.56	..
Percentage of internal expenditure on R&D financed by other domestic sources	USA	4.36	4.76	5.43	5.39	5.52	5.68	..
	Germany	0.26	0.41	0.29	0.36	0.35	0.31	..
	Poland	2.11	2.12	3.19	2.46	0.42	4.33	6.96
	EU27	1.82	2.13	2.36	2.60	2.49	2.64	..
Percentage of internal expenditure on R&D financed by foreign entities	USA	..	..	..	..	..	..	..
	Germany	1.83	2.15	3.75	3.82	4.01	4.01	..
	Poland	1.69	1.82	5.74	7.04	6.72	5.42	5.50
	EU27	6.63	7.15	8.90	8.59	8.95	8.73	..
Percentage of internal expenditure on R&D financed by the industry	USA	60.23	69.42	64.33	65.34	66.17	67.27	..
	Germany	60.03	66.04	67.57	68.29	68.12	67.27	..
	Poland	35.96	29.51	33.35	33.05	34.26	30.46	27.10
	EU27	52.00	55.29	53.79	54.74	54.60	54.20	..
Share of Business Enterprise R&D (BERD) in GDP	USA	1.76	2.02	1.80	1.86	1.92	2.02	..
	Germany	1.45	1.73	1.72	1.77	1.77	1.86	1.88
	Poland	0.24	0.23	0.18	0.18	0.17	0.19	0.19
	EU27	1.02	1.11	1.08	1.11	1.12	1.15	1.16
Households with Internet access as percentage of all households	USA	..	41.50	..	..	61.70	..	68.70
	Germany	..	16.40	61.64	67.11	70.67	74.88	79.07
	Poland	..	..	30.44	35.94	40.98	47.60	58.59
	EU27	..	..	48.38	49.18	54.15	60.39	65.17
Investments in venture capital as % of GDP	USA	0.04	0.27	0.04	0.04	0.05	0.05	0.05
	Germany	0.01	0.08	0.01	0.01	0.02	0.02	0.02
	Poland	..	0.02	0.00	0.00	0.00	0.00	0.00
	EU27	..	..	..	..	..	..	..

(..) data not available

The most general index that describes the innovative capacity of the economy is the share of total internal R&D expenditure in gross domestic product (GDP). The value of this index in Poland in the years 1995–2009 ranged from 0.56 to 0.68% and was notably lower than in the USA, Germany, and the entire EU. In these economies, this value went up to 2.79% of GDP in 2008 in the USA, 2.78% of GDP in 2009 in Germany, and 1.9% of GDP in 2009 in the EU. In all these economies, the share of R&D expenditure in GDP rose notably in the analyzed period. In Poland, it is difficult to identify an upward or downward trend.

R&D expenditure is not a sufficient basis for increasing the innovativeness of the economy. As regards facilitating technological development, the capacity of economies to transform research results into a specific market offer is most significant. One of the factors determining this capability is the intensity and quality of interrelations between R&D and industry. These are (indirectly) reflected by the R&D expenditure financing structure, especially by the proportion between the level of financing this expenditure by the government and by enterprises. Increasing government expenditure on R&D does not always trigger an increase in the innovativeness of the economy and does not always translate into economic growth. Economic growth is dependent on whether enterprises effectively use research outcomes. In the above context, it is very interesting to observe the changes in indexes reflecting the share of various sources of R&D financing, including government expenditure, other domestic sources, foreign programs (including mainly EU programs), and enterprises. It follows from the data presented in the above-mentioned table that in Poland, the share of government expenditures in R&D outlays amounts to 60% of the total R&D expenditure. To compare, in the USA or Germany, in 2008 this share achieved the level of 27.05 and 28.4% of the total R&D expenditure, respectively. In turn, the share of enterprises in R&D expenditure in Poland ranged from 27.1 to 35.96% and was notably lower at the end of the analyzed period than at its beginning. In the same period, the share of companies in R&D expenditure in the USA and Germany in 2008 achieved the volume of 67% and notably went up in the years 1995–2008. Significant differences in the R&D financing structure between Poland and highly developed countries, however, did not result from very high public outlays on research (as their relation to GDP was lower than that in the above-mentioned highly developed economies). They resulted rather from a very low R&D expenditure incurred by the industry. This situation to a large extent reflected the development level of the Polish economy and Polish enterprises.

Promoting access to information technologies might serve as another index describing innovative capacity. The percentage of households with Internet access as a percentage of all households might be employed as a means of measurement of innovative capacity. In the analyzed period in Poland, notable progress was observed within this scope. In 2009, nearly 58.59% of Polish households had Internet access. In 2005, this rate amounted to a mere 30.44%. However, also in this area, there was a gap between Poland and the USA and Germany in 2009. This resulted from the fact that 68.7% of American households and 79.07% of German households declared that they had Internet access.

A significant drawback of Poland's economy, in comparison with the USA and Germany, was the unpopularity of venture capital investments. While the share of such investments in 2009 amounted to 0.05% in the USA, and to 0.02% in Germany, in Poland it was practically nonexistent.

7.1.2.2 Human Capital

Human resources are the underlying factor of innovativeness of the given economy, as such activity is always the effect of human thought. Human resources constitute a special factor. At the same time, people serve as resources in economic processes and function as entities that determine how all the resources that are being pro-

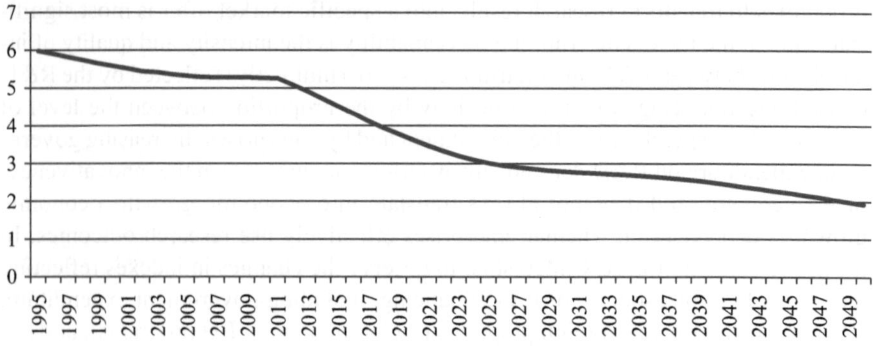

Fig. 7.2 Ratio of working-age population to post-working age population in the years 1995–2010 and forecast until 2050. (Source: Own study based on OECD data)

cessed into goods and services will be used. People also participate in the mutual exchange of resources and their products. The Polish economy has high potential in terms of population, and in the years 1995–2010, the number of people of working age was systematically increasing—from 25.61 million in 1995 to 27.24 million in 2010. However, the total population of Poland fell from 38.27 to 38.19 million in the analyzed period. This was the outcome of low population growth rates and negative net migration. The number of working-age population rose in previous years, because persons born during the 1980s baby boom entered the market.

Unfortunately, the above data do not allow an optimistic outlook. The maintenance of low population growth rates in Poland might have a very negative impact on the economy and stability of public finance. As follows from OECD forecasts, owing to the low fertility rates persisting in Poland, in the coming decades, the Polish society will start aging quite rapidly. This phenomenon is illustrated by the forecast proportion of the working-age population to the post-working-age population until 2050, in comparison with the data for the years 1995–2010, presented in Fig. 7.2. It follows from this figure that in the 2010–2020 decade, this relation will fall from 5.27 to 3.46, and in 2050—if current population growth trends are maintained—it will go down to 1.19. This means that the percentage of persons of working age in the total population will fall drastically, and this development will affect the economic growth and innovativeness.

Capital resources are the second most important factor influencing economic innovativeness, after human resources. Human capital refers to the quality of human resources, and it reflects especially the level and structure of education. In order to evaluate these resources, one has to assess their size, structure, development process, accessibility, and form.

One of the basic indexes describing the level of human capital is the structure of the working-age population by education level. This structure is presented in Table 7.14. The analysis of the data provided in this table leads to the conclusion that from 1995 to 2007, the structure of labor force education in Poland improved notably. The share of persons with primary education fell from 21.5 to 10.2%. The

Table 7.14 Structure of working-age population by education level. (Source: World Bank)

	1995	2000	2005	2006	2007
Percentage of labor force with primary education	21.5	15.7	11.7	10.7	10.2
Percentage of labor force with secondary education	64.7	68.7	69.4	68.9	68.3
Percentage of labor force with higher education	13.8	15.6	18.9	20.4	21.5

share of persons with secondary education went up from 64.7 to 68.3 %, similarly as the share of persons with higher education—from 13.8 to 21.5 %.

The above trends were also reflected by the increase in the average number of years of education of persons aged 25 or more in Poland—from 9.1 years in 1995 to 9.9 in 2010. In comparison with Germany, the USA, or Sweden, in 2010 the average years of education of persons aged 25 or more in Poland was still lower. Moreover, in all analyzed economies this index improved in the years 1995–2010. This means that the above countries increased investments in their human capital. Poland, despite the rise it witnessed, still has a gap to bridge to catch up with the most developed countries.

Not only the population potential, described by means of demographic and human capital indexes, is of significance for the economy, but also the level of utilization of human resources in the process of generating domestic income is important. In the years 2000–2010, the employment rate in Poland went up from 55 % to more than 59 %, which is a positive development (Fig. 7.3). Against the backdrop of EU member states and the USA, the employment rate in Poland is still relatively low. To compare, this rate exceeded 70 % in Germany, Austria, Sweden, Denmark, and the Netherlands. However, one should note that in comparison with the mid-1990s, the years 2000 and 2005, in 2009 the unemployment rate remained at a relatively low level in Poland.

Labor productivity can also serve as very important indicator of human resources and capital utilization in the generation of domestic income. These indicators constitute the relation of, successively, the GDP to total population, the number of employed persons, and hours of labor in the economy. Their analysis points to the source of the discrepancy between economic development levels. The data concerning this issue are also presented in Table 7.15. Their analysis shows that, in the years 1995–2010, Poland was gradually catching up with the economic level of European states. In 2010, the development divide between Poland and the EU27 average amounted to 37 %, while in 1995 it already amounted to 57 %. The gap between the productivity of one employed person was even lower and amounted to 33.2 % in 2010, while in 1995 this ratio equaled as much as 54 %. However, the above fact also means that a wider disparity between the levels of GDP per capita than between productivity per one employed person resulted from employment rates lower than in EU27, which indicates a low utilization of human resources in the economy. Data that are even more interesting can be obtained from the index describing productivity rate per working time. Within this scope, the divide between productivity levels in 2010 amounted to 46.1 %, while in 1995 it equaled 60.17 %. Notably lower

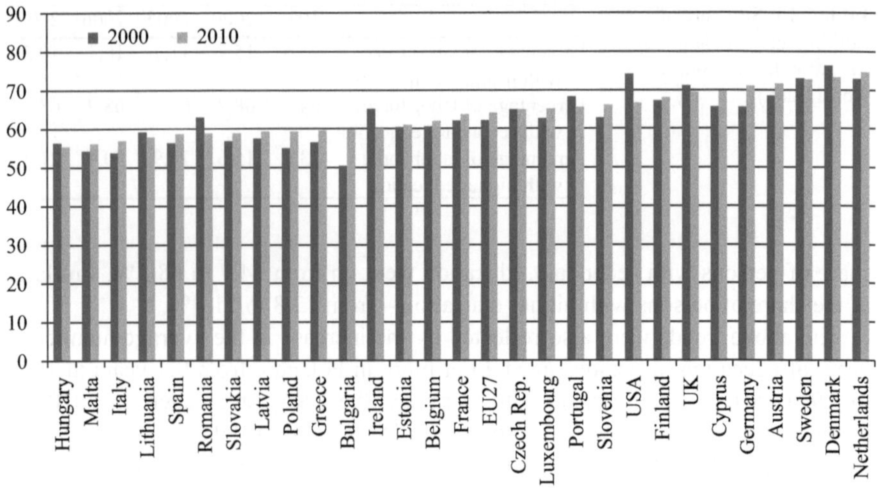

Fig. 7.3 Employment rate in Poland in comparison with selected economies in the years 2000 and 2010. (Source: Own study based on EUROSTAT data)

Table 7.15 Labor productivity in Poland in comparison with EU27 productivity (EU27 = 100). (Source: EUROSTAT)

	1995	2000	2005	2010
GDP per capita	43.0	48.0	51.0	63.0
Labor productivity per employee	46.0	55.5	61.7	66.8
Labor productivity per working time	38.3	45.5	49.7	53.9

productivity indicators per working time in comparison with productivity per one employee indicate that the average working time of each employee in Poland was markedly higher than the working time of other employees in EU27. The latter index also illustrates the actual productivity per hour of each employee in Poland. It is still much lower than the EU27 average, which, to a large extent, is the outcome of organizational and technological factors. However, Poles close this productivity gap by having a higher average working time. Still, lower employment levels exacerbate the disparities in GDP per capita.

Apart from labor productivity, labor costs are also significant. The changes in labor costs are presented in Table 7.16. Its analysis allows for drawing the conclusion that in 2009, the remuneration level in Poland amounted to approximately 50 % of the German remuneration level and about one-third of US remunerations. At the same time, in comparison to the mid-1990s, relative remunerations in Poland increased in comparison to the two above-mentioned economies. Interestingly, in the years 2000–2009 in Poland the relative hourly rate of remuneration fell in comparison to the USA and Germany, and in 2009 in Germany the hourly remuneration rate constituted 307 % of remunerations paid out in Poland. US remunerations amounted to 346 % of Polish salaries. However, in 2005, i.e., before the most recent crisis, this

Table 7.16 Labor costs in Poland in comparison with the USA and Germany. (Source: OECD)

Rate	Economy	1995	2000	2005	2009	2010
USD						
Annual renumeration per employee according to USD PPP	Germany	30,295.14	33,044.26	36,760.15	40,100.27	..
	Poland	10,192.94	14,366.00	15,815.94	20,073.36	..
	USA	35,912.94	44,931.05	52,959.33	59,049.21	..
Remuneration per hour according to USD PPP	Germany	21.07	24.04	27.42	30.95	..
	Poland	..	8.54	7.69	10.07	..
	USA	20.15	25.02	30.60	34.84	..
Labor costs per production unit	Germany	0.72	0.72	0.68	0.70	0.68
	Poland	0.63	0.65	0.55	0.55	0.57
	USA	0.67	0.68	0.65	0.64	..
Poland = 100						
Remuneration per employee according to USD PPP	Germany	297.2	230.0	232.4	199.8	..
	Poland	100	100	100	100	..
	USA	352.3	312.8	334.8	294.2	..
Remuneration per hour according to USD PPP	Germany	..	281.5	356.6	307.3	..
	Poland	..	100	100	100	..
	USA	..	293.0	397.9	346.0	..
Labor costs per production unit	Germany	114.3	110.8	123.6	127.3	119.3
	Poland	100.0	100.0	100.0	100.0	100.0
	USA	106.3	104.6	118.2	116.4	

(..) data not available

disproportion was even higher, to the detriment of remunerations in Poland. If one converts labor costs in relation to production units, one will obtain the index of actual labor costs, and these are not much lower in Poland than in the USA or Germany. As follows from the data presented in the above-mentioned table, these indexes in the USA and Germany amount to 127.3 and 116.4 %, respectively, of labor costs per production unit in Poland. This means that the remuneration level in all these economies is the reflection of the actual level of achieved labor productivity.

7.1.2.3 Characteristic Features of Science, Innovation, and Education Policies in the Years 2005–2010

In the second half of the 2000–2010 decade, the science, innovation, and education policies underwent significant changes resulting from the political cycle and changes in the political system. A significant element systemizing Poland's policy toward the above areas was the adoption of the 2009 *Plan for Organizing the Development Strategy* by the Council of Ministers. The passing of this document was aimed at organizing the priority documents in the form of the development strategies and programs that had been drafted until then. In this document, the drafting of nine development strategies implementing the Medium-term (and Long-term) National Development Strategy was proposed. This proposal embraced innovation and economic productivity strategy, human capital development strategy, strategy for transport development until 2020, energy safety and environment, efficient state program, social capital development strategy, national strategy for regional development 2010–2020, regions, cities, and rural areas, strategy for the development of the national security system of the Republic of Poland, and strategy for sustainable development of rural areas, fishing and fishery. Among the listed nine strategies, two refer directly to the NIS, i.e., the innovation and economic productivity strategy and the human capital development strategy.

Another program document is linked to the above strategies, namely the *National Reform Program: Europe 2020.* According to the provisions of this document, two of the three priority areas connected with broadly understood infrastructure and innovation are of key importance for the development of the NIS. In the context of the above-mentioned disadvantages of the Polish NIS connected with the low funds allocated to R&D and very weak links between enterprises and universities and public R&D centers, one should assume that the following points are of utmost importance to strengthen the Polish NIS: 1.6. R&D infrastructure development and modernization; 2.4. intellectual capital for innovativeness; 2.5. closer cooperation between science and economy; 2.6. informed management of the innovation drive; and 2.3. new directions for innovation development. In many cases, however, it is difficult to assess the effectiveness of the planned activities at the current stage. Initial comments on the above-mentioned policy areas are presented in Table 7.17.

To sum up the above deliberations, the main strengths and weaknesses of the Polish NIS, as well as the opportunities and threats this system might have to face may be assessed (Table 7.17).

7.1.3 Interrelations Between Innovation Level, Human Capital Accumulation, and Advantages in Foreign Trade

As indicated in the foreign trade analysis presented above, the main positive trends observed in the period of 1995–2009 comprise mainly the very dynamic growth of foreign trade after 2002 and the relatively fast revival of exports after its collapse during the 2009 crisis.

Apart from these general trends, also a wide range of positive structural changes were noted in Polish exports in the years 1995–2009. The following developments are worth mentioning:

- decrease in the share of commodities produced by LT and MLT industries;
- increase in the share of commodities produced by HT and MHT industries;
- decrease in the share of commodities produced by LI and CI industries;
- increase in the share of goods produced by RDI, MDI, and MM industry;
- decrease in the share of commodities produced by industries based on medium-qualified, blue-collar workers and semiqualified employees; and
- increase in the share of goods produced by industries based on highly qualified employees and medium-qualified workers;

Table 7.17 SWOT matrix: evaluation of the Polish national innovation system. (Source: Own study)

Opportunities	Threats
Greater access to external sources of R&D financing, including EU funds	Deteriorating demographic situation that might affect future economic growth
Advantages resulting from the increase in competitiveness and opportunity of joining the EU single market, as well as inflow of foreign investments	Instability of the global and European economy that might impact Poland's economy and the possibility of financing research from public funds and financial resources of enterprises
Opportunity of developing international relations and international cooperation within the EU	Insignificant progress in implementing structural and regulatory reforms in the economy
Polish enterprises have more possibilities of pursuing international expansion	Draining of highly qualified labor force Persisting unfavorable education structure
Strengths	Weaknesses
Increasing share of innovative goods in exports and the shifting of advantages in the direction of innovative goods	Weak ties between enterprises and universities and state research centers
Improved education structure of the population	Low participation of enterprises in R&D financing
Relatively good economic situation allowing for incremental increases in expenditure	Relatively low share of innovative goods in exports and advantages in trading of goods that are not very innovative
Relatively good standing of Polish enterprises that allows for production development and increases in capexI	Low effectiveness of universities and public R&D centers in commercialization of research
Improvement of the condition of infrastructure in Poland and gradual rise in expenditure on research—although their nominal value is still low	Low innovativeness of the economy in terms of patents Faulty structure of professional qualifications Inertia of the public R&D system

Despite these positive trends visible in the changes in foreign trade structure, one ought to note that Polish exports are characterized by many negative features, including among other things the fact that in 2009 still more than half of the value of Polish exports was constituted by exports of LT and MLT. In addition, more than half of the value of Polish exports was constituted by exports of commodities manufactured by industries based on medium-qualified, blue-collar workers and semiqualified employees.

When analyzing the results of Polish exports against the backdrop of the features of the Polish NIS, one ought to indicate that the increase in the value of Polish exports achieved in the most recent decade, including the change in export structure, was to a large extent possible, thanks to the inflow of direct foreign investments. According to Radło (2011), the industries that noted highest increases in exports in the recent decades were the industries which had witnessed a very high inflow of direct foreign investments. This influx was connected with many features of institutional nature, including accession to the EU, a high accessibility of qualified labor force in Poland, and persisting differences in labor costs. Here, it is possible to prove that there is a positive relationship between human capital and exports.

In the light of the above facts, it is not an easy task to draw conclusions on the basis of research conducted on whether the NIS led to the improvement of the competitiveness of Polish exports. Furthermore, when considering the disadvantages of this system—and especially the weak bonds between enterprises, universities, and public research centers—one might state that most probably the role of the NIS in developing competitive advantages in Polish foreign trade was marginal. However, one ought to add that the improvement in human capital accumulation, including a notable improvement in the education level of the Polish population, had a major impact on increasing the attractiveness of the Polish economy in terms of attracting foreign investments.

Nevertheless, when formulating conclusions such as those presented above, one should bear in mind that economies are based on various competitiveness factors depending on their development stage. This also means that the Polish economy—at the current development stage—built its competitive advantages by attracting foreign investments and reviving its innovation potential. Moving on to higher levels of economic development and competitiveness will mean that developing competitive advantages of national enterprises will gain in importance, similarly as the development of competitive advantages on the basis of the national economic potential. The fact that in recent years Polish companies have been increasingly interested in investing abroad is indicative of Poland's economy shifting toward that stage of developing national competitiveness. More and more Polish companies are competitive enough to export as well as invest equity in foreign markets.

For the above reasons, it should be indicated that in the nearest future we ought to expect an increase in the significance of the NIS in the area of creating competitive advantages of Polish enterprises. Therefore, it ought to be assumed that introducing improvements to the functioning of the NIS is of great importance. All the more, as in its current condition this system has numerous disadvantages. Its weak points concern mainly the weak ties between enterprises and the low commercialization of R&D activity results.

7.2 Case Study: Hungary

Marcin Gomułka

Hungary is a country with a small economy and, thus, a great capacity for adjusting and managing its competitiveness in a dynamic way. During the transition to the free market economy, Hungary managed to launch a wave of growth in exports by the inflow of foreign direct investments.

In the recent years, the Hungarian GDP growth rate has slowed down, as a result of the global financial and economic crisis. Unlike Poland, Hungary did not manage to avoid the negative growth in 2009. The decline in the GDP was significant and amounted to −6.8 %, while in the previous 2 years it was close to zero. Over the last 2 years (2010–2011), it was possible to slightly increase the growth rate to over 1 % per year. All in all, the years 2007–2011 were a period of a considerable slowdown of Hungary, which should be regarded as a symptom of a lack of macroeconomic competitiveness.

Table 7.18 Geographical structure of Hungary's exports (USD M and %). (Source: Own study)

	1995	2000	2005	2010	Structure in 2010 (%)
Germany	3,633	10,471	18,734	23,854	25.2
Italy	1,042	1,654	3,445	5,236	5.5
Great Britain	383	1,156	3,162	5,126	5.4
Romania	347	574	2,335	5,074	5.4
Slovakia	199	288	1,766	4,901	5.2
France	514	1,471	3,187	4,745	5.0
Austria	1,264	2,443	3,482	4,534	4.8
Poland	316	605	2,021	3,465	3.7
Russia	803	455	1,171	3,370	3.6
The Czech Republic	204	465	1,900	3,312	3.5
The Netherlands	366	1,522	2,406	3,029	3.2
Spain	119	523	1,998	3,013	3.2
The U.S.	394	1,475	1,896	1,930	2.0
Total exports	12,452	28,092	62,272	94,693	100.0

7.2.1 *Evaluation of Hungary's Competitiveness in Foreign Trade*

The most important export partner of Hungary is Germany, which in 2010 accounted for 25% of Hungarian exports. Germany was the main trading partner of Hungary already in 1995. Since then, its dominant position has not changed.

Attention should be paid to the fact that the economies of the United States, Russia, and China, which are large in terms of not only economics but also population, have small significance as export markets for Hungary. As an export partner, Poland is more important than Russia, the Netherlands, or Spain, and, in turn, count more than the United States. As for China as an export market, it is far behind other more significant countries.

To sum up, the majority of exports is directed to highly developed countries of the Western Europe. The comparison of data on the geographical structure of exports is contained in Table 7.18.

In the import structure, it is possible to observe both some similarities and some significant differences. Germany is Hungary's most important import partner, with a similar share in imports and in exports amounting to about 25%. Two other import partners are Russia and China, with shares equaling to 8 and 7%, respectively. In the case of Russia, it is undoubtedly related to Russia's position as a country exporting energy (gas and oil) to the European market. The case of China is connected with that country's considerable involvement in the export of processed manufactured goods.

As for the remaining countries, the import structure does not significantly vary from the export structure. Other important trading partners of Hungary are not only neighboring countries such as Austria, Poland, and Slovakia, but also more distant ones such as Italy, the Netherlands, and France. A direct neighbor, the Czech Republic, is only the tenth trading partner.

It is interesting that some of Hungary's neighbors do not figure on the list of its main trading partners. It concerns Ukraine, Slovenia, Croatia, and Serbia, as well as Romania in the case of imports.

Table 7.19 Geographical structure of Hungary's imports (USD M and %). (Source: Own study)

	1995	2000	2005	2010	Structure in 2010 (%)
Germany	3,602	8,213	18,211	22,272	25.0
Russia	1,754	2,589	4,868	6,826	8.0
China	126	939	3,588	6,177	7.0
Austria	1,654	2,366	4,342	5,537	6.0
The Netherlands	459	708	2,868	3,770	4.0
Poland	247	651	2,500	3,720	4.0
Italy	1,215	2,407	3,213	3,663	4.0
Slovakia	363	575	1,490	3,407	4.0
France	606	1,401	3,262	3,219	4.0
The Czech Republic	358	646	1,891	2,887	3.0
Belgium	388	721	1,467	1,985	2.0
Japan	338	1,701	2,249	1,895	2.0
UK	441	1,017	1,640	1,665	2.0
The U.S.	474	1,224	1,111	1,573	2.0
Total imports	15,186	32,080	65,920	87,356	100.0

All in all, the geographical structure of imports is, just like in the case of exports, focused on the trade with the Western Europe, with the addition of raw materials or LI goods imported from Russia and China (Table 7.19).

Before moving on to more advanced measures of competitiveness of the Hungarian economy in foreign trade, it is worth analyzing the goods structure of the trade.

Today (in 2010), in the case of Hungary, the goods structure of both exports and imports is dominated by group 7 of the SITC, i.e., machinery and transport equipment. Exports, to the tune of USD 53 billion, are, to a large extent, offset by imports equal to USD 39 billion. In terms of the trade structure, it means that as much as 56 % of exports and 45 % of imports belong to this category.

The second important position in Hungary's trade is SITC 6 manufactured goods classified chiefly by material. In terms of both quantity and of share, the import of raw materials dominates to some extent over the export in a ratio of USD 8.8 billion of exports to USD 10.7 billion of imports (9–11 % in terms of the share).

As for imports, the considerably dominant category in foreign trade is SITC 3, i.e., mineral fuels, lubricants, and related materials. These goods constitute 11 % of imports, but only 3 % of exports. The presence of this imbalance is understandable and related to the previously discussed geographical structure of foreign trade. It is connected with the import of fuels and other derivatives of hydrocarbons primarily from Russia. Hungary does not have its own substantial oil reserves, so it is largely dependent on its import.

Other trade categories have low importance. To sum up, the main characteristic of Hungary's foreign trade from the structural perspective is the need for equivalent exchange in the field of machinery, equipment, and transport equipment, as well as the need to find output markets for other goods categories of exports, in order to acquire the means for the import of fuel. In this context, the competitiveness understood as the ability to sell on world markets is the factor necessary for the efficient functioning of the economy.

Table 7.20 Export structure according to the groups of the SITC classification in USD M and %. (Source: Own study)

SITC Code	Goods	1995	2000	2005	2010	Structure in 2010 (%)
7	Machinery and transport equipment	3,047	15,798	35,865	53,367	56.0
6	Manufactured goods classified chiefly by material	2,248	3,080	6,334	8,806	9.0
5	Chemicals and related products	1,489	1,759	4,997	8,423	9.0
8	Miscellaneous manufactured articles	2,157	4,066	6,575	7,748	8.0
0	Food and live animals	2,254	1,798	3,339	5,934	6.0
9	Commodities and transactions not classified elsewhere in the SITC	7	461	2,242	5,386	6.0
3	Mineral fuels, lubricants, and related materials	379	451	1,619	2,611	3.0
2	Crude materials, inedible, except fuels	560	520	1,016	1,802	2.0
4	Animal and vegetable oils, fats, and waxes	48	47	127	294	0.0
1	Beverages and tobacco	259	103	151	267	0.0
	Total exports	12,448	28,083	62,265	94,638	100.0

From the dynamic perspective, it is interesting that, in the years 1995–2010, the value of export of SITC 7 machinery, equipment, and transport equipment increased from USD 3 billion to USD 53 billion. Attention should be paid to the fact that, while in 1995 the import of machinery was greater than the export, over 15 years the situation had reversed (Tables 7.20 and 7.21).

It is worth extending the analysis of the goods structure according to the SITC to include other structure profiles of exported and imported goods. A very important profile is the technological structure that allows the division of exports and imports into different levels of technological advancement. An increase in the share of technologically advanced goods is a symptom of growing competitiveness of exports of an analyzed country, whose companies increase their ability to sell on world markets through innovativeness.

The growth rate of Hungary's import structure indicates that important changes took place in the years 1995–2000, while the period of 2000–2010 was characterized by structural stability. Thus, the changes that took place until 2000 are the most important. They consisted in an increase in the share of HT goods in imports from 9 to 15 %. On the other hand, the share of LT goods decreased from the level of 17 % to the level of 10 %.

The export structure had been continuously improving throughout the 1995–2010 period. The share of HT goods had been slowly increasing, from the level of 7 % up to the level of 22 %. The share of the export of LT goods had been decreasing and in 2009 amounted to only 9 %, while at the beginning it was 22 %.

The change in the structure took place at the same time as the increase in absolute levels of import and export of all technology categories. These two situations could occur simultaneously due to the fact that the export and import of the HT goods had been increasing very fast, while the export and import of LT

Table 7.21 Import structure according to the groups of the SITC classification in USD M and %. (Source: Own study)

SITC Code	Goods	1995	2000	2005	2010	Structure in 2010 (%)
7	Machinery and transport equipment	4,463	15,771	31,481	39,168	45.0
6	Manufactured goods classified chiefly by material	3,587	5,376	9,198	10,747	12.0
3	Mineral fuels, lubricants, and related materials	1,811	1,530	6,675	9,290	11.0
5	Chemicals and related products	2,174	2,779	5,845	8,501	10.0
9	Commodities and transactions not classified elsewhere in the SITC	1	1,372	3,544	7,936	9.0
8	Miscellaneous manufactured articles	1,703	3,710	5,562	6,114	7.0
0	Food and live animals	715	799	2,281	3,526	4.0
2	Crude materials, inedible, except fuels	611	625	950	1,473	2.0
1	Beverages and tobacco	79	68	285	349	0/0
4	Animal and vegetable oils, fats, and waxes	41	47	97	237	0.0
	Total imports	15,186	32,078	65,918	87,340	100.0

Table 7.22 Structure of imports according to the technology levels of the OECD classification (in %). (Source: Own study)

	1995	2000	2005	2010
High technology	9	15	15	15
Medium-high technology	37	40	41	43
Medium-low technology	37	32	32	33
Low technology	17	12	11	10

Table 7.23 Structure of exports according to the technology levels of the OECD classification (in %). (Source: Own study)

	1995	2000	2005	2010
High technology	7	19	20	22
Medium-high technology	31	40	42	43
Medium-low technology	40	28	29	26
Low technology	22	13	10	9

goods had been increasing at a moderate pace. Attention should be paid especially to the increase in the export of HT goods in the years 1995–2010. All in all, the structural data on exports and imports according to the technology levels of the OECD classification indicate a very high improvement in Hungary's competitiveness (Tables 7.22 and 7.23).

In the case of Hungary, changes in the import structure indicate a decrease in the share of the import of LI goods from 14 to 7 %. At the same time, the share of technology-driven (TD) goods in imports increased from 23 to 41 %. It means that the TD goods achieved the biggest share in imports.

Table 7.24 Structure of imports according to the *Österreichisches Institut für Wirtschaftsforschung Wien* (WIFO) classification based on the extent of use of tangible and intangible expenses in the production process (in %). (Source: Own study)

Industries	1995	2000	2005	2010
Mainstream	28	25	24	24
Labor intensive	14	10	9	7
Capital intensive	25	20	20	20
Marketing driven	10	9	9	8
Technology driven	23	37	38	41

Table 7.25 Structure of exports according to the WIFO classification based on the extent of use of tangible and intangible expenses in the production process (in %). (Source: Own study)

Industries	1995	2000	2005	2010
Mainstream	22	18	20	20
Labor intensive	24	15	10	7
Capital intensive	21	13	15	15
Marketing driven	17	8	7	7
Technology driven	16	45	47	50

At the same time, a similar growth rate can be observed in the field of exports. In the years 1995–2010, the share of LI goods decreased from the level of 24 % to the level of 7 %. A very high increase in the share of TD goods in exports was recorded, from 16 to 50 %, although most of the changes took place in the years 1995–2000.

These changes have to be assessed very positively. They are a symptom of an increase in innovativeness of the Hungarian economy in the domain of export competitiveness. Fifty percent of the exports is already constituted by TD goods, while the share of LI goods in exports has been considerably reduced. The absolute value of each of these categories has been growing, although the growth rate has been varied. As for the TD goods, throughout 15 years their export has noted a 70-fold increase, which is an extraordinary achievement.

To sum up, the foregoing data also indicate a strong relation between innovativeness and international competitiveness of Hungary (Tables 7.24 and 7.25).

Another issue analyzed is the impact of human capital on the competitiveness of the Hungarian economy. It can be examined from the angle of the use of factor of labor of different quality in the field of export (or import) production. The growing share in the export of goods manufactured by highly skilled workers can indicate that the human capital has a positive influence on the competitiveness.

In the case of Hungary, it is possible to observe an increase in the share of goods manufactured by industries hiring mainly highly skilled and semiskilled white-collar workers in exports. Together, these two categories of products comprise 65 % of exports. As for the category of goods produced by low-skilled workers, its share in exports decreased from 22 to 9 % in the analyzed period of 15 years. The absolute value of the export of goods produced by low-skilled workers tripled, while the export of goods manufactured by industries hiring highly skilled workers recorded a 20-fold increase.

In the domain of imports, it is possible to notice a similar trend of structural changes. However, attention should be paid to the fact that the export structure has a qualitative advantage over the import structure in relation to goods produced by the

Table 7.26 Export structure of goods according to the WIFO classification based on the skill level of labor force (in %). (Source: Own study)

Export structure	1995	2000	2005	2010
Industries hiring highly skilled workers	7	19	20	22
Industries hiring semiskilled white-collar workers	31	40	42	43
Industries hiring semiskilled blue-collar workers	40	28	29	26
Industries hiring low-skilled workers	22	13	10	9

Table 7.27 Import structure of goods according to the WIFO classification based on the skill level of labor force (in %). (Source: Own study)

Import structure	1995	2000	2005	2010
Industries hiring highly skilled workers	9	15	15	15
Industries hiring semiskilled white-collar workers	37	40	41	43
Industries hiring semiskilled blue-collar workers	37	32	32	33
Industries hiring low-skilled workers	17	12	11	10

most highly skilled workers in a ratio of 22 to 15%. It also means that, in Hungary, as far as absolute values are concerned, the export of goods manufactured by highly skilled workers is higher than the import.

To sum it up, the export and import structures divided into categories according to the skill level also show a strong relationship between the human capital and the competitiveness of the Hungarian economy (Tables 7.26 and 7.27).

A standard measure of comparative advantages in trade is the RCA index. The index can be used in relation to the classification dividing trade into different subgroups. In relation to the competitiveness, the index assumes that certain categories of foreign trade can be considered more valuable than others. From the point of view of the competitiveness of a country, which translates into the welfare of its inhabitants, it is preferable to gain a competitive advantage in the domains of exports that are considered valuable. The export of HT goods can without a doubt be numbered among them.

This study uses the division of exports into categories according to the three above-mentioned classifications. The approach to the structure of trade taking into account its division into different technology levels (Table 7.28) indicates that, in the years 1995–2010, Hungary's advantage in the field of HT and MHT industries dynamically increased. The strongly growing competitiveness was confirmed by the fact that in the domain of HT industries the change was the biggest (an increase of RCA of 0.67), and the final level of the comparative advantage was the highest among all the four analyzed categories.

At the same time, the comparative advantage related to LT industries recorded the biggest decline (an increase of RCA by 0.38). However, at the end of the examined period, the lowest comparative advantage was noted in the field of MLT industries.

Table 7.28 Revealed comparative advantage (RCA) in Hungary's foreign trade according to the OECD classification based on the technology level. (Source: Own study)

	1995	2000	2005	2010
High-technology industries	−0.23	0.08	0.17	0.44
Medium-high-technology industries	−0.19	−0.10	−0.09	0.05
Medium-low-technology industries	0.06	−0.23	−0.22	−0.18
Low-technology industries	0.29	−0.08	−0.20	−0.09

Table 7.29 Revealed comparative advantages (RCAs) in Hungary's foreign trade according to the WIFO classification based on the extent of use of tangible and intangible expenses in the production process. (Source: Own study)

Industries	1995	2000	2005	2010
Mainstream	−0.19	−0.33	−0.14	−0.13
Labor intensive	0.57	0.45	0.08	0.08
Capital intensive	−0.12	−0.44	−0.25	−0.25
Marketing driven	0.56	−0.08	−0.25	0.01
Technology driven	−0.31	0.19	0.24	0.26

These data indicate a very positive trend in the economic and pro-export growth of Hungary. The shift in favor of HT industries is the evidence of a significant improvement in the competitive position of the country.

As for the RCA index calculated in relation to industries categorized according to the extent of use of tangible and intangible factors, it is more difficult to interpret (Table 7.29). First of all, it is impossible to indicate a determined trend in the domain of mainstream industries. The index shows a slight noncompetitiveness in this field. In turn, the revealed competitive advantage concerning the goods manufactured by LI industries had considerably declined. As for the goods produced by CI industries, a noncompetitiveness is observed, which is an unfavorable symptom. It means that Hungary is not fully capable of ensuring an adequate capital base for these industries or that the industries are less competitive than foreign industries. The export of goods produced by MD industries had been undergoing strong fluctuations, although at the end of the analyzed period of 1995–2010 the RCA index indicated a balance between exports and imports.

The most important and a very positive observation is the one concerning the change in RCA index for TD industries from the level of −0.31 to the level of 0.26. It indicates a considerable growth in the competitiveness of Hungary in the years 1995–2010 in the domains related to modern technologies. In the context of the subject of the present research, i.e., the relation between innovativeness and competitiveness, it is a sign of a strong increase in competitiveness in the areas that depend on innovativeness.

The data analyzed from the perspective of the division of industries according to the skill level of their workers indicate a growing competitiveness of Hungary measured by comparative advantages observed in relation to industries requiring the highest skill level (Table 7.30). Industries hiring highly skilled workers increased the advantage measured by the RCA index from the level of −0.23 (no advantage) to the level of 0.44 (strong advantage). A positive increase, although weaker, was recorded also in the field of industries hiring semiskilled white-collar workers. The comparative advantage

Table 7.30 Revealed comparative advantages (RCAs) in Hungary's foreign trade according to the WIFO classification based on the skill level of labor force. (Source: Own study)

Industries	1995	2000	2005	2010
Industries hiring highly skilled workers	−0.23	0.08	0.17	0.44
Industries hiring semiskilled white-collar workers	−0.19	−0.10	−0.09	0.05
Industries hiring semiskilled blue-collar workers	0.06	−0.23	−0.22	−0.18
Industries hiring low-skilled workers	0.29	−0.08	−0.20	−0.09

Table 7.31 The intensity of intra-industry trade in the trade of goods according to the OECD classification based on the technology level. (Source: Own study)

Industries	1995	2000	2005	2010
High technology	0.79	0.98	0.94	0.75
Medium-high technology	0.81	0.88	0.93	0.94
Medium-low technology	0.93	0.82	0.86	0.95
Low technology	0.95	0.89	0.87	0.99

related to goods produced by low-skilled workers noted in 1995 had, in turn, been eliminated (a decrease from the level of 0.29 to the level of −0.09).

The foregoing results indicate a strong improvement in Hungary's competitiveness concerning the use of human capital to achieve benefits of foreign trade. Thus, for this country, the research hypothesis of the study is confirmed.

As stated by J. Misala in Chap. 1, it is worth "preferring solutions that favor the consolidation of the international division of labor of intra-industry nature," because "the development of this subdivision of work visibly stimulates inventiveness and innovativeness." Thus, a high level of IIT in every field of foreign trade is a positive phenomenon. It seems, however, that in the case of foreign trade related to goods based on natural advantages, e.g., advantages in the domain of raw materials, a low level of IIT should not be criticized. The intra-industry division of labor has to occur in the domains in which a given country has no raw material basis for activity.

On the basis of the available data, it can be stated that the Hungarian economy is characterized by a very high intensity of IIT in goods manufactured by LT industries and other industries, up to the MHT industries. As for the HT industries, a full intensity was observed in the years 2000–2005, but recently it has declined to a lower level. However, it is not a disadvantage, because, as shown by the data concerning absolute levels of trade in this category, in Hungary, the export of HT goods is higher than the import (Table 7.31).

The analysis of the intensity of foreign trade conducted on the basis of the extent of use of tangible and intangible expenses indicates an increase in the intensity in all categories of the classification taken into consideration. Only the category of goods produced by TD industries constitutes an exception, as its intensity had increased in the years 2000 and 2005, but then it decreased in 2010 to the level of 0.83. In the case of other categories, the intensity of trade was between 0.91 and 0.97. The intensity of foreign trade in goods manufactured by TD industries in the years 1995 and 2010 has a positive interpretation, because it was related to export surpluses in this trade category (Table 7.32).

Table 7.32 Intensity of intra-industry trade (IIT) according to the classification of industries based on the extent of use of expenses. (Source: Own study)

Industries	1995	2000	2005	2010
Mainstream	0.81	0.77	0.90	0.97
Labor intensive	0.82	0.84	0.99	0.92
Capital intensive	0.84	0.72	0.85	0.91
Marketing driven	0.82	0.89	0.85	0.96
Technology driven	0.75	0.97	0.91	0.83

Table 7.33 Intensity of Hungary's intra-industry trade (IIT) in the trade of goods according to the WIFO classification based on the skill level of labor force. (Source: Own study)

Industries	1995	2000	2005	2010
Industries hiring highly skilled workers	0.79	0.98	0.94	0.75
Industries hiring semiskilled white-collar workers	0.81	0.88	0.93	0.94
Industries hiring semiskilled blue-collar workers	0.93	0.82	0.86	0.95
Industries hiring low-skilled workers	0.95	0.89	0.87	0.99

The dynamics of the intensity of IIT in the section related to the skill level of labor force does not show any unusual relationships between the intensity and the quality of labor force, nor a visible trend of change. As for the industries hiring highly skilled workers, it is possible to observe a transitory process of intensification in the years 2000–2005. Over this period, the intensity of trade had increased from the level of 0.79 to the level of 0.98–0.94, but in 2010 it decreased again to the level of 0.75. However, the process cannot be interpreted negatively, as both in 1995 and in 2010 the surplus of exports over imports in the domain of goods produced by industries hiring highly skilled workers was recorded in Hungary. Thus, the observed decline in the intensity has a rather positive interpretation (Table 7.33).

7.2.2 *Characteristics of Hungary's NIS*

The Hungarian NIS has inherited its traits from the economic system of the real socialism. This is the reason why its structure is similar to that of the Polish system. One of its characteristics is, among others, the existence of the Academy of Sciences. The system has evolved as a result of two important phenomena of transformation: economic transformation after the collapse of communism in 1989 and the adjustment to the scheme of the innovation policy of the EU after 2004.

The structure of Hungary's NIS can be divided into the following three levels:

1. the political level, including governmental bodies forming the science, innovation, and education policies;
2. the operational level, including intermediate bodies; and
3. the executive and research level, including academic institutions, research institutes, and private entities.

The innovation policy is a part of the general policy of the Hungarian government. In the domain of government's activities, the National Research, Innovation, and Science Policy Council (NKITT) is the most important government body, which takes care of these actions, debates under the chairmanship of the Prime Minister, and is responsible for science, technology, and innovations. The key role of the Council consists of coordinating and approving proposals for legislative changes that directly affect the operation of the national system of innovation in Hungary. NKITT is a continuation of other bodies with a similar objective, with which the Science and Technology Policy Council (TTPK) had been cooperating for the longest time.

The Ministry for National Economy oversees many of the government initiatives for innovativeness. It is the supervisor of the National Innovation Office (NIH).

The Ministry of National Resources (MNR) is not a common type of ministry. In 2010, it took over the responsibilities of the Ministry of Education. Moreover, it is responsible for health, social affairs, and sports. MNR supervises the formulation and the implementation of the science and education policies. It oversees the entire education system, from primary to higher education. Thus, it has a decisive influence on the formation of human capital.

The NIH is in charge of the implementation of the technology and innovation policies. It prepares R&D programs and innovative programs. It also governs the international research cooperation on behalf of the government. It reports to the Prime Minister.

The National Development Agency (NFU) manages the operational programs for the Hungary Development Plan (2007–2013) and the Research and Technology Innovation Fund (KTIA). It prepares medium- and long-term plans aimed at using the EU Cohesion and Structural Funds.

In the case of financing research and innovation, the largest sources are the KTIA and various operational programs of New Hungary Development Plan. The Hungarian Scientific Research Fund (OTKA) funds research conducted on a smaller scale. The biggest projects are cofinanced from Structural Funds.

The Hungarian Scientific Research Fund (OTKA) aims at supporting the development of research infrastructure and of research itself. It operates as an autonomous agency providing grants, which focuses on basic research. It independently determines the rules of allocation of means and announces competitions for funding.

Furthermore, there is the Hungarian Economic Development Centre (MAG Zrt.), which implements the innovation policy by administrating structural funds of the EU and national funds allocated to activities supporting the technology and innovation policy. The main recipients of the MAG funds are companies. The owner of MAG is the Hungarian Development Bank PLC.

The Hungarian Patent Office (MSZH) plays an important role in the field of protection of industrial property, the preparation and implementation of the related government strategy, policy recommendations, and pan-European cooperation in the domain of intellectual property protection.

Research is conducted by both private companies and public institutions. The strong point of the Hungarian innovation system is a significant share of companies in the employment (about a half in 2010) and in the research funding (60 % of GERD in 2010).

Other research contractors are academic institutions, which conduct about 20 % of research (GERD expenditure in 2010).

The third group of contractors is the Hungarian Academy of Sciences (MTA), which has ten research centers and five research institutes at its disposal. This organization is self-governing and public. In 2012, it was subjected to a strong consolidation reform. It had been conducting about 12 % of research (of GERD expenditure in 2010).

7.2.2.1 Assessment of Innovative Capacity and Position

The analysis of the innovative capacity and position of Hungary aims at answering the following two questions:

1. Does Hungary's economy have the necessary innovative capacities?
2. Does Hungary's economy use the capacities properly when translating them into the innovative position?

The key measure of the innovative capacity of NISs is the so-called "internal expenditures on Research and Development in proportion to the GDP." This relative measure is the most capacious index of how big the research activity is in a given country. In the case of Hungary, the share of these expenditures in the GDP increased from 1995 to 2009 from the level of 0.72 % of the GDP to the level of 1.15 % of the GDP. In comparison with the average recorded in the EU27, the index is about half of that, which means that the innovative activity is about two times smaller. A favorable trend, however, is the relative increase from the level of about 43 % to the level of 60 % in the share of the research intensity in the EU27.

The share of government internal expenditure on R&D (GOVERD) in the GDP indicates that the country has a considerable supportive role, but not the decisive role in the innovative activity. As for the percentage of internal expenditure financed by the government, it decreased from the level of 51 % to the level of 42 %. It shows an increase in the share of private sector in the research activity funding, which is a very positive sign. Nevertheless, the level is still higher than the average noted in the EU, which amounts to about 20 %.

The increase in the percentage of expenditures financed by foreign countries from 5 to 10 % is also very important. This change took place already in 2000 and, since then, the share of foreign countries remains at the same level. It is a positive phenomenon, because international cooperation allows the exchange of knowledge and experience, which improves the quality of research.

In the domain of the Internet access of households, a significant improvement has been recorded only in the recent years. Currently, 55 % of households have such access. The index can be used to illustrate the degree of development of the information society and the ability to use IT infrastructure for the creation of IT innovations.

Human resources for science and technology as a percentage of the working population is the total number of people who are currently or potentially involved in the work related to the creation, development, dissemination, and application of scientific and technical knowledge. In the case of Hungary, the index is about 30 % and is rising slowly. The value of the index is 15 % lower than the average recorded in the EU. However, it is a slight difference, and it could not constitute a developmental problem.

Investments in venture capital as percentage of GDP are at a very low, barely measurable level of 3–4 per mille of the GDP. As a result, they cannot have considerable impact on innovativeness of the Hungarian economy. The share of business internal expenditure on R&D (BERD) in GDP constitutes about 30 per mille of the GDP, which amounts to only 30 % of the EU standard. The percentage of internal expenditure on R&D financed by industry had been increasing throughout the examined period and is now approaching the EU average, in relation to which it is only 14 % smaller (Table 7.34).

In the domain of Hungary's innovative position, three groups of indexes explaining innovativeness are distinguished. First, innovativeness can be related to achieving high employment in innovative sectors. In other words, the desired result of the economy's competitiveness is the creation of the so-called good jobs. Second, the country may have a high innovative position in creating intellectual property. Third, the country may specialize in the export of HT goods, which is a measure of its competitiveness.

The indexes available for Hungary indicate a slowly increasing share of employment in HT and MHT industries, from 8 to 9.3 % of the total employment in 2009. The share of employment in knowledge-intensive service sectors also increased from 26.5 to 28.7 %.

As for the formation of the intellectual property capital, a progressive decline has to be pointed out. The patent intensity calculated by the number of submitted patent applications per USD 1 billion of funds allocated to R&D had declined, which means a decrease in the efficiency of innovation expenses. In the years 1995–2009, the patent intensity measured by the number of patent applications per million inhabitants also declined.

It seems that the innovative competitiveness of exports had stabilized. The share of the export of HT goods in total exports indicates that, in the years 1995–2000, an improvement was noted in this field. Afterward, however, it came to a standstill from the structural perspective. The Hungarian economy gains about 20 % of exports, thanks to HT products, and it is unlikely that the share will continue to grow. At the same time, total exports are increasing in terms of absolute values, so only a structural stagnation, and not absolute stagnation, can be considered (Table 7.35).

7.2.2.2 Assessment of Human Capital Resources

Hungary is characterized by a high level of human capital. The resources of labor force in the years 1995–2010 was increasing. The share of labor force with higher

Table 7.34 Hungary's innovative capacity. (Source: Own study)

Indexes of innovative capacity	1995	2000	2005	2009
Share of total internal expenditure on Research and Development (R&D) (GERD) in GDP	0.72	0.79	0.95	1.15
Share of government internal expenditure on R&D (GOVERD) in GDP	0.18	0.21	0.26	0.23
Percentage of internal expenditure on R&D financed by the government	0.53	0.49	0.49	0.41
Percentage of internal expenditure on R&D financed by other national sources	0.47	0.31	0.32	0.69
Percentage of internal expenditure on R&D financed by foreign countries	4.85	10.63	10.67	10.90
Households with access to the Internet (as % of all households)	n/a	n/a	22.12	55.07
Human resources for science and technology as a percentage of the working population	n/a	28.20	30.70	32.50
Investments in venture capital as % of GDP	n/a	0.003	0.004	0.001
Share of business internal expenditure on R&D (BERD) in GDP	0.31	0.35	0.41	0.66
Percentage of internal expenditure on R&D financed by industry	38.38	37.76	39.45	46.43
Relative level of the innovative capital in comparison with the EU27				
Share of total internal expenditure on Research and Development (R&D) (GERD) in GDP (%)	43	45	54	60
Share of government internal expenditure on R&D (GOVERD) in GDP (%)	66	83	108	88
Percentage of internal expenditure on R&D financed by the government (%)	134	140	141	121
Percentage of internal expenditure on R&D financed by other national sources (%)	26	14	14	26
Percentage of internal expenditure on R&D financed by foreign countries	73	149	120	125
Households with access to the Internet (as % of all households)	n/a	n/a	46	84
Human resources for science and technology as a percentage of the working population (%)	n/a	89	86	85
Share of business internal expenditure on R&D (BERD) in GDP (%)	30	32	38	57
Percentage of internal expenditure on R&D financed by industry (%)	74	68	73	86

education was also increasing, which constitutes a positive symptom of the competitiveness of the human capital. The share of labor force with primary education was decreasing.

The decline in expenditure on higher education as percentage of GDP per capita seems to be a problem. It decreased from the level of 36% of GDP per capita to the level of only 25% of GDP per capita. On the other hand, the expenditure on secondary education had been rising. It means that an increase in the support for secondary rather than higher education was recorded. It may also indicate a shift

Table 7.35 Hungary's innovative position. (Source: Own study)

Indexes of innovative position	1995	2000	2005	2009
Share of employment in high and medium-high-technology industries in the total employment (%)	n/a	8.07	8.34	9.26
Share of employment in knowledge-intensive service sectors in relation to the total employment (%)	n/a	26.50	28.22	28.73
Number of patent applications per USD 1 billion of funds allocated to R&D	n/a	0.89	0.49	0.39
Number of patent applications per million inhabitants	106.11	79.33	69.89	68.52
Share of export of high-technology goods in total exports	4.76	23.11	19.69	20.33
Relative level of the innovative capital in comparison with the EU27				
Share of employment in high and medium-high-technology industries in the total employment (%)	n/a	109%	127%	138%
Share of employment in knowledge-intensive service sectors in relation to the total employment (in %)	n/a	87%	87%	87%
Share of export of high-technology goods in total exports	n/a	108%	105%	122%

of the higher education structure toward the less cost-intensive education. The data show an increase in the number of years of education of the adults from the level of 10.4 to 11.7 years in 2010.

In the field of the use of human capital, it is possible to observe a continuously high level of unemployment. Only in the 2000–2005 period, the unemployment rate decreased from the level of 6.4 and 7.2. Until 2000, higher education was virtually a guarantee of lack of unemployment among the graduates of academic institutions. It turns out that, since 2005, higher education has no longer been such a guarantee. In recent years, in particular, the unemployment among the young has increased, which constitutes a serious social problem.

The data indicate also a continuous increase in the unit labor costs. It is not a significant problem, as long as it does not cause a loss of competitiveness in foreign trade. The previously analyzed data show that such phenomenon does not occur (Table 7.36).

7.2.3 Interrelationship Between the Level of Innovation and Human Capital Accumulation, and the Advantages in the Foreign Trade

It turned out that an important element of the increase in Hungary's competitiveness was the transition to the export of HT goods and a significantly advanced and progressive discontinuation of export of LT goods, which, in 1995, had a considerable share in this activity. In order to make such change possible, it is necessary to have appropriate human capital resources and an adequate level of innovativeness.

Table 7.36 Hungary's human capital resources and their use. (Source: Own study)

	1995	2000	2005	2010
Total labor force resources	4,175,453	4,162,530	4,268,806	4,278,423
Share of labor force with higher education (%)	n/a	16.4	20.2	n/a
Share of labor force with secondary education (%)	n/a	65.2	64.5	n/a
Share of labor force with primary education	n/a	18.4	15.3	n/a
Expenditures on higher education as % of GDP per capita	36	31	25	n/a
Expenditures on secondary education as % of GDP per capita	19	26	23	n/a
Average number of years of education of people aged 25+	10.4	11.2	11.5	11.7
Unemployment rate	10.2	6.4	7.2	10
The unemployed with higher education as % of the total number of the unemployed	4.1	4.1	7.6	n/a
Unemployment among the young (% of the unemployed aged 15–24)	18.6	12.7	19.4	26.5
Unit labor costs (ULC)	6.4	7.2	10.4	11.7

It has to be indicated that Hungary disposed of these resources since the times of the transformation. Although it would be possible to increase, for example, the number of engineers in response to the needs, it would considerably slow down the process of changing the export structure. In this context, the necessary elements have to be provided before they are used.

The data concerning the RCA show a very strong shift in favor of HT industries, which is a symptom of the creation of significant advantages in foreign trade. It should definitely be emphasized that Hungary has a comparative advantage in the field of innovative and technologically advanced domains of foreign trade.

The case study of Hungary indicates that the advantage results from having high-quality human capital at one's disposal. The dynamic increase in the internal expenditures on R&D (GERD) could also be significant. The comparative advantage related to innovativeness and human capital manifests itself especially in the increase in the export of HT products in total exports.

Nevertheless, it is worth mentioning that the decreasing number of registered patents is in contradiction with positive achievements related to the development of HT and TD goods.

Chapter 8
International Competitiveness of Countries with Unbalanced Innovation Systems

Krzysztof Falkowski, Günter Heiduk, Agnieszka McCaleb and Beata Michorowska

Unbalanced innovation systems are considered to be poorly developed. They can be found in as many as 19 global economy states. They are characterized by gross disparities in terms of the indexes that reflect both innovative capacity and position. Moreover, in the last decade, one could observe notable fluctuations in the innovation and human capital indexes, as well as the exacerbation of existing dissimilarities. Some states with unbalanced innovative systems, such as China, are heading for gradual elimination of divergence, while other systems—for instance, Russia—are experiencing an escalation of dissimilarities (Weresa 2012). This is the reason why these two countries have been selected to serve as examples of unbalanced innovation systems and undergo in-depth analysis. The third case study discussed in this chapter is the case of Mexico, which stands out from the unbalanced innovative systems thanks to its relatively high outlays on education (Weresa 2012). In this chapter, the development and evolution of competitive advantages of these countries have been evaluated, and the role of human capital and innovativeness in this process has been indicated.

K. Falkowski (✉) · G. Heiduk · A. McCaleb · B. Michorowska
World Economy Research Institute, Warsaw School of Economics,
Al. Niepodleglosci 162, 02-554 Warsaw, Poland
e-mail: kfalkow@sgh.waw.pl

G. Heiduk
e-mail: gheidu@sgh.waw.pl

A. McCaleb
e-mail: agnieszka.mccaleb@gmail.com

B. Michorowska
e-mail: bz26456@sgh.waw.pl

M. A. Weresa (ed.), *Innovation, Human Capital and Trade Competitiveness*, Innovation, Technology, and Knowledge Management, DOI 10.1007/978-3-319-02072-3_8,

8.1 Case Study: Russia

Krzysztof Falkowski

At the turn of the twenty-first century, Russia was struggling with serious consequences of the 1998 financial crisis. The economic policy launched at the time—aimed mainly at stabilizing the macroeconomic situation—as well as the favorable developments in global economy—especially changes in prices on international energy resources markets—spurred dynamic economic growth in Russia in a relatively short period of time. In the context of the economic growth that has been visible in Russia for some time now (apart from the 2008 crisis), one might ask how competitive Russia is on an international scale, and what the sources of its competitiveness are, whereby it would be worth discussing the role of innovations and human capital. These issues are the object of the analysis conducted in this section.

8.1.1 Russia's Competitiveness in Foreign Trade in the Years 1996–2010

In the years 1996–2010, Russian foreign trade was generally displaying an upward trend, except for two periods, i.e., 1998–1999 and 2009. Declines in these years resulted from the 1998 crisis and the global 2008 crisis. Moreover, the drop in the prices of energy resources on global markets, especially of oil, had a negative effect on Russian foreign trade, which translated into a decrease in the value of Russian exports.

In the years 1996–2010, Russian exports had a notably higher share in total Russian foreign trade than imports (Fig. 8.1).

The highest value of exports in Russia was noted in 2008, when they went up to US$468 million, whereas the lowest value was observed in 1998 and 1999, and amounted to US$72.2 million and US$72.8 million, respectively. Russian imports achieved their peak value in 2008 (US$267 billion). Russia's foreign trade had a positive trading balance in the whole analyzed period from 1996 to 2010.

When analyzing the geographic structure of Russian exports in the years 1996–2000, one ought to underline the lack of stabilization of Russia's main output markets. Only two countries (Germany and China) retained their position among Russia's top five exports partners. Moreover, it points to the relatively lower significance of Germany as output market for Russian goods. In 2000, the share of Germany in Russian exports amounted to 8.9 %, in 2005 to 8.2 %, and in 2010 to a mere 4.3 %. What is more, it is becoming obvious that China holds a high and stable position as Russia's exports market. In the years 1996–2010, the share of China in Russian exports oscillated around 5 %. Italy holds a strong position among Russia's main partners as it takes second or third place with a share ranging from 7.1 % in 2000, through 8.2 % in 2005 to 6.5 % in 2010. Also, the Netherlands has gained in importance as Russia's export partner, as in 2010 their share in Russian exports amounted to 14.3 % (Table 8.1)

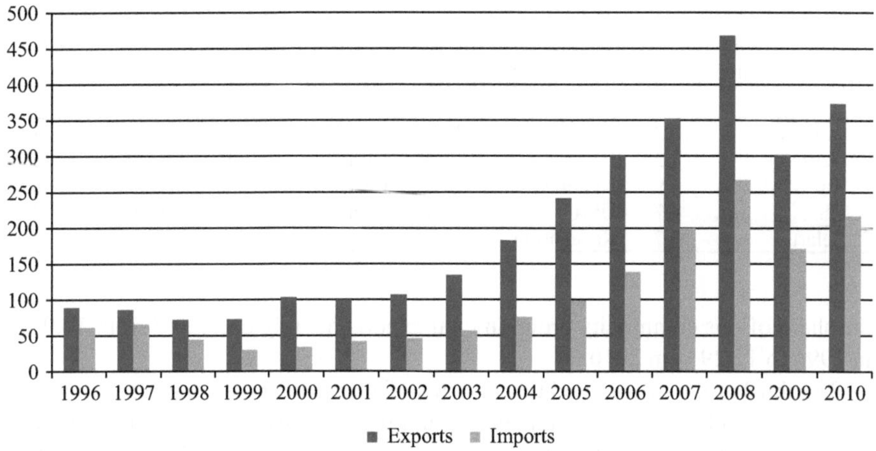

Fig. 8.1 Value of Russian exports and imports in the years 1996–2010 (billion US$). (Source: Own study on the basis of data obtained from the United Nations Commodity Trade Statistics Database)

Table 8.1 Russia's main export partners in the selected years of the period from 1996 to 2010 (% share in total exports). (Source: Own study on the basis of data obtained from the United Nations Commodity Trade Statistics Database)

1996		2000		2005		2010	
1. Ukraine	8.6	Germany	8.9	The Netherlands	10.2	The Netherlands	14.3
2. Germany	8.0	Italy	7.1	Germany	8.2	Italy	6.5
3. USA	5.7	Belarus	5.4	Italy	7.9	China	5.3
4. China	5.4	China	5.0	China	5.4	Belarus	4.8
5. Switzerland	4.5	Ukraine	4.8	Ukraine	5.1	Germany	4.3

The significance of CIS countries among Russia's main imports partners in the period from 1996 to 2010 fell markedly—it decreased threefold. To illustrate, Ukraine became a less important player in Russian imports (its share fell from 10.3 % in 1996 to 6.4 % in 2010), similar as Belarus (share in Russian imports dropped from 10.9 % in 2000 to 3.9 % in 2010). However, the significance of China as supplier of goods to Russia clearly went up. In 2010, approximately one-fifth of total Russian imports (17.9 %) stemmed from China, which translates into an increase in China's share by as much as 10.5 p.p., from 7.4 % in 2005.

Moreover, Germany is traditionally an important import partner of Russia. In 2010, Germany took second position among the most important imports market of Russia, with a share of 12.2 % (Table 8.2).

In the years 1996–2010, the major part of Russian exports was constituted by mineral products, especially mineral fuels (oil, natural gas), lubricants and related materials (group 3 according to the SITC classification). Their share in Russian exports went up from 43.11 % in 1996 to 69.03 % in 2010. The second position was taken by goods belonging to SITC 06, mainly so-called manufactured goods classified by material (including, among others, metals and metallurgic goods). However,

Table 8.2 Russia's main import partners in the selected years of the period from 1996 to 2010 (% share in total imports). (Source: Own study on the basis of data obtained from the United Nations Commodity Trade Statistics Database)

1996		2000		2005		2010	
1. Ukraine	10.3	Germany	11.5	Germany	13.5	China	17.9
2. Germany	8.5	Belarus	10.9	Ukraine	7.9	Germany	12.2
3. Kazakhstan	4.9	Ukraine	10.6	China	7.4	Ukraine	6.4
4. USA	4.7	USA	8.0	Japan	5.9	Japan	4.7
5. Belarus	4.7	Kazakhstan	6.5	Belarus	5.8	Italy	4.6

the share of this commodity group in total exports notably decreased from 19.94 % in 1996 to 12.29 % in 2010.

The third group of commodities in terms of share in Russian exports was constituted by so-called nonclassifiable goods and transactions, whereby this group lost significance. In 1996, its share in Russian exports amounted to 15.81 %, and in 2010 to a mere 4.88 % (Table 8.3).

In the years 1996–2010, the commodity structure of Russian imports was characterized by a much higher level of diversification than the exports structure. Machinery and transport equipment are most important here. Their share in Russian imports increased by as much as 20.7 p.p., from 20.03 % in 1996 to 40.73 % in 2010. SITC group 09, i.e., transactions and commodities, not classified elsewhere in the SITC, witnessed a drop in the share in imports in the years 1996–2010 by as much as 28.93 p.p. Interestingly, apart from group 0 (food and live animals), 1 (beverages and tobacco), and 3 (mineral fuels, lubricants and related materials), whose share in imports also fell (by 2.5, 0.92, and 1.04 p.p., respectively), an increase in the share in imports was observed in all other SITC groups. The highest growth (apart from group 09) was noted in the case of group 08—miscellaneous manufactured articles—as well as group 05—chemicals and related products (Table 8.3).

To sum up, the analysis of changes in the commodity structure of Russian foreign trade indicated that semiprocessed commodities, i.e., mineral fuels and related materials, are dominant in Russian exports, whereby in the case of imports to Russia, highly processed goods, i.e., machinery and transport equipment have taken the lead.

From the point of view of international competitiveness of the economy, it is worth taking a closer look at the changes in the commodity structure of its foreign trade, including the technology factor. Medium-high technology (MHT) commodities are visibly dominant in Russian exports. Their share in total Russian imports in the years 1996–2010 exceeded the volume of 70 %, and reached its peak value in 2010 (nearly 77.5 %). This share increased from 72.12 % in 1996 to 77.49 % in 2010 (in terms of quality, this translated into almost 3.5-fold growth). In the case of the remaining categories of industrial commodities, i.e., high-technology (HT), MHT, and low-technology (LT), their share in Russian exports fell. However, in terms of quality, their value increased twofold. It ought to be mentioned that HT commodities have a minor share in exports. It oscillated between 3.62 % in 1996 and a mere 2.37 % in 2010 (Table 8.4).

Table 8.3 Commodity structure of Russian exports in selected years of the period from 1996 to 2010 according to SITC (%). (Source: Own study on the basis of data obtained from the United Nations Commodity Trade Statistics Database)

		Exports				Imports			
SITC group	Name	1996	2000	2005	2010	1996	2000	2005	2010
0	Food and live animals	1.01	0.91	1.30	1.77	14.79	15.61	12.75	12.29
1	Beverages and tobacco	0.22	0.08	0.18	0.18	2.50	3.34	2.43	1.58
2	Crude materials, inedible, except fuels	5.41	4.50	4.39	3.36	2.05	4.30	2.48	2.12
3	Mineral fuels, lubricants and related materials	43.11	50.58	61.77	69.03	2.78	4.13	1.75	1.74
4	Animal and vegetable oils, fats and waxes	0.04	0.08	0.08	0.17	0.61	1.12	0.79	0.72
5	Chemicals and related products	5.97	6.05	4.20	4.44	8.84	14.49	13.66	13.83
6	Manufactured goods classified by material	19.94	17.75	14.71	12.29	10.43	13.57	12.82	12.69
7	Machinery and transport equipment	7.04	6.35	4.21	3.23	20.03	24.90	39.59	40.73
8	Miscellaneous manufactured articles	1.45	1.50	0.72	0.65	6.97	7.13	7.52	12.23
9	Commodities and transactions not classified elsewhere in the SITC	15.81	12.20	8.43	4.88	31.01	11.40	6.21	2.08
	Total	100.00	100.00	100.00	100.00	100.00	100.00	100.00	100.00

Table 8.4 Commodity structure of Russian exports in selected years of the period from 1996 to 2010 according to the OECD industry classification based on technological level. (Source: Own study on the basis of data obtained from the United Nations Commodity Trade Statistics Database)

Industries	Value (billion US$)				Share in total exports (%)			
	1996	2000	2005	2010	1996	2000	2005	2010
HT	3.09	3.49	4.47	6.45	3.62	3.65	2.56	2.37
MHT	14.35	18.41	25.73	37.83	16.83	19.27	14.77	13.88
MLT	61.51	67.06	132.99	211.23	72.11	70.21	76.33	77.49
LT	6.35	6.56	11.05	17.08	7.44	6.87	6.34	6.26

In the years 1996–2010, MHT and medium-low-technology (MLT) commodities were of greatest significance in the structure of Russian imports. At the beginning of that period, i.e., in 1996, the highest share in Russian imports was held by MLT commodities (36.82%), and the share of MHT amounted to 34.1%. At the end of that period, i.e., in 2010, MHT commodities were dominant in Russian imports with a share of 42.38%, and the share of MLT amounted to a mere 31.34%. In comparison to the situation of Russian exports, HT commodities have a higher and an increasing share of 11.03% in 2010, as compared to 8.72% in 1996. However, one ought to underline that this share was lowest in comparison to other commodity

Table 8.5 Commodity structure of Russian imports in selected years of the period from 1996 to 2010 according to the OECD industry classification based on technological level. (Source: Own study on the basis of data obtained from the United Nations Commodity Trade Statistics Database)

Industries	Value (billion US$)				Share in total imports (%)			
	1996	2000	2005	2010	1996	2000	2005	2010
HT	7.02	5.84	20.63	48.02	8.72	10.18	11.20	11.03
MHT	27.42	20.68	78.06	184.61	34.10	36.06	42.41	42.38
MLT	29.62	21.23	59.12	136.50	36.82	37.02	32.12	31.34
LT	16.37	9.60	26.27	66.43	20.36	16.74	14.27	15.25

Table 8.6 Commodity structure of Russian exports in selected years of the period from 1996 to 2010 according to the WIFO taxonomy of manufacturing industries based on factor inputs. (Source: Own study on the basis of data obtained from the United Nations Commodity Trade Statistics Database)

Industries	Value (billion US$)				Share in total exports (%)			
	1996	2000	2005	2010	1996	2000	2005	2010
Mainstream manufacturing (MM)	11.49	11.30	23.98	28.09	19.93	18.84	18.03	14.22
Labor-intensive (LI)	3.81	4.10	5.85	8.26	6.61	6.84	4.40	4.18
Capital-intensive (CI)	37.09	38.34	94.00	148.02	64.31	63.93	70.69	74.97
Marketing-driven (MDI)	2.18	2.92	3.74	7.20	3.78	4.87	2.82	3.65
Technology-driven (TDI)	3.10	3.31	5.40	5.87	5.37	5.52	4.06	2.98

groups. Another interesting issue is that of the gradual decrease in the significance of LT commodities in Russian imports (Table 8.5).

To sum up, in the years 1996–2010, MLT commodities were of greatest significance in the structure of Russian exports, while MHT commodities were dominant in imports. Taking into account the growth rate and trends of shifts in the significance of various commodity groups in Russian foreign trade, one ought to emphasize that in the case of exports, the key share of the abovementioned MLT commodities was additionally boosted, while in relation to imports, MHT went up at the cost of MLT commodities.

The analysis of the commodity structure of Russian foreign trade in the years 1996–2010 from the point of view of the significance of commodities with various degrees of utilization of material and nonmaterial outlays in the production process shows that CI commodities are dominant in Russian exports. Their share increased from 64.31% in 1996 to 74.97% in 2010. The value of exports of these commodities increased from approximately US$37 billion to approximately US$148 billion. Such a strong position of CI commodities in Russian exports resulted from the fact that Russia's exports are based on raw materials.

Moreover, it is worth noting the very low and decreasing share of commodities whose production process is based on marketing and research. In 2010, these shares amounted to 3.65 and 2.98%, respectively (Table 8.6). This is strongest evidence that the Russian economy was not capable of providing an attractive export offer, which is tantamount to the statement that it was completely uncompetitive in that scope.

Table 8.7 Commodity structure of Russian imports in selected years of the period from 1996 to 2010 according to the WIFO taxonomy of manufacturing industries based on factor inputs. (Source: Own study on the basis of data obtained from the United Nations Commodity Trade Statistics Database)

Industries	Value (billion US$)				Share in total imports (%)			
	1996	2000	2005	2010	1996	2000	2005	2010
MM	12.26	8.64	28.31	63.24	27.03	28.11	28.80	28.16
LI	4.29	2.24	8.52	26.69	9.45	7.29	8.67	11.89
CI	10.06	7.35	17.23	37.76	22.17	23.92	17.53	16.81
MDI	11.25	6.13	16.33	36.07	24.80	19.93	16.61	16.06
TDI	7.51	6.38	27.90	60.83	16.55	20.75	28.39	27.08

The commodity structure of Russian exports according to the taxonomy based on factor inputs is much more varied. In this case, it is very difficult to discern one dominant commodity group. Still, the share of MM commodities was highest and oscillated around 28 %. However, this commodity group did not surpass the second highest group, i.e., MDI, by much (MDI had a share of 27 % in 2010).

When analyzing the changes that occurred in this structure in the years 1996–2010, one ought to emphasize the growth of RDI commodities (most impressive growth, from 16.55 % in 1996 to 27.08 % in 2010), and the much lower increase in LI commodities (from 9.45 % in 1996 to 11.89 in 2010). Moreover, an increase in MM commodities occurred (from 27.03 % in 1996 to 28.16 % in 2010). In the case of MDI and CI commodities, a drop in their share in Russian imports was noted (from 24.8 % in 1996 to 16.06 % in 2010 and from 22.17 % in 1996 to 16.81 % in 2010, respectively—Table 8.7).

To sum up, in the years 1996–2010, CI commodities played a dominant role in the structure of Russian exports. MM and MDI commodities were of much greater significance in imports, which was much more diversified in terms of imports.

Another attempt at analyzing the commodity structure of Russian foreign trade incorporates the classification by the qualification level of the labor force necessary to manufacture commodities that are the subject of foreign trade both in exports and in imports. The analysis of such trading structure indicates that in the years 1996–2010, commodities manufactured by medium-qualified blue-collar employees were dominant in exports. Their share was very high and ranged from 70 to 80 %. When comparing the level of shares of all commodity groups in 2010 with the 1996 data, Russia witnessed an increase in the share of only this very group. In the case of the remaining commodity groups, however, slight drops in shares took place, though in quality terms exports are still on the rise. One ought to also note the very low share of goods that require highly qualified employees in Russian exports—in 2010 it amounted to only 2.37 % (Table 8.8).

Commodities manufactured by industries based on medium-qualified white-collar employees were dominant in Russian imports. In 2010, they scooped up 42.38 % of Russian imports. Also, commodities manufactured by industries employing medium-qualified blue-collar workers were of relatively high significance, as their share amounted to 31.34 % of total Russian imports in 2010.

Table 8.8 Commodity structure of Russian exports in selected years of the period from 1996 to 2010 according to the WIFO industry taxonomy based on labor force qualifications. (Source: Own study on the basis of data obtained from the United Nations Commodity Trade Statistics Database)

Industries based on	Value of exports (billion US$)				Share in total exports (%)			
	1996	2000	2005	2010	1996	2000	2005	2010
Highly qualified white-collar employees	3.09	3.49	4.47	6.45	3.62	3.65	2.56	2.37
Medium-qualified white-collar employees	14.35	18.41	25.73	37.83	16.83	19.27	14.77	13.88
Highly qualified blue-collar employees	61.51	67.07	133.00	211.23	72.11	70.21	76.33	77.49
Semiqualified blue-collar employees	6.34	6.56	11.05	17.08	7.44	6.87	6.34	6.26

Table 8.9 Commodity structure of Russian imports in selected years of the period from 1996 to 2010 according to the WIFO industry taxonomy based on labor force qualifications. (Source: Own study on the basis of data obtained from the United Nations Commodity Trade Statistics Database)

Industries based on	Value of imports (billion US$)				Share in total imports (%)			
	1996	2000	2005	2010	1996	2000	2005	2010
Highly qualified white-collar employees	7.02	5.84	20.63	48.02	8.72	10.18	11.21	11.03
Medium-qualified white-collar employees	27.42	20.68	78.06	184.61	34.10	36.06	42.41	42.38
Highly qualified blue-collar employees	29.62	21.23	59.12	136.50	36.82	37.02	32.12	31.34
Semiqualified blue-collar employees	16.37	9.60	26.27	66.43	20.36	16.74	14.26	15.25

Similarly, as in the case of exports, in imports also the share of commodities manufactured by industries employing highly qualified staff was at its lowest (11.03 % in 2010) in the analyzed period. In the years 1996–2010, the significance of commodities manufactured by industries hiring medium-qualified white-collar employees (from 34.1 % in 1996 to 42.38 % in 2010) and industries based on highly qualified employees, however, in this case—to a notably lesser extent (from 8.72 % in 1996 to 11.03 % in 2010). However, the drop in the share of other commodity categories in Russian imports, i.e., commodities originating from industries based on medium-qualified blue-collar workers, as well as semiqualified employees (36.82 % in 1996 to 31.34 % and 20.36 % in 1996 to 15.25 % in 2010, respectively—Table 8.9).

In conclusion, taking into account the fact that the commodities, which are the object of Russia's foreign trade in the analyzed period of time, originate from industries based on various labor force qualification levels, it ought to be underlined that although Russian exports were dominated by commodities manufactured by means of medium-qualified blue-collar workers, it was the commodities manufactured by industries based on medium-qualified white-collar employees that played the major and increasingly significant role in imports.

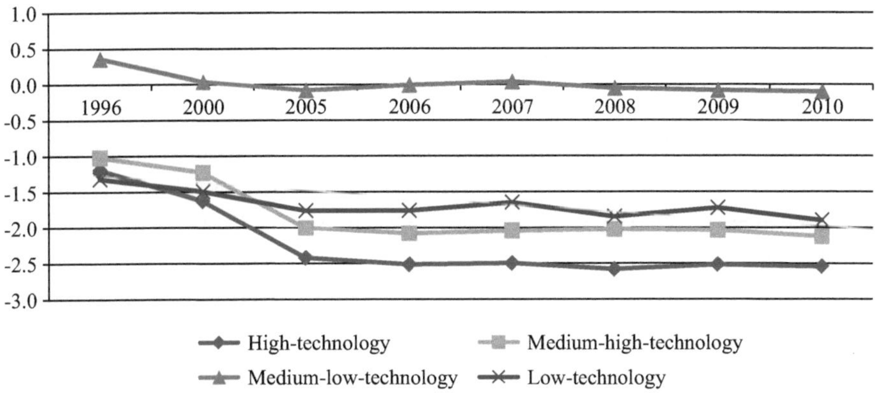

Fig. 8.2 Comparative advantages (RCA) in Russian foreign trade in selected years from 1996 to 2010 according to the OECD industry sector classification based on technological level. (Source: Own study on the basis of data obtained from the United Nations Commodity Trade Statistics Database)

The above analysis of the commodity structure of Russia's foreign trade from 1996 to 2010 leads to the following general conclusions:

1. Russian exports to foreign markets were dominated by MLT commodities, CI commodities, and commodities manufactured by medium-qualified blue-collar workers. Mineral products (fuels, i.e., oil and natural gas), lubricants, and related materials undoubtedly had the highest share in Russian exports
2. Russian imports were dominated by MHT, MM, and RDI commodities that were manufactured by industries based on medium-qualified white-collar employees. Machinery and transport equipment had the highest share.

The evaluation of Russia's competitiveness in foreign trade will be carried out by means of the revealed comparative index (RCA).

In the commodity groups classified according to their level of technological advancement in selected years from 1996 to 2010, Russia enjoyed comparative advantages in MLT trading. These advantages were not found in any of the remaining groups. What is more, Russia's competitiveness in trading HT, MHT, and LT commodities from 1996 to 2010 clearly deteriorated (Fig. 8.2).

Taking into account the commodity groups discerned on the basis of input factors, one ought to pay special attention to the fact that comparative advantages are present in only one group of Russian foreign trade, i.e., CI commodity trade (that as a matter of fact have the highest share in Russian exports, as has been mentioned earlier). In the analyzed period, refined petroleum products exhibited the highest level of competitiveness on the international market in this group of Russia's trade.

In the case of other commodity groups differentiated on the basis of input factors in the years 1996–2001, the RCA index assumed negative values, which indicates that there were no comparative advantages in this scope (Fig. 8.3).

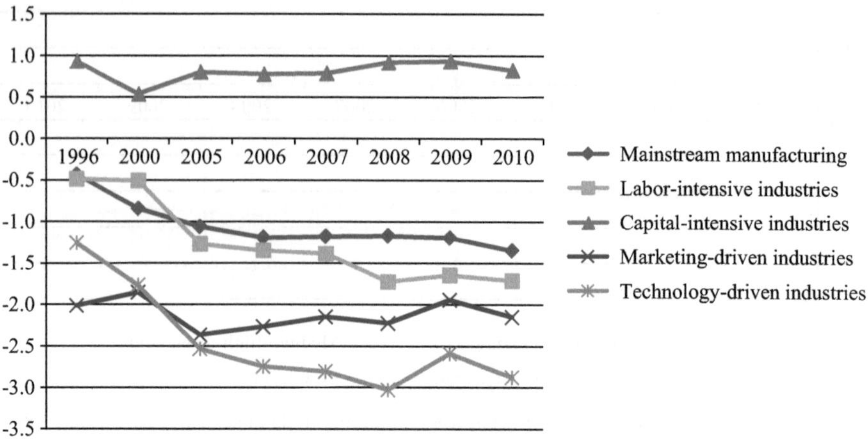

Fig. 8.3 Comparative advantages (RCA) in Russian foreign trade in selected years from 1996 to 2010 according to the WIFO taxonomy based on input factors. (Source: Own study on the basis of data obtained from the United Nations Commodity Trade Statistics Database)

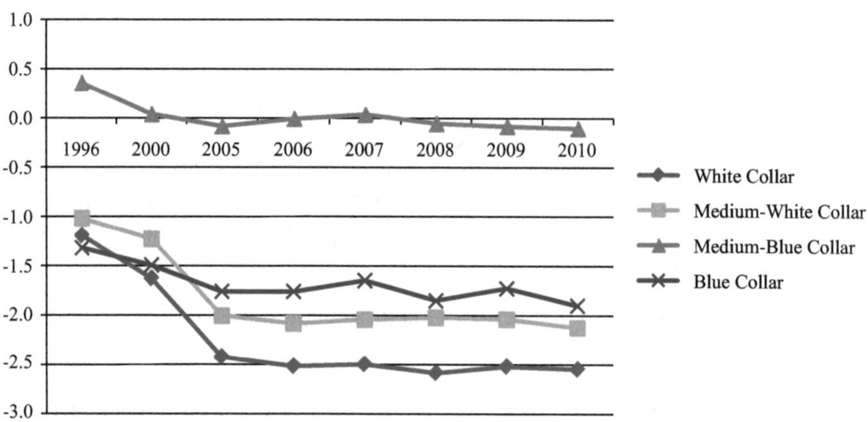

Fig. 8.4 Comparative advantages (RCA) in Russian foreign trade in selected years from 1996 to 2010 according to the WIFO taxonomy based on labor force qualification level. (Source: Own study on the basis of data obtained from the United Nations Commodity Trade Statistics Database)

The situation in commodity group trading classes according to the labor force qualification level necessary for its production was also unfavorable in Russia. The value of the RCA index for particular commodity groups assumed negative values and, moreover, decreased gradually each year (Fig. 8.4).

What also attracts attention here is Russia's gradual loss of competitiveness in the trading of commodities manufactured by medium-qualified blue-collar workers.

To sum up, Russian economy is characterized by a lack of comparative advantages in modern input-factor-intense industries. What is worse, Russia's competitive-

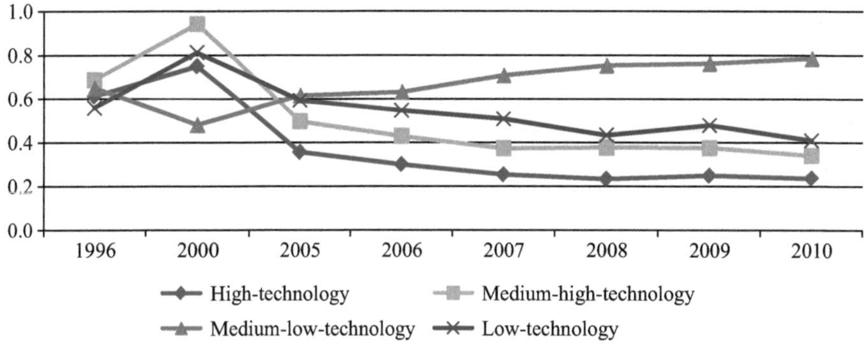

Fig. 8.5 IIT intensity indexes in Russian foreign trade from 1996 to 2010 according to the OECD industry taxonomy based on technological level. (Source: Own study on the basis of data obtained from the United Nations Commodity Trade Statistics Database)

ness is not only very low, but is also deteriorating. In terms of the country's further development, also in the face of the potential depletion of resources, this is a threat.

In contemporary global economy, specialization and intra-industry trade (IIT) play an increasingly important role. From this perspective, it is important and interesting to take a closer look at the answer to the following question: "was it possible to observe changes in IIT in Russia in 1996–2010, especially in relation to commodities classed as modern input-factor-intense industries?"

When analyzing Russia's foreign trade according to the level of technological advancement, one ought to state that in the analyzed period IIT was most intense among MLT commodities. Moreover, since 2000, IIT has been rising only within this group. At the same time, HT, MHT, and LT commodities experienced a dynamic fall. This most notable and at the same time most spectacular drop occurred in MHT commodities (Fig. 8.5).

Similarly, drops in IIT intensity in foreign trade in Russia from 1996 to 2010 occurred in most commodity groups discerned on the basis of input factors. The most spectacular and, at the same time, the most notable drop in this intensity took place in the group of LI and RDI commodities. Especially, the latter phenomenon is very unfavorable in relation to the global competitiveness of the Russian economy.

CI commodities had a fairly stable situation, though in the analyzed period the IIT ratio remained at a relatively low level (Fig. 8.6).

IIT intensity has clearly been decreasing since 2000. The affected groups are: commodities manufactured by industries hiring medium-qualified white-collar employees, medium-qualified blue-collar employees, and—especially—highly qualified employees. IIT intensity increased throughout the period from 2000 to 2010 only in the case of commodities manufactured by medium-qualified blue-collar workers. In this area of trade, Russia has managed to achieve relatively the highest level of international competitiveness (Fig. 8.7).

In conclusion, in the years 1996–2010, IIT in Russia was generally decreasing. The above statement does not apply to solely one group: CI MLT commodities man-

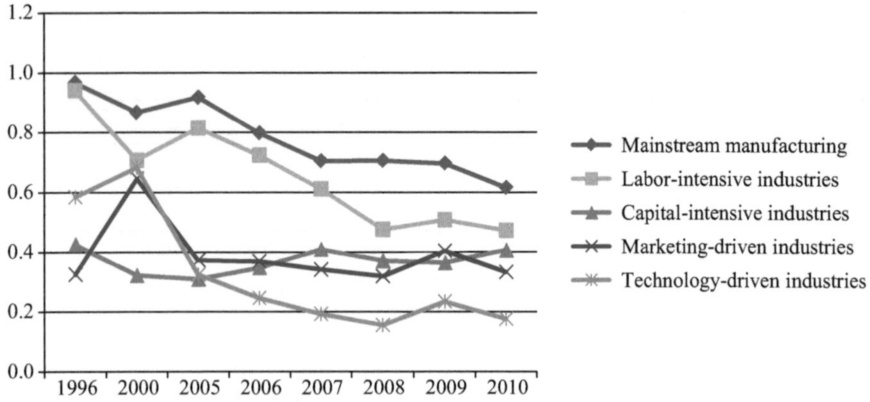

Fig. 8.6 IIT intensity indexes in Russian foreign trade in selected years of the period from 1996 to 2010 according to the WIFO taxonomy based on input factors. (Source: Own study on the basis of data obtained from the United Nations Commodity Trade Statistics Database)

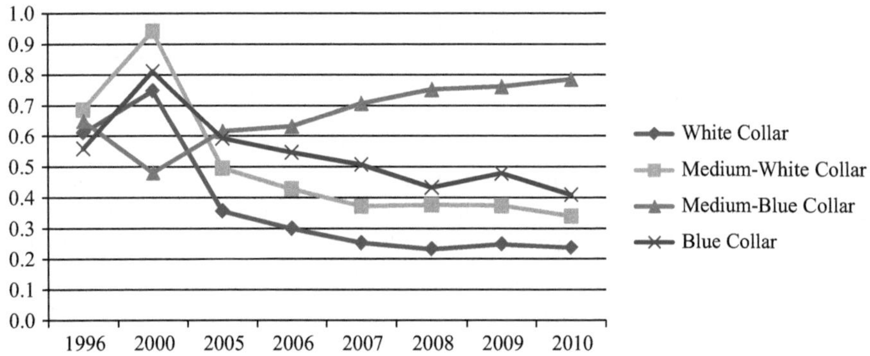

Fig. 8.7 IIT intensity indexes in Russian foreign trade in selected years of the period from 1996 to 2010 according to the OECD industry taxonomy based on labor force qualifications. (Source: Own study on the basis of data obtained from the United Nations Commodity Trade Statistics Database)

ufactured by medium-qualified blue-collar workers. In this area, Russia maintained relatively the highest level of competitiveness in the analyzed period.

8.1.2 National Innovation System in Russia from 1996 to 2010

The main entities constituting the Russian national innovation system (NIS) are: the federal government, regional state administration bodies, as well as the extremely important financial and investment network that creates network infrastructure enabling the operation of regional and cooperative centers for innovation, sector and coordination centers for innovation, sector management bodies, and regional and

Table 8.10 Number of R&D institutions in Russia according to type in the years from 2000 to 2009. (Source: Russian Center for Science Research and Statistics)

Institution type	2000	2003	2005[a]	2007	2008	2009
Research institutions	2,686	2,564	2,115	2,036	1,926	1,878
Project organizations	318	228	489	497	418	377
Construction projects and extractive organizations	85	68	61	49	42	36
Experimental enterprises	33	28	30	59	58	57
HEI	390	393	406	500	503	506
Industrial enterprises	284	248	231	265	239	228
Other	303	268	234	551	480	454
Total	4,099	3,797	3,566	3,957	3,666	3,536

[a] In 2005, the methodology of institution classification was altered

sector innovative structures. However, the task of the abovementioned financial and innovation network is to monitor the results of the activity of existing institutions that operate as part of the created network.

In Russia, the number of R&D institutions decreased by 14 % at the turn of the previous decade (Table 8.10). The most notable drop, by one-third, concerned the number of research institutions. A notable improvement was visible only in the case of higher education institutions (HEI), whose number increased by nearly 30 %.

When analyzing the share of particular types of institutions in the total number of R&D institutions in Russia, the dominant position of state research institutions attracts attention. According to the data presented by the Russian Center for Science Research and Statistics, they account for as much as 53 % of all organizations. Second most numerous are higher education units (14 %) and the third position is taken by enterprises (approximately 6 %).

The Russian Academy of Sciences, which associates leading scientists and researchers from all over the globe, is undoubtedly the most important public institution tasked with innovation creation. The research priorities of the Russian Academy of Sciences involve projects in mathematics, chemistry, and physics that consume more than half of this institution's budget. Almost 68 % of the scientific and research activity of the Russian Academy of Sciences is financed from public funds (OECD 2011, p. 156). The low level of internationalization of research studies, as well as the low level of information exchange on an international scale, is assumed to be the main culprit behind the relatively low effectiveness of the innovation-related enterprises performed by the Russian Academy of Sciences. The OECD estimates that almost half of the employees of the Academy do not participate actively in research projects, which means that the available human resources are not fully used. This results mainly from relatively low salaries, which prompts employees to seek additional forms of professional activity. Moreover, every third employee of the Russian Academy of Sciences is 60 or more years of age (OECD 2011).

When drafting a general characteristic of the Russian NIS, it is worth paying attention to the fact that there are special economic zones (SEZ) in this country, and these zones contribute to the increase in the innovativeness of the economy.

In Russia, there are six types of SEZ: innovation, production, tourist, harbor, and "old zones" of the 1990s, as well as innovation zones. Innovation zones are usually located in the vicinity of large agglomerations. Two can be found in the capital—Moscow. More innovation zones can be found in two large cities, i.e., Saint Petersburg and Tomsk. A different specialization was assigned to each zone. One of the Moscow zones—Dubna, is the center for information and nuclear technology development. The second zone—Zelenograd, specializes in micro- and nanotechnology. The SEZ in Saint Petersburg was established to develop analytical instruments, whereas the Tomsk SEZ specializes in industrial electronics development and biotechnology (Liuhto 2009, p. 8).

Universities play an important role in the functioning of every NIS. In the 1990s, after the fall of the Soviet Union, the number of universities in Russia started increasing. As the transforming economy provided new possibilities, the number of academic employees fell. A revival of R&D projects carried out at universities took place as late as in 2000. From then on, the number of universities conducting research activities notably increased in relation to the total number of universities, from 9.6 % in 2000 to 14.3 % in 2009. In the same period, to a much lesser extent, the percentage of researchers among general academic staff increased (from 4.9 % in 2000 to 6.5 % in 2009), similarly as the percentage of university expenditure on research in comparison to total domestic expenditure (from 4.5 % in 2000 to 7.1 % in 2009; Dezhina 2011, p. 8).

Currently, Russia is implementing new projects aimed at boosting universities' participation in R&D projects. As a result, the so-called "mirror laboratories" were established at a few universities. Their design was based on Western examples of academic research laboratories. This enables fulfilling their purpose, i.e., participation in international projects. One of the main goals of their functioning is to boost student exchange and cooperation aimed at obtaining and processing information on a global scale. However, it is worth underlining that scarce scientific infrastructure might thwart Russia's plans. Moreover, the government has yet to present long-term plans concerning development and financing of the abovementioned research laboratories (Dezhina 2011, pp. 18–19).

8.1.2.1 Russia's Innovative Capacity and Position from 1996 to 2010

When analyzing the innovative capacity of Russia, it is worth paying special attention to the funds allocated to R&D, as this is one of the main sources of own innovative solutions. Expenditure on R&D in Russia from 1996 to 2009 was not high, as its value ranged from 1 to 1.2 % of GDP. Taking into account the economic and financial potential of Russia, especially in periods of favorable price conditions on global energy resource markets, this value should therefore be assumed to be vastly inadequate for the capabilities and needs of the Russian economy. In this context, it is peculiar that from 2003 to 2008, when Russia witnessed higher and higher budget surplus, the percentage of funds allocated to R&D was gradually decreasing. In

2009, this percentage went up. However, this was the result of a decrease in GDP, which was triggered by the global economic crisis that did not spare Russia.

In the case of Russia, the R&D sector is undoubtedly dominated by the state, in terms of both finance and organization. In 2009, budget funds accounted for almost 66.5 % of total R&D expenditure. This was the highest volume in the analyzed period. The year 1999 was a hinge date here, as the level of public R&D financing was relatively the lowest in the analyzed period (approximately 51 % of GDP). This was connected with the incredibly difficult financial situation of the entire state resulting from the onset of the 1998 financial crisis. However, one ought to emphasize that since 1999 the percentage of budget funds allocated to R&D increased—slowly, but steadily.

In connection with such significant financing of R&D in Russia from budget funds, as much as 77 % of employees in this sector are hired by institutions and enterprises that are state bodies in the strict sense.

When analyzing the R&D financing structure in Russia, one ought to note that the percentage of own funds of Russian enterprises in general R&D financing is gradually decreasing, while public funds play an increasingly important role here. The above trend is proof that the state budget is playing a more and more important role of donator in the Russian R&D sector. Moreover, this increases the responsibility of state authorities for the condition of financing of this area, as it is a very important field in terms of the innovativeness and competitiveness of the entire economy.

In the context of financing the Russian R&D sector, it is also worth noting the percentage share of foreign entities in total financing, which was not significant. In 1996, it amounted to 5.58 % of total funds allocated to the financing of the R&D sector in Russia, while in 2009 the same index equaled 6.46 %.

As a result, it is extremely important that foreign investors are capable of vesting their trust in Russia and investing their funds in the development of its innovativeness. However, it is worth noting that many Russian entrepreneurs (48 % of large- and medium-sized enterprises according to the Russian Innovation Survey 2009–2010) cooperated with foreign partners in the field of technology and innovation over the period of 3 years (2006–2009). The main partners of Russian entrepreneurs within this scope were mainly companies based in Germany (36 %), the USA (23 %), China (16 %), and other CIS countries (10 %; Bauman Innovation and OPORA—Russian Innovation Survey, 2009–2010).

The innovativeness of the given country is the resultant of R&D expenditure and the level of effective utilization of available human capital (especially, employees with specialist and technical education).

One of the main determinants of the innovativeness of a given state is the number of patent applications. In the case of Russia, the number of patent applications per million residents from 1995 to 2007 clearly went up, which was very beneficial and favorable, since—as explained above—the competitiveness of the Russian economy in international trade is low. In 1995, this value amounted to 118.5, and at the end of the period, in 2007, it already went up to 193.5. Nevertheless, despite this improvement, there is still a chasm between Russia and countries such as the USA or Germany as regards this index (Fig. 8.8).

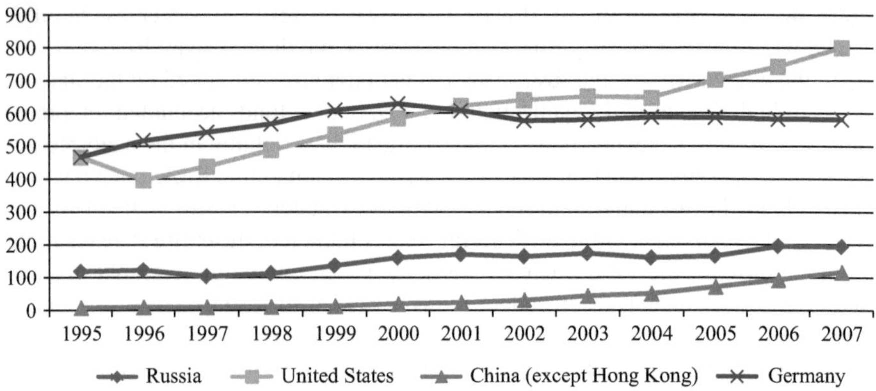

Fig. 8.8 Number of patent applications per million residents in Russia in comparison to selected countries from 1995 to 2007. (Source: Own study based on WIPO data)

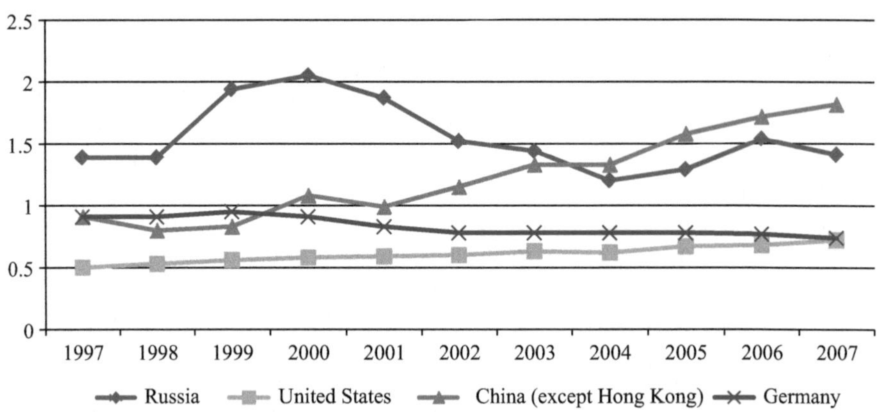

Fig. 8.9 Number of patent applications per US$1 million of expenditure on R&D in Russia in comparison to selected countries in the years 1997–2007. (Source: Own study based on WIPO data)

When we compare the number of patent applications in Russia and the financial outlays on R&D, it is evident that this financing system is highly effective as it yields many patent applications. From 1997 to 2007, it was markedly higher than in the USA or Germany.

In the case of Russia, the highest level of financing effectiveness in the R&D sector was achieved in 2000, when the number of patent applications per million residents was at its highest and equaled 2.05. Since then, it decreased. However, afterward it remained above 1.

The above might be indicative of a situation where funds are concentrated on those areas that generate most patent applications, i.e., computer software, nanotechnology, nuclear technology, aviation and aerospace industry, as well as armament and military in the broad sense (Fig. 8.9).

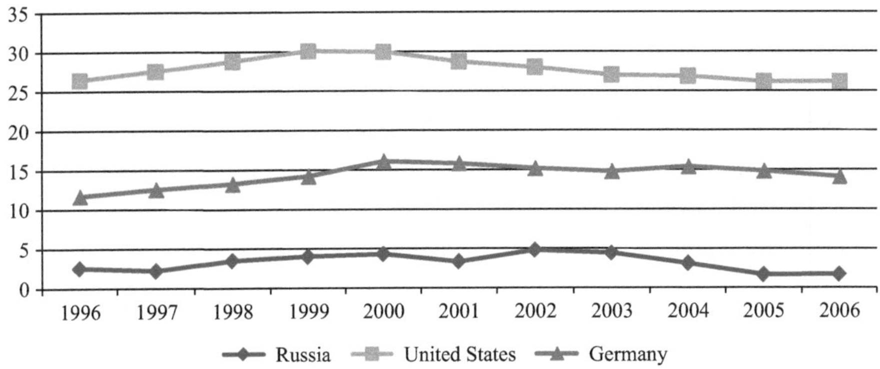

Fig. 8.10 The share of HT export products in total Russian exports as compared to the USA and Germany in the years 1996–2006 (%). (Source: Own study based on EUROSTAT data)

The above-specified growth in the number of patent applications filed in Russia from 1995 to 2007 influenced the growth in the share of HT commodities in Russian exports. It could already be observed from 1997 to 2002, and afterward this trend was reversed, which was a direct result of the increased significance of energy resources in Russian exports, connected with a dynamic increase in the prices of these resources on the global market. Still, the share of HT commodities in Russian exports from 1996 to 2006 was incomparably lower than the same index in Germany or the USA (Fig. 8.10).

8.1.2.2 Human Resources and Their Quality in Russia from 1996 to 2010

When analyzing the accumulated human resources as one of the potential determinants of Russia's long-term competitive advantages in foreign trade, it is worth starting with taking a closer look at the size and quality of this resource in Russia.

In the case of Russia, one can observe a permanent downward trend in the population of this country. Upon the fall of the USSR and the establishment of independent Russia at the end of 1991, according to official statistics, the country was inhabited by 148.9 million people. In 2000, the population equaled 146.8 million, in 2005—143.8 million, and in 2010—only 142.8 million. According to the UNCTAD forecast, in 2020 Russia's total population will amount to 141 million people. This downward trend is the result of a very high negative population growth rate and intensifying emigration tendencies, especially among youth.

When discussing changes in the Russian population, one ought to note the number of persons of productive age. The general trend has been convergent with that observed among the whole Russian population. However, a very notable drop in the number of persons in productive age was visible in Russia until 1998, i.e., the financial crisis. In 1996, the percentage of people in productive age in total population in Russia was still at the level of 50.8 %. In 1998, the same ratio amounted to a mere 45.9 %. The main consequence of the above very significant drop in this index

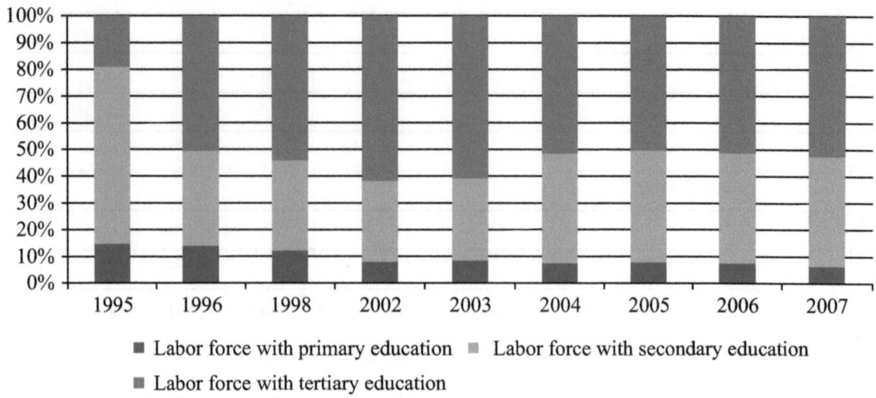

Fig. 8.11 Russian labor force structure according to education level in selected years of the period from 1995 to 2007 (% of total). (Source: Own study based on World Bank data)

was constituted by migration after the fall of the Soviet Union and the proceeding economic downturn that culminated with the abovementioned 1998 financial crisis.

Since 1999, the percentage of the professionally active population in Russia started to grow dynamically. However, according to UNCTAD forecasts, starting from 2012, it will unfortunately decrease.

The trends discussed earlier, characteristic of the changes in human capital in Russia, are especially unfavorable in terms of the possible improvement of the competitiveness of Russian economy. This is because these trends indicate that it is necessary to increase the effectiveness of utilizing the available, decreasing, human resources, and at the same time to improve their broadly understood quality. In this context, one ought to consider the broadly understood level of qualifications of the Russian labor force. One of the ratios used here is the structure of labor force by education level (Fig. 8.11).

In the years 1995–2007, the quality of Russian labor force was high in terms of education. In the whole period, the percentage of persons with higher and secondary education was higher than 85 %, and since 2002—even 90 % of total labor force in the country. In this period, one could observe the positive trend consisting in the gradual decrease of the percentage of persons with primary education.

An interesting phenomenon that could have been observed in Russia from 1995 to 2007 was that the percentage of persons with higher education increased at the cost of those with secondary and primary education. This was the consequence of, among others, the increase in the number of university graduates in Russia. However, in terms of innovativeness and the possibility of generating modern technological solutions, graduates of exact and technical sciences are especially needed, and within that scope a certain stagnation was observed. The notable increase in university graduates in Russia concerned especially the humanities (Crane and Usanov 2010, p. 98). Ipso facto, currently Russian university graduates do not, as a matter of fact, meet the requirements of the market, which is exhibiting high demand mainly for IT specialists, engineers, and production technologists.

In this context, it is also worth underlining that Russia is allocating more and more funds to higher education year after year. In 1995, the share of this expenditure in the GDP amounted to 9.8 %, in 2000—already as much as 10.9 % of GDP, and in 2005—12.6 % of GDP and in 2010—14.8 % of GDP.

In reference to human capital utilization level in Russia in the studied period, one ought to note that, starting from 1999 and afterward, a clear improvement was observed. However, when analyzing statistical data concerning the labor market in Russia, one has to take into account the vast "gray zone" of illegal labor.

In terms of the competitiveness of the given economy, in the context of utilizing (hiring) the available labor resources, it is extremely important that the best educated citizens of the given country are effectively employed. Within this scope, the situation in Russia is improving clearly. In the mid-1990s, the unemployment rate of persons with higher education amounted to approximately 40 %, and in 2010—to 29 %. This still meant that almost one-third of the unemployed in Russia were persons with higher education, however, their employment in the economy was on the rise, which should be classed as a positive trend.

8.1.2.3 Russia's Policy on Science, Innovation, and Education

Government activities performed as part of the broadly understood economic policy are by all means significant in terms of stimulating the development of innovativeness and competitiveness of each country. Apart from direct activities, their overriding role is to create appropriate investment conditions for other entities.

The first steps toward supporting innovation development in the Russian Federation in the indicated areas were taken in the 1996 federal law "Science and the Scientific and Technological Policy." In the course of time, interest in investing in R&D transformed from a written provision into an attempt at performing specific activities.

Special tasks were assigned to the Ministry of Economic Development. The goals and basic lines of the Russian modernization policy have been included in the long-term socioeconomic development concept for Russia until 2020. This concept foresees switching to an economy of innovative and society-oriented growth. The objective is to achieve this by modernizing traditional economy sectors, i.e., metallurgy, chemical production, construction, and agriculture.

Improving broadly understood infrastructure is one of the planned means of stimulating development. The plans also foresee investments in the development of new sectors such as research engineering, medical services, and education that are supposed to improve the competitiveness of the Russian economy. The activities of the Ministry of Economic Development are supposed to concentrate on supporting innovative enterprises in existing companies and assisting new and developing enterprises. To this end, systemic activities have been undertaken. The objective is to improve legislation in the area of innovativeness and stimulate demand for products provided by this sector. On 4 August 2009, the government adopted an action plan for supporting the activity of innovative enterprises. Moreover, every year the

Ministry of Economic Development organizes a conference on the improvement of the image of Russian brands and boosting their competitiveness on the domestic and foreign market.

As part of the abovementioned Ministry, a special Corporate Supervision Department was established. This Department is focused on the activity of development of institutions and state enterprises. The Department coordinates their activities and ensures that Ministry employees participate in the works of development companies and supervisory boards. The Department concentrates on improving the effectiveness of state enterprises and implementing modern management methods derived from Western models. Currently, the Department believes that development and venture capital financing promotion is its priority.

One of the first institutions whose goal is to support the innovativeness of the Russian economy, and as a result—its international competitiveness, is the special Government Commission on High Technology and Innovation (reformed in 2010). This Commission is a standing body tasked with the coordination of activities of federal executive bodies and cooperation between executive and regional bodies, institutions, and public and scientific associations to implement and formulate the policy on development on the NIS. Its chairperson is the Prime Minister of the Russian Federation, and is composed of, among others, selected ministers, ministry employees, presidents of enterprises, university deans and research institutions. The main goals of the Commission involve the supervision of activities of federal authorities in the area of innovation policy, formulating long-term scientific and technological forecasts, development of nanotechnology and nanoindustry, and creating a market for its products (OECD 2011, pp. 182–189).

Moreover, in 2009, the President established a special Commission on Modernization and Technological Development of the Economy. Its main tasks include: Developing the basis of a new government program on modernization and technological development of the Russian economy, coordinating the activities of state authorities' organs participating in the process of innovation implementation, as well as specifying the priorities of Russia's technological development.

Moreover, five main priorities for the Commission's work were selected, namely: nuclear technologies, space technology, medicine, energy, and information and telecommunications technologies.

When describing Russia's NIS from the perspective of the implemented innovation policy of the Russian authorities, one ought to note the clear concentration of both funds and efforts on the development of the nanotechnology sector that was classed as a strategic scientific research area. The funds that have been allocated to nanotechnology already amount to US$15 billion. According to estimates, in the next 5–6 years the value of this sector will go up by US$2 billion. Moreover, it is expected that until 2015, production will increase by RUB900 billion thanks to the implementation of nanotechnology solutions (today, production does not exceed RUB4–5 billion).

Therefore, the first Russian institution that aims at implementing the national objectives within the scope of nanotechnology—the Russian Corporation of Nanotechnologies (RUSNANO)—was established in September 2007. The main objective of the above institution is the commercialization of nanotechnology projects.

Another important project conducted under the current federal authorities' policy aimed at improving the innovativeness and competitiveness of the Russian economy is the construction of the Skolkovo Innovation Center. This Center is to follow the example of the American Silicon Valley and produce modern inventions and solutions for the industry. The Center is to be composed of a modern university campus and a range of institutions, such as technological parks, specialist laboratories, as well as financial and legal centers.

This enormous undertaking, which will cost the Russian government US$5 billion, does not have much in common with other cluster enterprises in the world. The Center is actually a country within a country—it has a powerful self-government, its own courts that are concerned with intellectual property protection, its own laws, and even its own police force. All this is supposed to eliminate the barriers that inhibit innovation development (such as red tape and corruption) in Russia as soon as possible by means of the advantages of its location (mainly access to educated labor force).[1]

To sum up, the advantages of Russia's NIS, especially in terms of its international competitiveness, include (Peltola 2008):

- High economic growth that makes it possible for the federal government to create conditions conducive to stimulating and developing proinnovation activity.
- Existence of a few strategic documents concerning economy innovativeness.
- Existence of special institutions that coordinate proinnovation activities at the federal level.
- Growing number of institutions and agencies that support proinnovation activities at the regional level.
- Improvement of legal safeguards within the scope of intellectual property, in compliance with international standards.
- Paying more attention to the monitoring and evaluation of the innovation policy, including international benchmarking.
- The disadvantages of Russia's NIS that inhibit its international competitiveness include (Peltola 2008):
- Limited inclusion of business into the process of formulating priorities for the development of the R&D sector and financing the development of that sector.
- Issues connected with the low technological advancement of the industry are poorly translated into provisions of the innovation policy, and enterprises are poorly included in the international knowledge transfer.
- Lack of solutions that would enable knowledge transfer and diffusion of technology among economy sectors.
- Low level of support for innovativeness of small- and medium-sized enterprises, especially at the stage of initial development.
- Low share of funds allocated to supporting the innovativeness of enterprises in the budget, mainly lack of financing for innovative increased-risk enterprises.
- Lack of effective formal principles of monitoring of innovation and evaluation policy tools.

[1] www.pi.gov.pl/parp/chapter_86365.asp?soid=65F3718B5F4A4A618C4AB4279455A2A4, (accessed on 20 May 2012).

Despite the specified disadvantages of the Russian NIS, it is possible to indicate opportunities for further growth, which would be especially desired in terms of improving the innovativeness and competitiveness of the Russian economy. The abovementioned opportunities might result mainly from:

- Gradually increasing cooperation between various institutions in developing and implementing the innovation policy.
- Developing the strategy and special institutions responsible for stimulating the innovative development of both enterprises and the whole economy.
- Orientation toward innovative strategic support for key sectors from the perspective of economy development.
- And, first and foremost, gradual growth in innovative awareness in Russia, both in political and business environments.
- In turn, the main challenges facing the Russian NIS are:
- Lack of further development of broadly understood innovative infrastructure.
- Lack of coordination of the activities of state institutions that are responsible for the innovation policy in the Russian Federation.
- Lack of attempts at facilitating access to cheap and secure financing of innovative activity.
- Emigration of highly qualified specialists.
- Lack of an effective system ensuring the application of innovative solutions to the economy.
- Lack of correlation between the R&D system and real socioeconomic needs of the country, also from the perspective of the competitiveness of its economy.

8.1.3 Interdependencies Between the Level of Innovation, Accumulation of Human Capital, and Advantages in Russia's Foreign Trade

Economic growth in Russia is based on exports of energy resources and is paradoxically dependent on the fluctuations of prices of these resources on the global market. The Russian economy cannot be called a knowledge-based economy, mainly because it displays a low level of human capital accumulation and innovation.

In reference to the quality and level of available human capital utilization in Russia, there is no doubt that the political transformation process, as well as the narrow specialization of the Russian economy that is limited to resource sectors, makes it impossible to fully use the potential of the human capital. Moreover, the structure and quality of human resources meet the requirements of the economy only to a small degree.

When analyzing the commodity structure of Russian exports in the years 1996–2010, one ought to underline that commodities manufactured by medium-qualified blue-collar workers are dominant. Their share was in Russian exports was very high and ranged from 70 to 80%. It is also worth emphasizing that industries hiring highly qualified employees have a very low share in Russian exports. In 2010, the share of these commodities was almost nonexistent and amounted to a mere 2.37 %. In turn, commodities manufactured by industries hiring medium-qualified white-collar employees traditionally play an important and increasing key role in Russian imports.

In the case of innovativeness and its role in creating competitive advantages of the Russian economy, one ought to underline that the innovativeness level of Russian enterprises is generally much lower than in other countries that display a similar level of socioeconomic development, and is additionally very diversified depending on the sector where these enterprises operate. From the point of view of the future potential improvement in the level of innovativeness of the Russian economy, it is alarming that the R&D area that is responsible for the creation of new innovative solutions is financed practically in whole from public funds that de facto reach large state enterprises. On the other hand, the low innovative awareness of Russian SMEs and relatively low funds allocated to the purchase of modern technological solutions do not allow the assumption that the innovation level that translates into a notable increase in Russia's competitive advantages in international trade has gone up recently.

MHT commodities are decidedly dominant in Russian exports. Their share in total Russian imports in the years 1996–2010 exceeded 70 %. It grew from 72.12 % in 1996 to 77.49 % in 2010.

When discussing the commodity structure of Russian exports, it is worth paying attention to the very low presence of HT commodities. It oscillated from 3.62 % in 1996 to a mere 2.37 % in 2010.

To sum up, Russia's economy is generally characterized by a lack of comparative advantages in modern input-factor-intense industries. Human capital and innovations do not play a major role in the process of developing advantages in Russian foreign trade. What is worse, Russia's international competitiveness is not only very low, but is also deteriorating. In terms of the country's further development, also in the face of the potential depletion of resources, this should be considered alarming.

8.2 Case Study: China

Günter Heiduk, Agnieszka McCaleb

China is the most impressive example of those economies that were able to tap the potential of the advantage of backwardness by building its development on exploiting its comparative advantages extensively and securing its development path through gradual, pragmatic, but still forward-oriented transformation (Lin 2011, p. 4). Many studies highlight capital accumulation, export orientation, and manufacturing concentration as the driving forces of China's growth (Felipe et al. 2010). In the case of China, this frequently observed development pattern has some special features. China's transformation approach is characterized by learning, first, through attracting foreign capital and knowledge and second, through educating human resources abroad. The result is a fast and drastic structural change. The catching-up of the manufacturing sector in favor of the agricultural sector has got its most visible and influential result in China's high export-led growth. The backward service sector could be considered as a future source of growth.

8.2.1 China's Competitiveness in Foreign Trade

The evolution of China's trade in the last three decades does neither fit any trade theories nor the development of other countries' trade. Very superficial observations deliver two significant features of China's trade, namely the extraordinary growth measured by the share in world trade and the drastic change in the factor intensity of its exports. China's share in world trade increased from 1 % in 1980 to 9.7 % in 2010 (WTO 2011). In the early 1980s, raw materials and agricultural products accounted for over 50 % of China's exports, whereas now manufactured goods make up over 90 % of its exports. The factor intensity in manufactured exports changed rapidly from LI products (textiles, apparel, footwear, toys) to technology-intensive goods (telecommunication equipment, office machines, electrical machinery). On the import side, technology-intensive products have increased steadily and make up approximately one-quarter of China's imports. Especially, in the last 5 years, shares of imported raw materials increased. China's export bundle as well as its level of IIT reached a degree of sophistication far above its level of development, which only recently passed the threshold from a low-income country to a middle-income country (Liang 2008, p. 100; Rodrik 2006, p. 4). From a global perspective, China achieved the highest rank in centrality[2] in its total trade as well as in its trade in capital goods and also in consumer goods, whereas it ranks third in intermediate goods (IMF 2011a, p. 5). In the last decade, the number of countries where China is the number one or two as a trading partner has increased from 13 to 78 (IMF 2011a, p. 3).

Despite the ongoing debate on the measurement of China's trade shares (e.g., Anderson 2007), there is well-accepted evidence for the growing share of China's trade in GDP as well as for the changing trade pattern, especially with respect to the export composition. A large number of studies aim to explain these developments. Among the most popular findings are the reduction of tariffs and nontariff barriers (e.g., Feenstra and Kee 2005).[3] Evidence for the relatively high productivity growth in the export sector is attributed to the gains from spatial fragmentation. Falling trade costs allow more efficient exploitation of these gains. China's integration into the East Asian production network is considered a main driving force of China's growing trade shares as well as its fast-growing IIT (Tang and Zhang 2009). The deepening fragmentation of production processes seems to explain the increasing

[2] The centrality is measured by the eigenvector centrality, which weighs the size of exports with linkages to other large players (IMF 2011c, p. 5).

[3] Low or no tariffs on imported intermediate products motivate firms with low productivity to engage in processing trade (Yu 2001, p. 2). See China's current tariff profile at http://stat.wto.org/TariffProfile/WSDBTariffPFView.aspx?Language=E&Country=CN.

skill intensity in China's exports (Amiti and Freund 2008). Several studies analyze the relationship between IFDI, ownership and export performance, concluding that foreign ownership and export performance are positively correlated (Fu et al. 2010). The higher the technology content of IFDI, the better is the export performance. Trade growth as well as the IFDI-driven increasing technological content of China's exports generate an increasing demand for education and result (with a time lag) in a higher supply of skilled labor (Harris et al. 2010, p. 18).

Last but not least, China's export performance is analyzed as a dependent variable of the RMB (Morrison and Labonte 2011). Even the export performance of competing developing countries is significantly depending on changes of the RMB (Mattoo et al. 2012). Countries with a high degree of export similarity with China (e.g., Taiwan, Korea, Mexico) experienced decreasing losses in export market shares when comparing 2008 (after RMB appreciation) with 2005 (before RMB appreciation; IMF 2011a, p. 5). Growing interest is devoted to the effects of an appreciation of the RMB on China's processing trade.[4] The results of the latest studies range from considerable negative effects on China's processing exports (e.g., Thorbecke 2011) to negligible (Xing and Deterts 2010) or even positive effects (Cheung et al. 2010).

The share of China's top 10 export partners in its total exports has steadily decreased over 15 years, from 78.13 % in 1995 to 60.39 % in 2010. After 2000, China's exports to developing countries increased at an average of 25.23 % and imports from developing countries have grown at an even faster pace of an average 42.68 % (Lu and Li 2010), which is due to China's attempt at diversifying its export markets as well as to its fast-growing imports of natural resources mainly from developing countries (Gong 2011).

Chinese exports are majorly divided between three trading blocs of East Asia, the North American Free Trade Area (NAFTA) and the European Union (EU). From 1995 until 2010, the EU share of Chinese exports rose from 13.3 to 20.5 % (Li et al. 2012), while the US share rose and fell (from 16.62 % in 1995 through around 21 % in 2000 and 2005 and ending at 17.99 % in 2010) and the share of Japan dropped from 19.13 to 7.67 %. From 1995 until 2010, China's top three export destinations were invariably USA, Hong Kong and Japan followed by other developed countries, especially EU member states such as Germany, Netherlands, United Kingdom, Italy (the EU is the most important market for Chinese exports), as well as South Korea, India, Russia and Singapore. The reason for that is China's specialization in LI manufacturing for the markets of developed countries. It is also the result of direct investment by MNCs from EU, USA, Japan, and Korea that assemble imported parts and components in China and export them mainly to developed markets, thus fostering China's exports (Li et al. 2012; Gaulier et al. 2007). Trade with ASEAN countries is to a high degree resulting from regional production networks where components are produced for assembly in China (Li et al. 2012).

[4] Processing trade are imports of intermediate inputs that are imported duty-free for assembling to processed exports. These imports and exports are not allowed to enter the domestic markets. Trade without special tariff treatment is called "ordinary trade." These products are allowed to enter the domestic market.

In 2010, China's top import categories from USA were electrical machinery and equipment (US$11.5 billion), power generation equipment (US$11.2 billion), oil seeds and oleaginous fruit (US$11 billion), aircraft and spacecraft (5.8 %), optics and medical equipment (5.2 %). Most of them saw significant increases over 2009 with power generation equipment and optics and medical equipment of above 30 %. Interestingly, vehicles, excluding rail, saw annual growth of 134 %. Leading export categories from China to the USA in 2010 were: electrical machinery and equipment (US$90.8 billion; 24.5 % change over 2009), power generation equipment (US$82.7 billion; 32.5 % change over 2009), apparel (US$28.8 billion), toys, games and sports equipment (US$25 billion), furniture (US$20 billion), footwear (US$15.9 billion).[5] Data shows that China serves as assembly of components in electrical machinery and power generation equipment that are imported from the USA and final products are exported back to the American market.

Between 1995 and 2010, Hong Kong remained at the top of China's export destinations, although the share in total exports has been slowly decreasing (from 24.19 % in 1995 to 13.84 % in 2010). However, approximately two-thirds of Hong Kong's imports from China are reexported to other markets (Li et al. 2012).

Japan, South Korea and other Asian economies are China's top partners in terms of its imports. MNCs from Asian economies invest in China, being mainly driven by cost optimization, unlike their counterparts from the USA and EU (which are interested in China's domestic market as well); their operations are focused on assembly of products, which are then exported to developed economies or their home markets. These MNCs import parts and components for assembly in their affiliates in China (Gaulier et al. 2007); and thus they contribute to the growth of Chinese exports to their home countries. Exports to and imports from Japan are related with high intensity of IIT coming from Japanese foreign direct investment (FDI) in China (Chen et al. 2011). Japan imports from China many agricultural and mineral products due to their scarcity in Japan (Chen et al. 2011, p. 37). Japan is the most important supplier of China's intermediate goods. Japanese MNCs in China import a lot of intermediate products from Japan and export large part of them back to Japan.

East Asian production networks have given rise to a "triangular trade pattern:" Japan and NIEs export capital goods and sophisticated intermediate goods (especially parts and components) to the less developed countries of the region (ASEAN* and China), which process them for exports destined to the USA and Europe (Gaulier et al. 2007).

EU is China's number one export partner, accounting for 20.1 % of its total exports in 2010. In 2011, 95.4 % of Chinese exports to the EU were manufactures composed mainly of machinery and transport equipment (49.1 %; 31.5 % of it being office and telecommunication equipment), textiles and clothing accounting for 13.3 %, and other manufactures for 19.1 %. Machinery and transport equipment have seen increases in Chinese exports to the EU27.[6]

---

[5] https://www.uschina.org/statistics/tradetable.html.

[6] http://trade.ec.europa.eu/doclib/docs/2006/september/tradoc_113366.pdf.

In 2011, Chinese imports from the EU comprised mainly machinery and transport equipment that accounted for 60.2 % (transport equipment seeing growth from 17.8 % in 2007 to 24.6 % in 2011), chemicals and related products for 10.9 %, and manufactured goods classified chiefly by material for 9.5 %.[7] According to Li et al., EU27 might be expected to specialize in and export machinery, mechanical appliances, vehicles, and spacecraft to China (Li et al. 2012, p. 139). "As far as China's trade with the EU is concerned the evidence suggests that intra-industry trade predominates and that there is a high degree of complementariness: China assembles labor-intensive commodities, such as computers (8471), line and cordless telephones and transmission apparatus (8517 and 8525), for export to the EU27, while the EU27 exports capital-intensive commodities, such as motor cars (8703), aircraft and spacecraft (8802), machines and mechanical appliances (8479) to China." (Li et al. 2012, p. 141).

The share of top 10 China's suppliers in its total imports has steadily decreased from 75.6 % in 1995 to 64.3 % in 2010. The EU is China's second largest source of imports after Japan (Li et al. 2012). China's imports from the EU, the USA, and Japan are large but decreasing (Li et al. 2012). China's manufacturing exports and domestic infrastructure investments have been sizeable and significant determinants of China's resource imports (Roberts 2011). China's top import partners can be divided into suppliers of parts and components for assembly in China (East and Southeast Asia) and suppliers of natural resources (Australia, India, Brazil).

China was Australia's second largest trading partner in 2005 and Australia was China's 13th largest trading partner (Sheng and Song 2008). Trade between the two countries is complementary. Australia exports mineral resources and agricultural products. China exports to Australia mainly LI consumer goods (textiles and clothing, electronic goods, general machinery; Sheng and Song 2008).

In 2010, Indian government announced its aim to double its bilateral trade with China within 4 years (Dolla 2011). Chinese exports to India are dominated by value-added products, mainly machinery. Indian and Chinese companies from TDI, which aim at becoming multinational players, contribute to the growth of Chinese exports of technology to India. Chinese firms also conduct infrastructural projects in India and import relevant machinery and telecommunications equipment (Dolla 2011). The structure of bilateral trade moves from primary sectors to manufactured and service products. Projects in renewable energy, infrastructure, and finance have seen significant growth in the last few years (Dolla 2011). However, in 2011 India's exports to China were dominated by "ores, slag, and ash" (49 % of total exports; 60 % in 2010) followed by cotton, accounting for only 10 % of total exports.[8]

China imports from Brazil are mostly basic commodities (iron ore, soya beans, crude oil) accounting for 83.7 % of China's imports from Brazil in 2010 and exports manufactures amounting to 97.5 % of China's exports to Brazil (Brazil-China Factsheet 2011; Economist 2012).[9]

[7] http://trade.ec.europa.eu/doclib/docs/2006/september/tradoc_113366.pdf.

[8] Indian Embassy in China; http://www.indianembassy.org.cn/DynamicContent.aspx?MenuId=3&SubMenuId=0.

[9] http://www.economist.com/node/21542780.

Manufactured goods comprised 70.26% of Thailand's total exports to China in 2008, followed by agricultural products that accounted for 16.84% (Manarungsan, 2009). Computers and components (29.78% of total exports), rubber (11.97%), finished oil (7.32%), and plastic pellets (6.64%) are primary export products to China that record a rising trend, especially finished oil recorded 77.92% growth in 2008 over 2007, and rubber products, 39.09%. China's exports to Thailand comprise capital goods in 39.21%, raw materials in 38.94%, and consumer products in 19.11%.

From 1995 to 2010, China's export structure has changed significantly. The analysis of China's export structure based on the SITC classification shows that exports of simple LI products declined in favor of higher-technology products. During that period, exports of semimanufactured goods, such as leather, rubber, textiles, metals, wood, paper, and more LI products, such as prefabricated buildings and construction fixtures, furniture, travel goods, apparel, footwear, photographic apparatus, watches, and clocks, have significantly decreased; from 22.06 and 37.75% in 1995 to 15.82% and 25.32% in 2010, respectively. Exports of machinery and transport equipment grew from 19.52% in 1995 to 48.07% in 2010, reflecting China's movement up the value chain. However, within this group, China relies on imports of technology-intensive components (Li et al. 2012). Moreover, higher competitiveness in machinery was not accompanied by a drop in China's competitiveness in textiles and clothing; in 2008, they recorded net export value of 25% and thus ranked second after machinery and transport equipment (Li et al. 2012). However, China's competitive advantage in the latter is not homegrown, as according to Li et al. foreign-financed enterprises in China accounted for 69.3% (US$494.4 billion) of the export value in machinery and electronic products in 2009.

Because of high growth, Chinese industry relies more and more heavily on imports of natural resources, especially energy, metals, and raw materials. Since 1995, imports of crude materials (crude rubber, cork and wood, textile fibers, crude fertilizers and crude minerals, metalliferous ores and metal scrap) have doubled, from 7.5% in 1995 to 15% in 2010. Imports of mineral fuels and lubricants (coal, petroleum, gas, electric current) in that period have more than tripled, from 3.89% in 1995 to 13.42% in 2010. In 2008, mineral products amounted to 71% of net import value (Li et al. 2012). Owing to China's increasing global consumption of natural resources, it influences global prices of such products as iron ore or coal that increased in 2009 and 2010 (Li et al. 2012). Machinery sector maintains the largest and, over the period analyzed, steady share in Chinese imports that amounted to 39.25% of total imports in 2010.

The commodity structure of China's exports analyzed based on the OECD classification of technological level of exported goods has also changed over the last two decades, showing growing importance in exports of HT and MHT products, from 7.65% and 23.7% in 1995 to 15.23% and 33.25% in 2010, respectively. The decrease in exports of LT products almost doubled, from 29.99% in 1995 to 16.85% in 2010. More technologically advanced operations of domestic and foreign firms in China are subject to "agglomeration effects of industrial clusters" and country's position in international production networks (Li et al. 2012; Sheng and Song 2008).

China's imports show increasing importance of MHT products, recording growth from 38.57% in 1995 to 47.62% in 2010; the share of LT goods decreased from 14.08% in 1995 to 6.52% in 2010. Imports of HT products were floating, from 11% in 1995, rising to above 15% in 2000 and 2005, and dropping to the level from 1995 in 2010. The fact that China still imports most sophisticated components means that transfer of know-how comes from FDI and joint ventures with foreign partners and technologically advanced skills are not yet possessed by domestic enterprises (Li et al. 2012).

When applying WIFO classification of tangible and intangible inputs in production to China's foreign trade, we can observe that China's entry into the WTO at the end of 2001 resulted in a sudden growth of value of Chinese exports of products from all types of industries, especially of TDI, LI, and MM industries. In 2005, the traditionally leading role of LI industries in Chinese exports gave in to TDI, which in 2010 accounted for one-third of Chinese exports. This marks China's transition from a country solely relying on its huge labor endowment and low labor costs (exogenous location-dependent comparative advantages) to a country whose growth is based on developed endogenous factors created by intangible investments in R&D. Combined with the previous OECD technological classification, it can be concluded that China's structural upgrading is deep-rooted. Although China's comparative advantage in LI industries is still strong (25% of total exports). The trade of MM evolves toward average for exports and imports, characteristic for triad countries (25% of export among triad countries in 1990s, growing trend for China that reached 23% in 2010; 15% of import among triad countries and declining trend for China accounting for 19.6% in 2010). After 2000, the share of CI industries remained stable at around 12%. MDI decreased their share from 16% in 1995 to 8.7% in 2010.

Chinese imports reveal decreasing importance of all types of industries except for TDI one, which grew dramatically from 19.2% in 1995 to 45% in 2010. It is caused by previously described imports of technologically advanced components for assembly in China. Imports of CI goods remain large, although they recorded a slump from 31.8% in 1995 to 26% in 2010. The share of MM in imports fell from 29.2% in 1995 to 19.6% in 2010; LI industries fell from 12.25% in 1995 to 4.99% in 2010; and MDI fell from 7.6% in 1995 to 4% in 2010.

Since China's opening up, the changes in the job profile and qualification structure of its exports and imports clearly reflect the development path from a low-income, labor-abundant country to a middle-income, human capital-building, and demanding country. At first sight, two features are striking (Table 8.11):

1. First, the sudden upheaval of the trade level in all four categories between 2000 and 2005, which may mainly result from China's accession to the WTO in December 2001.
2. Second, the highest growth of trade in medium white-collar/high-skill technology products, suggesting that China is leapfrogging one or more phases of a gradual spread of technology.

Table 8.11 Commodity structure of exports and imports, WIFO classification, selected years (in %). (Source: Own calculation based on WIFO, Economic Statistics)

	Exports				Imports			
	1995	2000	2005	2010	1995	2000	2005	2010
White-collar/high-skill technology	7	11	17	15	11	16	16	6
Medium white-collar/high-skill technology	24	28	31	33	39	38	40	11
Medium blue-collar/high-skill technology	39	36	33	35	36	34	34	48
Blue-collar/low-skill technology	30	25	19	17	14	12	10	35

The almost bisection of the share of blue-collar/low-skill products in China's exports from 30% in 1995 down to 17% in 2010 underlines the technological leapfrogging. The higher technology level in the production of intermediate products may cause the decreasing import demand for blue-collar/low-skill products. This seems to hold even with rising domestic wages in MHT and HT industries. Migration from rural to urban (industrial) areas allows the establishment of domestic blue-collar/low-skill production plants, which substitute the respective low-skill imports. The new comparative advantage in more sophisticated products and the still continuing advantage in low-skill products accelerate growth and rural–urban inequality, thus, further encouraging rural workers to move. According to Lewis (1954), the turning point in economic development is reached when the rate of emigration from rural areas exceeds the natural increase of the rural working-age population. Empirical studies deliver evidence that China has already passed the turning point or is approaching it (e.g., Garnaud and Huang 2006; Du and Wang 2010; Cai 2010; Yao and Zhang 2010). The one-child-policy and the increasing wages in rural and urban areas seem to explain the evidence, whereas the recently observed reverse migration seems to contradict the perception of the already surpassed turning point. Regarding the consequences of China's comparative advantage in international trade, it is argued that there will be a shift toward CI, white-collar/high-skill technology products (Garnaud 2010, p. 188). China's broader-based comparative advantage will result in new challenges for developed countries to adjust their industrial structures. Developing countries will lose pressure from China at their domestic, but also third markets. There is evidence that less-developed countries, especially in Southeast Asia, will take over China's former role as the manufacturer of low-skill products (e.g., Zhao 2011).

Despite the different trade dynamics of the four product groups, the common feature is the dominance of exports over imports in each group and each year, except in white-collar/high-skill technology products and medium white-collar/high-skill technology products in 1995, 2000, and 2005 (Table 8.12). The reason for this exception has not yet been explored.

Contrary to other developing/emerging countries, China is keeping its comparative advantage in low-skill manufacturing even when gaining new comparative advantage in high-skill manufacturing. The explanation of the double comparative advantage that has been revealed in China's evolution of its trade pattern in the last

Table 8.12 China's exports and imports according to WIFO classification based on the level of labor force, selected years (US$ million). (Source: Own calculations)

	Export				Import			
	1995	2000	2005	2010	1995	2000	2005	2010
White-collar/high-skill technology products	24,022.31	63,172.58	285,734.17	520,631.30	27,873.75	67,790.93	229,473.77	223,832.90
Medium white-collar/high-skill technology products	74,412.39	151,341.65	515,570.77	1,136,545.01	97,622.39	161,408.66	550,140.70	946,485.88
Medium blue-collar/low-skill technology products	121,472.09	197,409.46	550,224.20	1,185,152.20	91,983.85	147,511.66	471,565.53	687,749.11
Blue-collar/low-skill technology products	94,206.56	135,448.10	318,535.37	575,898.16	35,646.92	50,419.67	135,335.78	129,536.09

decade rests on the size and economic structure of the country. Three provinces/municipalities (Guangdong, Jiangsu, Shanghai) account for about 60% of China's exports. However, Guangdong Province, as the powerhouse for LI manufacturing, is steadily losing the share in China's exports, in fact from 40% in 1999 to approximately 30% in 2009 (Dyck and Levinger 2010, p. 4). The gradually decreasing comparative advantage in LI manufacturing is closely related to the decreasing IFDI in this province. The average annual growth of IFDI in Guangdong Province dropped from 24% in the 1990s to 5% in the last decade (Orr 2012). In contrast, FDI inflows to inland provinces (e.g., Shaanxi, Sichuan) or remote provinces (Yunnan, Heilongjiang) recently increased at higher rates than the highly export-oriented provinces, partly resulting from the government's "Go West Initiative" (Lu and Deng 2011). The government's ongoing strong efforts to improve the infrastructure in China's off-coastal provinces are reducing comparative disadvantages of these regions. The accelerating growth in Southeast Asian countries (especially Myanmar and Vietnam) offers comparative advantages resulting in increasing trade of the southern border provinces Yunnan and Guangxi.

The comparison of the development of provincial RCA indexes suggests that China's comparative advantage in LI manufacturing is moving from eastern to western regions and new comparative advantage in technology-intensive production is emerging in coastal regions.[10] The relatively low share of the population in the west that received higher education (UNESCO 2011, p. 17) is one of the most important current fundamentals for the comparative advantage of the these regions in LI manufacturing. The fast-growing share of postgraduates in the population in the east creates the basis for the emergence of technology-intensive comparative advantages.

The RCA indexes in the period 1980–2000 confirm the following evolution and pattern (Yue and Hua 2002, p. 293):

- With the gradual liberalization of trade and exchange rate regime, China's overall trade pattern shifted toward its comparative advantage of a labor-abundant country. The intensity of this shift differed significantly among the provinces.
- Provinces achieving a high congruency between their comparative advantage and their export pattern experienced higher and faster increasing exports.

Apart from the Ricardo- and Heckscher-Ohlin-type comparative advantage, the evolution of "institutional comparative advantages" (Hall and Soskice 2001) plays an important role in transition countries that open up to international markets. These advantages are shaped by the availability, type, and access to financial, human, and social capital. The concrete patterns of China's change from institutional comparative disadvantages toward advantages will be unconsolidated as long as China's transformation process is ongoing. The challenges identified so far are firstly the improvement of the innovative capacity of the economy and secondly the increase of levels of institutionalized trust in Chinese society (Redding and Witt 2008, p. 23).

[10] The Tianjin Binhai New Area reported 2775 new S&T-intensive enterprises in 2011 increasing the total number to 6,650 (http://www.tjbhnews.com/eng/system/2012/01/13/010102386.shtml).

An in-depth assessment of the development of the China-World RCA in 1996–2008 concludes that "China's booming export sectors are still largely scale-driven: the expansion of trade volumes are largely of relatively low value-added assembly-based and/or energy-intensive commodities, which demand the importation of raw materials and energy. There are however signs of industrial upgrading: capital and technology-intensive sectors are growing at an even faster pace than labor-intensive sectors, with strong increases in indictors of revealed comparative advantage (RCA) (RCA-I and RCA-II) in machinery and a decline in the relative importance of textiles and clothing sectors (RCA-II). An examination at the 4-digit level of HS (Harmonization System) commodity classification trade data revealed that the increase in the technological content of Chinese exports is largely a result of the growth of computers, office equipment and electronics and telecommunications equipment and that China remains dependent on imports of some of the most sophisticated components" (Li et al. 2012, p. 141).[11]

In the last decade, China's trade with EU27 is mirroring in most categories its trade with the world. China's RCA indexes for LI exports to the EU27 (textiles, clothing, footwear, headwear, furniture, toys) showed the highest but gradually decreasing scores, whereas EU27 RCA indexes scored high in machinery, mechanical appliances, vehicles, and spacecraft exports to China (Li et al. 2012, p. 140). However, China's RCA indexes for technology-intensive products' exports to the EU27 (especially machinery, vehicles, aircraft) continue to improve. The EU decision to launch antidumping, antisubsidy, and safeguard probes to certain Chinese ITC exports is considered as evidence for Chinese industrial policy measures (e.g., local content requirements) that assist the technological upgrading in manufacturing (Li et al. 2012, p. 140). The high dynamics of China's mean RCA are underlined by the fact that this score doubled and the standard deviation declined between 1996 and 2008.

The comparison of China's RCA with other countries of the BRIC group shows abundantly clear advantages in machinery, toys, games, and sports requisites (Singh et al. 2011, p. 10). The advantages in the first product group are increasing, whereas the latter three product groups experience slightly declining advantages. China's export pattern matches best with Brazil's import pattern suggesting high and even gradually increasing trade complementariness.

When breaking down the HS-2 digit level to HS-6 digit commodity level, China's maximum number of commodities with comparative advantage in the world market at the beginning of the last decade had been 129 products in electrical and electronic equipment, closely followed by organic chemicals (118) and articles of apparel (111) (Batra and Khan 2005, p. 11).

When focusing on a higher level of aggregation, three different methods of characterizing industries according to their input level allow economically insightful findings regarding China's revealed comparative advantages.

[11] RCA-I: Net exports as a percentage of gross exports plus imports; RCA-II: Share of a commodity in a country's exports relative to the share of the same commodity in total world exports (ibid p. 133).

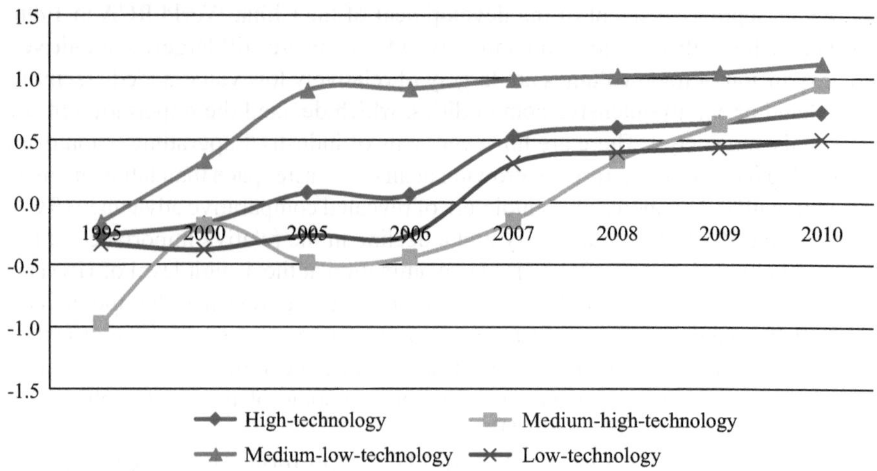

Fig. 8.12 China's RCA indexes of industries based on the technology level, 1995–2010. (Source: Own calculation based on OECD, STAN Database)

For the period of 1995–2010, China's comparative advantage in international trade according to the OECD technology level-based classification showed the following dynamic pattern (Fig. 8.12):

- The RCA index of MLT industries scored the highest comparative advantage, but since 2005 it has been losing growth dynamics. The improvement had been particularly high between 1995 and 2005. Since then, the trend flattened to a modest level of this RCA index.
- Since 2000, the development paths of the RCA indexes for HT and LT industries showed an almost parallel run. The former scored around 0.2 higher. The differences with the higher MLT RCA decreased in 2006/2007 and amounted since then to around 0.4 and 0.6, respectively.
- The most dynamic trajectory exhibited the MHT RCA index reducing the negative gap to the MLT RCA index from 1.38 in 2005 to 0.17 in 2010, and surpassing the LT RCA index in 2008, and 1 year later, also the HT index.
- The only RCA index scoring higher than 1 had been observed in MLT industries since 2008. If the trend of the RCA index in MHT industries continued in 2011, then its score should have surpassed the threshold of 1.

In the same period of time (1995–2010), the evolution of the comparative advantage according to the WIFO classification based on the extent of utilization of tangible and intangible *factor inputs* in the manufacturing process showed the following characteristics (Fig. 8.13):

- LI industries scored the highest RCA index with a stable, but modest upward trend.
- The development of the RCA index of mainstream industries almost matched the former trend at a lower level with an average difference of around −1.4.

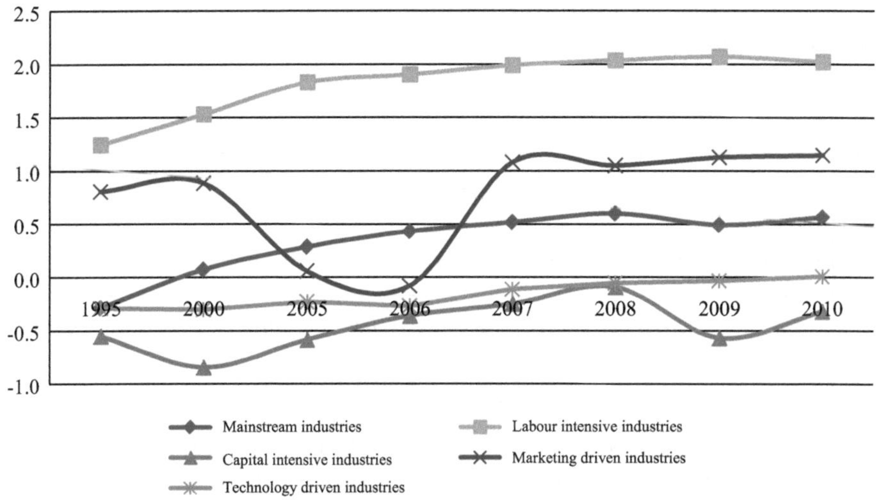

Fig. 8.13 China's RCA indexes of industries based on their factor input, 1995–2010. (Source: Own calculation based on OECD, STAN Database)

- In 1995, TDI exhibited almost the same RCA index as basic industries but developed with a lower growth rate resulting in a gap of −0.44 in 2010.
- As expected, the RCA index of capital industries scored lowest with largest differences to the RCA index of TDI in 2000 (−0.54) and 2009 (−0.53).
- The development of the RCA index of MDI did not match the trends of the other industries. It showed a sinus curve run between 1995 and 2007, with peaks in 2000 (0.88) and 2007 (1.08) and a slightly increasing trend afterward. Except 2005 and 2006, MDI's RCA scored higher than basic industries, but over the whole period of time, fairly lower than LI industries.

The third method to explore China's comparative advantage in international trade is based on the WIFO classification of industries according to their share of different skill levels of labor (Fig. 8.14):

- It is not surprising that industries with a high share of low-skilled labor exhibited the highest RCA scores. The development between 1995 and 2010 did not take a stable course. The steep rise in 2006/2007 was preceded by a gradual decline from 2000 onward, and an almost stable RCA index from 2008 to 2010.
- A quite similar trend at a lower level could be observed in industries with a high share of medium-skilled, blue-collar labor.
- The pattern of industries with high share of medium-skilled, white-collar labor showed a relatively stable, slightly upward trend.
- The strongest upward trend occurred in industries with a high share of high-skilled labor with a short minor decline in 2005/2006. Since the beginning of the last decade, the RCA indexes of these industries surpassed those of the medium-skilled, white-collar industries and since 2005 also those of the medium-skilled, blue-collar industries.

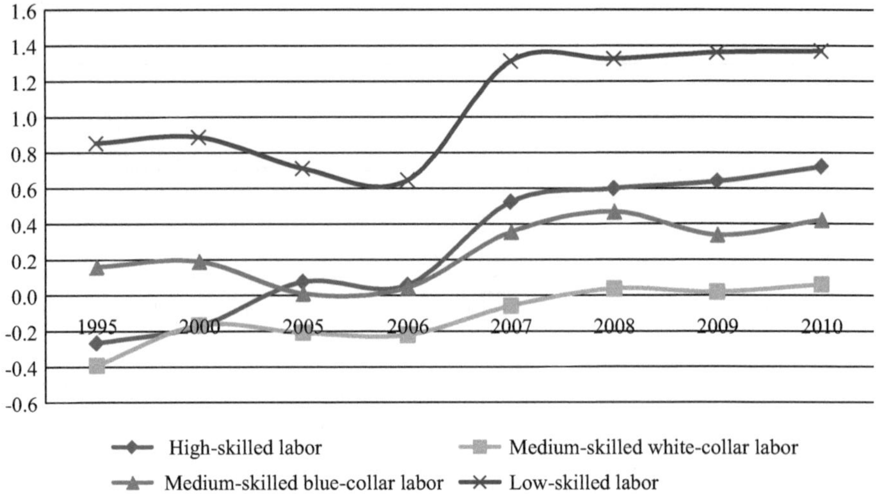

Fig. 8.14 China's RCA indexes of industries based on their skill levels of labor, 1995–2010. (Source: Own calculation based on WIFO, Economic Statistics)

- Only industries with a high share of low-skilled labor achieved RCA indexes higher than 1. If the trend in the industries with a high share of high-skilled labor continues, the RCA index should reach 1.00 in 2012.

The main message of the different levels and structures of product- and industry-based analysis of China's comparative advantage is the confirmed knowledge that low-skilled labor is still its most important source. Despite improving comparative advantages in technology-intensive, high-skill products and industries, China will be able to further rely on its labor-based comparative advantage with a gradually increasing skill level. Differently to theoretical predictions and examples in other emerging countries, the rise of comparative advantage in HT, high-skilled labor industries does not correspond to the decline in the comparative advantage of LT, LI industries. China might preserve this pattern in the near future if the government succeeds in transforming the labor market toward efficiency-guided functioning (Yang et al. 2010, p. 30).

Finally, it should be pointed out that the aforementioned results do not hold for any breakdown to bilateral country and/or product level (e.g., Ju et al. 2010).

The analysis of China's trade suggests that in many SITC product categories exports and imports rise at the same time. This increasing IIT goes along with the increasing integration of China into the world economy as well as with the sophistication of its trade. In order to explain the determinants of IIT, the literature suggests to analyze country-specific characteristics (e.g., factor endowment) or/and industry-specific features. Furthermore, additional insights regarding the competitiveness in IIT should be extracted from dividing IIT into horizontal and vertical IIT (HIIT, VIIT) (Hummels et al. 2001; Fontagné et al. 2005). The latter is two-way trade of products with different quality, which is predominantly attributed to the human capital intensity. Exports and imports of goods in the same category differ according to the stages of processing. HIIT mainly results from product differentiation, differences in preferences, and economics of scale. It should be noted that the total amount of IIT strongly depends on the level of classification of goods. At the one-digit SITC level, the IIT used to be larger than on more detailed levels.

The relatively deep value chain in capital goods (BEC Rev. 3, Chap. 4) and the central role of China as the top exporter of capital goods have motivated a large number of studies on the evolution of China's IIT. The country is the largest exporter and second largest importer in this product category. Based on the division into final and intermediate capital goods (BEC Rev. 3, Chap. 4.1 and 4.2), China is also the largest exporter and second largest importer in the former category and the largest exporter and importer in the latter category (Zhu 2009, p. 5). Early studies already suggest that IIT played an important role in China's trade with developing and developed countries where VIIT was clearly dominating HIIT (Hu and Ma 1999, p. 5). Large differences in IIT across industries seem to result rather from unique country characteristics, such as malfunctioning of the market mechanism, "consumption ahead of time," lack of modern infrastructure in traffic, communication, and logistic. China's accelerated integration into the world economy after the accession to the WTO has changed the structure of its IIT toward the already dominating VIIT. The higher the similarity with partner countries, the lower is the VIIT. Contrary to HIIT, factor endowment did not play an important role in determining VIIT. In general, IIT is larger with countries that match China's size measured by population (Zhu 2009, p. 14).

The currently available studies on China's IIT, VIIT, and HIIT do not provide sufficient answers to their respective determinants. With respect to the methodology, the main reason is the ongoing search for the best method that allows a clear separation of HIIT from VITT. The different approaches to measuring differences in product quality result in different shares of HIIT and VIIT in total IIT, leading to a "rather fuzzy set of conclusions" (Azhar et al. 2008, p. 337). The method applied in this study suggests a surprising result that China's exports to Malaysia, Thailand, and the Philippines are of considerably lower quality than their imports of products from the same industry (Azhar et al. 2008, p. 343). Former studies seem to underestimate China's HIIT. Another weakness is the analysis of the determinants of changes in the dispersion of quality over time.

The high share of VIIT in China's IIT and even in its total trade results to a high degree from its policy to attract IFDI. In the last decade, foreign-invested enterprises (FIEs) have contributed an average of 81.5% of China's processed trade (China Daily 2011). Assuming that a considerable part of VIIT stemming from IFDI is at least of MLT, China might lose its comparative advantage LI, low-skilled workers VIIT to countries in Southeast Asia if these countries succeed in timely quality improvements in product categories where producers in China are no more interested in allocating resources.

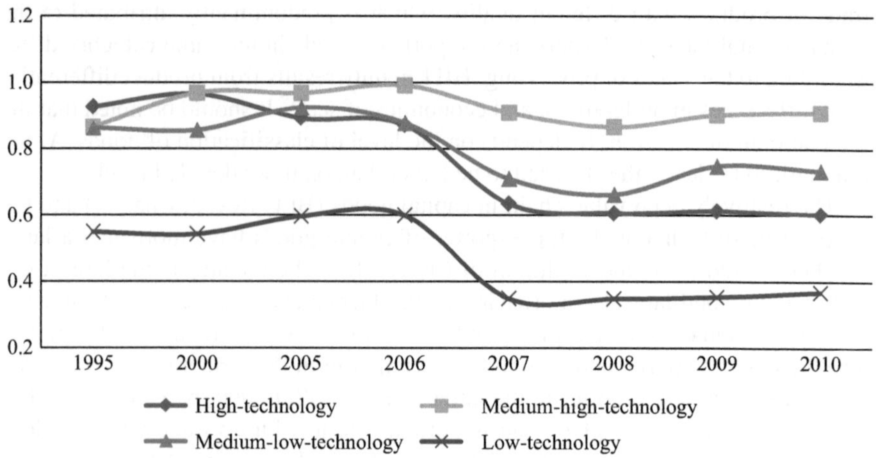

Fig. 8.15 China's IIT classified by the technology level, 1995–2010. (Source: Own calculation based on OECD, STAN Database)

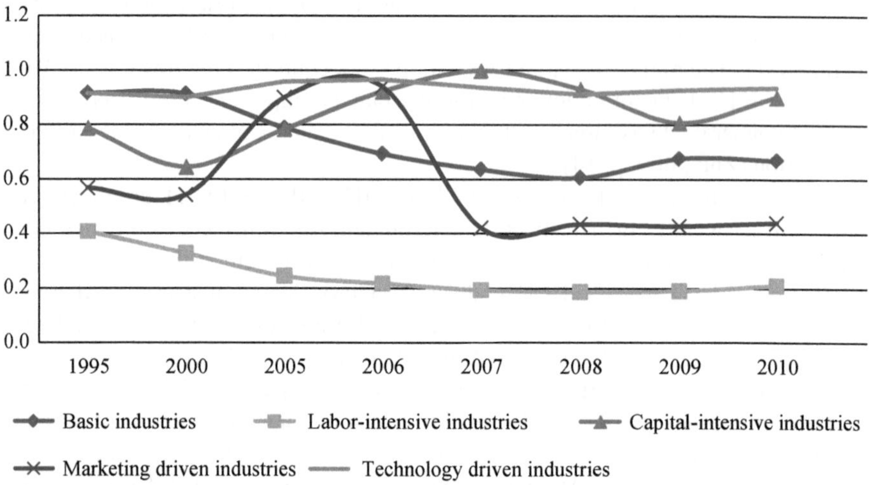

Fig. 8.16 China's IIT classified by factor inputs,1995–2010. (Source: Own calculation based on OECD, STAN Database)

In addition to the conventional break down of IIT into HIIT and VIIT, the following categorization allows to shed light on the technology, factor and human capital content of China's IIT from 1995 to 2010 (Figs. 8.15, 8.16 and 8.17):

- The common feature in all categories is the type of the correlation between labor and the amount of the IIT. The more important labor and the lower its quality in China's trade, the lower its IIT is. This is not surprising because China, as a labor-abundant country with a large rural population of circa 700 million, is able

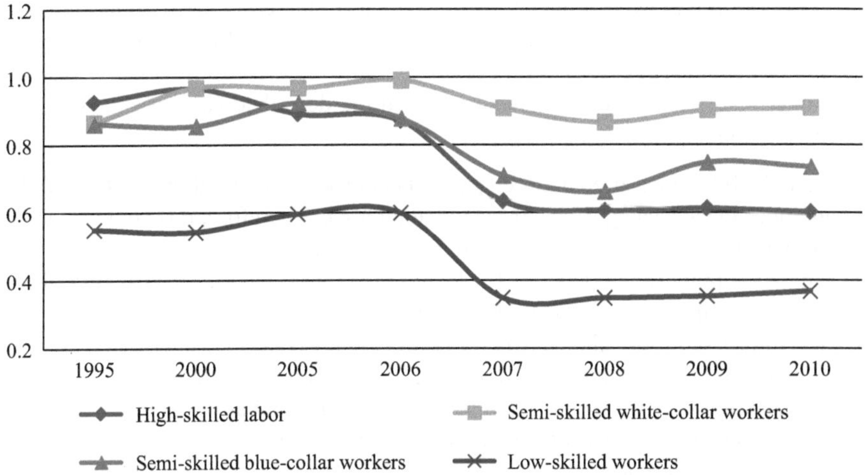

Fig. 8.17 China's IIT measured by skill levels of labor, 1995–2010. (Source: Own calculation based on WIFO, Economic Statistics)

to produce most of these products domestically. The successful integration of a large part of the rural population into the manufacturing industries is reflected in the downward trend of this IIT, as well as in its sharp drop in 2005–2007.[12]

- The categorizations according to the technology level and the skill level of the labor force show similar trends and shapes of the curves. The falling trend shows a clear drop in 2005–2007. Medium-technology IIT followed by semiskilled white-collar workers IIT were less exposed to this drop. Since 2000, both classified IIT achieved the highest level in their respective category.
- The differences between HT IIT and MLT IIT are relatively low. Both categories of IIT are fairly larger than LT IIT, but smaller than MLT IIT, especially after 2005. This also holds when categorizing IIT according to the skill level of the labor force. High-skilled labor IIT and semiskilled blue-collar workers IIT are similar by trend and scope, as well as by their respective difference from low-skilled IIT and semiskilled white-collar workers IIT.
- The aforementioned characteristics are less clear when categorizing IIT according to the factor inputs. The trend of IIT in basic industries matches the IIT in the LI industries at a higher level. IIT in TDI showed a relatively stable trend at a higher level than IIT in basic industries. Between 2000 and 2008, the gap between both types was growing. In general, the IIT in CI industries was the only one with an increasing but fluctuating trend. A drop from 1995 to 2000 was followed by an increase until 2007, another drop until 2009 and a recovery in 2010. The level is approaching the highest level of IIT in TDI. Between 1995 and

[12] According to estimations, the number of migrant workers in urban areas increased by 2.5 million between 2005 and 2008 (Herd et al. 2010, p. 17).

2000, the IIT in MDI dropped, then had sharply increased until 2006, dropped again sharply in 2007, and stabilized since then at a level in between IIT of LI and basic industries.

The interpretation of China's IIT classified in the described categories allows the following conclusions:

- Usually, IIT is measured between pairs of countries. China's above-described IIT includes in each category its respective total trade. Therefore, it is not possible to draw conclusions regarding the similarity with other countries.
- China's relatively low level of LI, low-skilled labor, LT IIT suggests that there are few competitors with a similar specialization pattern.
- The higher scores in all other categories of China's IIT can be interpreted as clear evidence for China's integration into the global production networks.
- The relatively high MHT IIT and semiskilled white-collar workers IIT describe China's position as one of the countries in the upper-medium segment of the global production networks.
- HT IIT and high-skilled workers IIT are the second lowest IIT, indicating that China's innovativeness seems to be relatively low. The high IIT in TDI does not necessarily contradict this interpretation. China's processing exports in telecommunication and office machines rely on high imports of parts and components. The Chinese value-added is relatively low, which is suggested by several empirical studies on China's IIT (e.g., Xing 2010).
- In order to get deeper insights into China's competitive position in global IIT, a breakdown to a combination of country- and industry-specific IIT seems to be necessary.
- The most general conclusion that can be drawn from the three types of categorization of IIT is China's gradual moving away from LI processing trade.
- It has to be noted that the basic G–L index does not provide information about the relation between exports and imports in the same category. This may modify the aforementioned interpretations of China's IIT.
- The IIT analysis should be complemented and related to China's processing trade and the ownership structure of this part of trade. Even the analysis of global IIT delivers indications toward the importance of processing trade in explaining IIT (Brühlhart 2009, p. 457).

An in-depth study on the correlations between China's IIT and its processing trade reached the following conclusions (Koopman et al. 2008, p. 24): The share of foreign content in China's manufacturing exports increased from close to 50 % in 1997–2002 to 60 % in 2007. Low-skilled, LI exports tend to have high shares of domestic content, whereas high-skilled, technology-intensive exports (e.g., computers, electronic devices, telecommunication equipment) exhibit notably low shares of domestic content. The exports of foreign-invested companies (wholly owned foreign companies and Sino-foreign joint ventures) usually contain a relatively low share of domestic content. This result is in line with the formerly expressed presumption that China's innovative competitiveness in trade is foreign-owned.

8.2.2 Characteristics of China's NIS

Before 1999, China's national system of innovation (NSI) was characterized by the leading role of the government where public laboratories were centers of research and universities focused on teaching with a very limited role in R&D activities (Sun and Liu 2010). In 1999, in order to prepare the country for the challenges of the economic globalization, the knowledge-based era and its WTO entry, the Chinese leadership announced a goal of creating China's own system of scientific and technological innovation (Sun and Liu 2010). In fact, China's NSI remains centrally planned with a very limited role of the industry in shaping science, technology, and innovation policies. The state sets up country's R&D goals through different programs, provides support through various science and technology (S&T) measures (mainly financial), and promotes interactions among key actors. It determines S&T targets and mobilizes human capital and material resources for their realization (Sun and Liu 2010; Springut et al. 2011). Besides strategic goal setting, China's NSI is characterized by development of chosen key technologies. Since 1999, the priority has been the development of knowledge-intensive industries and promotion of HT industries. Main S&T programs include: National Key Technologies R&D Program, National High Tech R&D Development Program (863 Program), and National Basic Research Program of China (973 Program). National Key Technologies R&D Program was launched in 1982 and aimed at providing technical support to industrial restructuring, introducing technical innovation, and applying high and new technologies. It focused on agriculture, IT, environmental protection, medicine, and public welfare technologies (Swissnex 2009). The 863 Program was established in 1986 and promotes applied research and acceleration of HT development. It focuses on key projects that include: "IT, biotechnology and advanced agricultural technology, advanced materials technology, advanced manufacturing and automation technology, energy technology as well as resource and environment technology."[13] The 973 Program started in 1993 and supports basic research, especially multidisciplinary, in areas of agriculture, energy, information, resources and environment, population, and health and materials (Swissnex 2009). Most of the projects within the program have international character.

The *National Medium- to Long-Term Plan for the Development of Science and Technology (2006–2020)* (MLP) defines NSI as "a social system which is led by the state, enabling the market to play the basic role of resources allocation, where the main actors of the innovation system are closely connected and can effectively interact with each other." (Sun and Liu 2010). China's policymakers aim at creating an enterprise-centered NSI, with the state investing in S&T infrastructure such as laboratories, R&D centers, other organizations, or buildings. According to MLP, China's NIS comprises "four main sub-systems: the technological innovation

[13] Swissnex, A Quick Overview of the Science and Technology System in China, http://www.swissnexchina.org/background/1090821-a-quick-overview-of-the-st-system-xiaoyin-2-.pdf.

system; the knowledge innovation system; the national defense innovation system; and the regional innovation system." (Sun and Liu 2010).

Since 2000, different reforms have taken place, such as transformation of public research institutes (PRIs) into, for example, nonprofit organizations with a larger government budget; and "delinking" of university research institutes (URIs) from their affiliated enterprises (Zhong and Yang 2007). Many policy measures were adopted to promote technology innovation such as "Regulations on the Establishment of Venture Investment Mechanism" and "Measures to Stimulate the Development of Software Industry and Integrated Circuit Industry." Science Parks have been given more attention as a key component of the NIS. The 10th Five-Year Program (2001–2005) addressed the goal of promoting commercialization of S&T achievements; development of university science parks; and enhancing innovation capacities, especially in integrated circuit and software technology (Zhong and Yang 2007).

Since China's leaders became aware that government-sponsored technology development programs were not efficient enough in technological catching up, they decided to open more to FDI. Through FDI, not only technology was accessed (MHT and HT sectors), but through foreign MNCs China was integrated into global production networks of HT goods (Naughton 2007). China's strategy for obtaining technology was the policy of allowing foreign MNCs privileged access to China's market in exchange for technology sharing. However, very few projects brought anticipated results and although Chinese government still tries to use its bargaining power when it feels it has it, such as in case of nuclear power facilities suppliers, now Chinese policy makers prefer free competition among many players (Naughton 2007).

China's continually growing R&D expenditures reflect the country's ambitions to move up the technological ladder. China's spending on R&D in 2011 grew by 21.9 % year-on-year to US$139.7 billion and constitutes 1.83 % of the country's GDP (0.9 % in 2000; 1.3 % in 2005), which is close to the level of spending on R&D by developed countries (Naughton 2007; China Daily 2012; OECD 2010a). It was predicted that China would surpass Japan in 2011 and become the largest source of R&D spending in the world after the USA (Springut et al. 2011). In 2011 alone, China spent 39.6 billion yuan (approximately US$5.88 billion) on basic scientific research projects, and the country constructed 130 national engineering research centers and 119 national engineering laboratories (China National Bureau of Statistics, quoted in China Daily 2012).

From the beginning of reforms in 1978 until the end of 1990s, the aim of China's education system was to expand basic education to the whole population. Since the beginning of the twenty-first century, the aim has been to expand higher education (OECD 2010a).

Education along with economic development is China's top priority as China needs well-trained, qualified engineers with managerial skills for the success of its S&T goals. In spite of impressive numbers that Chinese education has been generating (new university intakes growth from 2.7 million in 2001 to 6.1 million in 2008; graduates growth from 1 million in 2001 to 5.1 million in 2008; total enrollment growth from 7.2 million in 2001 to 20.2 million in 2008; science and engineering degrees constituting almost 40 % of all degrees in 2005; OECD 2010a; Zhao and Zhu 2010), it faces many problems. Among them are: passive learning; teaching, preparing for test-taking, and not fostering creative and entrepreneurial skills; producing graduates with skills inadequate for market needs; inequalities in terms of access and quality of education (due to introduction of tuition fees and partial privatization of education; *hukou* system that hinders allocation of the best students to the best universities); corruption (nepotism, bribery, exchange of favors); and plagiarism that negatively impacts quality of education and research in China (Schwaag Serger and Breidne 2007).

The educational reform has been an ongoing project that aims at strengthening primary education, developing education of higher learning and vocational training, pushing forward quality of education and developing a life-long learning education system; encouraging universities to spin off technology-based enterprises (Hung 2009). The weaknesses of education are reflected in the main programs of the *Ministry of Education* (*MOE*) such as Project 211 and Project 985. The former was implemented in 1993 and its main goal was the improvement of teaching quality, level of scientific research and administrative efficiency of HEI. Project 985 was introduced in 1998 and it provides funding for key universities to become first-class universities in the world (Swissnex 2009). Improvement of the quality of education and especially that of higher education is also the main task of the recent *National Medium- to Long-Term Reform and Development Plan for Education (2010–2020)* (Shi, 2011).

It should be noted that an important source of high-quality human capital for Chinese universities and companies are Chinese students obtaining their tertiary education overseas, especially in the USA, Japan, and the EU. Chinese college students constituted the largest group among foreign students studying in the USA in 2010–2011 academic year (22 %, an increase of 23 % compared with 2009) (China Daily 2011). The number of students returning is rather low. Therefore, since 1999 China's government encouraged students to return voluntarily and the latest global financial crisis caused many to look for jobs back home as China's economy maintained stable growth. However, students who do not return connect domestic scientists and engineers to international networks of research and innovation (Naughton 2007).

The higher education sector is the key performer of R&D in China. Since 2000, more than 50 % of S&T and R&D funding has come from the government. The government goal is to transform key Chinese universities with relatively strong research capacity in a few key subjects, into world-class research institutions. Therefore, R&D activities are concentrated in a few large universities and focus on a few key disciplines in natural sciences and engineering. In 2005, R&D expenditure by the top 50 universities accounted for 66 % of total R&D expenditure in natural sciences and engineering in the higher education sector (Schaaper 2009). Higher education sector focuses on basic and applied research and is an important contributor to S&T outputs in the form of scientific publications and patent applications (Schaaper 2009). The higher education sector plays an important role in S&T

diffusion and commercialization of R&D results, which takes place through direct participation in the technology market and cooperation with the business sector (by 2005, 50 national university S&T parks had been established, containing 6,075 start-up firms, hosting 110,200 entrepreneurs) (Schaaper, 2009).

Public education spending as a share of GDP has been fluctuating (from 2.3 % in 1990 to 3.2 % in 2003 and 1.8 % in 2007) (Dahlman et al., 2007). The plan to increase the spending to 4 % of GDP (target set in 1993) was unsuccessful and postponed to 2012 (included in the MLP) (China Daily, 2010).

Some researchers are concerned whether universities are able to manage two significant tasks of improving teaching quality comparable with international standards and developing internationally competitive capacities in basic research (Kroll and Schiller 2010).

8.2.2.1 Main Actors of the China's NIS

Government Agencies The State Council Steering Committee of S&T and Education, NRDC, and Ministries of Science and Technology and Education are the most important institutions creating China's NSI policy.

The State Council "is the highest executive organ of State power, as well as the highest organ of State administration."[14] It issues major strategies and guidelines related with S&T policy. The *State Council's Steering Committee on S&T and Education* coordinates national S&T policy. It comprises the leaders of the major science agencies.

The *National Development and Reform Commission (NDRC)* is an economic planning body under the State Council. It ensures compatibility of S&T planning with macroeconomic policy. It is in charge of strengthening innovation in SMEs, devising a plan for special projects promoting independent innovation capabilities, including such tasks as developing guidelines for increasing the recognition of Chinese brands, and developing guidelines for building national engineering laboratories (McGregor 2010). The *Ministry of Science and Technology (MOST)* needs to cooperate closely with NDRC when promoting technological R&D and innovation, and developing measures to encourage S&T investment and support innovative SMEs (OECD 2007).

The MOST plays a leading role in developing national science policy and in working out and implementing many of the national funding programs. It allocates funding for setting up and strengthening incubators, research laboratories, mega-projects, science parks, and HT development zones (McGregor 2010). Its tasks include: formulating strategies, priority areas, policies, laws, and regulations for S&T; promoting the building of the NIS; conducting research on major S&T issues related to economic and social development; guiding reforms of the S&T system; formulating policies to strengthen basic research, HT development, and industrialization;

---

[14] The Central People's Government of the PRC, http://english.gov.cn/2008-03/16/content_921792.htm.

designing and implementing programs to fund basic and applied research, to induce firms to innovate, and to create science parks and incubators; developing measures to increase S&T investments; allocating human resources in S&T and encouraging S&T talents; and promoting international S&T cooperation and exchanges (OECD 2007), providing S&T-related information, data, and statistics.

The leading institutions in the complex government structure of the Chinese S&T system are the State Council Steering Committee of S&T and Education, the MOST, the MOE, the *Ministry of Finance* (MOF) (Springut et al. 2007).

The MOF was assigned to design fiscal incentives aimed at increasing R&D and innovation in enterprises, and drafting policies for driving innovation through government procurement incentives (McGregor 2010).

The Ministry of Education (MOE) is in charge of promoting basic research and developing human resources in S&T policies together with MOST, the Chinese Academy of Engineering (CAE), the Chinese Academy of Sciences (CAS), and the Natural Science Foundation of China (NSFC). It also provides support for R&D in universities and science parks (OECD 2007).

"*The State Commission of Science, Technology and Industry for National Defense* is responsible for research policy in defense-related research."[15]

Intermediary Institutions The NSFC mainly supports basic research in S&T, identifies and fosters talented researchers, and manages the National Natural Science Fund. NSFC cooperates with the MOST to formulate the principles, policies, and plans for the development of basic research in China. NSFC supports other Chinese foundations in natural sciences (NSFC 2012). The main problem of NSFC is its limited number of staff and an increasing size of funds it manages (OECD 2007).

The success of Silicon Valley is a driving force behind establishment of science parks in China. They were to "attract high tech manufacturing firms for the purpose of jump-starting economic development" (Walcott, Xiao quoted in Macdonald, Deng 2004). By 1998, there were 53 national science parks, and by 2007, 548 innovation centers. Four of the former ones are located in municipalities controlled by the central government: Shanghai, Beijing, Tianjin, and Chongqing. The rest are located along the Eastern coast, which reflects distribution of industrial resources, technological capabilities, and skilled human resources (Zhang and Sonobe 2011). The Zhongguancun Science Park is the largest and oldest science park in China, established in 1988, located near Beijing.

China's government wanted to upgrade its export structure and thus it incorporated science parks into its strategy. As a result, China's parks produced a disproportionate number of HT goods for exports, becoming a production base rather than a center of innovation. Moreover, in order to achieve that, they had to rely on FDI rather than on its own technological, innovative capabilities (Yu et al. 2009). So far, China's science parks and business incubators were not successful in terms

[15] European Commission, Private Sector Interaction in the Decision Making Processes of Public Research Policies Country Profile: China, http://ec.europa.eu/invest-in-research/pdf/download_en/psi_countryprofile_china.pdf.

of reforming the NIS (constituting a platform for innovation and synergies among local entities) (Yu et al. 2009).

The *Chinese Association for Science and Technology (CAST)*. CAST is the largest national nongovernmental organization of scientific and technological workers in China.[16] CAST transfers research results of public research institutions to the industry sector. It cooperates closely with CAS and the industry sector; provides public funding for technological research; and promotes industrial research, applications in technological development, and risky R&D undertakings. Its mission also includes improving science literacy of the general public, encouraging academic exchange, science popularization, and providing consultancy on technology-related issues. CAST develops international cooperation with similar organizations, and since 2004 it has had a consultative status with the Economic and Social Council of the United Nations (CAST 2012).

In addition to these intermediary institutions, several field-specific research institutions (e.g., CAS, CAE) deliver background information, policy advice, and recommendations. Educational institutions such as the top Chinese universities ("Chinese Ivy League") emerged to important pools of talents.

8.2.2.2 China's Innovative Capacity

The coaction of the three elements that shape the innovation capacity—innovation infrastructure, cluster-specific conditions, and the quality of linkages (Porter and Stern 2001; Furman et al. 2002)—is hard to quantify and to compare between countries, especially when it comes to assessing the capacity of communication and cooperation between scientific, applied, and commercial research. It is well known that China's enforcement of IPR protection does not meet international standards, which in turn may disturb the incentives for indigenous innovative research. But the main reason why China should be considered a special case when evaluating and comparing its innovative capacity is the dominance of foreign know-how as the basis for developing domestic innovation clusters. Even the offshoring of R&D of the US and European MNCs to China (and India) may not result in growing capacity of the total innovation production chain in China. In fact, this observation may be interpreted as a newly emerging pattern of the international division of labor. It is argued that the greater stock of knowledge and the lower costs for R&D inputs in China (and India) will shift foreign investment from manufacturing to R&D and contribute to gradually building up the national innovation competence and capacity in China. If the US and European MNCs specialize in their home countries in entrepreneurship resulting in commercializing innovations, then this pattern is expected to be beneficial for the "old" and the "new" innovation leaders (e.g., Bhide 2008). In any case, "China, whose export prowess is based on manufacturing, appears to be building its innovation system through major investments in R&D spending, attracting foreign firms and advanced education. While it has a major initiative to spur

[16] www.cast.org.cn.

indigenous innovation, it is still too early to tell how fruitful these efforts will be" (Hira 2008, p. 16). Recent studies pick up this observation and examine, through a model-based empirical analysis, the effects of international technology diffusion by social networks on the national innovation capacity (e.g., Huang et al. 2010). As already elaborated in part 3 regarding China's competitiveness in trade, the label "national" often is pretending to the perception that China is a homogenous country where mechanisms or political measures lead to successful equalization of any differences. In contrast, China is one of the best examples where regions always matter. China's regions differ in their capacity to innovate. The aforementioned Furman–Porter–Stern concept "cannot be adjusted to China directly because of the uniqueness of China's context. Therefore, studies on RIC (*Regional Innovative Capacity)* are needed" (Zhou et al. 2010, p. 18). Although this study emphasizes the impact of the coexistence of different actors in the multilevel institutional governance system as reasons for the regional perspective on China's innovative capacity, the question arises whether China's regional innovative capacity is correlated with the cluster-based stock of IFDI. At present, the results of empirical studies are controversial, reaching from statistically insignificant correlation between IFDI and regional innovative capacity (Chen 2007) to significant contribution of IFDI to overall regional innovative capacity (Fu 2008). The Guangdong Province is rated highest in regional knowledge innovative capability among China's provinces resulting to a high degree "from national and regional innovation inputs, the structure of local intellectuals, and the situation of foreign finance utilization" (Ma 2010, p. 29). A case study on the Wuxi semiconductor industry cluster concludes that, besides the well-known innovation drivers, the transnational Chinese technical community is more important than IFDI (Chou et al. 2011). Another study on the effect of global-local networks via foreign-invested enterprises (FIEs) in the information and communication industry in three megacity regions confirms the ambiguous results so far that foreign-invested R&D does not generally strengthen China's local innovative capacity (Wei et al. 2012). Especially, in Chinese SOEs in HT industries, the import of foreign technology does not effectively promote innovation unless it is coupled with investment in R&D. "By contrast, the assimilation and utilization of domestic technology is less demanding for absorptive capacity" (Li 2011, p. 1247). The fast-growing S&T intermediaries are an important supporting component for Chinese innovating enterprises. But at the present stage, this industry does not fully match the demand of Chinese enterprises (Zhao and Zheng 2011, p. 44). From the perspective of innovation commercialization, enterprises are indeed the final beneficiaries of upgrading China's innovative capacity, but the public sector is the pivotal point in moving China toward an important creator of innovation. "The outstanding puzzling feature of China's innovative capacity to date is the less than stellar performance of its public sector in providing a means to steer the overall technological development of the country through its catch-up stage" (Hu and Mathews 2008, p. 1478). Shortcomings that moderate the building of China's innovative capacity are resulting from the "old" and partly obsolete patent stock, inefficiency of public R&D expenditures, shortage of S&T manpower, insufficient IP protection, lack of venture capital, and extremely high specialization in electronics area (Hu and Mathews 2008, p. 1472).

China's high decade-long economic growth may lead to the gradual erosion of comparative advantage in LI production. If China follows the development path of many advanced countries, its future position in the globalized world economy will depend on its capability to innovate. This requires a new direction of thinking in politics, business, and science. Instead of exploiting given comparative advantages, all actors need to focus on creating man-made competitiveness. There is common sense that the key to success lies in an early transformation from a country's late-comer to a forerunner position with respect to sustainable growth that ensures innovation. There is sufficient evidence that China is on track in this direction. A number of widely accepted indicators (Malerba et al. 2011) provide information on China's path toward innovative competitiveness. Compared to other emerging economies, China can rely on a considerable and still growing stock of advanced knowledge incorporated in FDI inflows. The model of clustering public–private–academic partnership that proved successful in many countries seemed to be imitated by China. It goes without saying that a large and geographically, as well as socially, fragmented country is barely able to offer a nationwide favorable environment for innovation. Given the ongoing transformation, the sectoral concentration of FDI in ICT-industries and their regional/local concentration, it is to be expected that even the analysis of China's path toward innovation since 1995 does not allow the assessment of its full range of innovative capabilities. Even if China-specific factors are disregarded, the measurement of national innovative capacity has to cope with the causality between inputs that are supposed to result in innovation and comprehensively measured outputs. Furthermore, empirical research faces difficulties with differentiating the drivers of innovation according to their importance, strength, and sustainability.

As a result of the growing pressure on advanced economies to maintain sustainable growth in an increasingly competitive global environment, empirical research on drivers of productivity growth intensified at the beginning of this century, resulting in a number of proposals to measure the national innovation capacity by complex indexes. In 2001, Porter and Stern (2001, p. 4) calculated China's innovation capacity index at 18.1, which ranked China 43rd out of 75 investigated countries.[17] It is striking that in four subindexes (proportion of scientists and engineers, innovation policy, cluster innovation environment, and linkages) China achieved almost the same rank, which is quite unusual for advanced and emerging economies. Based on a regression line of 75 countries, which shows the relation between innovative capacity and GDP per capita, China belonged to a group of countries where the innovative capacity grew ahead of the sophistication of the overall economy, promising a sound growth path in the future (Porter and Stern 2001, p. 15).

The trend of the main indicators seems to confirm this optimistic estimation. The gross domestic expenditure on R&D (GERD) increased relatively steadily from 0.57 % of GDP to 1.70 % of GDP in 1995–2009. On a lower level, the trend of the business enterprise expenditures on R&D (BERD) showed an almost parallel development reaching 1.25 % of GDP in 2009. The government intramural expenditure

[17] See Furman et al. 2002, for the concept of national innovative capacity.

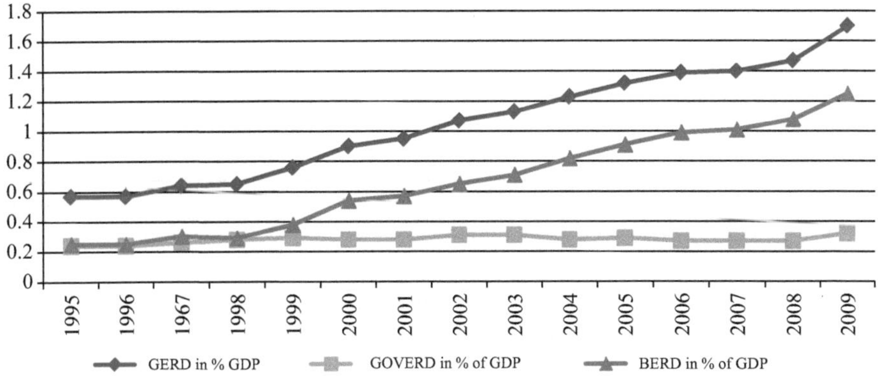

Fig. 8.18 GERD, GOVERD, and BERD as percentage of GDP, 1995–2009. (Source: Own calculation based on OECD data)

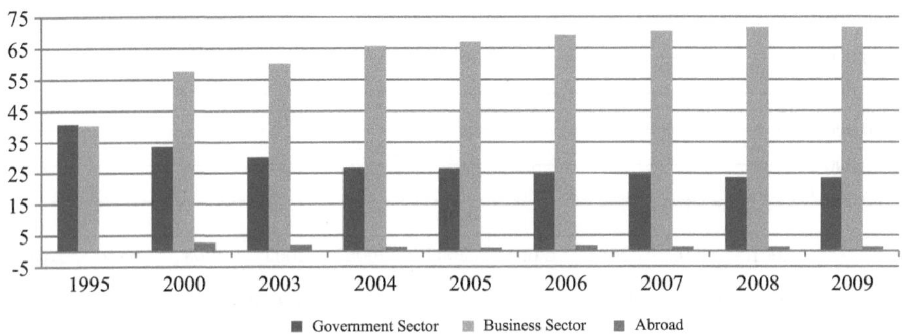

Fig. 8.19 Percentage of GERD financed by government, business, foreign sector, selected years. (Source: Own calculation based on OECD data)

on R&D (GOVERD) remained on a low level, fluctuating slightly between 0.24% of GDP in 1995 and 0.32% in 2009 (Fig. 8.18).

The structural development of the GERD differs considerably over time as well as by source. The percentage of GERD financed by government decreased steadily from over 30% at the beginning of the last decade to less than 25% in 2009, whereas the business sector increased its share from around 60 to 72%; the share of foreign-financed R&D on GERD decreased from 2.7% in 2000 to 1.35% in 2009 (Fig. 8.19).

The gap between China's national innovation capability and the current innovative position is illustrated by comparing China's rank in R&D generators, R&D users, and R&D importers (Dutta 2011, p. 98): In the first category, China does not appear under the top 10 countries, whereas it is ranked fifth in the second and first in the third category. More important than a yearly snapshot, is tracing the past trend and extracting indicators for China's future innovative position. The number of resident patents filings measured both by R&D expenditure and by population grew disproportionately since the middle of the 1990s (Fig. 8.20).

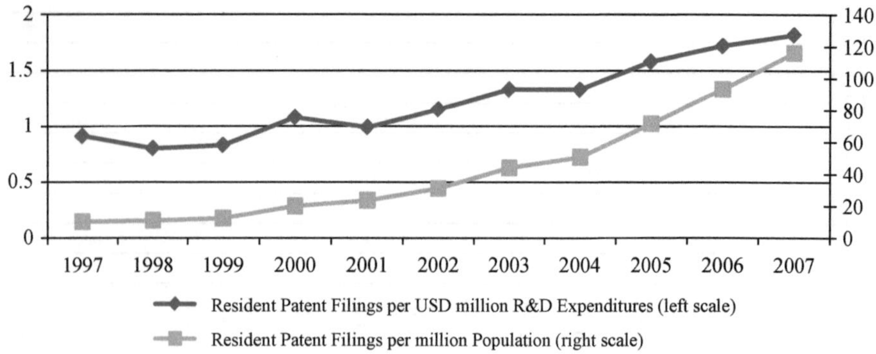

Fig. 8.20 Number of resident patent filings per US$ million R&D expenditures and per million population, 1997–2007. (Source: Own calculation based on WIPO)

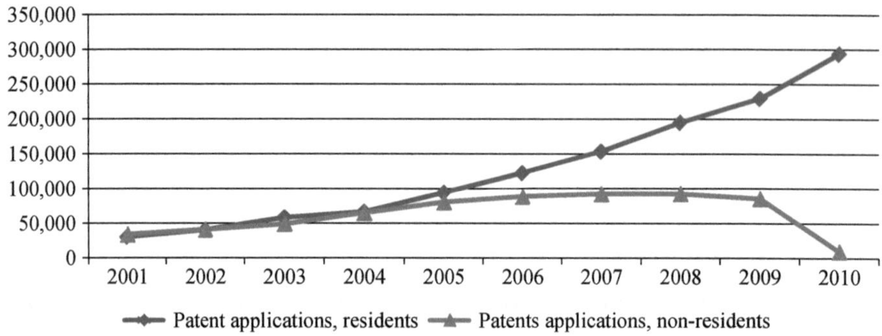

Fig. 8.21 Number of patent applications, residents and nonresidents, 2001–2010. (Source: Own calculation based on The World Bank data)

China's annual average percentage change in relative and absolute GERD is approximately 10 times higher than that of the USA and EU27 (Shelton and Foland 2011, p. 3). The comparison of China's innovative output with the USA and Japan—usually measured by the number of published patent applications—shows that the gap was closed in 2011, allowing the prediction that China will lead the global patent information landscape from 2012 onward (Thomson Reuters 2011, p. 5). The increase is due to the growing share of domestic patent applications in total patent applications from 51.9 % in 2006 to 72.7 % in 2010, which is at least partly due to the absolute decline of nonresident patents (Fig. 8.21). The top five industrial sectors are electrical machinery/apparatus/energy, measurement instruments, digital communication, computer technology, and pharmaceuticals. The latter is China's leading sector measured by its share of almost 40 % in global patent applications.

At first glance, these trends seem to confirm the positive effects of the input in R&D on the innovative output. But many studies express doubts whether the impressive increase in input factors, such as R&D, results in significant improvements of China's innovative position. In pharmaceuticals, a large proportion of the patent applications cover traditional Chinese medicine (Fig. 8.22).

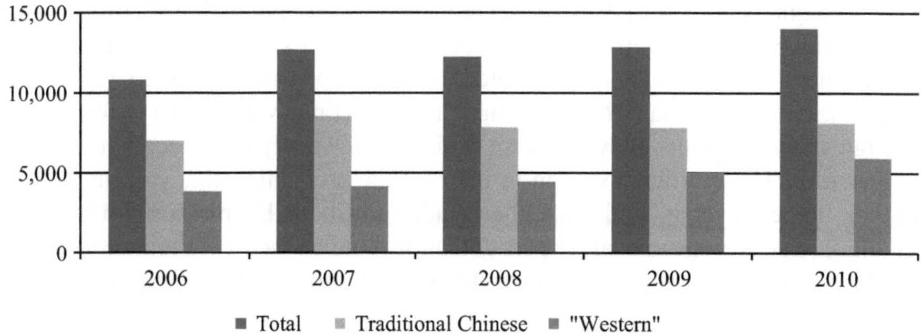

Fig. 8.22 Number of domestic Chinese patent applications in traditional and "western" pharmaceuticals, 2006–2010. (Source: Own calculation based on Thomson Reuters 2011, p. 15)

Studies analyzing the innovative value of Chinese patents point to the relatively low share of invention patents. In 2011, around 32 % of 1.63 million patent application cases accounted for invention patents, whereas two-third belonged to the "new design" and "innovative utility model" categories (Intellectual Property Protection in China 2012). The extremely low number of internationally recognized patents (approximately 3 % of the number of patents filed by the USA or EU or Japan) could be considered as an indicator for the low value of Chinese patents (Gupta and Wang 2009, p. 112). The "innovation-by-the-number-mentality" of the government may stimulate the low-value patents and thus increase considerably the total number of patent filings (New York Times 2011).

8.2.2.3 Human Capital in China

The commitment of the Chinese government to continued high growth—stated especially in the 10th and 11th 5-Year Plans—was put into action, among others, by several investment programs in the education sector. Unlike in other countries at a similar level of development, the government's education policy gave top priority to the transformation of higher education. In terms of quantity, the number of full-time teachers in regular institutions in higher education as well as the number of graduates, postgraduates, and students studying abroad increased considerably. At the same time, qualitative improvements in teaching and research had been encouraged by additional funding of national key research projects and institutions. This enabled the traditional "national champions" in higher education, such as Peking University and Tsinghua University, to climb up in the international rankings. In Times Higher Education World University Rankings 2011–2012, the former is ranked 49th and the latter, 71st (Thomson Reuter 2012). The national landscape of universities changed drastically through consolidation efforts, especially by merging small industry-specific universities into comprehensive universities in order to receive, in addition to municipal/provincial funding, also central government funding. Many universities cooperate closely with foreign universities by offering their study programs, partly regularly taught by foreign scholars (e.g., Tsinghua–

MIT strategic partnership). Besides several hundred foreign-affiliated colleges and universities with cooperative teaching programs, the experiment of establishing independent foreign universities resulted in the opening of the University of Nottingham, Ningbo, China (UNNC) campus in 2004 (Ennew and Fujia 2009). Sino-foreign study programs allow Chinese students a relatively easy access to foreign universities. These programs seem to increase the probability that students return to their home university and finish their studies in China. Despite a sharp increase in tuition, families spend a larger part of their income on their children's education.

With increasing income per capita, the Chinese households' education expenditures increased disproportionately and exceed many families' ability to pay for higher education, especially in rural areas (Dong and Wan 2012, p. 5). This may contribute to the still evident educational backwardness in the West. Despite this barrier, the enrolment in tertiary education institutions increased in the last decade from 10 % to over 25 %.

The trend in the number of graduates in natural sciences and engineering is considered an important indicator for a country's innovative competitiveness. Whereas in China the number of first university degrees in natural sciences and engineering approximately quadrupled between 1999 and 2008, the increase in technologically leading countries was modest, in Japan even negative.

Even the high growth of China's economy—mainly caused by the expansion of the manufacturing sector and the technological upgrading in core industries—that is generating around 2.5 million new white-collar jobs annually, is not enough to absorb the fast increasing number of graduates. Despite the fact that in 2010 the labor demand returned to the precrisis level, an increasing number of graduates form a new, young, and overqualified generation of migrant workers (China Labor Bulletin 2012, p. 18).[18] The employment rate of skilled migrant workers without degree increased in 2011 to 97 %. However, the employment rate of college graduates was just less than 90 % indicating that the unemployment rate of around 10 % is much higher than the average urban jobless rate of 4.1 % (Social Sciences Academic Press 2011). This is another example of the already highlighted regional and sectoral mismatch in China's economy. Whereas the demand for blue-collar workers in manufacturing and construction industries is still high, there exists an oversupply in urban high-qualified labor and at the same time a shortage of the latter in western part of China. At an estimated average annual growth of 3 % in 2012–2015, it is to be expected that in 2012 around 6.8 million college and university graduates will search for jobs. As a consequence, the MOE is currently intervening into the enrolment procedures in universities, aiming to restrict the number of students in fields with low employment perspectives.

At first view, the R&D output measured by Chinese patent applications (Fig. 8.21) suggests China's catching-up in innovation. The growing number of

[18] The "old" type of migrant workers moved from rural to urban areas without being registered in their urban place of residence. They lack the "*hukou*" status. Migrant workers contributed considerably to the structural change between rural and urban population. According to the Census 2010, approximately 100 million workers migrated in the last decade from rural to urban areas (Cox 2011).

researchers in R&D is in line with this development. The low number of researchers in basic research may explain why only around 16 % of patents applications are granted for inventions (China Statistical Yearbook 2011). Another indicator may be the stagnating number of national invention prizes as well as the minor increase in publications on S&T.

As already mentioned, academic institutions are the major source of technological innovation (Hu and Mathews 2008, p. 1468). Improving the domestic innovative capacity needs first of all a solid basic research. In the last decade, China's expenditures on basic research represented an annual average of approximately 5 % of total R&D expenditures, which is far lower than in industrialized countries (OECD and Main Science&Technology Indicators 2012). Around 90 % of these R&D expenditures are allocated to government research and academic institutions. The bottleneck is rather the lack of highly qualified scientific and technical personnel than the total number of researchers or the amount of funds for R&D. Furthermore, the utilization of the results of basic research for applied research often lacks an efficient transfer process (Zhang and Zheng 2011, p. 131). But there is evidence that universities contribute considerably to innovation in the private sector, being the source of spin-offs (Hu and Mathews 2008, p. 1476).

By using a simulation model, a recently published study estimated the contribution of education to human capital building and finally economic growth by comparing 15 countries, among others, China (Sun et al. 2012). Compared to the USA, Germany, France, UK, and Australia, China's economic contribution of education is lower, but higher than in countries such as Canada, New Zealand, and South Korea. China's position in the middle level is mainly attributed to the government's policy to give education high priority in its development strategy. The gap between countries with higher economic contribution of education results in the priority to meet the demand for low-skilled and medium-low-skilled workers in manufacturing industries. Surprisingly, the study also identified the relative lack of education funding as a major bottleneck. It is unquestioned that human capital contributed significantly to China's economic growth, according to Whalley's and Zhao's (2010) estimation of 38.1 % for the period of 1978–2008. The negative contribution of TFP to growth between 1999 and 2008 (−7.03 %) "Negative TFP growth, along with a high contribution of physical and human capital to economic growth, seems to suggest that there has been a decrease in the efficiency of input usage in China and misallocation of physical and human capital" (Whalley and Zhao 2010, p. 30). In order to better exploit the large potential of human capital, substantial inequalities in access to higher education have to be addressed and impediments to labor mobility have to be reduced (Heckman and Yi 2012).

8.2.2.4 Characteristics of scientific, innovative, and educational policy in the years 2005–2010

China's current policy for science, technology, and innovation is formulated in the MLP. The plan calls for China to "enter the ranks of innovative countries by 2020" and a world leader in S&T by 2050 (developing "indigenous innovation" (*zizhu chuang and xin*); capable of challenging even the most advanced nations for technological preeminence (Springut et al. 2011). China's Prime Minister, Wen Jiabao, summarized the plan's ultimate goals: to develop technologies related to energy and water resources and environmental protection; to master core technologies in information technology (IT) and production technology; to catch up with the most advanced nations in selected areas within biotechnology; to raise the pace of development in space and aviation technology as well as oceanology; and to strengthen both basic and strategic research (Schwaag Serger and Breidne 2007).

The plan sets a goal of reducing reliance on foreign technology to only 30 % by 2020 (OECD 2007). At the same time, it provides for strengthening basic research, which is underdeveloped as it receives only a 5 % share of national expenditure on R&D (Kroll and Schiller 2010). Another target is raising the ratio of R&D to GDP to 2 % by 2010 and to 2.5 % or more by 2020.

It encourages Chinese enterprises to establish overseas R&D centers. It calls for "'establishing the nation's credibility and image in international cooperation' and 'to perfect the nation's intellectual property rights system.' It also sets goals for expanded cooperation with foreign universities, research centers and corporate R&D centers" (McGregor 2010).

The plan received some criticism. On the one hand, the plan is called unrealistic due to bureaucratic control and corruption as well as an inadequate education system based on rote learning. On the other hand, it raises fears of technonationalism and technology theft as it "defines indigenous innovation as 'enhancing original innovation through coinnovation and reinnovation based on the assimilation of imported technologies'" (McGregor 2010).

The State Intellectual Property Office of China has issued the *National Patent Development Strategy (2011–2020)*. Its overall goal is to enhance China's capacity to create, utilize, protect, and administer patents and thus encourage and protect innovation, support economic transformation, and promote social and economic development. It recognizes that patented technology is strategic resource of the country's core competitiveness. The Strategy aims at improving legal environment, efficiency and capacity of the patent system so that it ranks among the advanced countries, providing adequate number of patent professionals working in the administration and private sector, increase public awareness of importance of patents through education and media. The goals for 2015 include: two million of patent filings (invention, utility models, and design patents); being among the world's top two countries in terms of annual invention patents granted to domestic applicants; improvement of quality of patents; doubling the number of patents per one million people and number of patent filings by Chinese residents in other countries; based on national strategic demands obtain patent rights of a number of core technologies that support development of emerging industries in China and its industrial restructuring; enhance benefits of utilizing patents (establishing patent transaction services in major cities); increase significantly the share of patent-intensive products and the share of exported products with patent rights; foster a number of dominant enterprises that own core patents and are internationally competitive; establish 10 model cities that comprehensively utilize the patent system and have excellent intellectual property market environment; double the number of patent examiners to approximately 9,000 (in the USA, there are currently 6,300 examiners); and certified patent agents should reach 10,000. According to the US Patent and Trademark Office, the recent filings by Chinese entities reflect priorities of China's industrial strategy such as solar and wind energy, IT and telecommunications, and battery and manufacturing technologies for cars (Lohr 2011; New York Times 2011).

In 1995, reforms and restructuring of China's education system were introduced under four strategies: "joint development," "restructuring," "merger," and "cooperation." These reforms have continued until today. In 2004, the central government, in order to help inland regions to develop their higher education, picked one university in each of 12 provinces for "joint development," which meant increased funding (Zhao and Zhu 2010). The 2003–2007 "Action Plan for Invigorating Education" is a blueprint for implementing strategies of "Rejuvenating China through Science and Education." The main goals are to extend 9 years compulsory education to rural areas and eradicate illiteracy in the backward western regions; improve quality of compulsory education in Eastern areas. The Plan extends 211,985 and "High-Level Innovative Talents" projects that aim at developing China's leading universities into world-class ones, improving research in selected fields to compete internationally and overall contribute to enhancing China's innovation system (Schwaag Serger and Breidne 2007; Zhou 2004). Moreover, within the plan there are number of projects such as "Plan for the Innovation in Postgraduate Education," the "Project for Scientific and Technological Innovation in Higher Education Institutions," and the "Project for Social Science Prosperity in Higher Education Institutions." Their goal is an overall coordination of development of various disciplines, talent training, S&T innovation, the establishment of human resources in teaching and research, and international cooperation and exchanges (Zhou 2004).

In terms of international cooperation, the priorities include: developing Confucius institutes, increasing the number of international students coming to China and the number of degree programs taught in English in China, and encouraging Chinese universities to collaborate with foreign universities and to establish branches abroad (Embassy of Switzerland 2010).

China's National Plan for Medium and Long-Term Education Reform and Development (2010–2020) addresses issues not resolved by former plans, such as the quality of teaching, and problems not tackled before, such as preschool education, migrant workers' children's equal access to education, as well as enabling them to pass exams in host cities (Embassy of Switzerland 2010). The plan states that, during compulsory education, admissions must be based on geographic location without the requirement for any certificates or other entry tests. The Nation-wide Teacher Training Program was introduced, aimed at improving equal access to education resources. All teachers in primary and secondary education were to participate in at least one training every 5 years. In 2010, the program obtained US$81.2 million funding. The Plan also emphasizes "education for all," as for decades, the priority was educational development in selected fields (Embassy of Switzerland 2010).

The *National Talent Development Plan (2010–2020)* aims at transforming China from "an investment-driven economy to a talent-driven economy" (Wang 2010). The main goal is to increase the overall talent number from nearly 114 million in 2008 to 180 million in 2020; the number of people in labor force with higher education from 9.2% in 2008 to 20% in 2020; the ratio of human capital investment as a share of GDP shall increase from 10.75% in 2008 to 15% in 2020. The plan will support development of human resources in six categories: political leaders and officials; business entrepreneurs; technical professionals; highly skilled talents; practical talents for rural areas and agriculture; and professional social workers (Wang 2010).

It is predicted that by 2020 around 5 million skilled people will be needed in: equipment manufacturing, IT, biotechnology, new materials, aeronautics and astronautics, oceanography, finance and accounting, international business, environmental protection, energy resources, agricultural technology, and modern traffic and transportation. Also, more than 8 million professionals will be needed in the fields of education, political science and law, medicine and health, publicity and cultural information, and disaster prevention (Wang 2010).

The plan also supports China's soft power strategy and provides support for research in fields such as philosophy, social science, news and publishing, culture and art, heritage protection, and related social and humanities areas. The goal is to foster a group of highly respected and talented people who would have wide influence. It also plans to attract outstanding overseas talents to work in China in the next 5 to 10 years (Wang 2010).

8.2.3 Interdependencies Between the Level of Innovation, Human Capital, and Advantages in Foreign Trade

The summary of strengths and weaknesses of China's NSI as well as threats and opportunities it faces is presented in Table 8.13. Theoretical approaches suggest a two-way dependency between innovation, human capital and trade. Trade can affect human capital building and innovation through learning-by-doing and technology transfer. But innovation and human capital can also create advantages in trade. Both types of dependencies are considered as an important source of economic growth. In the first case, trade is the vehicle that is necessary for creating dynamic growth. In the second case, trade is allowing the increase of impact of human capital and innovation on growth. In most industrialized countries, both directions of dependencies are contributing to growth, especially if trade is only one part of a complex globalization strategies of MNCs. The latter benefit from national R&D programs, transfer innovation first into firm-owned advantages and make innovation via trade-embedded knowledge transfer to an international public good. It should be mentioned that a high degree of openness is positively influencing the two-way dependencies.

The case of China differs considerably, mainly because of its implemented model of development. The opening-up strategy in the 1980s was targeted first and foremost on attracting foreign capital ("invite-in strategy"). The combination of foreign capital with unskilled, low-wage labor developed China's East into the "factory of the world." The export boom—fueled by public investment in physical infrastruc-

Table 8.13 China's NSI SWOT analysis

Strengths	Weaknesses
Big and growing R&D funding	Low quality of human resources for S&T and inadequate related infrastructure (OECD 2008) Weak institutions that inhibit market-led innovation (e.g., corporate governance, enforcement of intellectual property rights; OECD 2008) Interactions and linkages between actors are lacking and thus China's NSI is compared to "archipelago" of "innovative islands" (OECD 2008) Insufficient coordination of the public research system (Kroll and Schiller 2010) No equal treatment of enterprises in terms of support policies (not all enterprises are eligible to apply for support for cooperation with the domestic public research sector; Kroll and Schiller 2010) Lack of managerial know-how (OECD 2007) Lack of incentives encouraging to take the risk of innovating (OECD 2007) Government support system rather crowds out than encourages business investment in risky ventures (OECD 2007) Not supportive financial system (OECD 2007) Inefficient when comparing inputs with outputs Not adequate level of basic research (due to problems with raising funds and acquiring the best scientists) Lack of a mechanism, which allocates best talents to best universities (barrier: *hukou* system)
Opportunities	Threats
Fast expansion of China's resources of research scientists and engineers	Insufficient economic growth
Construction of many research facilities (Springut et al. 2011)	Political instability
Importing, transferring, and investing in foreign technology (high currency reserves, M&A strategies of Chinese MNCs)	Aging population
Encouraging the return of the Chinese educated abroad	Shift of student's preferences from natural sciences and engineering to humanities and social sciences Growing international conflicts in China's violation of IPR Allocating S&T resources to a "prestige project" (e.g., astronautics/cosmonautics)

ture—contributed considerably to China's economic growth in the 1990s. During the first two decades of China's opening-up, human capital and innovation did not play a prominent role as a trade-enhancing factor. Anyway, production facilities of foreign companies contributed to a gradual evolution of a medium-low-skilled stock of human capital, but were mostly restricted to the Special Economic Development Zones. The link between human capital, innovation, and trade intensified at the beginning of the last decade. China's entry into the WTO, further facilitations for foreign companies, and heavy investment in higher education created incentives for foreign MNCs to invest in production facilities in MHT and HT sectors. The direct contribution of the former (and still existing) export of LT, LI products to technology and knowledge transfer might be rather small. But as a "forerunner phase" the indirect contribution might be considerable. China learned how internationalization functions and how to set positive incentives for foreign investment in more sophisticated industries. Foreign MNCs started to put China on their landscape of globalization of production and R&D. The first wave of Chinese graduates and postgraduates who studied natural sciences or engineering abroad and returned to China created a small but highly motivated work force in advanced technology product industries.

China was and still is a country, which benefited the most from the "innovation" of MNCs' fragmentation of production processes and the resulting global production networks. IFDI in MHT and HT industries, followed by increasing imports of parts and components, created the basis for China's new trade pattern. The "old" one-way trade in textiles, leather products, and toys is complemented by two-way trade in industries with higher level of technology and human capital such as electronics. The Chinese value-added is often relatively low because foreign MNCs hesitated with R&D investments in China and therefore did not stimulate demand for high-skilled workers on a relevant macroeconomic level, at best in state-level HT economic development zones. China's export growth in this second phase of opening-up cannot primarily be attributed to indigenous innovation and intensive input of human capital. The new pattern of IFDI in medium- and HT industry segments and the resulting two-way trade delivered an increasing contribution to China's growth except in 2008/2009. In 2005, the contribution of net exports to China's growth accounted for 2.3 % up from 0.6 % in 2004. After a sharp drop during the crisis years 2008/2009, the share of trade in China's growth rate is recovering but did not even reach the 2004 level (IMF 2012). The large stimulus package in 2009 increased drastically the contribution of capital formation to China's growth.

There is already evidence for shifts to a third pattern of trade. The new determinants are foreign MNCs' investment in R&D on the one hand and first results of China's change in its development model from a "capturing value strategy" toward a "value creating strategy" on the other hand. As demonstrated in part 4.1, China's government is paying increasing attention to improving its NSI, aiming to make full use of its potential of human resources. The inputs in R&D and human capital building can compete with technologically leading countries. Progress has been achieved in several output categories such as the fast-growing number of graduates and postgraduates in natural sciences and engineering, the increase in the number of scientific publications, and the sharp rise of overall patents. These achievements could not hide the fact that in terms of innovativeness the output cannot compete with technological leaders, which is best demonstrated in China's low number of patents classified as inventions. But it has to be noted that the time lag between input and output in R&D as well as in high-level human resource building is longer than in LT manufacturing. With the emergence of Chinese MNCs, this time lag is expected to shrink. Domestic as well as external developments will influence the emergence of China as a competitor to the innovation leaders and finally determine whether this way creates a sustainable basis for growth. The adoption of the 12th 5-Year Plan signaled that Chinese leaders consider exports as a growth-supporting factor, which allows the importing of growth-sustaining resources. Turning the focus on indigenous innovation, which should improve domestic living conditions, may open opportunities for a next pattern of trade where domestic innovation determines exports of HT final products with a high domestic value added.

In the modern history of industrialization, China is the first country where long-term benefits from trade did not result from the process of transforming Ricardian comparative advantages into indigenous innovative competitiveness via R&D and human capital upgrading. China obviously needs an additional development phase in between where the growth-creating effects of trade are based on a low-value but high-volume participation in global production networks via IFDI and knowledge transfer.

8.2.4 Conclusions

Altogether, the following conclusions are common sense (e.g., Herrero and Chen 2012, p. 1). First, China is a main manufacturing center of HT products but, generally, only at the low end of the value chain of global HT industries. Second, China's efforts to improve its innovation capacity is documented by fast-increasing inputs, such as R&D expenditures, building up a huge human capital pool, improving the enforcement of the protection of intellectual property rights, as well as by innovation-stimulating government measures. Third, the output of the innovation process looks impressive in terms of quantitative results but far less in terms of "relevance."

Assessing China from the perspective of its transformation from a manufacturing hub ("Made in China") to an innovation hub ("Made by the Chinese") needs a comprehensive analytical concept, which is based on the awareness that both are complementary yet contending (Cheung 2009, p. 5). China's specific conditions in terms of size, human resources, geography, and culture may allow enduring interactions between both concepts.

8.3 Case Study: Mexico

Beata Michorowska

The main objective of this subsection is to establish the relationship between the development of comparative advantages in foreign trade and the characteristic features of the Mexican NIS.

The basic macroeconomic indexes for Mexico's economy indicate that it is in pretty good shape. In the years 1996–2010, Mexico's economy was characterized by a moderate increase in GDP. The only exception was observed in 2009, when this parameter decreased by 6.2. The development of the Mexican economy is to a

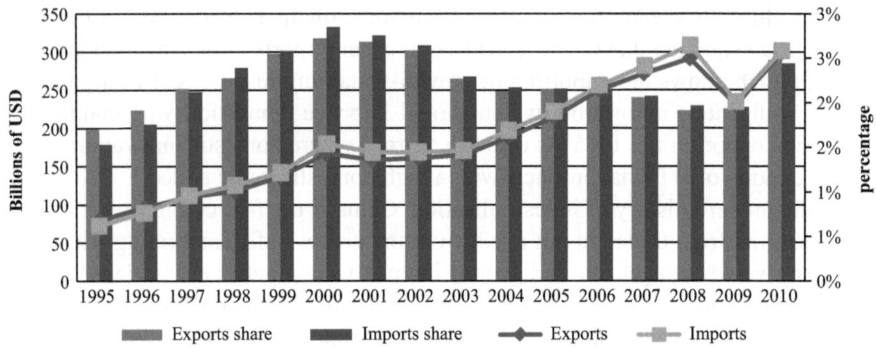

Fig. 8.23 Volume of Mexico's trading (billions of US$) and its share in global exports and imports (in %). (Source: Own elaboration based on the UN Comtrade database)

large extent dependent on the standing of the US economy—the main trade partner of Mexico.

8.3.1 Evaluation of Competitiveness of Mexico in Foreign Trade

Mexico's trade rose almost uninterruptedly in the analyzed period, i.e., from 1995 to 2010. Only in the years 2001–2003, one could observe a slowdown, and in 2009—a sudden drop in Mexico's trading. The 2009 deceleration resulting from the recession, affecting both exports and imports, however, was worked off in 2010 (Fig. 8.23).

Mexico's international trade is dominated by the USA—Mexico's main trade partner. In 1995, nearly 86% of Mexican exports were channeled to two economic partners—NAFTA members: the USA and Canada. This share decreased only slightly in 2010—to nearly 84%, while the share of Canada went up from 2.5 to 3.6%, and the volume of trading with the USA went down from 83.4 to 80.1%. Mexico's remaining export partners are of lesser importance. The most significant countries belonging to this group are: China, Brazil, and Spain (although each and every one of these states purchased less than 1.5% of Mexican export goods in 2010). In comparison with 1995, the significance of the Japanese market fell notably as Mexican exports recipient. The volume of trading with Japan fell by 50%: from 1.2% in 1995 to 0.6% in 2010. Generally, however, not much has changed in the geographical structure of Mexican exports over the analyzed period.

The geographical structure of Mexican imports indicates notable changes between 1995 and 2010. This is mainly the result of the decrease in the significance of the USA from 74.5 to 48.2%) and the rise of China as Mexico's potent import partner (from 0.7 to 15.1%). Moreover, South Korea increased its share in the Mexican imports market (from 1.1 to 4.2%), similarly as Canada (from 1.9 to 2.9%). One can also observe a greater diversification of import directions, as other

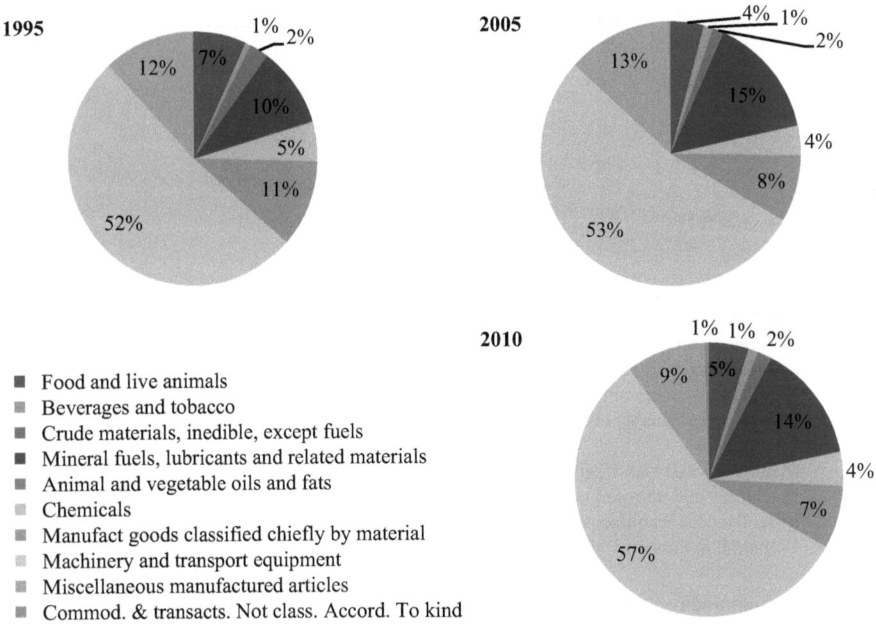

Fig. 8.24 Commodity structure of Mexican exports in selected years of the period from 1995 to 2010 according to SITC. (Source: Own elaboration based on the UN Comtrade database)

partners gained in importance. Although the unit volume of these countries is not high, together this group constitutes more than 20 % of Mexican imports—in 1995 this amounted to 12 %.

Generally, in 2010, the trade structure of Mexico's foreign trade was similar to that of 1995. The most important commodity group of Mexican exports is constituted by machinery and transport equipment. The significance of this group even went up in the analyzed period from 52 % in 1995 to 57 % in 2010. In 1995, raw materials (except for fuels) were the second most important commodity group of Mexican exports—12 %. These were followed by industrial commodities classified by raw material—11 %. Nevertheless, in the following years, mineral fuels started gaining in significance. In 2005, they assumed the volume of 15 % of Mexican exports, which was only slightly lower in 2010—14 %. Other commodity groups are of lesser importance, and their share in Mexico's exports remains at a similar level. This means that in 2010 Mexican exports were composed of: machinery and transport equipment (nearly 60 %) and raw materials and goods classified by material (30 %), which leaves only 10 % to commodities belonging to different groups (Fig. 8.24).

The structure of Mexican imports is similar to the country's exports structure. Although the importance of the group: machinery and transport equipment is slightly lower, it went up from 42 % in 1995 to 47 % in 2010. Manufactured goods classified by material and chemicals and related products (14 and 11 % in 2010, respectively)

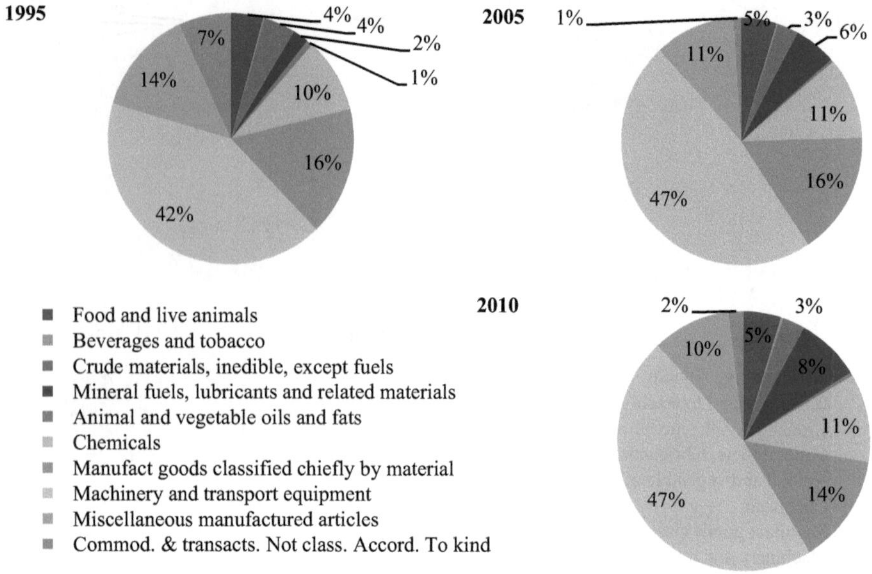

Fig. 8.25 Commodity structure of Mexican imports in selected years of the period from 1995 to 2010 according to SITC. (Source: Own elaboration based on the UN Comtrade database)

are of greater importance in the case of exports. Mineral fuels are less significant here, although this group is witnessing a marked increase in imports: from 2% in 1995 to 8% in 2010). Similar export and import commodity structures indicate high importance of intra-industry trade (IIT) in the pursued international trading (Fig. 8.25).

The commodity structure of Mexican trading according to the OECD classification based on technological level also presents a similar structure in the case of both exports and imports. No significant changes have occurred in their structure in recent years (between 1995 and 2010). An advantage is visible in MT commodities, mainly in MHT goods. To a lesser extent, trading involves HT and LT products. Exports are more technologically advanced than imports. In 2010, LT commodities constituted only 7% of exports (10% in 1995). However, HT products accounted for 17% of exports (Fig. 8.4). HT commodities include the most important group of Mexican export goods: electronics and communication (nearly US$65 billion in 2010). In the dominant group of MHT commodities (43% in 2010), the key role is played by commodities belonging to the group of motor vehicles (almost US$70 billion in 2010), research instruments (nearly US$44 billion), electric devices (more than US$34 billion), and nonelectric devices (more than US$47 billion in 2010) (Fig. 8.26).

The commodity structure of Mexican imports indicates higher significance of LT and MLT goods (Fig. 8.5). In the case of MLT, the group embracing refined petroleum products (US$60 billion in 2010) is of key importance, while these products themselves are exported at a notably smaller scale (more than US$17 billion in 2010). In the HT commodity group, more electronic and communication goods are

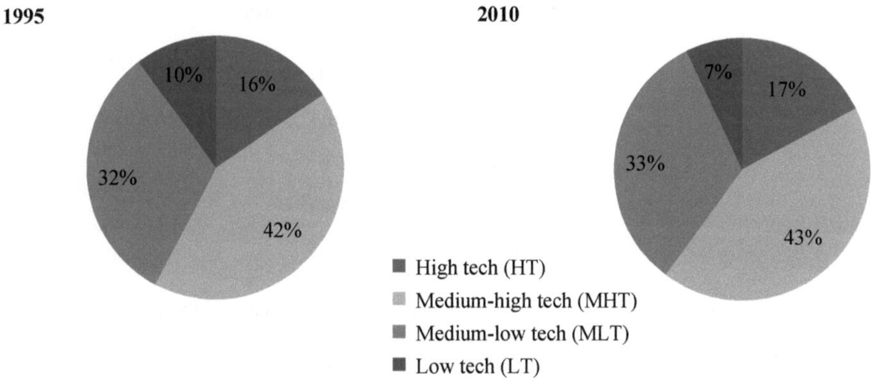

Fig. 8.26 Commodity structure of Mexican exports in selected years of the period from 1995 to 2010 according to the OECD industry classification based on technological level. (Source: Own elaboration based on the UN Comtrade database)

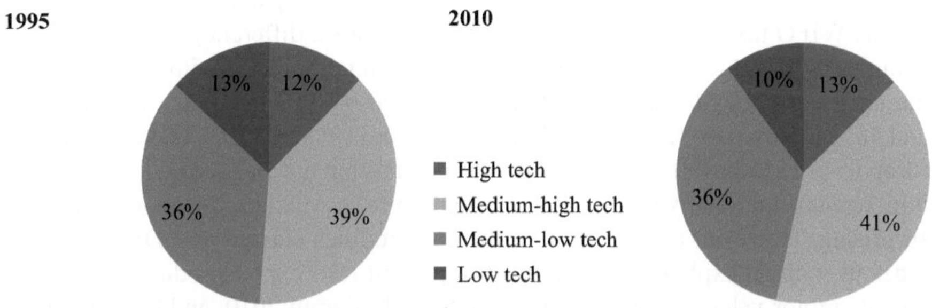

Fig. 8.27 Commodity structure of Mexican imports in selected years of the period from 1995 to 2010 according to the OECD industry classification based on technological level. (Source: Own elaboration based on the UN Comtrade database)

imported (nearly US$50 billion in 2010), while in the MHT group—motor vehicles are equally important (almost US$41 billion in 2010), as well as research instruments (close to US$48 billion), electric devices (US$42 billion) and nonelectric devices (more than US$50 billion in 2010). Currently, imports of chemicals are more important than exports (US$53 billion as compared to US$19 billion in 2010) (Fig. 8.27).

The structure of Mexico's international trade according to the WIFO taxonomy based on labor force skills is similar to the commodity structure of international trade classified by technological level. Both classifications indicate that MT products are dominant in Mexico's international trade. They also point to the fact that the contribution of MW employees is highest in the production of goods that are the object of international trade. Both in exports and imports, products manufactures by MW employees are dominant. In this group, goods produced by MW employees are superior in numbers: they account for approximately 40%. The second largest group is constituted by products based on MB workers—approximately one-third.

Fig. 8.28 Commodity structure of Mexican exports in selected years of the period from 1995 to 2010 according to the WIFO taxonomy of manufacturing industries based on input factors. (Source: Own elaboration based on the UN Comtrade database) (Data not available: *MM* textile finish, *LI* forging, pressing, embossing, and rolling of metals; founding of metals; and metalworking and coating, *MDI* reproduction of recorded data media (only 2010))

Among export commodities, goods based on employees with high skills are the most important group in the area of imports (17 % as compared to 13 % in 2010).

The WIFO taxonomy based on input factors offers a different perspective of the commodity structure of Mexico's foreign trade. In compliance with this approach, the largest part of Mexican exports is constituted by TDI products (40 % in 1995 and 46 % in 2010), followed by MM commodities (19 % in 2010), CI (18 %) and LI (drop from 15 % in 1995 to 10 % in 2010), and, last but not least, MDI (7 %). Taking into account the relatively low R&D expenditure in Mexico, this result might be surprising. However, a quick look at the commodities classed as RDI goods provides us with an explanation. The most important RDI export products of Mexico include motor vehicles (MHT)—nearly US$40 billion in 2010, and household appliances (in total, nearly US$39 billion)—which means that exports are not generated by Mexican companies, but by numerous international companies representing this industry. The importance of the vehicle sector is emphasized by the fact that in this group of CI products, car parts and accessories are most important (almost US$17 billion in 2010) (Fig. 8.28).

The specification based on input factors is more precise when detecting the differences between exports and imports, although these differences were more visible in 1995. It also better presents the changes that occurred in Mexican imports between 1995 and 2010 (Figs. 8.28 and 8.29).

In Mexican imports, MM commodities are of relatively higher importance, although their significance fell from 31 % in 1995 to 24 % in 2010. RDI products gained in importance, as their share increased from 27 to 36 % in the analyzed period. Also, the share of CI industries went up from 22 to 27 %. Refined petroleum products (below US$21 billion in 2010) and basic chemicals (US$18 billion in 2010), as well as car parts and accessories (nearly US$18 billion in 2010) were the most important elements of this group. However, in the group of RDI products, the following subgroups were dominant: household appliances (in total, just under US$38 billion in 2010), office equipment and computers (US$15 billion in 2010),

Fig. 8.29 Commodity structure of Mexican exports from 1995 to 2010 according to the WIFO taxonomy of manufacturing industries based on input factors. (Source: Own elaboration based on the UN Comtrade database) (Data not available: *MM* textile finish, *LI* forging, pressing, embossing, and rolling of metals; founding of metals; and metalworking and coating, *MDI* reproduction of recorded data media (only 2010))

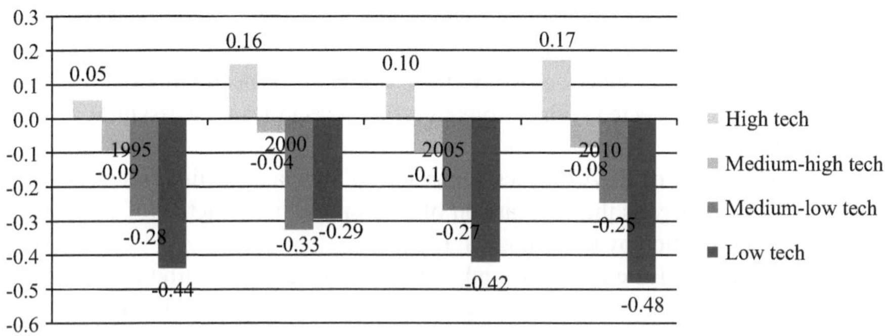

Fig. 8.30 Comparative advantages in Mexico's foreign trade (RCA) exports in selected years of the period from 1995 to 2010 according to the OECD industry classification based on technological level. (Source: Own elaboration based on the UN Comtrade database)

production of electron lamps and other electronic elements (US$16 billion in 2010), and motor vehicles (less than US$13 billion in 2010) (Fig. 8.29).

Despite the dominance of MT industries in Mexico's international trade, these areas do not display comparative advantages—when measured by means of the revealed comparative advantage (RCA) index (Fig. 8.30). RCA indexes are negative both for MHT and MLT, though the lack of comparative advantages is most visible in LT industries in the analyzed years. Interestingly, only Mexican HT industries exhibit relatively weak comparative advantages. In this group, the aviation and aerospace industry achieves the best results: in 2010, RCA equaled 0.36. Industries connected with household appliances production have slightly lower advantages. Their RCA amounts to 0.28. Pharmaceuticals have the worst results among HT industries: in 2010, their RCA amounted to −1.19. In the MHT sector, despite the general negative results, the car industry is witnessing major comparative advantages. It's RCA=0.53. The chemical industry displays lowest values here: RCA=−1.01.

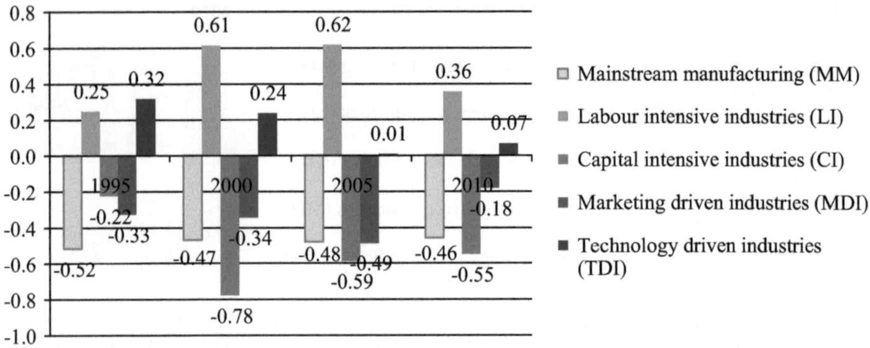

Fig. 8.31 Comparative advantages in Mexican foreign trade (RCA) in selected years from 1995 to 2010 according to the WIFO taxonomy based on input factors. (Source: Own elaboration based on the UN Comtrade database)

Two sectors that are generating significantly high comparative advantages in the MLT industries deserve attention: production of commodities from nonmetal materials and metal production (RCA = 0.47 and 0.52, respectively). On the other hand, industries connected with petroleum refinery seem least successful with their RCA at the level of − 1.21.

The comparison of RCA indexes according to the classification based on labor force skills, similarly as the comparison of the commodity structure of trading, reflects the specification by technological level.

Mexico has relative comparative advantages in the group of products manufactured by HS employees. RCA indexes become less favorable as the skills of employees decrease. In the group of products manufactured by MW employees, only the production of motor vehicles exhibits relative comparative advantages. The biggest disadvantages refer to those products that are manufactured by LS workers.

Similarly, the results of the RCA index analysis according to the WIFO taxonomy based on input factors are surprising. Mexico's highest comparative advantages are vested in LI industries, and TI industries exhibit scant advantages as well (Fig. 8.31). MM and CI industries that constitute a major part of Mexican trading, similarly as MDI sectors, are devoid of significant comparative advantages in international trade.

In the group of LI industries, the highest comparative advantages are displayed by the furniture industry (RCA = 0.99 in 2010) and the sector that produces bricks, tiles, and construction-related products (RCA = 0.86 in 2010). It is worth taking a closer look at technology-based industries, especially the automotive sector (RCA = 1.14 in 2010), aviation and aerospace (RCA = 0.71 in 2010) and medical equipment (RCA = 0.81 in 2010). In the group of CI industries, few comparative advantages are noted by the ceramic tile industry (RCA = 0.86 in 2010), cement, calcium and gypsum (RCA = 1.32 in 2010) and the sector that produces precious and nonferrous metals (RCA = 0.53 in 2010).

Sectors belonging to MM industries have been dominated by negative comparative advantage indexes in international trade. The only positive examples of sectors

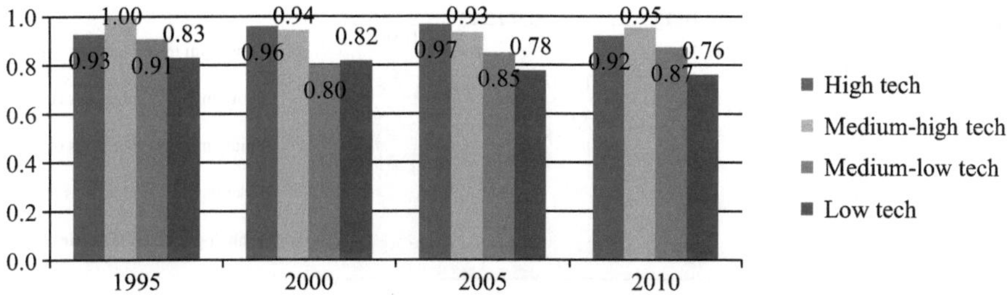

Fig. 8.32 IIT in Mexico in selected years of the period from 1995 to 2010 by the OECD classification based on technological level. (Source: Own elaboration)

that managed to generate comparative advantages are: production of concrete and gypsum goods (RCA=1.86 in 2010) and production of household equipment (RCA=1.69 in 2010).

Mexico's highest comparative advantages are visible in industries concerned with the manufacturing of tobacco products (MDI). In 2010, the RCA index for this industry amounted to 2.51.

Mexico's international trade is to a large extent based on IIT. This concerns mainly MHT industries. In 1995, trading in this group was almost solely of intra-industry nature (IIT=1). HT commodity trading is also based on IIT to a very large extent (IIT=0.92 in 2010). To a slightly lesser extent, this also concerns trading in MLT products (IIT=0.87 in 2010), and to the smallest extent—trading in LY commodities (IIT=0.76 in 2010). In the analyzed period, no radical changes in the shares of particular producer groups in IIT occurred. It is only possible to notice a slight decrease in IIT indexes (Fig. 8.32).

When analyzing particular sectors, one should note that the lowest volume of IIT can be observed in those Mexican industries that lack significant comparative advantages. In 2010, the IIT index for the pharmaceutical sector equaled only 0.46 (in 1995, IIT=0.70). At the same time, the chemical industry achieved a result amounting to 0.53, while the petroleum refinery sector obtained the outcome of 0.45.

This dependency is not mutual, as the industries that cater for Mexico's comparative advantages in international trade are not very focused on IIT. These sectors are characterized by a relatively average IIT index volume (aerospace—0.83, electronics-communication—0.87, cars—0.74, nonferrous metals—0.75, products from nonmetal materials—0.78).

Mexico's IIT presented on the basis of labor force skills also displays similar trends as in the case of the specification according to the technological advancement of products. The lowest IIT index is exhibited by industries based on LS workers, while the highest value of this index is displayed by industries based on MW employees and HS employees.

More diversity in the IIT index may be observed when Mexico's trade is organized by the degree of utilization of tangible and intangible outlays in the production process.

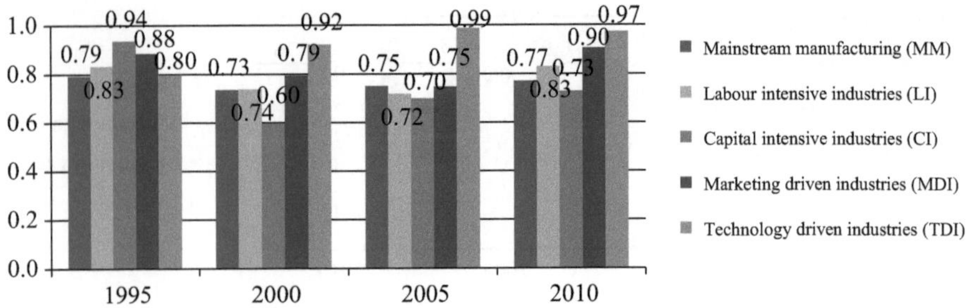

Fig. 8.33 Mexico's IIT in selected years from 1995 to 2010 according to the WIFO taxonomy based on input factors. (Source: Own elaboration)

In 2010, IIT was most robust in the RDI industry (IIT = 0.97). In 1995, this index was much lower, i.e., 0.80. On the other hand, trading in CI industries in 1995 was based on IIT to a larger extent than in 2010 (IIT = 0.94 in 1995 and 0.73 in 2010) (Fig. 8.33).

In the group of MM industries, the intensity of IIT exceeds average values in sectors that manufacture machinery (especially general-purpose machinery—IIT = 0.96 and machinery for forestry and agriculture—IIT = 0.94) and electric motors, generators, and transformers (IIT = 1) or insulated wires and cables (IIT = 0.91). In the group of CI industries, in 2010 the highest IIT index was noted for the sector concerned with car parts and accessories and motors (IIT = 0.98). At the same time, one ought to note the increase in the IIT index in the automotive sectors that constitutes part of the RDI industry group. In that sector, the intra-industry index went up from 0.15 in 1995 to 0.49 in 2010. The highest intra-industry indexes in this group are generated by medical and surgical equipment and orthopedic devices production sectors (IIT = 0.92), measurement, control, research and navigation equipment and instruments production (IIT = 0.97), and office equipment and computer production (IIT = 0.97).

Industries with highest RCA indexes also achieve above-average IIT indexes (cars: 0.49, concrete and gypsum goods production: 0.27, household appliances production: 0.32, cement, calcium and gypsum: 0.43, beverages: 0.48, and in the case of tobacco products: only 0.15).

In conclusion, it is worth noting that Mexico is witnessing most intense trading in: MT goods, mainly RDI commodities, but also MM products, and these goods are mainly produced by employees with medium skills. In commodity categories, this means that Mexican exports are based on machinery and transport equipment. North America is Mexico's main output market. On the other hand, imports are more diversified and China is gaining in importance in this field.

Mexico's competitiveness in international trade is the resultant of the competitiveness of particular economy sectors. Some industries have a positive influence on competitiveness, and some have a negative impact on the country's competitive edge. If one assumes that relative comparative advantages determine competitiveness, then the competitiveness of the Mexican economy is driven by the results of HT, LI, and RDI industries that use mainly HS labor force.

On the other hand, the competitiveness of the economy is hindered by the results of sectors that produce LT products, as well as CI, MM, and MDI industries. These sectors are based on employees with medium and low skills.

8.3.2 Characteristic Features of Mexico's NIS

Mexico's NIS is to a large extent based on public entities: institutions established by the administration and public research centers. This is a direct reflection of the structure of financing R&D activity in Mexico, as the share of the private sector is very low here.

The government is the legislative body that bears responsibility for Mexico's R&D system, and the General Council for Scientific Research and Technological Development headed by the President of the Republic is the government entity tasked with drafting and implementing appropriate policies. However, the most important government institution operating in the sector of S&T is the National Council for Science and Technology (CONACYT). This office is responsible for ensuring coordination of activities at the ministerial level, including the drafting and implementation of the research and technological policy of Mexico (Erawatch, p. 11). Moreover, the council is composed of the Minister of Finance, representative of eight sector Ministries (Education, Internal Affairs, Economy, Health, Energy, Environment, Agriculture, and Communication and Transport)—that have access to funds allocated to supporting programs promoting S&T, as well as from members of the Scientific and Technological Advisory Forum (FCCyT)

Financing of the scientific and technological policy at the technological level has two main channels: budget funds assigned to sector ministries that finance research institutions, and funds or programs allocated to science and technologies and budget funds allocated to funds, programs, and research centers financed by CONACYT.

FCCyT is the main intermediary body that was formally established to provide advisory within the scope of science policy. The Forum was created in 2002 as part of the S&T Law in order to ensure that representatives of scientific and business environments are involved in processes connected with the implementation of the science, technology, and innovation policy (STI). FCCyT constitutes a platform for discussion and exchange of opinions on STI-related issues that enables the interested parties to reach an agreement with regard to key issues. The diagram of Mexico's NIS has been presented in the following section.

Research is mainly conducted in research institutions that operate under the supervision of Sector Ministries at 27 research centers subordinated to CONACYT, public and private HEI, and institutions of the private sector (research centers, universities, and single enterprises[19]; Erawatch, p. 13).

[19] The group of most important research units encompasses institutes operating as part of: the National Autonomous University of Mexico, the Center for Research and Advanced Studies, the Metropolitan Autonomous University, and the National Polytechnic Institute.

This means that research is focused on the public sector, which receives the majority of public R&D funds. As a result, the public sector develops most innovations. Moreover, research activity is concentrated on a few locations in the country. The most important area is the capital, Mexico City, (38% of researchers belonging to the National Scientists System[20]) operates in the Federal District—Distrito Federal).

Apart from central authorities and the institutions they establish, the national research system is complemented by private enterprises, private research centers, public and private HEI, and regional institutions, i.e., State Scientific and Technological Councils and six regional CONACYT offices. The Interministerial Budget Committee analyzes the criteria for granting funds established by the ministries and agencies responsible for the implementation of STI programs. The Interministerial and Integration Committee reviews and promotes offers for creating programs that would involve a number of ministries and other supporting programs that encompass administration at the federal and state level. The National Science and Technology Conference promotes collaboration at the federal and state level within the scope of issues connected with policy priorities, shaping, and implementing programs.

8.3.2.1 The Innovative Capacity and Position of Mexico

One of the basic determinants of the innovative potential is constituted by the outlays on R&D activity. According to OECD data, in Mexico, the share of R&D expenditures in GDP is low and does not exceed 0.44%. The EU average in the analyzed period oscillated between 1.7 and 1.9% of GDP, while in the USA, these outlays amounted to 2.8% in 2009. Although in Mexico this index went up in the period from 1995 to 2009, this trend is not especially dynamic.

Year 2002 was a turning point, as it was then that R&D expenditures were to a larger extent utilized by the enterprise sector (BERD), and not by government institutions (GOVERD).

However, these funds are still mainly provided by government sources (53.17% in 2009), and enterprises rank second here (39.06% in 2009). Other sources; higher education sector, nonprofit organizations, or foreign institutions, carry little weight. What is important is that in the analyzed period, one should observe the trend directed at increasing the significance of enterprises at the expense of government sources. The importance of foreign sources is also higher, but nonprofit organizations still lag behind.

Human resources for S&T are another important element that is necessary to enable the provision of conditions conducive to fostering innovation. In 2010, only 14.2% of Mexico's population aged over 18 could be classified as members of this group (INEGI-CONACYT data).

[20] Conacyt data, *Base de datos del SNI*, 2001–2010.

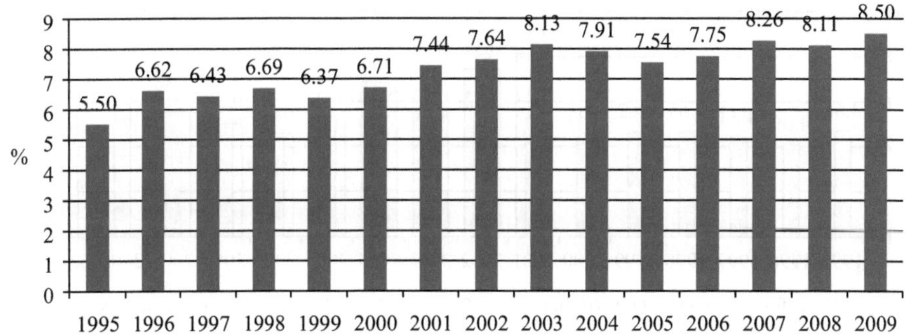

Fig. 8.34 Population with higher education (ISCED 5 or higher) involved in scientific and technological activity as percentage of total number of professionally active population. (Source: CONACYT, Informe General del Estado de la Ciencia y la Tecnología)

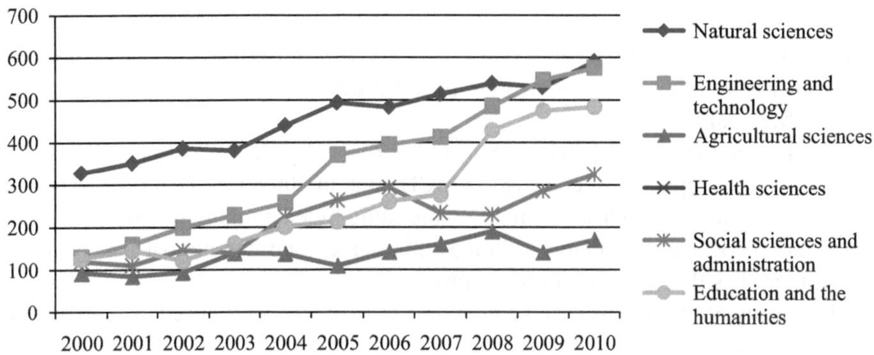

Fig. 8.35 Graduates of Ph.D. studies by faculty, 2000–2010. (Source: Conacyt, Encuesta de Graduados de Doctorado 2010)

This percentage did not increase notably (only by 0.6 p.p.) in the years 2003–2010. For comparison, in countries with a high level of technological advancement, such as Switzerland, more than 50 % of the professionally active population constitutes human resources for S&T.

Low availability of Human Resources in Science and Technology (HRST) in Mexico affects its utilization in the economy. In 2009, only 8.5 % of the professionally active population was constituted by persons who finished their education at higher level (ISCED 5 or higher) and were involved in scientific and technological activity. Although this percentage increased from 5.5 % in 1995, nevertheless, these results confirm Mexico's low innovation capacity (Fig. 8.34).

The majority of university students opt for humanities and social sciences. This trend is reflected by the structure of graduates of Ph.D. studies (Fig. 8.35). The group of graduates of social and administrative faculties is largest and is growing most dynamically. Every year, the number of graduates of Ph.D. studies in natural

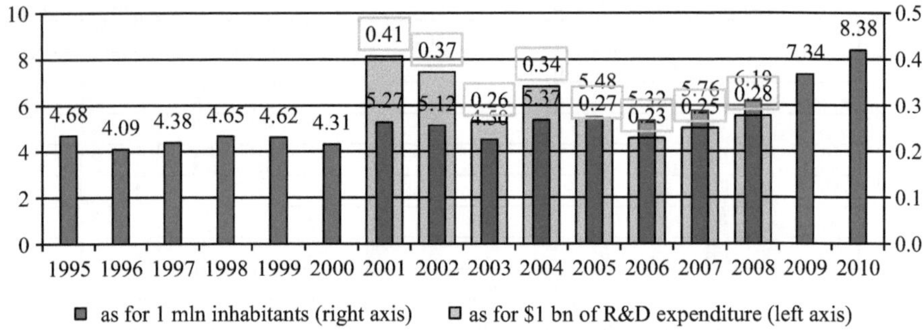

Fig. 8.36 Number of submitted patent applications per US$1 billion of R&D expenditures in comparison with the number of patent applications per one million citizens. (Source: WIPO Statistics Database and World Bank 2011)

sciences, engineering, and technology increases by almost 200. The increase in the latter group has been significant since 2000—575 in 2010, as compared to 159 in 2000. At the same time, in almost all faculties, a rapid growth in the number of graduates is visible. This is indicative of the growing interest in enhancing skills among Mexicans.

The low percentage of households with Internet access is indicative of Mexico's low innovative capacity. According to OECD data, in 2009 only 18.4 % of Mexican households had Internet access. In 2009, the same index amounted to 68.7 % in the USA and 79.1 % in Germany. Although this percentage has almost tripled since 2001, there is still a lot of room for improvement.

The group of indexes based on patent data is most frequently used to describe the innovative position of the country. Patent indexes indicate low innovative intensity in Mexico. Although the number of patent applications per one million inhabitants almost doubled since 1995, in 2010 it amounted to a mere 8.38. On the other hand, innovative intensity of R&D expenditures decreased by more than 30 % between 2001 and 2008, from 0.41 to 0.28 of a patent per US$1 billion (Fig. 8.36), which points to the fact that the effectiveness of R&D expenditures is falling.

8.3.2.2 Human Capital in Mexico

One of Mexico's key competitiveness factors is its numerous population and labor force that is still relatively cheap. In Mexico, the population itself has not only increased over the past decades, it is also expected to go up to 123 million by 2041 (from the current 112 million). What is more, human capital utilization in the Mexican economy increased over the years. Labor force resources increased in the years 1995–2010 from 34.5 to 46.8 million. This meant that 43 % of the population was active on the labor market in 2010—5.5 p.p. more than in 1995.[21] The employ-

[21] In 2010, Mexico's population amounted to 112.3 million (INEGI. Population and Housing Census 2010).

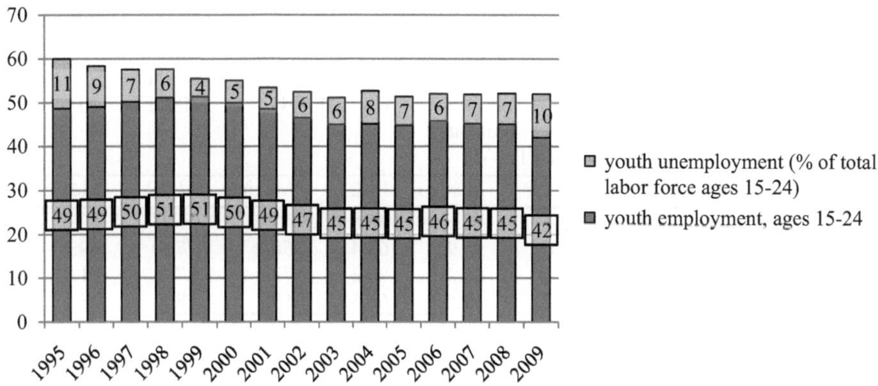

Fig. 8.37 Youth on the Mexican labor market from 1995 to 2009 (data in %). (Source: World Bank data)

ment rate in Mexico remained at a high level almost throughout the period from 1995 to 2010. It was not until 2009–2010 that the employment rate growth slowed and the number of the unemployed went up. In 2010 in Mexico, the total labor force equaled 46.8 million, and 44.4 million persons were employed.

Unemployment statistics in Mexico are very favorable, though one has to bear in mind that they do not include persons of working age who remain outside the labor market. In the years 1998–2008, unemployment in Mexico remained at a level similar to natural unemployment and amounted to approximately 2.5–3.5 %. In 2010, the unemployment rate was even lower than in 1995 (although when calculated in absolute value, the number of the unemployed exceeded the 1995 level in 2010). Nevertheless, the exceptionally good 1999 result (unemployment at the level of 2.5 %, i.e., 938,000 persons) is followed by growth both in the number and percentage of the unemployed. In 2010, unemployment in Mexico amounted to 2,467,000, i.e., 5.3 %.

Low unemployment among youth, which is a problem for many highly developed European countries, should be deemed a positive feature of the Mexican economy. However, also in this area negative trends may be observed—in 2009: a drop in employment and increase in unemployment, however, these data are still more favorable than in the case of many competitive European economies,[22] as well as the USA (youth unemployment amounting to 18 %) and Canada (15 %). The positive undertone of these data was altered by the fact that approximately 18.4 % of 15–19-year olds and 30 % of 25–29-year olds in Mexico do not participate in the education process, are unemployed, or are even outside the labor market (OECD 2011) (Fig. 8.37).

The quality of the state's human capital is indicative of the education level of its employees. Although in recent years more and more persons in Mexico improve their qualifications, as they decide to continue education to the secondary or even higher level, however, the level of participation in secondary and higher education

[22] For example, in 2009 in Spain youth unemployment amounted to 38 %, in Italy—to 25 %.

Table 8.14 Employment structure according to highest obtained education level, Q3 2010 (in %), EU—Mexico. (Source: EUROSTAT, Mexico-EU—basic statistical indicators)

Age group	Total		Incomplete primary, primary, lower secondary (ISCED 0–2)		Higher secondary (ISCED 3–4)		Higher (ISCED 5–6)	
	EU27	Mexico	EU27	Mexico	EU27	Mexico	EU27	Mexico
15–24	25.2	43.1	22.4	44.6	45.8	41.7	56.1	40.6
15–74	57.7	58.5	38.6	55.9	63.4	58.2	77.0	69.2
25–49	78.3	70.0	63.2	65.2	80.2	73.5	87.0	81.3
25–74	61.8	64.2	43.0	59.4	66.6	69.3	78.3	78.1

is the lowest among the OECD states. It is also notably lower than in the European Union. This concerns mainly the elderly.

The structure of unemployed persons by education level is a negative economic signal indicating low effectiveness of the education system. The Q4 2010 data show that persons with secondary education were the largest unemployed group in Mexico—36.6%. This group was followed by persons with higher education (34.6%). Generally, the lower the education level, the smaller the unemployment rate. These data show that the educations system, both at the secondary and the higher level, is not adjusted to the labor market.

The employment structure in reference to education shows which employees are sought after and, indirectly, what the technological advancement of the economy is. The comparison of Mexico and the EU reveals significant differences. To start with, the employment rate of persons with the lowest education is much higher in Mexico. This concerns mainly young persons (44.6% in Mexico, and 22.4% in the EU). A much higher level of employment of persons with the lowest education level is observed also in the 15–74 age group (55.9% in Mexico and 38.6% in the EU), comparable in the 25–49 age group (65.2 and 63.2%, respectively). The structure of employment is notably different in the case of persons with secondary and higher education. Both young persons (15–24) and middle-aged persons (25–49) are more likely to find employment in the EU than in Mexico. The position of Mexicans with higher and higher secondary education in the labor market is worse than that of persons with lower education (Table 8.14).

Such a high level of employment of labor force with low education results in labor costs in the Mexican economy being at relatively low levels. Labor costs in Mexico, the USA, and Canada are presented in Fig. 8.38. Both the hourly wages (US$6/h in 2009) and the annual salary (US$12,528 in 2009) are markedly lower than in other NAFTA member states—especially the USA (in 2009: US$35/h and US$59,049 per annum), but also lower than in most EU Member States. The rates paid out in Mexico are similar to the non-OECD average.

Unit labor costs (ULC) are lower by almost one-third in Mexico in comparison with the USA (Fig. 8.39). Industry is the most competitive element of the Mexican economy (in terms of ULC), and the construction sector seems to be least competitive. Other trade and services sectors assume similar ULC levels.

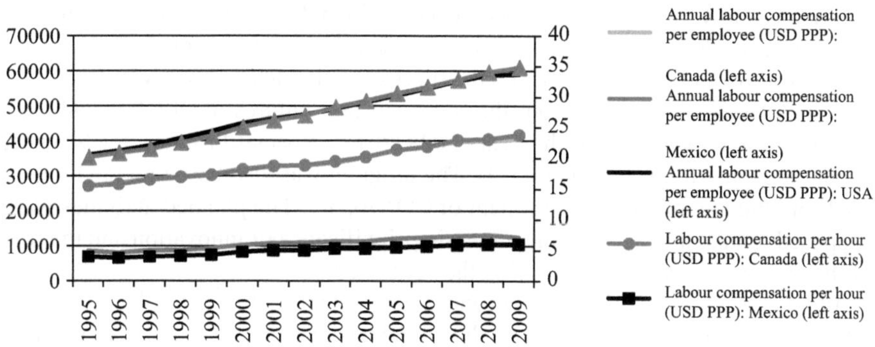

Fig. 8.38 Remuneration in Canada, Mexico and the USA. (Source: Own elaboration based on OECD data)

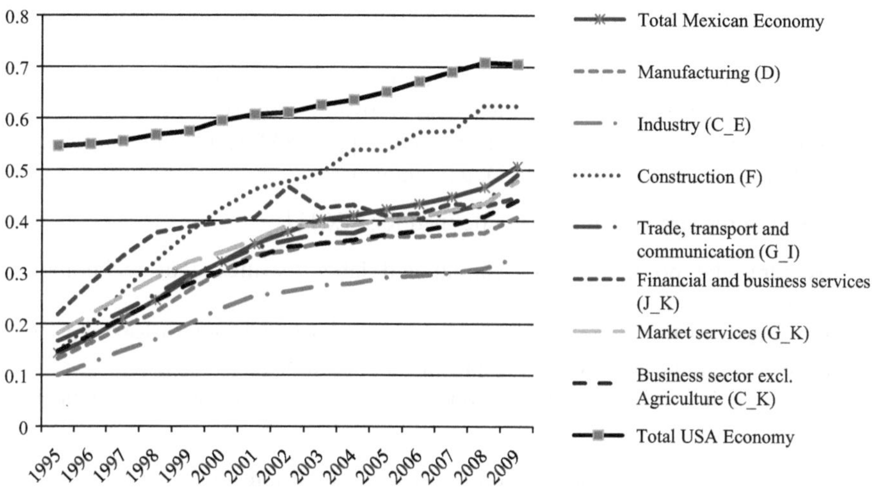

Fig. 8.39 ULC in the Mexican and US economy, and in selected sectors of the Mexican economy. (Source: Own elaboration based on OECD data)

8.3.2.3 Characteristic Features of the Science, Innovation, and Education Policy in Mexico in the Years 2008–2012

In the years 2001–2006 in Mexico, the CONACYT implemented the Special Program for Science and Technology (PECYT). As part of the program, numerous initiatives aimed at improving and implementing Mexico's STI. However, the evaluation of the program showed that despite certain successes, the main objectives were not fulfilled. Structural disadvantages rendered the NIS ineffective (OECD 2009, pp. 9–11). It remains unclear, among other things, how institutions responsible for policy creation were differentiated from those responsible for its implementation. The wide range of supporting instruments, where some were poorly financed and referred to various strategic documents, additionally burdened with ineffective management, weakened the activities undertaken by the government. As a result, the policy mix had limited impact on the success of the Mexican innovation system.

Conclusions were drawn from the successes and failures of PECYT, and used to create a new program for the years 2008–2012—Special Program for Science, Technology and Innovation (PECITI). The program was also based on the Law on S&T and the law on the establishment of CONACYT. The planned activities were supposed to take into account the increasing significance of innovation and the role of the enterprise sector in developing the S&T sector.

PECITI is a framework document and an instrument enabling the planning of the STI policy in Mexico. It is based on the guidelines of the 2007–2012 National Development Plan that defines five priorities within the scope of STI (Erawatch, p. 14):

1. Developing short-, medium-, and long-term national policies that strengthen the chain: education—basic and applied research—technology and innovation.
2. Promoting decentralization of STI activities in order to support regional development.
3. Financing basic and applied research, technology, and innovation.
4. Investing in research, technological, and innovation infrastructure.
5. Promoting responsible public investment to support STI development: high-quality human capital, training, R&D, and innovation.

Moreover, three strategic lines were defined for PECITI:

1. Investing the productivity of capital and labor by means of technological innovation.
2. Boosting investments in infrastructure, thanks to which enterprises and employees will gain access to appropriate production tools.
3. Increasing investment in people through investment in enhancing education quality.

This program is the key contribution to creating conditions enabling the development of a knowledge-based economy in Mexico. PECITI discerns the following science- and technology-related areas as priorities: biotechnology, medicine, energy, environment, industrial technologies (industrial processing), textiles, nanotechnologies, information and communication technologies, applied mathematics, modeling and priority sectors: food, aviation, automotive and car parts, electric and electronic, pharmaceuticals and medical services, metallurgy, metalworking and capital goods, and chemical and petrochemical.

As part of PECITI 2008–2012, new programs aimed at stimulating the entrepreneurship of researchers, state research centers, and the private sector were established. Moreover, programs supporting intermediary institutions that support connections between science and industry were developed. Policies; supportive of training (e.g., CONACYT—Scholarship program for graduates), supporting HRST growth and accumulation (e.g., National award for member of the Researcher System, Institutional Consolidation Program), are believed to be especially successful elements of the policy mix (Erawatch, p. 3).

The most important programs (in terms of budget size) are; Sector funds, R&D, and innovation-stimulating package and mixed funds (Erawatch, p. 15).

In Mexico, there are 17 sectoral funds, which are financed and implemented jointly by CONACYT and sector ministries to support STI in accordance with the strategic goals. The funds allocated to their implementation are, however, limited. In 2009, the total of €120 million, 13% of which was constituted by CONACYT funds, was at the disposal of all the funds.

The R&D and innovation-stimulating package involves three various supporting programs aimed at strengthening R&D and innovativeness by focusing especially on SME and microenterprises and the cooperation between enterprises, as well as collaboration between enterprises, public research centers, and HEI. In 2009, the additional amount of €88 million was allocated to this program. The car, the health, and the food industries benefited most from this program.

Mixed Funds were introduced in 2001 and are still the most effective instrument of promoting applied research and/or innovative capacity at the state and municipal level. Their goal is to provide support in expressing federal and regional STI policies and programs. In 2009, Mixed Funds amounted to €71 million. The financial contribution of CONACYT varied depending on the project, however, it has never fallen below 50%. Mixed funds accounted for approximately 4% of the CONACYT budget in the period from 2000 to 2006.

The evaluation of Mexico's NIS is summed up by means of the SWOT analysis (Table 8.15).

As presented in the above table, the human capital of Mexico is the source of most weaknesses and provides most opportunities for further growth. The young population constitutes invaluable capital. However, it requires undertaking many actions that will enable developing this group in a manner that will be beneficial for the entire economy—and guarantee that it becomes a source of innovation and competitiveness.

8.3.3 Interdependencies Between Innovation and Human Capital Accumulation in Mexico and the Country's Advantages in Foreign Trade

Mexico belongs to the group of countries where competitiveness is based on traditionally low labor costs. Currently, this country is not among the leaders of innovation. The level of social understanding and involvement in innovative processes in low. This is the result of poor education. Despite the improvement observed in recent years, the innovation-supporting policy is still managed ineffectively, as its programs are excessively fragmented and underfinanced.

Despite the favorable economic situation (relatively high economic growth, low inflation, low unemployment rate), R&D and innovation expenditures in Mexico are limited. This is caused by the insufficient involvement of mainly the enterprise sector. As a result, R&D activity and innovation remain the domain of the public sector.

Table 8.15 SWOT analysis of six key areas of the NIS. (Source: Programa Nacional de Innovación, Comité Intersectorial para la Innovación México 2011)

Strengths	Weaknesses
1. Domestic and international market	
Large domestic market	Limited competition in some sectors
Increase in share of some industrial sectors in global market	Insufficient information provided to consumers
Regional and sector excellence clusters	Low penetration of international markets other than North America
Capability of attracting FDI	High concentration of exports in large enterprises
Preferential access to international markets	Low level of consumer adaptation of technologies
Large export capacity of large enterprises	Low awareness of environment-related issues among consumers
Opportunities	Threats
Population of young consumers	Growing competition from emerging economies
Geographical proximity of the USA	Accelerated expansion of key knowledge and technologies
Growing demand for RDI products	Weak ties with regions of dynamic economic, scientific, and technological growth
Shift of production and trade to goods and services that are more knowledge-intense	Regional concentration of innovation
Global interest in innovation-related issues (biological diversity, environment protection, health)	
Growing demand for social KI products	
Strengths	Weaknesses
2. Creating knowledge—strategic approach	
Group of HEI and highest quality public research centers	Low HRST mobility
Large group of skilled scientists	Shortcomings of the education system
Cultural differences as a source of creativity	Low labor force skills at the national level
Skilled and competitive labor force in industrial sectors at the global level and in some regions of the country	Insufficient technological infrastructure
	Definite majority of SMEs with low capability of technology absorption
	Preference for imported technologies
	Complicated law on knowledge development
	Ineffective information systems
	Weak connections between the National Scientists System and the private system

Table 8.15 (continued)

Strengths	Weaknesses
Opportunities	Threats
Young population	Accelerated expansion of scientific and technological border
Possibility of educating a large group of engineers	Intensification of the global struggle for talents
Including knowledge and global technological platforms into the network	Regional concentration of innovative capacity
Diffusion of technologies among multinational enterprises, together with the development of global value chains based on innovation	
Strengths	Weaknesses
3. Strengthening the innovativeness of enterprises	
Regional and sectoral excellence clusters	Most SMEs with low capacity to absorb technology
Vast and competitive database of industrial entrepreneurs	Scant knowledge of enterprises within the scope of opportunities offered by the intellectual property protection system
Large SME group that perceives knowledge as the basis of its activity	Limited competition in some sectors
Availability of public patent information systems	Imported technologies are deemed better than domestic ones
Highly global economy allows for accessing new technologies	Lack of innovation culture at the workplace
Opportunities	Threats
Shift of production and trade to goods and services that are more knowledge-intensive	Intensification of the global struggle for talents
Participation of SMEs in the innovation-based strategy	Weak regional concentration of innovative capacity
Diffusion of technologies among multinational enterprises, together with the development of global value chains based on innovation	
Using the knowledge contained in patents in Mexico	
Strengths	Weaknesses
4. Financing innovation	
Capability of attracting FDI to particular sectors of the economy	Low allocation of funds and insufficient political involvement in the implementation of the STI policy
Special public sectors tasked with supporting enterprises	Complex qualification procedures for support programs
	Insufficient knowledge within the scope of finance among entrepreneurs and investors
	Legal framework hindering the growth of private equity funds
	Financial markets are ill-adapted to innovation-related investments

Table 8.15 (continued)

Strengths	Weaknesses
Opportunities	*Threats*
Large population of potential entrepreneurs	International financial crisis
Utilization of private equity funds in Mexico	
Increase in financial savings	
Strengths	Weaknesses
5. Human capital	
Group of universities (public and private) and high-quality public and private research centers	Low level of cooperation between the private and the public sector aimed at ensuring appropriate professional training
Considerable number of skilled scientists	Low mobility of human resources within the STI
Cultural diversity as a source of creativity	Delayed and low absorption of practices concerning education quality
	Insufficient number of training specialists within the scope of intellectual property
	Lack of business and productivity culture
	Lack of professionalism and knowledge among persons responsible for connecting HEI, CPI, and enterprises with regard to innovations and utilization of IT
	Limited number of information systems that identify the skills required to perform a specific task in strategic sectors
	Lack of human resources with higher education
	Regional concentration of educational opportunities
	Low quality of primary education, especially in areas such as English and mathematics
	High percentage of persons leaving school at the secondary level

Table 8.15 (continued)

Strengths	Weaknesses
Opportunities	Threats
Young population	Accelerated expansion of science and leading technologies
Possibility of developing a large group of engineers	Regional concentration of innovative capacity
Integration with knowledge platforms and global technological platforms	Intensification of the global competition for talents
Differentiating production and trade by introducing more KI goods and services	
Increase in demand for skilled and specialist human capital	
Increasing investments made by enterprises in the human capital of its employees	
Strengths	Weaknesses
6. Legal and institutional framework	
Drawing from the experience of some public institutions to promote STI and economic growth	Ineffective management of the NIS
Mechanisms of intellectual property protection	Excessively bureaucratic procedures for the management of support programs
Existence of technology management standards	Conflicting provisions of law
	Inefficiency of public management
	Unbalanced implementation of the policy mix
	Shortcomings of the judiciary
	Lengthy patent application procedures
	Long-drawn adjudication of disputes concerning intellectual property rights
Opportunities	Threats
New regulatory frameworks concerning sustainable growth, environment pollution, biodiversity, and alternative energy sources	Incompliance with international regulations
Reform of the law on S&T	Lack of continuity in the government's policy resulting from constant changes
Revision of the law on industrial property	

Although Mexico has a numerous young population, HRST are limited. This is connected with the fact that Mexicans quit school relatively early, at the level of primary or lower secondary education. And, in the case of undertaking university studies, social sciences and humanities are the most popular, with engineering and technical studies being less in demand among students. As the majority of Mexico's labor force does not have high skills, it remains inexpensive. Consequently, ULC remain at relatively low levels, and the international competitiveness of the economy remains high. However, this means that Mexico still competes in the area of low costs of input factors—especially labor, rather than by promoting the innovativeness of its economy.

Currently, the trade structure—especially the main export channel that has been maintained for years—poses a serious obstacle to economic changes. Such a great dependence of the Mexican economy on the US economy leads to a situation where Mexico adapts its production to the needs of the large recipient (complementarity principle) and concentrates on the industries from which the USA withdraws. It does not compete with the HT industry, but provides MT products. For years, this situation was beneficial for Mexico and guaranteed growth (in periods when the USA generated growth), however, in times of intensifying global competition, Mexico's position has become unstable. Mexico's greatest rival is China, which entered the US market by offering goods that to a large extent were substitutes of Mexican commodities and adopting the same strategy of price competition. As Mexico does not stand a chance of winning a price war with China, it has to change its strategy and turn to manufacturing products that would be complementary also to Chinese products. Another consequence of these changes is the drop in the significance of the USA as an output market and the expanding of Mexico's exports to other states. Currently, the best export destinations are countries with high economic growth, as thanks to this the Mexican economy could pick up speed as well.

The current economic structure leads to a situation where Mexico's trading focused mainly on MT and RDI commodities. However, this does not mean that Mexico also has comparative advantages in these sectors. In the area of MT products, practically only manufacture of motor vehicles, metal working and nonmetal material processing are competitive. However, minimal comparative advantages are noted in RDI production (thanks to the automotive, medical equipment, and aviation industries). Importantly, Mexico is developing comparative advantages in HT commodities (apart from pharmaceuticals), although this area of trade is not as large-scale as the trading in goods that are not so advanced in terms of technology. Comparative advantages are also noted in LI production (the highest in furniture production, but also in textiles and construction-related products). In the field of MM production, household appliances and concrete, gypsum, and cement products are the most competitive.

These sectors can be classed at the specialization of Mexico's foreign trade. Until now, in the analyzed period of 1995–2010, no significant change occurred in the area of this specialization. However, one can note a drop in the competitiveness of such key sectors as manufacture of motor vehicles or textile-related industries.

In order to change the specialization of the economy and shift it toward HT, RDI, and MDI commodities, first, one would have to employ HS employees in the economy (which will be possible in the coming years, as the population of Mexico is young and numerous). Second, supply of knowledge and technology on the market should be increased, which is tantamount to increasing human resources for R&D and the level of financing research and innovation activity, and in the short term—continuing to attract foreign technologies, among other things, by means of FDI.

Summary and Conclusions Regarding Economic Policies Supporting Medium- and Long-Term Competitive Position of Domestic Goods on Foreign Markets

Marzenna Anna Weresa

The analysis of numerous and varied theoretical concepts in the area of international economy and the theory of competitiveness, as well as the empirical research whose results were presented in this book, have confirmed that human capital, innovativeness, and competitiveness are mutually dependent phenomena. What is more, this correlation is multidirectional. Human capital is of key significance in the process of developing knowledge and introducing it to the economy in the form of technological, organizational, and marketing innovation. It is the creative people—scientists, inventors, entrepreneurs, and consumers—and their knowledge, expertise, and inclination to take risks and experiment that constitute the source of innovative ideas and new applications of existing solutions. Therefore, it is impossible to separate innovativeness from human capital, as these two phenomena are intertwined. This is confirmed by the fact that measures describing the level of human capital development and quality are commonly used in order to assess economy innovativeness. That is why, in this monograph, when evaluating various national economies in terms of resources, human capital quality, and innovativeness, we attempted to integrate these two intertwined phenomena against the backdrop of the currently developing national innovation system concept. The innovation system of national economies is understood here in the broad sense as the combination of four groups of phenomena, such as: (1) the group and structure of domestic entities that participate in the development of new knowledge and its distribution (at home and abroad) and commercialization, and are involved in the creation of innovations and their practical applications; (2) all institutions that form the environment conducive to the development of science, technology, and entrepreneurship and their transformations; (3) interactions between these entities and institutions; and (4) knowledge resources accumulated in the given economy (Weresa 2012, p. 37).

This system also encompasses the technological and institutional aspects of innovation development, as well as accumulated human capital resources.[1]

[1] This does not exhaust all the complex aspects of human capital (e.g., problems concerning health or social ties), however, the authors wittingly limited themselves to the narrower definition of

M. A. Weresa (ed.), *Innovation, Human Capital and Trade Competitiveness*, Innovation, Technology, and Knowledge Management, DOI 10.1007/978-3-319-02072-3,

The analytical framework of the national system concept (despite certain limitations—c.f.: footnote) allowed for the joint analysis of human capital and innovativeness of the given country. Such a presentation allows for focusing on the relationship between these two phenomena and competitive advantages of countries in international trade, which is the core of the analysis undertaken by the authors of this book. This connection is multidimensional and heavily conjugated. As proven pursuant to the theory (Chap. 1 and 3), on the one hand, the characteristics of the national innovation system (i.e., innovativeness of the economy as well as human capital resources and quality) determines the competitive advantages of states and, what is more, might lead to improvements in the long run. However, in this case, we are dealing with impact channeled in the opposite direction. It is foreign trade that, as one of the carriers of technical progress and a channel of technology transfer, can influence the quality of human capital in the given country. Trading entails various types of learning, e.g., learning by doing or learning by interacting, and even enables the direct transfer of technological achievements from other economic areas (through direct purchase of patents, licenses, manufacturing technologies, and know-how). On the one hand, international trade influences the size and intensity of domestic resources use in the process of generating domestic income. On the other hand, the level of human capital determines the capacity of the economy to quickly adapt to modern (also foreign) technologies and organizational solutions. The current condition of the national innovation system determines in which direction this influence will be directed. These interrelations have been presented by means of the econometric model. Estimating its parameters confirmed the thesis that the force and direction of various factors on the competitiveness of countries in foreign trade varies depending on the characteristic features of the given economy. Econometric modeling does not fully reflect economic reality. Only case studies of selected countries representing various types and subtypes of innovation systems would allow for complementing econometric studies by additional subjects and describe its specificity and diversity of solutions, especially of the institutional ones, and their influence on the directions of changes that are becoming visible. Different approaches to developing innovative capacity and position adopted by various countries—leaders in innovativeness with dynamic innovation systems—can serve as an example here. However, strengthening their dynamics and forging them into competitive advantages in international trade is performed differently in Switzerland or Finland than in Ireland. The former two countries focus on domestic resources and their development mainly thanks to all their internal forces, and only support this process by encouraging integration with the global market. Ireland bases its innovative, i.e., competitive, advantages on internationalization, which translates into, among other things, a high influx of foreign direct investments. However, what

human capital that perceives all the acquired and useful abilities of all the inhabitants of a country as a part of capital (Schultz 1961, p. 2). A similar interpretation is also applied nowadays by assuming that human capital embraces the knowledge, skills, competencies, and attributes embodied in individuals that facilitate the creation of personal, social, and economic well-being (OECD 2001, p. 18) "…all of the acquired and useful abilities of all of the inhabitants of a country as a part of capital" (Schultz 1961, p. 2).

all these states have in common is the strong (financial and institutional) support for the education sector, as it is this sector that determines the qualitative changes in human capital resources and the long-term innovative capacity. This triggers an increase in competitiveness in foreign trade.

Two countries with so-called unbalanced systems provide another example of varied solutions within the same group of innovation systems: China and Russia. China, thanks to its active government policy on technology transfer and R&D, is gradually bridging the innovation gap and catching up with global leaders. This is confirmed by the continuous improvement of the structure of comparative advantages, which is shifting toward trading in technologically advanced commodities. Still, China's competitive advantages on international markets are founded on traditional factors, i.e., low labor costs. Nevertheless, it is possible to note positive changes and attempts to compete by means of innovation.

Russia represents a completely different path of developing international competitiveness. The innovation system of the Russian Federation is becoming increasingly unbalanced and comparative advantages in international trade stem from natural resources. These two extreme examples illustrate how differently particular states develop long-term competitive advantages in international trade, even if those countries represent the same national innovation system type. A broader spectrum of this issue is presented in the other case studies discussed in this book. In total, 12 countries from various parts of the world that represent different innovation system types were analyzed. What follows from their examples is that improving the international innovative position must be preceded by balancing entities, institutions, and interrelations within the innovation system, as well as by a gradual accumulation of knowledge in these states. The economic policy, including the innovative policy, pursued by the government is exceptionally important in this context.

The analyses conducted show that in order to actively utilize its advantages on international markets and develop competitive advantages in trade, a country needs to obtain a certain "critical mass," i.e., a conglomerate of human capital resources, research potential, and institutional solutions in the area of science, research, and education. Only some types of national innovation systems, e.g., dynamic innovation systems, have such a critical mass at their disposal. "Catching-up" and unbalanced systems have yet to develop one.

Certain general conclusions can be drawn from the analyses conducted.

First, innovation-based competition develops gradually over longer periods of time.

Second, there are no universal tools that support human capital development and innovation, and that bring about the same effect irrespective of the state in which they are implemented. The local context is significant, including, especially, the national innovation system model (type) that functions in the given country.

Third, institutional solutions that are well adapted to the development level of the innovation system are of key importance. Their scope cannot be narrowed down to encompass only innovative policy or science policy in the strict sense. It is necessary to assume an approach directed at releasing human energy, motivation to take action (and risk), and entrepreneurship.

Finally, education is the key to develop knowledge-based competitive advantages. This has been proven not only by quantitative research, but also by all the case studies.

It is worth completing these general conclusions concerning the economic policy by detailed recommendations for countries with various types of national innovation systems. Although one can assume that countries with dynamic and performing innovation systems are already basing their competitive advantages in international trade on human capital and innovations, and all they need to do is slightly modify (or correct) their existing policies, the situation is quite different in other states. The conclusions for countries with catching-up innovation systems might prove especially interesting. Poland is an example of such a system, as it is capable of autonomously developing its economic policy (including its innovation policy) only to a certain extent. This results from the fact that Poland, as an EU Member State, is obliged to adhere to certain supranational guidelines (e.g., the currently binding strategy *Europe 2020*).

It seems that in the case of a "catching-up" innovation system that is characteristic of Poland, the economic policy should be targeted at developing the education system. This concerns both the financial dimension that would guarantee constant growth in expenditures in this area and the institutional dimension that determines education quality. Also changes in the approach to the education process itself are of great importance. It is significant to target all activities at developing the skill of effective problem-solving and adapting the education offer to the needs of the labor market. Moreover, it is necessary to internationalize education to a greater extent, not only at the academic level, but also at the level of secondary and primary schools.

Moreover, as the examples of many states show, supporting innovation-related infrastructure development stimulates competitiveness. Such support involves the development of technology parks and other institutions tasked with mediating between scientific and business environments. Clusters are another underrated policy instrument whose potential has been slightly overlooked in the so-called "catching-up" innovation systems. If appropriate support is provided, these structures can develop into specific growth poles of the economy, and boost the export capacity of local enterprises.

The power of the science policy is underestimated in many countries (including Poland), although this policy plays an important role in translating knowledge into innovation and competitive advantages in foreign trade. First, it is necessary to postulate wider internationalization of science. Moreover, it is important to channel R&D activity to meet the market needs, facilitate cooperation with business, and stimulate the commercialization of research results. This goal could be achieved by introducing a system of incentives that would encourage cooperation between enterprises and universities in the scope of developing innovative research and implementation projects. Such joint projects could be financed from public funds. In the case of Poland, the science policy would be more effective if it entailed a wider promotion of EU programs targeted at scientists. Examples of good practices in this scope are provided by, e.g., Ireland and Austria.

One of the lines of the policy supporting the development of knowledge-based international competitiveness consists in the active support of external knowledge usage. External knowledge can be obtained through direct foreign investments and—in the case of Poland—EU structural funds. The objective is not only to attain foreign capital, but also to channel it into projects connected with education, science, R&D activity, and technology development. Therefore, it is recommended to ensure stronger relations between the innovation policy and foreign economic policy. A number of the countries whose experiences have been analyzed in this book developed an innovation system managed at central level by a body that bears responsibility for linking and coordinating various economic policy elements (including sectoral policies). The presence of such a government competitiveness council that coordinates various, sometimes incoherent, activities and regulations supports the process of drafting and implementing the national competitiveness strategy.

Editor and Author Bios

Andreas Bielig

Andreas Bielig, PhD, works as an Assistant Professor at the Warsaw School of Economics in Poland at the World Economy Research Institute and the Polish-German Academic Forum. Study of Economic Sciences at the University of Leipzig, PhD in Economics at the Technical University of Ilmenau, Research Fellow at the Helmholtz-Institute for Environmental Research (UFZ) in Leipzig, DAAD Fellow at the Warsaw School of Economics in Poland. Areas of his research include Intellectual Property, Economic Analysis of Law, Institutional Economics, International Economics. He received a DAAD long term Scholarship at the Warsaw School of Economics (2006–2012).

Ireneusz Bil

Ireneusz Bil, PhD, received his doctoral degree while studying the energy safety in the European Union, giving Germany as the example, from the Warsaw School of Economics, which is where he currently works in the World Economy Research Institute's team that is focused on conducting research on the German economy. His research interests include the European integration, cooperation between the Polish and German economies as well as the politics and energy safety. He is a member of the International Association for Energy Economics.

Ziemowit Czajkowski

Ziemowit Czajkowski is a PhD student at the Warsaw School of Economics and a member of the World Economy Research Institute's research team since the year 2006. Also at the Warsaw School of Economic, Mr. Czajkowski finished his master studies in economic with a thesis discussing the minimization of the dispersion for immunized bond portfolios.

M. A. Weresa (ed.), *Innovation, Human Capital and Trade Competitiveness*, Innovation, Technology, and Knowledge Management, DOI 10.1007/978-3-319-02072-3,

Krzysztof Falkowski

Krzysztof Falkowski, PhD, is a member of the World Economic Research Institute team since January of 2001. His studies focus on the Baltic economies as well as on the Eastern European countries.

Marcin Gomułka

Marcin Gomułka is a PhD student at the Warsaw School of Economics where he has been working as a teaching assistant since 2006 with the focus on classes with topics of banking. Mr. Gomułka received his Master of Science in Economics where his studies lead him to writing a thesis titled "Algorithmisation of the Shapley-Shubik Index. Application for the analysis of Poland's position in the EU-Constitution negotiations."

Günter S. Heiduk

Günter S. Heiduk, PhD, studied economics at universities in Erlangen-Nürnberg, Cologne and Aachen, Germany. After finishing his PhD dissertation as well as his habilitation he accepted in 1981 an appointment as Professor of International Economics at Duisburg-Essen University from where he retired in 2007. During his tenure he served as Dean of the Faculty of Business and Economics as well as Vice-President for International Academic Affairs. In 1983 he co-founded the East Asian Studies Programme. Since 1986 he is Chairman of the Asia-Pacific Economic Research Institute, a scientific institution at Duisburg-Essen University. He had been invited as Visiting Professor/Researcher at Kyoto University, Tokyo Keizai University, Fudan University, Wuhan University. He also worked as long- and short-term expert in several European Commission and German government financed projects in Albania, Armenia and China. Since December 2007 he had been appointed Professor and Head of the East Asian Center at Warsaw School of Economics, Poland. Furthermore, he holds a honourable position as Guest Professor at the International Business School, Northwest University, Xi'an, P.R. China. He published 25 books and more than 50 articles.

Oskar Kowalewski

Oskar Kowalewski, PhD, is an Associate Professor at the World Economy Research Institute at the Warsaw School of Economics (SGH) in Poland. He is also a Fellow of the Wharton Financial Institutions Center at the University of Pennsylvania. In 2009 he was awarded a grant by the Ministry of Science and Higher Education for a stay at the Wharton Financial Institution Center in the years 2009–2011. He was previously Assistant Professor in the Department of Finance and Director of the Polish—German Center for Economic Research at the Leona Kozminski University in Warsaw. In 2005, he received the Kosciuszko Postdoc Fellowship for the 1 year stay at the Wharton School at the University of Pennsylvania. In 2009 he was Visiting Professor at the Department of Economics at the University of Osnabruck in Germany. His primary research interest is on corporate finance, financial intermediation and international comparisons of financial systems. His current projects include a study of corporate governance in public owned companies and the relationship between knowledge based firms' and the structures of financial system across the countries.

Arkadiusz Michał Kowalski

Arkadiusz Michał Kowalski, PhD, works at the World Economy Research Institute, Warsaw School of Economics. His research and academic teaching focus on innovation systems, industrial clusters and international competitiveness. He received his PhD in Economics from the Collegium of Economic Analysis in the Warsaw School of Economics in 2006. Now, his habilitation procedure is carried out on the basis of the monograph on the development of clusters in Poland and their role for the innovativeness of the economy. In 2006–2011, he was cooperating with the Ministry of Economy as an expert in the fields of innovation and clusters. In 2009–2011, he was representing Poland in "Transnational Alliance of Clusters Towards Improved Cooperation Support" (TACTICS) Reflection Group. In 2013, he cooperates as an external expert with the Ministry of Regional Development on preparing programmes supporting clusters in new financial perspective 2014–20. Other experiences include visiting scholarships at the University of Macedonia in Thessalonica (1999–2000) and Oxford (2000) (during master studies), University of Southampton (2004–05) and University of Glasgow (2005–06) (during Ph.D. studies), and University of Coimbra, Portugal (2010), ICN Business School, Nancy, France (2011, 2013), University of Richmond, USA (2011–12) and INSEEC Business School, Paris & Bordeaux, France (2013).

Agnieszka McCaleb

Agnieszka McCaleb is a Research Fellow at the East Asian Center, Warsaw School of Economics and a PhD candidate at the English language Doctoral Programme in Management and Economics at Warsaw School of Economics. McCaleb graduated from Warsaw School of Economics, major in International Economic and Political Relations. She simultaneously achieved her M.A. in Sinology, obtained at Warsaw University. From 2006 to 2009, she was Marketing Manager for Chinese markets at Selena Group, one of the largest Polish investors in China. McCaleb's research interests focus on Chinese multinational companies.

Beata Michorowska

Beata Michorowska, PhD, received her PhD degree from the Warsaw School of Economics after conducting her research on the topic of the transfer of technology enterprises that arises as a result of foreign direct investment inflows to Poland over a decade of 1998–2008. Her work experience is derived from positions in the Polish Agency for Enterprise Development and ECORYS Poland.

Józef Misala

Józef Misala, PhD, has been a full professor of economics at the Warsaw School of Economics since 1993 where he has been working since 1978 (known then as the Main School of Planning and Statistics) while, since 1999, also being the director of Chair for International Economics and Regional Economic Integration at the Technical University of Radom. Outside of academia, Dr. Misala has been the advisor to Minister and deputy director of the Economic Strategy Department of the Polish Central Planning Office.

Tomasz M. Napiórkowski

Mr. Tomasz Napiórkowski is a PhD student at the Warsaw School of Economics where his research is focused on foreign direct investments and multinational enterprises. There he is a part of the World Economy Research Institute's research team and has been involved in five projects since his start in September of 2011, with over nine research projects completed overall. Mr. Napiórkowski received his Master of Business Administration and Master of Arts in Economic degrees from Old Dominion University in the USA where his studies focused on econometrics and its application in the local microeconomic environment. He also holds a Bachelor of Science in Business Administration degree with a double major in Decision Sciences and Marketing.

Józef Olszyński

Józef Olszyński, PhD, is a professor at the Warsaw School of Economics. He was a titular Ambassador and a Head of an Economic and Commercial Section in the Embassy of the Republic of Poland in Berlin. His research interests are: German integration concepts, German position in the European Union, Polish-German economic relations.

Mariusz-Jan Radło

Mariusz-Jan Radło, PhD, holds PhD in Economics from the Warsaw School of Economics (SGH 2003). He gained a Master's degree in Economics (1998) and Political Science (1999) at the Maria Curie Skłodowska University in Lublin. He also holds a postgraduate degree in European Integration from the National School of Public Administration (Poland) and the Ecole Nationale d'Administration (France). Since 2004, he has worked as a research fellow and lecturer at the Warsaw School of Economics. Since 2009 he is a managing partner at SEENDICO in Warsaw. Between 2004 and 2007, Dr. Radlo was the vice president for research of the Polish Lisbon Strategy Forum at the Gdańsk Institute for Market Economics, where he was responsible for the coordination of research projects investigating macroeconomic and sector reforms in Poland. Before 2004, he worked for the Ministry of Foreign Affairs and Office of the Committee for the European Integration. He has participated in many research and consulting projects concerning economic policy, strategy development, and sector and macroeconomic issues. He was a visiting professor and lecturer at the Aarhus School of Business in Denmark (2009 and 2010) and at the Wirtschaftsuniversität Wien in Austria (2011). In 2012 (June–September), he was a visiting research fellow at Corvinus University of Budapest in Hungary.

Marzenna Anna Weresa

Marzenna Anna Weresa, PhD, is a professor of the World Economy Research Institute at the Warsaw School of Economics (Poland), and the Institute's director. She holds D.Sc. in Economics (habilitation, 2003) and PhD degree in Economics (1995) from the Warsaw School of Economics in Poland. In 1999–2000 she worked as a research fellow at the University College London. She also was a Visiting Professor at the University of Leicester in the UK and at the Northeastern Illinois University in Chicago (USA). She has been working as an expert for the European Commission (projects on R&D, innovation and technological competitiveness) and conducted advisory projects for enterprises and governmental organizations. Her research focuses on innovation, technology transfer and foreign direct investment. She authored and co-authored over 80 books and scientific articles.

Grażyna Wojtkowska-Łodej

Grażyna Wojtkowska-Łodej, PhD, is a professor at the Warsaw School of Economics. She is the Head of the European Union Unit, which is a part of the Institute of Foreign Trade and European Studies in the Collegium of World Economy. From September 2005 to August 2012, she was the Dean for the Bachelor Studies. She is an author and a co-author of various publications and expertizes for public administration and for numerous businesses. Her major areas of scientific interest include: international economics, economic integration in Europe with special interest paid to issues of international trade, industry, energy and environmental policy.

Bibliography

Aghion, P., & Howitt, P. (1998). *Endogenous growth theory*. Cambridge: MIT Press.

Ahuja, A., Cheung, L., Han, G., Porter, N., & Zhang, W. (2010). *Are house prices rising too fast in China?* IMF Working Paper, WP10/274. IMF: Washington, DC.

Ainamo, A. (1997). Evolution of the Finnish system of innovation: The Contribution of Nokia. In B. Fynes & S. Ennis (Ed.), *Competing from the periphery: Core issues in international business* (pp. 423–439). Dublin: The Dryden Press.

Aiginger, K. (1998). A framework for evaluating the dynamic competitiveness of countries. *Structural Change and Economic Dynamics, 9*(2), 159–188.

Aiginger, K. (2006). Competitiveness: From a dangerous obsession to a welfare creating ability with positive externalities. *Journal of Industry, Competition and Trade, 6*(2), 161–177.

Aiginger, K. (2008), Wettbewerbsfähigkeit von Firmen, Regionen und Ländern. *Die Volkswirtschaft*, (3).

Aiginger, K. (2009). *Speed of change and growth of manufacturing*. Vienna: Austrian Insitute of Economic Research.

Alders, P. (2005). Human capital growth and destruction: The effect of fertility on skill obsolescence. *Economic Modelling, 22*(3), 503–520.

Ambroziak, Ł (2012). *Niemcy. Koniunktura gospodarcza świata i Polski w latach 2010–2013*. Warsaw: Instytut Badań Rynku, Konsumpcji i Koniunktur.

Amiti, M., & Freund, C. (2008). *The anatomy of China's export growth. Policy Research Working Paper, 4628*. Washington, DC: World Bank.

Anderson, J. (2007). *Is China export Led?* UBS Investment Research, Asian Focus. *Sept, 27,* 2007.

Anger, C. (2007). *Deutschlands Ausstattung mit Humankapital—Ergebnisse des IW-Humankapitalindikators*. IW-Trends. Cologne: Institut der deutschen Wirtschaft Köln.

APEC. (2000). *Towards knowledge-based economies in APEC*. Singapore: APEC Economic Committee.

Aquino, A. (1978). Intra-industry trade and intra-industry specialization as concurrent sources of international trade in manufactures. *Weltwirtschaftliches Archiv, 114*(2), 275–296.

Arrow, K. J. (1973). Higher education as a filter. *Journal of Public Economics, 2*(3), 193–216.

Asian Development Bank. *Statistical data base system*. https://sdbs.adb.org/sdbs/index.jsp.

Atkinson, R. D. (2012). *Enough is enough: Confronting Chinese innovation mercantilism*. Washington, DC: The Information Technology and Innovation Foundation.

Atzei, A., Groepper, P., & Novara, M. (1999). Innovations for competitiveness: European views on "Better-Faster-Cheaper". *Acta Astronautica, 44*(7–12), 745–754.

Baczko, T., Krzywina, E., & Pieńkowska, M. (2010). Główne wnioski. In T. Baczko (Ed.), *Raport o innowacyjności gospodarki Polski w 2009 roku*. Warsaw: Instytut Nauk Ekonomicznych Polskiej Akademii Nauk.

Bai, C.-E., Hsieh, C.-T., & Qian, Y. (2006). *The return to capital in China*. NBER Working Paper, 12755. Mass: Cambridge.

M. A. Weresa (ed.), *Innovation, Human Capital and Trade Competitiveness*, Innovation, Technology, and Knowledge Management, DOI 10.1007/978-3-319-02072-3,

Balassa, B. (1962). *Recent developments in the competitiveness of American industry and prospects for the future. Factors affecting the United States balance of payments*. Washington, DC: US Governmental Printing Office.

Balassa, B. (1965). Trade liberalization and revealed comparative advantage. *The Manchester School, 33*(2), 99–123.

Balcerowicz, L. (1980). Organizational structure of the national economy and technical innovation. *Acta Oeconomica, 24*(1–2), 151–167.

Balzat, M. (2002), *Theoretical basis and the empirical treatment of national innovation systems.* Discussion Paper Series No. 232, Universitaet Augsburg, Institute for Economics.

Barry, F. (2005). FDI, transfer pricing and the measurement of R & D intensity. *Research Policy, 34,* 673–681.

Barro, R. J. (1991). Economic growth in a cross section of countries. *The Quarterly Journal of Economics, 106*(2), 407–443.

Barro, R. J. (1998). *Notes on growth accounting*. NBER Working Paper Series, No. 6654 (July).

Barro, R. J., & Lee, J.-W. (1993). *International comparisons of educational attainment.* NBER Working Paper Series, No. 4349 (April).

Barro, R. J., & Lee, J.-W. (1996). International measures of schooling years and schooling quality. *The American Economic Review, 86*(2), 218–223.

Barro, R. J., & Lee, J.-W. (2000). *International data on educational attainment updates and implications*. NBER Working Paper Series, No. 7911 (September).

Barro, R. J., & Lee, J.-W. (2010). *A new data set of educational attainment in the world, 1950–2010*. NBER Working Paper Series, No. 15902 (April).

Batra, A., & Khan, Z. (2005). *Revealed comparative advantage: An analysis for India and China. Working Paper, No. 168*. New Delhi: Indian Council for Research on International Economic Relations.

Becker, G. S. (1964). *Human capital: A theoretical and empirical analysis, with special reference to education*. Chicago: University of Chicago Press.

Belitz, et al. (2011). *Die deutsche forschungsintensive Industrie in der Finanz- und Wirtschaftskrise im internationalen Vergleich*, Studien zum deutschen Innovationssystem No. 4-2011, Berlin.

Belke, A., & Schneider, F. (2005). *Privatization in Austria: Response to International and External Pressures*. CESifo DICE Report, No. 1/2005.

Bhide, A. (2008). *The Venturesome economy: How innovation sustains prosperity in a more connected world.* Princeton: Princeton University Press.

Bieńkowski, W., Weresa, M. A., & Radło, M. J. (Ed.). (2010). *Konkurencyjność Polski na tle zmian gospodarczych w krajach OECD. Ewolucja znaczenia czynników konkurencyjności*. SGH: Warsaw.

Bils, M., & Klenow, P. J. (2000). *Does schooling cause growth*? *American Economic Review, 90*(5), 1160–1183.

Bishop, J. (1992). The impact of academic competencies on wages, unemployment, and job performance. *Carnegie-Rochester Conference Series on Public Policy, 37*(1), 127–194.

Board of Governors of the Federal Reserve System. (2012). http://www.federalreserve.gov/aboutthefed/mission.html. Accessed Spring 2012.

Board on International Comparative Studies in Education, Commission on Behavioral and Social Sciences and Education, & National Research Council. (1995). *International comparative studies in education: Descriptions of selected large-scale assessments and case studies*. Washington, DC: National Academies Press.

Borowski, J. (2008). *Globalizacja, konkurencyjność międzynarodowa i strategie przedsiębiorstw*. Białystok: UwB.

Borst, N. (2011). *How should we measure investment in China?* China Economic Watch, Peterson Institute for International Economics, Aug 29, 2011.

Bossak, J. W. (1984). *Społeczno-ekonomiczne uwarunkowania międzynarodowej zdolności konkurencyjnej gospodarki* Japonii. Monografie i Opracowania No. 153. Warsaw: SGPiS.

Bossak, J. W. (2000). Międzynarodowa konkurencyjność gospodarki polskiej— ujęcie instytucjonalne. In H. Podedworny, J. Grabowiecki, & H. Wnorowski (Ed.), *Konkurencyjność gospodarki polskiej a rola państwa przed akcesją do Unii Europejskiej*. Białystok: Uniwersytet w Białymstoku.

Bossak, J. W. (2006). Teoria i metodologia. Krytyczna ocena stosowanych metod analizy. In M. Weresa (Ed.), *Polska. Raport o konkurencyjności 2006*. Warsaw: SGH.

Bossak, J. W., & Bieńkowski, W. (Ed.). (2001). *Konkurencyjność gospodarki Polski w dobie integracji z Unią Europejską i globalizacji*. Warsaw: SGH.

Bossak, J. W., & Bieńkowski, W. (2004). *Międzynarodowa zdolność konkurencyjna gospodarki i przedsiębiorstw. Wyzwania dla Polski na progu XXI wieku*. Warsaw: SGH.

Boston Consulting Group (BCG). (2008). *Creative Switzerland? Creating an innovation powerhouse*. Zurich: The Swiss–American Chamber of Commerce and The Boston Consulting Group.

Bożyk, P. (2008). *Międzynarodowe stosunki ekonomiczne*. Warsaw: PWE.

Bożyk, P., Misala, J., & Puławski, M. (1998, 2002). *Międzynarodowe stosunki ekonomiczne*. Warsaw: PWE.

Brander, J., & Spencer, B. (1985). Export subsidies and market share rivalry. *Journal of International Economics, 10*(18), 83–100.

Brandt, L., Ma, D., & Rawski, T. G. (2012). *From divergence to convergence: Re-evaluating the history behind China's economic boom. Working Papers, No. 158/12*. London: Department of Economic History, London School of Economics.

Breschi, S., & Malerba, F. (1997). Sectoral innovation systems: Technological regimes, Schumpeterian dynamics, and spatial boundaries. In C. Edquist (Ed.), *Systems of innovation: Technologies, institutions and organizations* (pp. 130–156). London–Washington: Pinter.

Brocka-Palacz, B., & Bil, I. (2010). Zmiana znaczenia czynników międzynarodowej konkurencyjności gospodarczej wynikająca z procesów globalizacji i integracji z Unią Europejską. Przypadek Niemiec. In W. Bieńkowski, M. A. Weresa, & M. J. Radło (Ed.), *Konkurencyjność Polski na tle zmian gospodarczych w krajach OECD*. Warsaw: SGH.

Brühlhart, M. (2009). *An account or global intra-industry trade, 1962–2006. The World Economy*, 32, 401–459.

Bruno, M. (1963). Domestic resource costs and effective protection clarification and synthesis. *Journal of Political Economy, 80*(1), 16–33.

Budnikowski, A. (2006). *Międzynarodowe stosunki gospodarcze*. Warsaw: PWE.

Budnikowski, T. (2011). Niemiecki rynek pracy w 20 lat po zjednoczeniu. In: T. Budnikowski (Ed.), *Gospodarka niemiecka 20 lat po zjednoczeniu*. Poznań: Instytut Zachodni.

Budnikowski, A., & Kawecka-Wyrzykowska, E. (Ed.). (2000). *Międzynarodowe stosunki gospodarcze*. Warsaw: PWE.

Bundesamt für Statistik. (2011a). *Neuchâtel*. http://www.bfs.admin.ch.

Bundesamt für Statistik. (2011b). *International aspects of Swiss research and development in 2008. From the internationalization of enterprises to international public cooperation*. Neuchatel: Federal Statistical Office (FSO).

Bundesministerium für Bildung und Forschung (BMBF). (2010). *Bundesbericht Forschung und Innovation 2010*. Berlin.

Bundesministerium für Bildung und Forschung (BMBF). (2012a). *Berufsbildungsbericht 2012*. Berlin.

Bundesministerium für Bildung und Forschung (BMBF). (2012b). *Bericht über die Umsetzung des Bologna-Prozesses in Deutschland 2009–2012*. Berlin.

Burenstam-Linder, S. (1961). *An essay on trade and transformation*. New York: Wiley.

Burzyński, W., Marczewski, K., & Wojnar, J. (1997). *Analiza cen osiąganych przez eksporterów polskich i konkurentów z Unii Europejskiej na rynku niemieckim w latach 1992–1995. In Zagraniczna polityka gospodarcza Polski 1996–1997*. Warsaw: IKCHZ.

Brülhart, M. (1994). Marginal intra-industry trade: Measurement and relevance for the pattern of industry adjustment. *Weltwirtschaftliches Archiv, 130*(3), 600–613.

CAE. (2012). http://www.cae.cn.

CAST. (2012). http://www.cast.org.cn.

Cai, F. (2010). Demographic transition, demographic dividend, and Lewis turning point in China. *Economic Journal, 3*(2), 107–120.

Cantwell, J. (Ed.). (1994). *Transnational corporations and innovatory activities (Vol. 17) UN-LTNC*. London: Routlege.

Cantwell, J. (2005). Innovation and competitiveness. In J. Fagerberg, D. C. Mowery, & R. R. Nelson (Eds.), *The Oxford handbook of innovation*. Oxford: Oxford University Press.

Carballo-Cruz, F. (2011). Causes and consequences of the Spanish economic crisis: Why the recovery is taken so long? *Panoeconomicus, 3,* 209–328.

Carlsson, B. (Ed.). (1997). *Technological systems and industrial dynamics*. Boston: Kluwer Academic Publications.

Castellacci, F. (2007). Innovation and the competitiveness of industries: Comparing the mainstream and the evolutionary approaches. *Technological Forecasting & Social Change, 75*(2008), 984–1006.

Centintas, H., & Barisik, S. (2009). Export, import, and economic growth: The case of transition economies. *Transition Studies Review, 15*(4), 636–649.

Cervantes, M. (2007). *Background note for panel 3: Implications for TIP and NESTI, Joint NESTI-TIP Workshop on innovation indicators for policy making and impact assessment*. http://www.oecd.org/dataoecd/20/13/38911869.pdf. Accessed Aug 2011: OECD.

Chamberlin, E. H. (1934). *The theory of monopolistic competition*. Cambridge: Harvard University Press.

Chen, Y. (2007). Impact of Foreign direct investment on regional innovative capability: A case of China. *Journal of Data Science, 5,* 577–596.

Chen, K. M., Rau, H. H., & Chiu, R. L. (2011). Determinants of China's exports to the United States and Japan. *The Chinese Economy, 44*(4), 19–41.

Cheng, M. Y., & Ghulam, R. (2007). Knowledge gap and earnings differential in the knowledge-based economy. Applied Economics Letters *14*(February), 219–221.

Cheung, G. C. K. (2009). Made in China vs. made by Chinese: Global identities of Chinese business—An Introduction. *Journal of Contemporary China, 18*(58), 1–5.

Cheung, K. Y. (2010). Spillover effects of FDI via exports on innovation performance of China's high-technology industries. *Journal of Contemporary China, 19*(65), 541–557.

Cheung, Y., Chinn, M., & Fujii, E. (2010). China's current account and exchange rate. In R. Feenstra & S.-J. Wei (Eds.), *China's growing role in world trade* (pp. 231–280). Chicago: University of Chicago Press.

Chilimoniuk, E. (2006). *Ocena skuteczności wybranych narzędzi polityki handlowej w warunkach oligopolu międzynarodowego, Kolegium Gospodarki Światowej*. Warsaw: SGH (niepublikowana praca doktorska).

China Daily. (2010). Government to increase spending on education. http://www.chinadaily.com.cn/china/2010-03/01/content_9515384.html.

China Daily. (2011a). 80 % China's processed trade from foreign firms. http://www.chinadaily.com.cn/china/2011-09/27/content_13804859.html.

China Daily. (2011b). China tops the list of overseas students in US. http://www.chinadaily.com.cn/china/2011-11/16/content_14101379.html.

China Daily. (2012). China's R & D spending surges 21.9 %. http://www.chinadaily.com.cn/bizchina/2012-02/22/content_14670805.html.

China Labour Bulletin. (2012). A decade of change: The worker's movement in China 2000–2010. http://www.clb.org.hk/en/sites*/default/files/File/research_reports/Decade%20of%20the%20Wor*kers%20Movement%20final.pdf.

Chou, T.-L., Ching, C.-H., Shu, S.-M., & Chang, J.-Y. (2011). Global linkages, the Chinese high-tech community and industrial cluster development: The semiconductor industry in Wuxi, Jiangsu. *Urban Studies, 48*(14), 3019–3042.

Ciesielska, D. A., & Radło, M.-J. (2011). *Identyfikacja potencjałów oraz barier rozwojowych polskich województw*, Raport z badania wykonanego na zlecenie Departamentu Analiz i Prognoz Ministerstwa Gospodarki, Warsaw.

CONACYT. *Informe General del Estado de la Ciencia y la Tecnología*, www.conacyt.gob.mx., INEGI-CONACYT. *Encuesta sobre Investigación y Desarrollo Tecnológico*, www.inegi.org.mx.

Conrad, K. (1992). Comment on D. W. Jorgenson and B. M. Fraumeni. Investment in education and U.S. economic growth. *The Scandinavian Journal of Economics, 94*(January), 71–74.

Cordes, A., & Gehrke, B. (2011). *Außenhandel, Strukturwandel und Qualifikationsnachfrage. Aktuelle Entwicklungen in Deutschland und im internationalen Vergleich*, Studien zum deutschen Innovationssystem No. 3-2011, Hannover.

Cox, W. (2011). China: Urbanizing and moving east, new geography. http://www.newgeography.com/content/002218-china-urbanizing-and-moving-east-2010-census. Accessed 5 April 2011.

Csaba, L. (2006). *How much trade and FDI theories help in analysing competitiveness—Related issues*, refrat wygłoszony na międzynarodowej konferencji. Łańcut: *New Europe*.

Czarny, E. (2002). *Teoria i praktyka handlu wewnątrzgałęziowego, Monografie i Opracowania, No. 496*. Warsaw: SGH.

Dagum, C., & Slottje, D. J. (2000). A new method to estimate the level and distribution of household human capital with application. *Structural Change and Economic Dynamics, 11*(1–2), 67–94.

Dahlman, C. J., Zeng, D. Z., & Wang, S. (2007). *Enhancing China's competitiveness through lifelong learning*. Washington, DC: World Bank Publications.

Daneke, G. A. (1998). Beyond Schumpeter: Nonlinear economics and the evolution of the U.S. innovation system. *Journal of Socio-Economics, 27*(1), 97–115.

Deardorff, A. V. (1980). The general validity of the law of comparative advantage. *Journal of Political Economy, 88*(5), 941–957.

Deardorff, A. V. (1982). The general validity of the Heckscher-Ohlin theorem. *American Economic Review, 72*(4), 683–694.

Deardorff, A. V. (2003). *Sources and implications of comparative advantages*. Michigan: University of Michigan, maszynopis powielony.

Deardorff, A. V. (1984). Testing trade theories and predicting trade flows. In R. W. Jones & P. B. Kenen (Eds.), *Handbook of international economics*. North Holland: Amsterdam-Washington-Oxford.

Deardorff, A. V. (2005). How robust is comparative advantage? *Review of International Economics, 13*(5), 1004–1016.

Deardorff, A. V. (2006). *Deardorff's glossary of international economics*. Washington, DC.

Deason, L. (2008). *Determinants of diffusion and downstreaming of technology-intensive products in international trade. Working Paper No. 2009-10-B*. Washington, DC: KS International Trade Commission.

Denison, E. F. (1967). *Why growth rates differ: postwar experience in nine western countries*. Washington, DC: Brookings Institution.

Department of Education and Skills Statement of Strategy 2011–2014. http://www.education.ie/. Accessed 30 April 2012.

Derick, J., Kraemer, K. L., & Linden, G. (2011). Who Profits from Innovation in global value chains? A study of the iPod and notebook PCs. *Industrial and Corporate Change, 19*(1), 81–116.

Deutschland im internationalen Wettbewerb. In *Erfolge im Ausland—Herausforderungen im Inland*, Sachverständigen Rat Jahresgutachten, Wiesbaden 2004.

Deutschland in Zahlen. Ausgabe 2011, Institut der Deutschen Wirtschaft, Köln 2011.

Deutschland in Zahlen. Ausgabe 2012, Institut der deutschen Wirtschaft, Köln 2012.

Exportweltmeister Deutschland—Titel auf Zeit?—Statistisches Bundesamt Wiesbaden, 2010.

Devonshire-Ellis, C. (2010). The RMB position and the mysteries of the china unemployment fund. *China Briefing*, Nov 12, 2010.

Dezhina, I. (2011). *Developing research in Russian universities* (pp. 18–19). Moscow: NIS Center.

Di Bello, G., & Andreta, E. (2012). *Boosting competitiveness through "Regional Innovation Strategies"*. Discussion Paper: Competitiveness and Smart Specialisation Strategies, APRE.

Dickmann, N. (2005). *Grundlagen der demographischen Entwicklung. In Perspektive 2050. Ökonomik des demographischen Wandels* (pp. 35–66). Köln: Institut der Deutschen Wirtschaft.
Dolla, V. S. (2011). *Technology trade in India-China relations: Divergent dynamics and implications. Journal of International Affairs, 64*(2), 125–141.
Dollar, D., & Wei, S.-J. (2007). *Das (Wasted) Kapital: Firm ownership and investment efficiency in China. NBER Working Paper, 1998*. Cambridge: National Bureau of Economic Research.
Dołęgowski, T. (2002). *Konkurencyjność instytucjonalna i systemowa w warunkach gospodarki globalnej*. Warsaw: SGH.
Dong, H., & Wan, X. (2012). Higher education tuition and fees in China: Implications and impacts on affordability and educational equity. *Current Issues in Education, 15*(1), 1–9.
Dornbusch, R., Fischer, S., & Samuelson, P. A. (1977). Comparative advantage. Trade and payments in Ricardian model with continuum of goods. American Economic Review, *67*(5).
Dosi, G. Freeman, Ch., Nelson, R. R., Silverberg, G., & Soete, L. (Eds.). (1988). *Technical change and economic theory*. London: Pinter.
Drucker, P. F. (1992). *The age of discontinuity: guidelines to our changing society*. Transaction Publishers.
Du, Y., & Wang, M. (2010). Discussion on potential bias and implications of Lewis turning point. *China Economic Journal, 3*(2), 121–136.
Dublin, L. I., & Lotka, A. J. (1977). *The money value of a man. Reprinted*. New York: Amo Press.
Dunning, J. (1958). *American investment in British manufacturing industry*. London: George Allen and Unwin.
Dunning, J. H. (1973). The determinants of international production (Vol. 25). Oxford Economic Papers.
Dunning, J. H. (1992). *The competitive advantage of countries and the activities of transnational corporations*, Transnational Corporations, No. 2.
Dunning, J. H. (1994). Re-evaluating the benefits of foreign direct investment. Reading University. http://www.unctad.org/en/docs/iteiitv3n1a3_en.pdf.
Dunning, J. H., & Narula, R. (1996). The investment development path revisited: Some emerging issues. In J. H. Dunning & R. Narula (Eds.), *Foreign direct investments and governments; catalysts for economic restructuring*. London: Routledge.
Durlauf, S. N., Johnson, P. A., & Temple, J. R. W. (2005). Chapter 8 growth econometrics. In *Vol. 1* (Part A; pp. 555–677). Elsevier.
Dutta S. (Ed.). (2011). *The global innovation index 2011*. Fontainebleau: INSEAD.
Dyck, S., & Levinger, H. (2010). *China's provinces*. Deutsche Bank Research, 25 Feb, 2010. Frankfurt a. M.
Eaton, J., & Grossman, G. (1986). Optimal trade policy and industrial under oligopoly. *Quarterly Journal of Economics, 101*(2), 383–406.
EC Trade (2012). China. http://trade.ec.europa.eu/doclib/docs/2006/september/tradoc_113366.pdf.
Ederer, P. (2006). *Innovation at work: The European human capital index*. Brussels: The Lisbon Council.
Ederer, P., Schuller, P., & Willms, S. (2002). *Wieviel Bildung brauchen wir? Humankapital in Deutschland und seine Erträge*. Frankfurt a. M.
Ederer, P., Schuler, P., & Willms, S. (2007). *The European human capital index: The challenge of central and eastern Europe*. Brussels: The Lisbon Council.
Ederer, S., Janger, J., Kaniovski, S., Kletzan-Slamanig, D., Berger, J., Fortin, I., Hofer, H., Paterson, I., Skriner, E., Schönpflung, K., Schuh, U., & Schwarzbauer, W. (2010). *Assessing the Lisbon strategy 2005–2010 and estimating expected effects from reaching the EU 2020 Gaals*. Vienna: Óesterreichisches Institut für Wirtschaftsforschung, Institut für Höhere Studien. December 2010.
Edquist C. (Ed.). (1997). *Systems of innovation: Technologies, institutions, organizations*. London–Washington: Pinter.
Ehmer, P. (2011). *Structural change in China.* Deutsche Bank Research, *16 February 2011*. Frankfurt/M: Deutsche Bank.

Eichgreen, B., & Tong, H. (2011). *The external impact of China's exchange rate policy: Evidence from firm level data.* NBER Working Paper, 17593. Cambridge: National Bureau of Economic Research.

Eisner, R. (1989). *The total incomes system of accounts*. Chicago: University of Chicago Press.

Embassy of Switzerland. (2010). *Science, technology and education news from China*. Number 73.

Engel, E. (1883). *Der Wert des Menschen*. Berlin: Verlag von Leonhard Simion.

Ennew, C. T., & Fujia, Y. (2009). Foreign universities in China: A case study. *European Journal of Education, 44*(1), 21–36.

Erawatch. (2010). *Country Reports 2010: Mexico*. http://erawatch.jrc.ec.europa.eu/.

Ethier, W. J. (1982). National and international returns to scale in the modern theory of international trade. *American Economic Review, 72*(3), 389–405.

Ethier, W. J. (2009). *Comparative advantages meets Alan Deardorff*. Pennsylvania: University of Pennsylvania, Department of Economics, mimeo.

Etzkowitz, H., & Leydesdorff, L. (2000). The dynamics of innovation: from national systems and "Mode 2" to a triple helix of university–industry–government relations. *Research Policy, 29,* 109–123.

Euromonitor International. Data exported: July 13, 2012.

European Commission. Private sector interaction in the decision making processes of public research policies. Country Profile: China. http://ec.europa.eu/invest-in-research/pdf/download_en/psi_countryprofile_china.pdf.

European Commission. (2001). *Innovation and enterprise creation: Statistics and indicators. Proceedings of the conference held at Sophia Antipolis, 23 and 24 November 2000. Innovation papers No. 18*. Luxemburg: Office for Official Publications of the European Communities.

European Commission. (2009). *INNO-Policy TrendChart—Innovation policy progress report. Spain 2009*. Brussel: Enterprise Directorate-General.

European Commission. (2010). *Europe 2020: A Strategy for smart, sustainable and inclusive growth.* Communication from the Commission. Brussels: 2010; *Basic figures of the EU, Spring 2012*. Eurostat.

European Commission. (2011). *Innovation union scoreboard 2011. The Innovation Union's performance scoreboard for Research and Innovation*. PRO INNO Europe, Inno Metrics UNU–UERIT.

European Commission. (2012). Innovation union scoreboard 2011, PRO INNO Europe, Inno Metrics. http://ec.europa.eu/enterprise/policies/innovation/facts-figures-analysis/innovation-scoreboard/index_en.html.

European Union. (2011a). *Innovation union scoreboard 2010. The innovation union's performance scoreboard for Research and Innovation*. Brussels: European Commission.

European Union. (2011b). *Innovation union competitiveness Report 2011. Zusammenfassung*. Brussels: European Commission.

EUROSTAT. (2010). *Mexico-EU—basic statistical indicators*, http://epp.eurostat.ec.europa.eu.

INEGI. (2010). *Population and Housing Census 2010*. http://www.inegi.org.mx.

Ewerhart, G. (2001). *Humankapital in Deutschland: Bildungsinvestitionen, Bildungsvermögen und Abschreibun-ben auf Bildung*. Nürnberg.

Expertenkommission für Forschung und Innovation (EFI). (2011). *Report 2011. Research, Innovation and Technological Performance in Germany*. Berlin.

Ezeala-Harrison, F. (1999). *Theory and policy of international competitiveness*. Westport: Praeger Publishers.

Ezeala-Harrison, F. (2005). On the competing notions of international competitiveness. *Advances in Competitiveness Research, 13*(1).

Fabre, A. (2012). *The German economic model: A strategy for Europe? Foundation Robert Schumann*. "European Issues" No. 237, 23 April 2012.

Farr, W. (1853). The income and property tax. *Journal of the Statistical Society of London, 16, 1*(March), 1–44.

Fast Company. (2010). Most innovative China companies. http://www.fastcompany.com/mic/2010/industry/most-innovative-china-companies.

Fast Company. (2012). The world's 50 most innovative companies. Top 10 China. http://www.fastcompany.com/most-innovative-companies/2012/industry/china.

Falvey, R. E., & Kierzkowski, H. (1987). *Product quality, intra-industry trade and perfect competition in international trade. Essays in honour of W. M. Corden*. Oxford: Macmillan.

Feenstra, R., & Kee, H. L. (2005). *Trade liberalization and export variety: A comparison of China and Mexico*. Policy Research Working Paper Series, 3213. The World Bank.

Feldstein, M. (2011). *The role of currency realignments in eliminating the US and China current account imbalances*. NBER Working Paper, 16674, Cambridge, Mass.

Felipe, J., Kumar, U., Usui, N., & Abdon, A. (2010). *Why has China succeeded—and why it will continue to do so*. Working Paper, 611, Levy Economics Institute at Bard College, Annandale-at-Hudson, NY.

Fiedor, B. (1979). *Teoria innowacji*. Warsaw: PWN.

Fiedor, B., & Czaja, S. (2003). *Kapitał ludzki oraz mechanizmy i główne sfery jego rozwoju w Polsce w okresie transformacji*. In J. Mujżel (Ed.), *Problemy wzrostu gospodarczego w warunkach ustrojowej transformacji w Polsce* (pp. 74–119). Warsaw: Instytut Nauk Ekonomicznych PAN.

Fischer, B. (2011). *China's long march to innovative success*. Forbes. http://www.forbes.com/sites/billfischer/2011/07/26/chinas-long-march-to-innovation-success/. Accessed 26 July 2011.

Florczak, W. (2007). *Kapitał ludzki a rozwój gospodarczy*. In I. Władysław Welfe (Ed.), *Gospodarka oparta na wiedzy*. Warsaw: PWE.

Flossmann, A., Piatek, R., & Wichert, L. (2007). *Going beyond returns to education: The role of noncognitive skills on wages in Germany*. Econometrics of education.

Foders, F., & Langhammer, R. J. (Eds.). (2006). *Labour mobility and the world economy*. Berlin-Heidelberg: Springer-Verlag.

Fontagné, F., Freudenberg, M., & Gaullier, G. (2005). *Disentangling horizontal and vertical intra-indstury trade*. CEPII Working Paper, No. 2005/10, Centre d'Etudes Prospectives et d'Informations Internationales, Paris.

Fontagne, L., Freudenberg, M., & Peridy, N. (1997). *Trade patterns inside the single market*. Document de Travail 97-07, Centre Etudes Perspectives et d'Informations Internationales, Paris.

Fraumeni, B. M. (2008). *Human capital: From indicators and indexes to accounts*. Turin: OECD, 4. November.

Fraumeni, B. M. (2011). *Human capital accounts: Choice of rates and construction of volume indices*. NBER Working Paper Series, No. 16895 (March).

Freeman, C. (1987). *Technology policy and economic performance: Lessons from Japan*. London: Pinter.

Freeman, R. B. (1981). *Career patterns of college graduates in a declining job market*. NBER Working Paper Series, No. 750 (September).

Frejtag-Mika, E. (Ed.). (2006). *Teoria i praktyka ekonomii a konkurencyjność gospodarowania*. Warsaw: Difin.

Fu, X. (2008). *Foreign direct investment, absorptive capacity and regional innovative capabilities: Evidence from China. Oxford Development Studies, 36*(1), 89–110.

Fu, D., Wu, Y., & Tang, Y. (2010). *The effects of ownership structure and industry characteristics on export performance*. Discussion Paper 10-09, Western Australia University, Perth.

Fujita, M., Krugman, P., & Venables, A. J. (1999). *The spatial economy: Cities, regions and international trade*. Cambridge (Mass): Cambridge University Press.

Furman, J. L., Porter, M. E., & Stern, S. (2002). The determinants of national innovative capacity. *Research Policy, 31,* 899–933.

Gagacka, M. (2007). *Kapitał ludzki i społeczny a innowacyjność mikroprzedsiębiorstw*. referat z III Konferencji Naukowej z cyklu "Wiedza i Innowacje" pt. Wiedza i innowacje w rozwoju gospodarki: siły motoryczne i bariery, Collegium Novum, Kraków.

Garnaud, R. (2010). Macro-economic implications of the turning point. *China Economic Journal, 3*(2), 181–190.

Garnaud, R., & Huang, Y. (2006). *Continued rapid growth and the turning point in China's economic development*. In R. Garnaud & L. Song (Eds.) (2006). *The turning point in China's development* (pp. 12–34). Australian National University: Asia Pacific Press.

Gaulier, G., Lemoine, F., & Unal-Kesenci, D. (2007). China's integration in east Asia: Production sharing, fdi & high-tech trade. *Econ Change, 40,* 27–63.

Gehrke, B., & Krawczyk, O. (2012). *Aussenhandel mit forschungsintensiven Waren im internationalen Vergleich*. Niedersächsisches Institut für Wirtschaftsforschung e. V. Hannover, Studien zum deutschen Innovationssystem, No. 11-2012.

Gemmell, N. (1996). Evaluating the impacts of human capital stocks and accumulation on economic growth: Some new evidence. *Oxford Bulletin of Economics and Statistics, 58*(1), 9–28.

Giesecke, H. (2005). Humankapital als Bildungsziel? Grenzen ökonomischen Denkens für das pädagogische Handeln. Neue Sammlung, (3), 377–389.

Glejser, H., Gossens, K., & Vanden Eede, M. (1979). *Inter*-industry and intra-industry specialization do occur in world trade. Economic Letters, *13*.

Główny Urząd Statystyczny. (2011). *Nauka i Technika w 2009 r*. Warsaw.

Godinho, M. M. (2007). *Towards a taxonomy of innovation systems*. Presentation to the Globelics Phd School, http://www.globelicsacademy.net/2007/presentations/2007 %20Globelics%20 Godinho.pdf.

Godinho, M. M., Mendonca, S., & Pereira, T. S. (2003). *Mapping innovation systems: A framework based on innovation surveys data*. Paper presented at the First Globelics Conference "Innovation Systems and Development Strategies for the Third Millennium". Rio de Janeiro, November, 2–6 Nov.

Godinho, M. M., Mendonca, S., & Pereira, T. S. (2006). *A taxonomy of national innovation systems, lessons from an exercise comprising a large sample of both developed, emerging and developing countries*. http://smartech.gatech.edu/bitstream/handle/1853/35988/Manuel%20 Mira%20Godinho.pdf.

Gong, M. (2011). Global specialization and the China-US economic imbalance. *The Global Studies Journal, 3*(4).

Görlich, D., & de Grip, A. (2007). *Human capital depreciation during family-related career interruptions in male and female occupations*. Kiel Institute for the World Economy.

Gorynia, M. & Łaźniewska, E. (Ed.). (2009). *Kompendium wiedzy o konkurencyjności*. Warsaw: PWN.

Gorynia-Pfeffer, N. (2011). *Das Konzept des nationalen Innovationssystems und seine Effektivität am Beispiel Deutschlands*. Working Papers IGŚ, SGH, Warsaw.

Government of Ireland. (2002). *National spatial strategy for ireland 2002–2020. People, places and potential*. Dublin: Stationery Office.

Government of Ireland. (2006). *Strategy for science, technology and innovation 2006–2013*. Dublin.

Government of Ireland. (2007). *Ireland: National development plan 2007–2013. Transforming Ireland: A better quality of life for all*. Dublin: Stationery Office.

Górzyński, M. (2005). *Przegląd wskaźników monitorowania systemów wspierania innowacyjności w krajach UE i wybranych krajach pozaeuropejskich—wnioski i rekomendacje dla Polski*. Warsaw.

Graham, J. W., & Webb, R. H. (1979). *Stocks and depreciation of human capital: New evidence from a present-value perspective*. *Review of Income and Wealth, 25*(2), 209–224.

Greenaway, D., Hine, R. C., & Milner, C. R. (1994). Country—Specific factors and the pattern of horizontal and vertical intra—industry trade in the UK. *Weltwirtschaftliches Archiv, 130*(1), 77–100.

Grömling, M. (2005). *Wirtschaftswachstum. Perspektive 2050. Ökonomik des demographischen Wandels* (pp. 67–96). Köln: Institut der Deutschen Wirtschaft.

Grubel, H. G., & Lloyd, P. J. (1975). *Intra-industry trade. The theory and measurement of international trade in differentiated products*. New York: McMillan.

Gu, B., & Cai, Y. (2009). *Fertility prospects in China*. United nations expert group meeting on recent and future trends in fertility. UN/POP/EGM-FERT/2009/6, 2–4 December 2009. New York.

Gundlach, E. (1995). The role of human capital in economic growth: New results and alternative interpretations. Weltwirtschaftliches Archiv, *131*(2) (1. January), 383–402.

Guordon, J. (2006). *Explaining trade flows: Traditional and new determinants of trade patterns.* Munich: Munich Personal RePEC Archive.

Gupta, A. K., & Wang, A. (2009). *Getting China and India right—Strategies for leveraging the world's fastest-growing economies for global advantage.* Hoboken: Wiley.

Gupta, S. D. (2007). Comparative advantage and competitive advantage: An economic perspective and synthesis. In S. S. Lodha, B. P. Bhatnager, & N. Sharma (Eds.), *Global competitive advantage. Concepts, approaches and perspectives.* Delhi: Abhijeet Publications.

Guzek, M. (2004). *Międzynarodowe stosunki gospodarcze. Zarys teorii i praktyki handlowej.* Poznań: Wydawnictwo Wyższej Szkoły Bankowej.

GWK. (2010). *Exzellenzinitiative geht in die dritte Runde. Pressemitteilung vom 12.3.2010.* Bonn: http://www.gwk-bonn.de/fileadmin/Pressemitteilungen/pm2010-02.pdf. Accessed 9 May 2012.

Hall, B., & Van Reenen, J. (2000). How effective are fiscal incentives for R & D? A review of the evidence. *Research Policy, 29,* 449–469.

Hall, P. A., & Soskice, D. (2001). *An introduction to varieties of capitalism.* In P. A. Hall & D. Soskice (Eds.), *Varieties of capitalism: The institutional foundations of comparative advantage* (pp. 1–68). Oxford: Oxford University Press.

Halper, S. (2010). *The Beijing consensus.* New York: Basic Books.

Hämäläinen, T. J. (2003). *National competitiveness and economic growth. The changing determinants of economic performance in the world economy.* Cheltenham: Edward Elgar.

Hanushek, E. A., & Kim, D. (1995). *Schooling, labor force quality, and economic growth.* NBER Working Paper Series, No. 5399.

Hanushek, E. A., & Kimko, D. D. (2000). Schooling, labor-force quality, and the growth of nations. *American Economic Review, 90*(5), 1184–1208.

Harris, R. G., Robertson, P. E., & Xu, J. Y. (2010). *The international effects of China's growth, trade and education booms.* Discussion Paper 10.04, University of Western Australia, Perth.

Harris, J., & Todaro, M. P. (1970). Migration, unemployment and development: A two-sector analysis. American Economic Review, *60*(2).

Hart, D. (2009). *Accounting for change in national systems of innovation: A friendly critique based on the U.S. case when talking about the Internet boom and bust. Research Policy, 38*(4), 647–654.

Hartog, J., & Oosterbeek, H. (2007). What should you know about the private returns toeducation? In J. Hartog & H. Maassen van den Brink (Eds.), *Human capital: Theory and evidence.* New York: Cambridge University Press.

Hatzichronoglou, T. (1996). *Globalisation and competitiveness: Relevant indicators.* OECD Science, Technology and Industry Working Paper, OECD, Paris.

Hatzichronoglou, T. (1997). *Revision of the white collar/high skill technology sector and product classification.* STI Working Papers 1997/2, OECD/GD(97)216, OECD.

Heckman, J. J., Stixrud, J., & Urzua, S. (2006). *The effects of cognitive and noncognitive abilities on labor market outcomes and social behavior.* NBER Working Paper Series, No. 12006.

Heckman, J. J., & Yi, J. (2012). *Human capital, economic growth, and inequality in China.* NBER Working Paper, 18100, Cambridge, MA.

Heijs, J. (2011). *Thematic report 2011 under specific contract for the integration of INNO policy trendchart with ERAWATCH (2011–2012).* European Commission DG Enterprise and Industry.

Heilemann, U., Lehmann, H., & Ragnitz, J. (2006). *Ländern-Rankings und internationale Wettbewerbsfähigkeit. Eine kritische Analyse.* Baden-Baden: Nomos.

Heitger, B., Schrader, K., & Stehn, J. (1999). *Handel, Technologie und Beschäftigung.* Tübingen: J.C.B.Mohr.

Herd, R., Koen, V., & Reutersward, A. (2010). *China's labour market in transition: Job creation, migration and regulation.* Economics Department Working Papers, No. 749, OECD, Paris.

Herrero, A. G., & Chen, F. (2012). *Innovation: Where does China stand?* Economic watch—China, BBVA Research, Banco Bilbao Vizcaya Argentaria, Bilbao.

Hicks, J. R. (1932). *The theory of wages.* London: Macmillan.

Hippel, E. von (2005). *Democratizing innovation*. Cambridge: MIT Press.
Hira, R. (2008). *The offshoring of innovation*. EPI Briefing Paper, #226, Economic Policy Institute, Washington, DC.
Hirsch, S. (1977). *Rich man's, poor man's and everyman's goods, Aspects of Industrialization*. Tübingen: J.C.B. Mohr.
Hoen, A. R., & Oosterhaven, J. (2006). On the measurement of comparative advantage. *The Annals of Regional Science, 40*(3).
Holmes, P., Michałek, J. J., & Smith, A. (1993). Comparative advantage changes in Poland's trade and the EC response. Russian and East European Finance and Trade, *29*(3).
Hołda, M. (2010). Dlaczego Austria (niemal) dogoniła Szwajcarię. In L. Balcerowicz & A. Rzońca (Eds.), *Zagadki wzrostu gospodarczego. Siły napędowe i kryzysy—analiza porównawcza*. Warsaw: C. H. Beck.
Houthakker, H. S. (1959). *Education and income*. The Review of Economics and Statistics, *41*(1), 24–28 (1. February).
Hsieh, C.-T., & Klenow, P. J. (2007). *Misallocation and manufacturing TFP in China and India*. NBER Working Paper, 13290, Cambridge, Mass.
Hu, M.-C., & Mathews, J. A. (2008). China's national innovative capacity. *Research Policy, 37*, 1465–1479.
Hu, X., & Ma, Y. (1999). International intra-industry trade of China. *Review of World Economics, 135*(1), 82–101.
Huang, H.-C., Shih, H.-Y., & Wu, Y.-C. (2011). Contagion effects of national innovative capacity: Comparing structural equivalence and cohesion models. *Technological Forecasting & Social Change, 78*(2), 244–255.
Human Development Report. (2011). *Sustainability and equity: A better future for all.*
Hummels, D., Ishii, J., & Yi, K. M. (2001). The nature and growth of vertical specialization in world trade. *Journal of International Economics, 54*, 75–96.
Hung, S. W. (2009). Development and innovation in the IT industries of India and China. *Technology in Society, 31*, 29–41.
IBnGR. (2012). *Stan i prognoza koniunktury gospodarczej. Kwartalne Prognozy Makroekonomiczne* (No. 73). Warsaw.
IMF. (2011a). *People's republic of China—Spillover report for the 2011 article iv consultation and selected issues*. IMF Country Report No. 11/193, Washington, DC.
IMF. (2011b). *People's republic of China—2011 article iv consultation*. IMF Country Report No. 11/192, Washington, DC.
IMF. (2012). *China economic outlook*. Washington, DC (February 06, 2012).
INEGI. (2012). *Krajowe badanie zawodu i zatrudnienia. Wskaźniki strategiczne*. http://www.inegi.org.mx.
INNOVATION UNION SCOREBOARD (IUS). (2010). *The innovation union's performance scoreboard for research and innovation*. Bruksela.
Intellectual Property Protection in China. (2012). *More patents owned by domestic holders in China*. http://www.chinaipr.gov.cn/policyarticle/policy/statistics/201205/1291622_1.html.
Jaffe, A. B. (1999). *The U.S. patent system in transition: Policy innovation and the innovation process*. NBER Working Paper No. 7280.
Jagiełło, M. (2003). *Wskaźniki międzynarodowej konkurencyjności gospodarki. Studia i Materiały, No. 80*. Warsaw: Instytut Koniunktur i Cen Handlu Zagranicznego.
Jagiełło, M. (2005). *Strategiczne budowanie konkurencyjności gospodarki*. Warsaw: Poltext.
Janc, K. (2009). *Zróżnicowanie przestrzenne kapitału ludzkiego i społecznego w Polsce*. Wrocław: Instytut Geografii i Rozwoju Regionalnego, Uniwersytet Wrocławski.
Jeong, B. (2002). *Measurement of human capital input across countries: A method based on the laborer's income*. *Journal of Development Economics, 67*(2), 333–349.
Johns, R. A. (1985). *International trade theories in the evolving international economy*. London: Macmillan.
Jones, L. E., & Manuelli, R. (1990). A convex model of equilibrium growth: Theory and policy implications. Journal of Political Economy, *98*(5), 1008–1038 (1. October).

Jones, R. W. (2000). *Globalization and the theory of input trade*. Cambridge: MIT Press.
Jorgenson, D. W. (1995a). *Productivity, Volume 1: Postwar U.S. economic growth*. The MIT Press.
Jorgenson, D. W. (1995b). *Productivity, Volume 2: International comparisons of economic growth*. MIT Press.
Jorgenson, D. W., & Fraumeni, B. M. (1989). *The accumulation of human and nonhuman capital, 1948–1984* (pp. 227–286). NBER Chapters. NBER, Inc.
Jorgenson, D. W., & Fraumeni, B. M. (1990). The output of the education sector. In Z. Griliches (Ed.), *Output measurement in the service sectors* (pp. 303–338). Chicago: University of Chicago Press.
Jorgenson, D. W., Gollop, F. M., & Fraumeni, B. M. (1987). *Productivity and U.S. economic growth*. Harvard University Press.
Jorgenson, D. W., & Griliches, Z. (1967). The explanation of productivity change. *The Review of Economic Studies, 34*(3), 249–283 (1. July).
Ju, J., Liu, Q., Ma, H., Qian, Y., & Wei, Z. (2010). *Anti-comparative advantage: A puzzle in US-China Bilateral Trade*. http://www.hkimr.org/cms/upload/seminar_app/sem_paper_0_408_Paper_2011-06-03.pdf.
Judson, R. (2002). Measuring human capital like physical capital: What does it tell us? *Bulletin of Economic Research, 54*(3), 209–231.
Kaufmann, F. (2004). *Kapitaldeckung versus Umlagefinanzierung in den sozialen Sicherungssystemen. Wie kann man umsteuern?* Frankfurt: http://wipog.de/app/download/2523482802/VortragFXKaufmann20041112.pdf. Accessed 12 Nov 2004.
Keller, W. (2003). *International technology diffusion*. University of Texas, maszynopis powielony.
Keller, W. (2009). *International trade, foreign direct investment and technology spillovers*. NBER Working Paper No. 15442, Cambridge, Mass.
Kendrick, J. W. (1976). *The formation and stocks of total capital*. NBER.
Kibritcioglu, A. (2002). *On the smithonian origins of "New" trade and growth theories*. Office of Research Working Paper, No. 02-0100, University of Illinois at Urbana-Champaign, Illinois.
Kiker, B. F. (1966). The historical roots of the concept of human capital. The Journal of Political Economy, *74*(5), 481–499 (1. October).
Kirsch, I. (2001). *The International Adult Literacy Survey (IALS): Understanding what was measured*. Princeton: Statistics & Research Division, Educational Testing Service (ETS).
Klodt, H. (1989). *Technologietransfer und internationale Wettbewerbsfähigkeit*, Kieler Arbeitspapier No. 351. Kiel: Institut für Weltwirtschaft.
Knell, M. (1999). *Foreign investment enterprises and productivity convergence in central Europe*. In G. Hunya (Ed.), *European integration through FDI: Making central European industries competitive*. London: E. Elgar.
Köppen, M. (1998). *Strukturelle Wettbewerbsfähigkeit von Volkswirtschaften. Ein Gegenentwurf zur Standortkonkurenz*. Marburg: Metropolis.
Korpi, T., & Tåhlin, M. (2009). Educational mismatch, wages, and wage growth: Overeducation in Sweden, 1974–2000. *Labour Economics, 16*(2), 183–193.
Kotowska, I. E., & Węziak, D. (2007). Pomiar kapitału ludzkiego i jego zróżnicowanie. In J. Czapiński & T. Panek (Eds.), *Diagnoza społeczna 2007*. Warsaw: Rada Monitoringu Społecznego.
Kotyński J. (Ed.). (1988). *Ocena metodologii i wyników badań konkurencyjności polskiego przemysłu przez zespół z Banku Światowego*. Warsaw: IKCHZ.
Kowalski, A. M. (2008). *Polityka regionalna Unii Europejskiej w Irlandii i Wielkiej Brytanii w latach 1989–2006*. Warsaw: Oficyna Wydawnicza Szkoły Głównej Handlowej.
Kowalski, A. M. (2009). Instytucje wspierające komercjalizację technologii. In M. A. Weresa (Ed.), *Własność intelektualna. Wybrane aspekty ekonomiczne*. Warsaw: SGH w Warszawie—Oficyna Wydawnicza.
Kowalski, A. M. (2009). Własność intelektualna w świetle polityki innowacyjnej. In M. A. Weresa (Ed.), *Własność intelektualna. Wybrane aspekty ekonomiczne*. Warsaw: SGH w Warszawie—Oficyna Wydawnicza.
Kowalski, A. M. (2010). Znaczenie klastrów w tworzeniu nowych form innowacji. *Optimum. Studia Ekonomiczne, 3*(47), 246–262.

Kowalski, A. M. (2010a). Studium przypadku—Irlandia. In W. Bieńkowski, M. Weresa, & M. J. Radło (Eds.), *Konkurencyjność Polski na tle zmian gospodarczych w krajach OECD*. Warsaw: SGH w Warszawie—Oficyna Wydawnicza.

Kowalski, A. M. (2010b). The role of clusters in enhancing ties between science and business. In M. A. Weresa (Ed.), *Poland: Competitiveness report 2010. Focus on clusters*. Warsaw: Warsaw School of Economics—Publishing.

Kowalski, A. M. (2011). Europejskie inicjatywy na rzecz zwiększania innowacyjności i konkurencyjności gospodarki poprzez internacjonalizację klastrów. *Studia Europejskie, 1*(57), 79–100.

Kowalski, A. M. (2011a). Wpływ kryzysu gospodarczego na procesy współpracy i konkurencji. *Master of Business Administration, 2*(109).

Kowalski, A. M. (2011b). Innowacyjność polskiego przemysłu przetwórczego. In M. A. Weresa (Ed.), *Polska: Raport o konkurencyjności 2011*. Warsaw: IGŚ, SGH w Warszawie.

Kowalski, A. M., & Szlachta, J. (2007). Wnioski dla Polski wynikające z doświadczeń polityki strukturalnej Unii Europejskiej w Irlandii i Wielkiej Brytanii. In M. Klamut (Ed.), *Polityka ekonomiczna—współczesne wyzwania*. Warsaw: Wydawnictwo Naukowe PWN.

Kraemer, K. L., Linden, G., & Dedrick, J. (2011). *Capturing value in global networks: Apple's iPad and iPhone*. PCIC Paper. Personal Computing Industry Center, University of California at Irvine, Irvine, CA.

Kroll, H., & Schiller, D. (2010). Establishing an interface between public sector applied research and the Chinese enterprise sector: Preparing for 2020. *Technovation, 30,* 117–129.

Krugman, P. R. (1990). *Rethinking international trade*. Cambridge: MIT Press.

Krugman, P. R., & Obstfeld, M. (2007). *Ekonomia międzynarodowa. Teoria i polityka*. Warsaw: PWN.

Krugman, P. R., & Venables, A. J. (1987). *The seamless world: A spatial model of international specialization*. CEPR Discussion Papers Series, London.

Krugman, P. A. (1984). Import protection and export promotion; international competition in the presence of oligopoly and economies of scale. In H. Kierzkowski (Ed.), *Monopolistic competition and international trade*. Oxford: Clarendon Press.

Kuciński, K. (2006). *Geografia ekonomiczna. Zarys teoretyczny*. Warsaw: SGH.

Kuntluru, S., Muppani, V. R., & Khan, M. A. (2012). Foreign direct investment and export performance of pharmaceutical firms in India: An empirical approach. *International Journal of Economics and Finance, 4*(5), http://www.ccsenet.org/ijef.

Kuźnar, A. (2008). *Międzynarodowa konkurencyjność polskich usług w latach 2000–2007*. In M. A. Weresa (Ed), *Polska. Raport o konkurencyjności 2008*. Warsaw: SGH.

Kyriacou, G. A. (1991). *Level and growth effects of human capital: A cross-country study of the convergence hypothesis*. C.V. Starr Center for Applied Economics, New York University.

Lampert, H. (1996). *Priorität für die Familie*. Berlin.

Lanjouw, J. O., & Schankerman, M. (2004). Patent quality and research productivity: Measuring innovation with multiple indicators. *Economic Journal, 114*(495), 441–465.

Lau, L. J., Bhalla, S., & Louat, F. F. (1991a). *Human and physical capital stock in developing countries -construction of data and trends*. Draft mimeo. World Development Report. The World Bank.

Lau, L. J., Jamison, D. T., & Louat, F. F. (1991b). *Education and productivity in developing countries: An aggregate production function approach*. The World Bank.

Le, T., Gibson, J., & Oxley, L. (2005). *Measures of human capital: A review of the literature*. Treasury Working Paper Series 05/10. New Zealand Treasury.

Leamer, E. E., & Levinsohn, J. (1995). International trade theory; The evidence. In G. Grossman & K. Rogoff (Eds.), *Handbook of international economics*. Amsterdam: Elsevier.

Lederborgen, M. (1991). *Produktdifferenzierung im internationalen Handel. Ein Beitrag zur Analyse von Aussenhandelsstrukturen*. Bochum: Universität Bochum.

Leitl, M. (2011). *Humankapital?* Harvard Business Manager, http://www.havardbusinessmanager.de.

Levine, R., & Renelt, D. (1992). A sensitivity analysis of cross-country growth regressions. *The American Economic Review, 82*(4), 942–963.

Lewis, W. A. (1954). Economic development with unlimited supplies of labor. *Manchester School of Economic and Social Studies, XXII,* 139–191.

Li, L., Dunford, M., & Yeung, G. (2012). International trade and industrial dynamics: Geographical and structural dimensions of Chinese and Sino-EU merchandise trade. *Applied Geography, 32,* 130–142.

Li, X. (2011). Sources of external technology, absorptive capacity, and innovation capability in Chinese state-owned high-tech enterprises. *World Development, 39*(7), 1240–1248.

Liang, Y. (2008). Why are China's exports special? *The Chinese Economy, 41*(6), 99–118.

Liberda, B. (2005). *Inwestycje w kapitał ludzki a stopa oszczędzania gospodarstw domowych w Polsce*. Ekonomista 4/2005, PTE, Warsaw.

Liberda, Z. B., & Maj, E. (2009). Idee i nowoczesny wzrost. In B. Fiedor (Ed.), *Nauki ekonomiczne wobec wyzwań współczesności*. Warsaw: PWE.

Lim, J. (2012). *Why China won't be innovative for at least 20 more years*. TechNode (March 26, 2012).

Lin, J. Y. (2010). *The China miracle demystified*. Paper prepared for the panel on perspectives on Chinese Economic Growth at the Econometric Society World Congress in Shanghai on August 19, 2010.

Lin, J. Y. (2011). *China and the global economy*. Remarks at the 20th Anniversary of the University of Science and Technology Hong Kong (March 23, 2011).

Lin, J. Y., & Li, Y. (2003). *Export and economic growth in China: A demand-oriented analysis*. Paper presented at the 4th International Conference of the Chinese Economy in Clermont-Ferrand on October 23–24, 2003, http://www.cerdi.org/uploads/sfCmsContent/html/197/JustinLi.pdf.

Lincen, G., Dedrick, J., & Kraemer, K. L. (2011). Innovation and job creation in a global economy: The case of apple's iPod. *Journal of International Commerce and Economics, 3*(1), 223–239.

Lipsey, R. E. (2002). *Home and host country effects of FDI*. NBER Working Paper No. 9293, Cambridge, Mass.

Lipsey, R. E. (2006). *Measuring the impacts of FDI on central and eastern Europe*. NBER Working Paper No. 12808, Cambridge, Mass.

Liuhto, K. (2009). *Special economic zones in Russia. What do the zones offer for foreign firms?* Turku: Pan-European Institute.

Lodha S. S., Bhatnagar, B. P., & Sharma N. (Eds.). (2007). *Global competitive advantage. Concepts, Approaches and Perspectives*. Delhi: Abhijet Publication.

Lohr, S. (2011). *When innovation is, too, made in China*. New York Times, 1st January 2011, http://www.nytimes.com/2011/01/02/business/02unboxed.html.

Lopez-Claros, A., & Mata, Y. N. (2011). Policies and institutions underpinning country innovation: Results from the innovation capacity index. In A. Lopez-Claros (Ed.), *The innovation for development report 2010–2011, Innovation as a driver of productivity and growth* (pp. 3–66). New York: Palgrave MacMillan.

Louri, H., Loufir, R., & Papanastassiou, M. (2002). Foreign investment and ownership structure: An empirical analysis. *Empirica, 29,* 31–45.

Lu, Z., & Deng, X. (2011). *China's western development strategy: Policies, effects, prospects*. MPRA Paper, 35201, Munich. *December, 05,* 2011.

Lu, X. D., & Li, R. L. (2010). South-South cooperation: Is there a foundation in trade? *Journal of Chinese Economic and Foreign Trade Studies, 3*(3), 221–234.

Lubiński, M., Michalski, S., & Misala, J. (1995). *Międzynarodowa konkurencyjność gospodarki. Pojęcie i sposób mierzenia*. Warsaw: Instytut Rozwoju i Studiów Strategicznych.

Lucas, R. E. (1988). On the mechanics of economic development. Journal of Monetary Economics, *22*(1), 3–42 (July).

Lundvall B.-A. (Ed.). (1992). *National systems of innovations*. London: Pinter.

Lynn, R., & Vanhanen, T. (2002). *IQ and the wealth of nations*. Greenwood Publishing Group.

Łosiewicz, M. (2009). *Wiedza i kapitał ludzki a konkurencyjność przedsiębiorstw w skali regionu*. Gdańsk: Wydawnictwo Uniwersytetu Gdańskiego.

Ma, G.-y. (2010). China's regional knowledge innovation capability assess: Based on a factor analysis methodology. *Management Science and Engineering, 4*(1), 26–33.

Maastricht Economic and Social Research and Training Centre on Innovation and Technology (UNU-MERIT). (2011). *Innovation union scoreboard 2011, The Innovation Union's performance scoreboard for Research and Innovation*. Brussel.

Macdonald, S., & Deng, Y. (2004). Science parks in China: A cautionary exploration. *International Journal of Technology Intelligence and Planning, 1,* 1–14.

Machlup, F. (1962). *The production and distribution of knowledge in the United States*. Princeton University Press.

Maj, A. J. (2012). Die Innovationsförderung in der EU am Beispiel Österreichs und Polens. *Eine landeskundliche Studie*, (25.07.2012).

Malerba, F. (1999). *Sectoral systems of innovation and production.* Prepared for a conference: DRUID Conference on: National Innovation Systems, Industrial Dynamics and Innovation Policy Rebild, June, 9–12.

Malerba, F., Salter, A., & Altelli, A. (2011). *Indicators for structural change*. Final Report of the expert group on the measurement of innovation, Bruxelles.

Manarungsan, S. (2009). Thailand-China cooperation in trade, investment and official development assistance, http://www.ide.go.jp/English/Publish/Download/Brc/pdf/01_thailandandchina.pdf.

Mankiw, N. G., Romer, D., & Weil, D. N. (1992). A contribution to the empirics of economic growth. *The Quarterly Journal of Economics, 107*(2), 407–37.

Mansfield, E. (1963). Size of firm, market structure and innovation. *Journal of Political Economy, 71,* 556–576.

Marczewski, K. (2002). *Zmiany kursu walutowego a ceny i reakcje przedsiębiorstw w handlu zagranicznym*. Warsaw: IKCZH.

Markusen, J. R., Venables, A. J., Konan, D. E., & Zhang, K. H. (1996). *A unified treatment of horizontal drect investment, vertical direct investment and the pattern of trade in goods and services*. NBER Working Paper, No. 5696. New York.

Markusen, J. R., & Venables, A. J. (1997). *Foreign direct investment as a catalyst for industrial development.* NBER Working Paper 6241. http://www.mber.ord/papers/w6241.

Marrewijk, C., & Hinloopen, J. (2004). *Dynamics of Chinese comparative advantage*, Tinbergen Institute Discussion Paper, No. 034/2.

Marshall, A. (1919). *Industry and trade*. London: Macmillan.

Marshall, A. (1930). *Principles of economics: An introductory volume (8th ed.)*. London: Macmillan.

Mattoo, A., Mishra, P., & Supramanian, A. (2012). *Spillover effects of exchange rates: A study of the Renminbi. IMF Working Paper, WP/12/88*. Washington, DC: IMF.

Marxt, C., & Brunner, C. (2009). *Innovationssystem Schweiz. Eine Bestandsaufnahme 2009*, Bern.

Mazurkiewicz, A. (2010). *Kapitał ludzki w procesie kształtowania sprawności organizacji*. Rzeszów: Uniwersytet Rzeszowski.

McGregor, J. (2010). *China's drive for indigenous innovation'. A web of industrial policies*, APCO Worldwide, 46. http://www.uschamber.com/sites/default/files/reports/100728chinareport_0.pdf.

Messinis, G., & Ahmed, A. (2009). *Human capital, innovation and technology diffusion. Working Paper*. Melbourne: Centre for Strategic Economic Studies, Victoria University.

Metcalfe, S. (1995). The economic foundations of technology policy: Equilibrium and evolutionary perspectives. In P. Stoneman (Ed.), *Handbook of the economics of innovation and technological change*. Oxford: Blackwell Publishers.

Michałek, J. (2002). *Polityka handlowa. Mechanizmy ekonomiczne i regulacje międzynarodowe*. Warsaw: PWN.

Midelfart-Knarvik, i inni. (2000). *The location of European industry. Economic Papers, No. 142 April*. Brussels: European Communities.

Miller, H. P. (1964). *Lifetime income and economic growth*. California: Institute of Government and Public Affairs, University of California.

Miller-Idriss, C. (2004). Challenge and change in the German vocational system since 1990. *Oxford Review of Education, 28,* 473–490.

Milliken Institute. (2012). *Innovation scoreboard—Country innovation profiles*. Santa Monica: Milken Institute.

Mincer, J. (1974). *Schooling, experience, and earnings. NBER;*. New York: Columbia University Press.

Mincer, J. (1958). Investment in human capital and personal income distribution. *The Journal of Political Economy, 66*(4), 281–302.

Ministerstwo Finansów. (2012). *Program Konwergencji Aktualizacja 2012*. Warsaw: Ministerstwo Finansów.

Ministerstwo Gospodarki Konfederacji Szwajcarskiej (2012). http://www.evd.admin.ch/ index.html?lang=en. Accessed Feb 15 2012.

Mira, M., & Liu, G. (2010). *The OECD human capital project: Progress report*. St. Gallen: OECD (26 August).

Misala, J. (1990). *Teorie wymiany międzynarodowej*. Warsaw: PWN.

Misala, J. (1992). *Analiza porównawcza rozwoju gospodarczego Hiszpanii i Polski*. Warsaw: IK-CHZ.

Misala, J. (2001). Socjalna gospodarka rynkowa a uczestnictwo Niemiec w międzynarodowym podziale pracy. In Małachowski W. (Ed.), *Współczesna gospodarka rynkowa. Doświadczenia w Niemczech. Implikacje dla Polski*. Warsaw: SGH.

Misala, J. (2001, 2003). *Współczesne teorie wymiany międzynarodowej i zagranicznej polityki ekonomicznej*. Warsaw: Oficyna Wydawnicza SGH.

Misala, J. (2004). Cechy i tendencje rozwojowe handlu zagranicznego Niemiec. In W. Małachowski (Ed.), *Polska—Niemcy a rozszerzenie Unii Europejskiej*. Warsaw: SGH.

Misala, J. (2005). *Wymiana międzynarodowa i gospodarka światowa. Teoria i mechanizmy funkcjonowania*. Warsaw: SGH.

Misala, J. (2007). *Międzynarodowa zdolność konkurencyjna i międzynarodowa konkurencyjność gospodarki narodowej*. Radom: Politechnik Radomska.

Misala, J. (2009). *Historia rozwoju teorii i polityki konkurencyjności międzynarodowej*. Warsaw: SGH.

Misala, J. (2010). Analiza stadium przypadku: Hiszpania. In W. Bieńkowski, M. A. Weresa, & M.-J. Radło (Eds.), *Konkurencyjność Polski na tle zmian gospodarczych w krajach OECD. Ewolucja znaczenia czynników konkurencyjności*. Warsaw: SGH.

Misala, J. (2010). Problemy związane ze sformułowaniem teorii międzynarodowej wymiany usług. *Ekonomiczno-Informatyczny Kwartalnik Teoretyczny, No. 24*.

Misala, J. (2010). *Rozwój wymiany zagranicznej i międzynarodowej konkurencyjności gospodarki Niemiec po II wojnie światowej*. materiał w druku przez Instytut Zachodni w Poznaniu: Instytut MSG, SGH.

Misala, J. (2011a). *Międzynarodowa konkurencyjność gospodarki narodowej*. Warsaw: PWE.

Misala, J. (2011b). *Zależności i współzależności między handlem międzynarodowym i międzynarodowymi przepływami kapitału w świetle teorii oraz dotychczasowych doświadczeń*. Warsaw, maszynopis powielony: Instytut Międzynarodowych Stosunków Gospodarczych, SGH.

Misala, J. (2011). Rozwój wymiany zagranicznej i międzynarodowej konkurencyjności Niemiec. In T. Budnikowski (Ed.), *Gospodarka niemiecka 20 lat po zjednoczeniu*. Poznań: Instytut Zachodni.

Misztal, P. (2010). *Zmiany kursu walutowego i dynamika cen w krajach o różnym poziomie rozwoju gospodarczego*. Warsaw: SGH.

Moenius, J. (2006). *Measuring comparative advantage: A Ricardian approach*. Redlands: University of Redlands, mimeo.

Mohr, D. (2011). *Ausverkauf der Deutschland AG*. http//www.Frankfurter Allgemeine Zeitung.net/. Accessed 11 May 2011.

Monkiewicz, J. (1981). *Międzynarodowy transfer wiedzy technicznej. Elementy teorii i praktyki*. Warsaw: PWE.

Moon, H. C., Rugman, A. M., & Verbeke, A. (1998). A generalized double diamond approach to the global competitiveness of Korea and Singapore. *International Business Review, 7*(2), 135–150.

Moran, T. H. (2011). *Foreign manufacturing multinationals and the transformation of the chinese economy: New measurements, new perspectives. Working Paper Series, 11-11*. Washington, DC: Peterson Institute for International Economics.

Morrison, W. M. (2011). *China's economic conditions*. CRS Report for Congress, RL33534. Washington, DC.

Morrison, W. M., & Labonte, M. (2011). *China's currency policy: An analysis of the economic issues*. CRS Report for Congress, RS21625. Washington, DC.

Motohashi, K., & Yun, X. (2007). China's innovation system reform and growing industry and science linkages. *Research Policy, 36,* 1251–1260.

Mowery, D. C. (1996). *The U.S. national innovation system: Recent developments in structure and knowledge flows*. Prepared for the OECD meeting on "National Innovation Systems," October 3, 1996.

Mulligan, C. B., & Sala-i-Martin, X. (1995). *A labor-income-based measure of the value of human capital: An application to the states of the United States*. NBER Working Paper Series, No. 5018 (February).

Muñoz, E., Espinosa, J., & Díaz, V. (2000a). *Innovation policy and the concept of national system of innovation in the Spanish context: Are they ghost images or real entities?* Grupo de Ciencia, Tecnología y Sociedad (CSIC), Working Paper 00–14.

Muñoz, E., Espinosa, J., & Díaz, V. (2000b). *Technology, innovation and economy in Spain: national and regional influences*. Grupo de Ciencia, Tecnología y Sociedad (CSIC), Working Paper 00–03.

Mytelka, L. K., & Smith, K. (2002). Policy learning and innovation theory: An interactive and co-evolving process. *Research Policy, 31*(8–9), 1467–1479.

National Bureau of Statistics of China. (2012). *China Statistical Yearbook—2011*. Beijing: National Bureau of Statistics of China.

National Science Foundation. (2012). *Science and engineering indicators 2012*. http://www.nsf.gov/statistics/seind12/c2/c2s4.html. Arlington.

Narula, R. (2003). *Globalization and technology: Interdependence, innovation systems and industrial policy*. Cambridge: Polity Press.

Narula, R. K., & Wakelin, K. (1996). Technological competitiveness, trade and Foreign direct investment. *Structural change and economic dynamics, 9,* 373–387.

Narula, R., & Guimon, J. (2009). *The contribution of multinational enterprises to the upgrading of national innovation systems in the EU new member*. http//www.oecd.org/dataoecd/20/16/44246353.pdf. OECD Global Forum on International Investment.

Naughton, B. (2007). *The Chinese economy. Transitions and growth*. Cambridge: MIT.

Naughton, B. (2008). A political economy of China's economic transition. In L. Brandt & T. G. Rawski (Eds.), *China's great economic transformation* (pp. 91–135). Cambridge and New York: Cambridge University Press.

Neary, P. J. (1995). Factor mobility and international trade. *Canadian Journal of Economics, 28,* 4–23.

Neary, P. J. (2003). Competitive versus comparative advantages. *The World Economy, 26*(4), 457–470.

Nehru, V., Swanson, E., & Dubey, A. (1993). A new database on human capital stock: Sources, methodology and results. *The World Bank.*

Nelson, R., & Winter, S. (1977). In search of a useful theory of innovation. *Research Policy, 6*(1), 36–76.

New York Times. (2011). *National patent development strategy (2011–2020)*. http://graphics8.nytimes.com/packages/pdf/business/SIPONatPatentDevStrategy.pdf.

Nijkamp, P., & Siedschlag, J. (Eds.). (2011). *Innovation, growth and competitiveness*. Heidelberg-Berlin: Springer.

Niosi, J., Saviotti, P., Bellon, B., & Crow, M. (1993). National systems of innovation: In search of a workable concept. *Technology in Society, 15*(2), 207–227.

North, D. C. (1984). Transformation costs, institutions and economic history. *Journal of Institutional and Theoretical Economics, 140,* 7–17.

North, D. (1995). *Institutions and economic development*, paper presented at the OECD (DSTI/Development Centre) Workshop on "Institutional Framework and Economic Development", Paris, November.

NSFC. (2012). *Main responsibilities*. http://www.nsfc.gov.cn/e_nsfc/desktop/zn/0101.html.
OECD. (1996). *The knowledge-based economy*. Paris: OECD.
OECD. (1998). *Human capital investment, an international comparison*. Paris: OECD.
OECD. (1999). *Managing national innovation systems*. Paris: OECD.
OECD. (2001). *The well-being of nations, the role of human and social capital*. Paris: OECD.
OECD. (2007). OECD reviews of innovation policy: China. Synthesis Report, 68.
OECD. (2008). *Executive summary. OECD reviews of innovation policy: China* (p. 7). Paris: OECD.
OECD. (2009). *Science, technology and industry outlook: Spain*. Paris: OECD.
OECD. (2010a). *Podręcznik Frascati. Proponowane procedury standardowe dla badań statystycznych w zakresie działalności badawczo-rozwojowe*. Warsaw: Ministerstwo Nauki i Szkolnictwa Wyższego.
OECD. (2010b). *Science, technology and industry outlook 2010*. Paris: OECD.
OECD. (2010c). *OECD economic surveys: China 2010*. Paris: OECD.
OECD. (2010d). *OECD Science, technology and industry outlook 2010*. Paris: OECD.
OECD. (2010e). *Shanghai and Hong Kong: Two distinct examples of education reform in China. In Strong performers and successful reformers in education: Lessons from PISA for the United States*. Paris: OECD.
OECD. (2011a). *Education at a glance 2011: OECD indicators*. Paris: OECD.
OECD. (2011b). *Reviews of innovation policy: Russian Federation*. Paris: OECD.
OECD. (2011c). *Reviews of innovation policy: Russian Federation* (pp. 182–189). Paris: OECD.
OECD. (2012a). OECD economic surveys Germany, overview February 2012. Economics Department Working Papers. http//www.oecd.org/eco/workingpapers.
OECD. (2012b). *Main science and technology Indicators—2012 (1st ed.)*. Paris: OECD.
OECD. (2012c). *OECD economic surveys: Poland*. Paris: OECD.
OECD, & World Bank Institute. (2001). *Korea and the knowledge-based economy. Making the Transition*. Paris: OECD and World Bank Institute.
OECD, & Eurostat. (2008). *Podręcznik Oslo. Zasady gromadzenia i interpretacji danych dotyczących innowacji*. Warsaw: Ministerstwo Nauki i Szkolnictwa Wyższego.
OECD, & Statistics Canada. (1995). *Literacy, economy and society: Results of the first International adult literacy survey*. OECD Publishing.
OECD, & Statistics Canada. (2000). *Literacy in the information age: Final report of the international adult literacy survey*. OECD Publishing.
OECD, Statistics Canada, & Human Resources Development Canada. (1997). *Literacy skills for the knowledge society: Further results from the International adult literacy survey*. OECD Publishing.
OECD. (2009). *Programa Nacional de Innovación, Comité Intersectorial para la Innovación México 2011* (pp. 9–11). Mexico: OECD Reviews of innovation Policy.
Olczyk, M. (2008). *Konkurencyjność. Teoria i praktyka*. Warsaw: CeDeWu.
Olszyński, J. (2012). *Implikacje kryzysu w Niemczech dla polsko-niemieckich stosunków gospodarczych i gospodarki Polski (No. 33)*. Warsaw: Zeszyty Naukowe KGŚ SGH.
Orr, G., (2012). *What's next for Guandong?* Harvard Business Review. http://blogs.hbr.org/cs/2012/02/what_next_for_guangdong.html. Accessed 15 Feb 2012.
Österreichischer Forschungs und Technologiebericht. (2012). BMWF, BMVIT, Wien.
Ozawa, T. (1992). Foreign direct investment and economic development. *Transnational Corporations, 1*, 27–54.
Pareto, V. (1897). *Cours d'Economie Politique*. Lousanne: H.Valloton Geux et Gie.
Peltola, K. K. (2008). *Russian innovation system in international comparison—Opportunities and challenges for the future of innovation development in Russia*. Turku: Turku School of Economics. http://www.tse.fi/FI/yksikot/erillislaitokset/pei/Documents/Julkaisut/Peltola%201108%20web.pdf. Accessed 20 May 2012.
Peneder, M. (1999). *The new WIFO taxonomy of manufacturing industries*. WIFO Working Papers No. 114, Vienna.
Penrose, E. T. (1959). *The theory of the growth of the firm*. Oxford: Oxford University Press.

Petty, W. (1690). *Political Arithmetik*. Reprint, 1899. Hull.

Pfeiffer, F., & Reuß, K. (2008). Age-dependent skill formation and returns to education. *Labour Economics, 15*(4), 631–646.

Piech, K. (2009). *Wiedza i innowacje w rozwoju gospodarczym. W kierunku pomiaru i współczesnej roli państwa*. Warsaw: Instytut Wiedzy i Innowacji.

Pitelis, Ch. N. (2011a). *The competitive advantage and catching-up of nations: A new framework and the role of FDI, clusters and public policy*. In P. Nijkamp & J. Siedschlag (Eds.), *Innovation, growth and competitiveness*. Heidelberg-Berlin: Springer.

Pitelis, Ch. N. (2011b). Knowledge: A learning-based perspective on foreign direct investment and the multinational enterprise. In P. Nijkamp & I. Siedschlag (Ed.), *Innovation, growth and competitiveness* (pp. 221–234). Heidelberg: Springer.

Plate, C. (2005). *Internationalisierung der Gütermärkte und internationale Wettbewerbsfähigkeit einer Volkswirtschaft*. Frankfurt a. M.: Peter Lang GmbH.

Plünecke, A. (2004). *Bildungsreform in Deutschland—eine Positionsbestimmung aus bildungsökonomischer Sicht*. Beiträge zur Ordnungspolitik, 2003/4, Köln.

Plünecke, A., & Seyda, S. (2005). *Bildung. Perspektive 2050. Ökonomik des demographischen Wandels* (pp. 121–144). Köln: Institut der Deutschen Wirtschaft.

Płóciennik, S. (2010). Systemowe konsekwencje globalnego kryzysu finansowego i gospodarczego dla Niemiec. In J. Olszyński (Ed.), *Polska i Niemcy w Unii Europejskiej. Gospodarki i przedsiębiorstwa w procesie integracji*. Warsaw: SGH.

Polasek, W., Schwarzbauer, W., & Sellner, R. (2010). *Human capital and regional growth in Switzerland*. Wiedeń: Institute for Advanced Studies (HIS).

Pontikakis, D., McDonnell, T., & Geoghegan, W. (2006). Ireland's national innovation system: An exploratory study of supporting institutions and dynamic actors. *Prometheus, 24*(1), 37–58.

Porter, M. E. (1990). *The competitive advantage of nations*. New York: Free Press.

Porter, M. E. (2001). *Porter o konkurencji*. Warsaw: PWE.

Porter, M. (2006). *Przewaga konkurencyjna. Osiąganie i utrzymywanie lepszych wyników*. Gliwice: Helion.

Porter, M. (2008). *On competition*. Boston: Harvard Business Review.

Porter, M. E., & Stern, S. (2001). *National innovative capacity*. Institute for Strategy and Competitiveness, Harvard Business School. http://www.isc.hbs.edu/Innov_9211.pdf.

Posner, M. V. (1961). *International trade and technical change* (Vol. 13). Oxford Economic Papers.

Poznańska, K. (2012). *Outsourcing prac badawczo–rozwojowych—doświadczenia krajów wysoko rozwiniętych*. In M. A. Weresa & K. Poznańska (Ed.), *Procesy tworzenia wiedzy, transferu osiągnięć naukowych i technologicznych do biznesu* (pp. 215–244). Warsaw: SGH.

Przybyszewski, R. (2007). *Kapitał ludzki w procesie kształtowania gospodarki opartej na wiedzy*. Warsaw: DIFIN.

Psacharopoulos, G. (1994). Returns to investment in education: A global update. *World Development, 22*(9), 1325–1343.

Psacharopoulos, G., & Arriagada, A. M. (1986). Educational composition of the labour force: An international comparison. *The International Labour Review, 125,* 561.

Psacharopoulos, G., & Arriagada, A. M. (1992). *The educational composition of the labor force: An international update*. The World Bank.

Psacharopoulos, G., & Patrinos, H. A. (2002). *Returns to investment in education : A further update*. The World Bank.

Ptak, K. *Innowacje w Rosji*. http://www.pi.gov.pl/parp/chapter_86196.asp?soid=21E6F290CA4B47D1B5FE89B179ED4F4C. Accessed 16 June 2012.

Purvis, D. D. (1972). Technology trade and factor mobility. *Economic Journal, 82,* 991–999.

Quaas, F. (2000). *Soziale Marktwirtschaft*. Bern: Verlag Paul Haupt.

Radło, M.-J. (2011). *Klasyfikacje sektorów*. Warsaw: SGH.

Radło, M.-J. (2011). Structural changes and the competitiveness of Poland's manufacturing sector. In M. A. Weresa (Ed.), *Poland competitiveness report 2011*. Warsaw: Warsaw School of Economics.

Radulescu, M., & Serbanescu, L. (2012). The impact of FDIs on exports, and export competitiveness in central and eastern European countries. *Journal of Knowledge Management, Economics and Information Technology*, (8).

Redding, G., & Witt, M. A. (2008). *China's business system and comparative advantage*. INSEAD Working Paper, 2008/14/EPS/EFE, Fontainebleau.

Reichel, R. (2002). *Ökonomische Theorie der internationalen Wettbewerbsfähigkeit von Volkwirtschaften*. Berlin: Deutscher Universitatis—Verlag.

Reiljan, J., Hinrikus, M., & Ivanov, A. (2000). *Key issues in defining and analysing the competitiveness of a country*. Faculty of Economics and Business Administration, Working Paper, No. 1, University of Tartu, Tartu.

Reinert, E. S. (1994). *Competitiveness and its predecessors—A 500-year cross national perspective*. STEP-report, STEP Group, Oslo.

Roberts, I., & Rush, A. (2011). Understanding China's demand for resource imports. *China Economic Review*.

Robinson, J. (1973). *Herezje ekonomiczne*. Warsaw: PWE.

Rodrik, D. (2006). *What's so special about China's exports*? *China & World Economy, 14*(5), 1–19.

Rogers, E. M. (2003). *Diffusion of innovations*. New York: Free Press.

Romer, P. M. (1986). *Increasing returns and long-run growth*. Journal of Political Economy, *94*(5), 1002–1037 (1. October).

Romer, P. M. (1989). *Human capital and growth: Theory and evidence*. NBER, Inc.

Romer, P. M. (1990). *Endogenous technological change*. *Journal of Political Economy, 98*(5), 71–102.

Roper, S. (2004). *Innovation policy: An effective way of reducing spatial disparities in small nations?* Innovation Lab Europe, Working Paper, Aston Business School.

Roszkowska, S. (2006). Kapitał ludzki a wzrost gospodarczy w Polsce w ujęciu regionalnym. In E. Okoń-Horodyńska & K. Piech (Ed.), *Unia Europejska w kontekście strategii lizbońskiej oraz gospodarki i społeczeństwa wiedzy w Polsce*. Warsaw: Intstytut Wiedzy i Innowacji.

Ruffin, R. J. (1984). *International factor movements*. In R. W. Jones & P. B. Kenen (Eds.), *Handbook of international economics* (pp. 237–288). Amsterdam-Washington: North Holland.

Rusangirsi, Y. (2003). *Inequality and welfare analyses in the Harris-Todaro model*. Nottingham: University of Nottingham.

Ruskin, J. (1905/1996). Munera Pulveris. In E. T. Cook & A. Wedderburn (Eds.), *The works of John Ruskin* (Vol. 17, pp. 131–132). Cambridge: Cambridge University Press.

Rybczynski, T. M. (1955). Factor endowment and relative commodity prices. Economica, *22*(1).

Rymarczyk, J. (2010). *Innowacyjność małych i średnich przedsiębiorstw*. Ekonomia. *Regionalizm i Globalizacja, 7*, 243–252. Prace Naukowe Uniwersytetu Ekonomicznego we Wrocławiu No. 132, Wyd. Uniwersytetu Ekonomicznego we Wrocławiu, Wrocław.

Rynarzewski, T. (2005). *Strategiczna polityka handlu międzynarodowego*. Warsaw: PWE.

Rynarzewski, T. (Ed). (2011). *Konkurencyjność międzynarodowa we współczesnej gospodarce światowej*. Poznań: Wydawnictwo Uniwersytetu Ekonomicznego w Poznaniu.

Rynarzewski, T., & Zielińska-Głębocka, A. (2006). *Międzynarodowe stosunki gospodarcze. Teoria wymiany i polityki handlu międzynarodowego*. Warsaw: PWN.

Sala-I-Martin, X., & Artadi, E. V. (2004). *The global competitiveness index. The global competitiveness report 2004–2005*. Hampshire: Palgrave Macmillan.

Saltykov, B. (1997). The reform of Russian science. *Nature, 388,* 16.

Sanz-Menéndez, L. (2003). *Public-private partnerships in innovation policy: The Spanish experience*. Unidad de Políticas Comparadas (CSI), Working Paper 03–01.

Schaaper, M. (2009). *Measuring China's innovation system: National specificities and international comparisons*. OECD Science, Technology and Industry Working Papers, 2009/1, OECD publishing, ©OECD, doi:10.1787/227277262447.

Schäfer, H., & Seyda, S. (2005). *Arbeitsmärkte. Perspektive 2050. Ökonomik des demographischen Wandels* (pp. 97–120). Köln: Institut der Deutschen Wirtschaft.

Schasse, et al. (2011). *FuE-Aktivitäten von Wirtschaft und Staat im internationalen Vergleich*. Studien zum deutschen Innovationssystem No. 2-2011. Hannover: Essen.

Schwaag Serger, S., & Breidne, M. (2007). China's fifteen-year plan for science and technology: An assessment. *Asia Policy, 4,* 135–164.
Shelton, R. D., & Foland, P. (2009). *The race for world leadership of science and technology: Status and forecasts*. Paper presented at the 12th International Conference Scientometrics and Informatics, Rio de Janeiro, July 2009.
Sheng, Y., & Song, L. G. (2008). Comparative advantage and Australia-China bilateral trade. *Economic Papers, 27*(1), 41–56 (March 2008).
Shi, J. (2011). The foundation and trends of undergraduate education reform in China's research universities. The case of Tsinghua University. *Chinese Education and Society, 44*(5), 67–83.
Scholz, C., Stein, V., & Bechtel, R. (2005). Zehn Postulate für das Human-Capital-Management. Personalwirtschaft, 2003/5. http:// www.personalwirtschaft.de.
Schultz, T. W. (1961). Investment in human capital. The American Economic Review, *51*(1), 1–17 (1. March).
Schultz, T. W. (1972). *Human capital: Policy issues and research opportunities* (pp. 1–84). *Economic research: Retrospect and prospect*. t. 6: *Human Resources*, NBER, Cambridge Mass, http://www.nber.org/chapters/c4126.
Schüren, V. (2012). Two TRIPs to innovation. Pharmaceutical innovation systems in India and Brazil. SFB-Governance, Working Paper Series, No. 37, Research Center (SFB) 700, June 2012, Berlin.
Schweizerische Arbeitskrafteerhebung (SAKE). (2010). Neuenburg.
Schweizerische Nationalbank (SN). (2011). *Direktinvestitionen 2010*. Zurich.
Schweizerischer Wissenschafts- und Technologierat (SWTR). (2007). *Empfehlungen für die schweizerische Bildungs-, Forschungs- und Innovationspolitik 2008–2011*. Bern.
Schweizerischer Wissenschafts- und Technologierat (SWTR). *Neun Empfhelungen zue Foerderung der Innovation in der Schweiz*. Zeszyt SWTR 3/2009, Bern.
Schydlowsky, D. M. (1984). A policymaker's guide to comparative advantage. *World Development, 12*(4), 439–449.
Scitovsky, T. (1958). *Economic theory and western European integration*. London: Allen & Unwin.
Siebert, H. (1999, 2007). *The world economy*. London: Roultledge (2007).
Siebert, H. (2000). *The paradigm of locational competition*. Kiel Discussion Papers, No. 367, Kiel Institute of World Economics, Kiel.
Siebert, H. (2005). *Jenseits des sozialen Marktes*. München: Deutsche Verlagsanstalt GmbH.
Siebert, H. (2006). Locational competition: A neglected paradigm in the international division of labour. *The World Economy, 29*(2), 137–159.
Siebert, H., & Klodt, H. (1999). *Towards global competition: Catalysts and constraints*. 21-st century economic dynamics, OECD, Paris.
Siek, E. (2009). *Międzynarodowe przepływy siły roboczej w procesie integracji Polski z krajami Unii Europejskiej*. Toruń: Wydawnictwo Adam Marszałek.
Singh, A., Dhania, R., Upasani, R., & Mathur, S. K. (2011). *Analyzing the trade flows for Brazil, Russia, India, China and South Africa (BRICS)*. http://www.scribd.com/doc/63710945/Analyzing-the-Trade-Flows-for-BRICS.
Sinn, H. W. (2005). *Die Basar-Ökonomie*. Econ.
Siyanbola, W., Abiodun Egbetokun, A., Boladale Abiola Adebowale, B. A., & Olumuyiwa Olamade, O. (2012). *Innovation systems and capabilities in developing regions. Concepts, issues and cases*. London: Gower.
Smith, A. (1776). *An inquiry into the nature and causes of the wealth of nations*. Oxford 1976: Clarendon Press.
Smits, R., & Kuhlmann, S. (2004). The rise of systemic instruments in innovation policy. *International Journal of Foresight and Innovation Policy, 1*(1–2), 4–32.
Social Sciences Academic Press. (2011). Annual report on the development on the development of chinese talents 2011, Beijing.
Södersten, B., & Reed, G. (1994). *International economics*. Houndmills-London: Macmillan Press Ltd.

Soete, L. (1981). *A general test of technological gap trade theory. Weltwirtschaftliches Archiv, 117*(4), 638–666.

Soete, L. (1990). Technical change theory and international trade competition. In J.de la Mothe & L. M. Ducharme (Ed.), *Science, technology and free trade*. London.

Soete, L., Verspagen, B., & Bas ter Weel (2010). Systems of innovation. *CPB Working Paper*, No. 138, CPB Netherlands Bureau for Economic Policy Analysis, Hague, February 2010. http://www.cpb.nl/eng/pub/cpbreeksen/discussie/138/disc138.pdf.

Solow, R. M. (1957). Technical change and the aggregate production function. *The Review of Economics and Statistics, 39*(3), 312–320.

Solleiro, J. L., & Castanon, R. (2005). Competitiveness and innovation systems: The challenges for Mexico's insertion in the global context. *Technovation, 25,* 1059–1070.

Sołdaczuk, J., Kamecki, Z., & Bożyk, P. (1983, 1987). *Międzynarodowe stosunki ekonomiczne. Teoria i polityka*. Warsaw: PWE.

Springer, K. (2000). *Do we have to consider international capital mobility in trade modes?* Kiel Working Paper No. 964, Kiel Institute of World Economy, Kiel.

Springut, M., Schlaikjer, S., & Chen, D. (2011). *China's program for science and technology modernization: Implications for American competitiveness* (pp. 1–143). Arlington: CENTRA Technology Inc.

Stern, S., Porter, M. E., & Furman, J. L. (2000). *The determinants of national innovative capacity*. NBER, Working Paper, No. 7876, September.

Stern, S., Porter, M. E., & Furman, J. L. (2002). The determinants of national innovative capacity. *Research Policy, 31,* 899–933.

Stifterverband für die deutsche Wissenschaft. (2010). *Forschung und Entwicklung facts*. Januar 2010, Essen.

Strategie der Bundesregierung für Forschung, Technologie und Innovation, Potenziale ausschöpfen, Dynamik steigern. Zukunft schaffen: Der Weg zum Innovation Leader, (25 July 2012).

Straubhaar, T. (2000). *Internationale Migration. Gehen oder Bleiben: Wieso gehen wenige und bleiben die meisten?* HWWA Discussion Paper, No. 111, HWWA, Hamburg.

Stroombergen, A., Rose, D., Nana ,G., & Pink, B. (2002). *Review of the statistical measurement of human capital*. Wellington: Statistics New Zealand.

Suarez-Villa, L. (1990). Invention, inventive learning, and innovative capacity. *Behavioral Science, 35*(4), 290–310.

Sulmicki, P. (1977). *Międzynarodowa wymiana gospodarcza*. Warsaw: PWE.

Sun, Y., & Liu, F. (2010). A regional perspective on the structural transformation of China's national innovation system since 1999. *Technological Forecasting and Social Change, 77,* 1311–1321.

Sun, W., Yang, X., & Xiao, G. (2011). Understanding China's high investment rates and FDI levels: A comparative analysis of the return to capital in China, the United States and Japan. *Journal of International Commerce and Economics, United States International Trade Commission, 3*(1), 157–188.

Sun, H., Guo, H., Hu, J., & Zhu, K. (2012). Comparative study on economic contribution rate of education of China and foreign countries based on soft simulation method. *Applied Soft Computing, 12,* 2106–2113.

Svensson, L. E. O. (1984). Factor trade and good trade. Journal of International Economics, *16*.

Swann, G. M. (2009). *The economics of innovation: An introduction*. Cheltenham: Edward Elgar.

Swissnex. (2009). *A quick overview of the science and technology system in China*. http://www.swissnexchina.org/background/1090821-a-quick-overview-of-the-st-system-xiaoyin-2-.pdf.

Szymański, A. (2008). *Austria. Prawie pełna kontynuacja (in:) Zmiana i kontynuacja. Polityka europejska wybranych państw Unii Europejskiej. In L. Jesień (Ed. naukowa)*. Warsaw: PISM.

Świerkocki, J. (2011). *Zarys ekonomii międzynarodowej*. Warsaw: PWE.

Tang, H., & Zhang, H. (2009). China reshapes the east Asian production network. *China Economist, 4*(2), 90–102.

Teichmann. E. (2010). Analiza studium przypadku: Finlandia. In W. Bieńkowski, M. A. Weresa, & M. J. Radło (Eds.), *Konkurencyjność na tle zmian gospodarczych w krajach OECD. Ewolucja znaczenia czynników konkurencyjności*. Warsaw: SGH.

Temple, J. (1999). The new growth evidence. *Journal of Economic Literature, 37*(1), 112–156 (1 March).

Tesch, P. (1980). *Die Bestimmungsgründe des internationalen Handels und der Direktinvestitionen*. Berlin: Duncker & Humblodt.

Thomson Reuters. (2011). *Chinese patenting—Report on the current state of innovation in China*. New York.

Thomson Reuters. (2012). *World university rankings 2011–2012*. http://www.timeshighereducation.co.uk/world-university-rankings/2011-2012/top-400.html.

Thorbecke, W. (2011). Investigating the effect of exchange rate changes on China's processed exports. *Journal of the Japanese and International Economies, 22*(4), 535–544.

Todaro, M. P. (1969). A model of labour migration and Urban unemployment in less developed countries. *American Economic Review, 59*(2).

Trabold, H. (1995). *Die internationale Wettbewerbsfähigkeit einer Volkswirtschaft (No. 5)*. Berlin: Deutsches Institut für Wirtschaftsforschung, Vierteljahreshefte zur Wirtschatfsforschung.

Trend Watching. (2012). *Made better in China*. http://www.trendwatching.com/trends/pdf/2012-06 %20MADE%20BETTER%20IN%20CHINA.pdf.

UNCTAD. (2012). *World investment report 2011. Non-Equity modes of international production and development. Country fact sheet-Switzerland*. New York.

United Nations. (2011). *World population prospects: The 2010 revision*. Washington, DC.

UN Comtrade. *United nations commodity trade statistics database*. http://comtrade.un.org/. Accessed Spring 2012.

UNDP. (2011). *Human development report 2011, China country profile*. http://hdrstats.undp.org/en/countries/profiles/CHN.html.

UNDP. (2011). *Human development report 2011—Sustainability and equity: A better future for all*. New York: Palgrave Macmillan.

UNESCO. (1993). *World education report*. Paris: UNESCO.

UNESCO. (2011). *World data on education*, 2010/2011 (7th ed.). IBE/2011/CP/WDE/CC, http://www.ibe.unesco.org/fileadmin/user_upload/Publications/WDE/2010/pdf-versions/China.pdf.

Usanov, A. (2008). *Special economic zone in Kaliningrad as a tool of industrial development: The case of the consumer electronics manufacturing*. Electronic Publications of Pan-European Institute.

U.S. Department of Commerce. http://www.commerce.gov/about-department-commerce. Accessed Spring 2012.

U.S. Department of Education. http://www2.ed.gov/about/what-we-do.html, http://www2.ed.gov/about/reports/strat/index.html?src=ln, http://www.ed.gov/recovery. Accessed Spring 2012.

Usher, A. P. (1929). *A history of mechanical inventions*. Cambridge: Harvard University Press.

Uyarra, E. (2010). What is evolutionary about regional systems of innovation? Implications for regional policy. *Journal of Evolutionary Economics, 20*(1), 115–137.

Varblane, U. (2006). *Does the broadening of EU detain its competitiveness on the world market*, referat przedstawiony na międzynarodowej konferencji. *The future competitiveness of the EU and its Eastern Neighbors* organizowanej w Turku przez Pan-European Institute, Turku.

Vernon, R. (1966). International investment and international trade in the product cycle. Quarterly Journal of Economics, *80*.

Vollrath, T. L. (1991). A theoretical evaluation of alternative trade intensity mesures of revealed comparative advantage. *Weltwirtschaftliches Archiv, 127*(2), 265–280.

Wang, H. (2010). *China's talent plan. Where will it lead China to*? Brookings Institution, http://www.brookings.edu/~/media/Files/events/2010/0920_china_talent/0920_china_powerpoint.pdfÄhnliche Seiten.

Warneke, P. (Ed.). (1995). *Wirtschaftsstandort Deutschland. Chancen und Gefahren*. Berlin: Duncker und Humboldt.

Weber, B., & Wolter, S. (2001). Wages and human capital: Evidence from Switzerland. In R. Asplund & P. T. Pereira (Eds.), *Returns to human capital in Europe* (pp. 325–350). Helsinki: ETLA.

Weber, S. (2008). *Human capital depreciation and education level: Some evidence for Switzerland. Materiały konferencyjne z Annual Conference of the European Association of Labour Economists (EALE)*. Amsterdam: September 2008.

Wei, D. Y. H., Zhou, Y., Sun, Y., Lin, & G. C. S (2012). Production and R & D networks of foreign ventures in China: Implications for technological dynamism and regional development. *Applied Geography, 32*(1), 106–118.

Weresa, M. A. (2001). The impact of foreign direct investment on Poland's trade with the European union. *Post-Communist Economies, 13*(1), 71–83.

Weresa, M. A. (2002). *Wpływ handlu zagranicznego i inwestycji bezpośrednich na innowacyjność polskiej gospodarki*. Warszawa: SGH.

Weresa, M. A. (2005). Does Foreign Direct Investment (FDI) facilitate innovation process in Poland? In A. Jasiński (Ed.), *Transition economies in the European research and innovation area: New challenges for their science and technology*. Warsaw: Warsaw University.

Weresa, M. A. (2009). Innowacje jako źródło przewagi konkurencyjnej w gospodarce opartej na wiedzy. In E. Mińska-Struzik & T. Rynarzewski (Eds.), *Szoki technologiczne w gospodarce światowej* (pp. 187–200). Poznań: *Wydawnictwo Uniwersytetu Ekonomicznego w Poznaniu*.

Weresa, M. A. (2012). *Systemy innowacyjne we współczesnej gospodarce światowej, Wyd*. Warszawa: Naukowe PWN.

Whalley, J., & Zhao, X. (2010). *The contribution of human capital to China's economic growth. NBER Working Paper, 16592*, Cambridge, MA.

White House. (2012). *Strategy for American innovation: Driving towards sustainable growth and quality jobs*. http://www.whitehouse.gov/administration/eop/nec/StrategyforAmericanInnovation and http://www.whitehouse.gov/innovation/strategy/executive-summary. Accessed Spring 2012.

Williamson, O. (1964). Innovation and market structure. *Journal of Political Economy, 73*(1), 67–73.

Statistics Database, W. I. P. O. http://www.wipo.int/ipstats/en/general_info.html.

WIPO. (2011). *World intellectual property report—The challenging face of innovation*. Geneva.

Woessmann, L. (2000). *Specifying human capital: A review, some extensions, and development effects*. SSRN eLibrary (October).

Woessmann, L. (2003). Specifying human capital. *Journal of Economic Surveys, 17*, 239–270.

Wolf, M. (2012). How to blow away China's gathering storm-clouds. *Financial Times*, *13* (21 March 2012).

Wolff, E. N. (2000). Human capital investment and economic growth: exploring the cross-country evidence. Structural Change and Economic Dynamics, *11*(4), 433–472 (December).

World Bank. http://data.worldbank.org.

World Bank. World Development Indicators (WDI) & Global Development Finance.

World Bank. (2012a). *China 2030—Building a modern, harmonious, and creative high-income society*. Conference Edition 67179, Washington, DC.

World Bank. (2012b). *World development indicators*. http://data.worldbank.org/indicator. Accessed 14 April 2012.

World Economic Forum. (2010). *The global competitiveness report 2010–2011*. Geneva.

World Economic Forum. (2011). *The global competitiveness report 2001–2012*. Geneva.

World Economic Forum. (2011). *The global competitiveness report 2011–2012*. Geneva.

WTO. (2011). *International Trade Statistics 2011*. Geneva.

Wysokińska, Z. (2004). Międzynarodowa konkurencyjność gospodarki. Zagadnienia teoretyczne. Wnioski dla Polski. In J. Staszewski (Ed.), *Konkurencyjność gospodarki Polski w warunkach integracji z Unią Europejską i globalizacji*. Warsaw: Wyższa Szkoła Ekonomiczna.

Wyżnikiewicz, B. (1998). Konkurencyjność przemysłu przetwórczego w świetle analizy DRC. In J. Winiecki (Ed.), *Konkurencyjność gospodarki polskiej*. Warsaw: Towarzystwo Ekonomistów Polskich.

Wziątek-Kubiak, A. (2001). Międzynarodowa specjalizacja a konkurencyjność. *Ekonomista, 4*.

Wziąte-Kubiak, A. (2003). *Zmiany konkurencyjności polskiego przemysłu w latach 1994–2000*. Warsaw: Bellona, INE–EAN.

Wziątek-Kubiak, A., Pęczkowski, M., & Balcerowicz, E. (2011). Occasional and persistent innovators in Poland: An empirical study of obstacles to innovation and their competitiveness. *Studia Ekonomiczne, 4*, 345–379.

Xing, Y., & Detert, N. (2010*). How the iPhone widens the United States trade deficitwith the people's republic of China.* ADBI Working Paper, 257, Asian Development Bank Institute, Tokyo.
Yang, D. T., Chen, V., & Monarch, R. (2010). *Rising wages: Has China lost its global labor advantages?* Discussion Paper No. 5008, Institute for the Study of Labor (IZA), Bonn.
Yang, D. T., Zhang, J., & Zhou, S. (2011). *Why are saving rates so high in China?* NBER Working Paper 16771, Cambridge Mass.
Yao, Y., & Zhang, K. (2010). Has China passed the lewis turning point? A structural estimation based on provincial data. *China Economic Journal, 3*(2), 155–162.
Yu, J., Stough, R. R., & Nijkamp, P. (2009). Governing technological entrepreneurship in China and the west. Public Administration Review, *69*(Special Issue), S95–100.
Yu, M. (2011). *Processing trade, tariff reductions, and firm productivity: Evidence from Chinese products*. Working Paper 302, Forum for Research in Empirical International Trade.
Yue, C., & Hua, P. (2002). *Does comparative advantage explains export patterns in China*? *China Economic Review, 13,* 276–296.
Zhang, H., & Sonobe, T. (2011). *Development of science and technology parks in China, 1988–2008*. Economics: The Open-Access, Open-Assessment E-Journal, 5, 2011–6, doi:10.5018/economicsejournal.ja.2011-6. http://dx.doi.org/10.5018/economics-ejournal.ja.2011-6.
Zhang, Z., & Zheng, X. (2011). *The study of technology innovation of chinese enterprises by regression analysis*. *Asian Social Science, 7*(2), 130–135.
Zhao, S. (2011). *China vs. Vietnam as the future workshop of the world. China Briefing*. http://www.china-briefing.com/news/2011/05/20/china-vs-vietnam-as-the-future-workshop-of-the-world.html. Accessed 20 May 2011.
Zhao, L., & Zhu, J. (2010). China's higher education reform: What has not been changed? *East Asian Policy, 2*(4), 115–125.
Zhao, X., & Zehng, Y. (2011). Development of Chinese science and technology intermediaries and their integration into the open innovation paradigm. *Technology Analysis & Strategic Management, 23*(1), 25–48.
Zhong, X., & Yang, X. (2007). Science and technology policy reform and its impact on China's national innovation system. *Technology in Society, 29,* 317–325.
Zhou, J. (2004). The full implementation of the 2003–2007. Action plan for invigorating education. China Education and Research Network, http://www.edu.cn/20040324/3102182.shtml.
Zhou, Y., Yang, X., Parker, R., & Steffens, P. (2010). *Regional innovative capacity in China: From the perspective of embedded autonomy*. Entrepreneurship, strategy, and international business, Paper 10, http://repository.usfca.edu/esib/10.
Zhu, Y. (2009). *A panel data analysis on China's intra-industry trade in the capital goods sector*. IWH Discussion Papers No. 18, November 2009, Halle Institute for Economic Research, Halle.
Zielińska-Głębocka, A. (Ed.). (2003). *Potencjał konkurencyjny polskiego przemysłu eksportowego w warunkach integracji z Unią Europejską*. Gdańsk: Wydawnictwo Uniwersytetu Gdańskiego.
Zielińska-Głębocka, A. (2009). Handel międzynarodowy w dobie światowego kryzysu. In J. Winiecki (Ed.), *Kryzys globalny: początek czy koniec* (pp. 61–84). Gdańsk: Regan Press.
Zorska, A. (1998). *Ku globalizacji*. Warsaw: PWN.
http://ec.europa.eu/enterprise/policies/innovation/files/ius-2011_en.pdf.
http://epp.eurostat.ec.europa.eu/portal/page/portal/statistics/search_database.
http://othes.univie.ac.at/6960/1/2009-10-06_0107672.pdf-praca.
http://stats.oecd.org/.
http://www.aba.gov.at;
http://www.ait.ac.at.
http://www.bmvit.gv.at.
http://www.fiw.at;
http://www.go-international.at;
http://www.ibwe.at.
http://www.pi.gov.pl/parp/chapter_86365.asp?soid=65F3718B5F4A4A618C4AB4279455A2A4.
http://www.WIFO_IHS_Lissabon_31032011.pdf.

Annexure: Sector Classifications

Mariusz-Jan Radło

Thanks to socioeconomic activity classification systems, it is possible to organize our knowledge of the economy and its structure. Official classifications—based on standards adopted at the international level—are used by statistical offices to collect data on the activity of business entities. Thus, they not only facilitate the process of describing the structure of the given economy and the changes occurring therein, but also allow for drawing comparisons at the international level.

In the studies on economic competitiveness discussed in this book, three activity classifications referring to the following riteria were employed:

- Technology level of processing industries.
- Utilization level of input factors.
- Skills of the labor force that forms the basis of the given industry.

Classification Based on Technology Level of Processing Industries

The classification based on the first above-mentioned criterion discerns four groups of processing industries (Table A1):

- High-technology (HT).
- Medium-high-technology (MHT).
- Medium-low-technology (MLT).
- Low-technology (LT).

The usefulness of this classification was confirmed mainly by the observation that companies that use technologies more intensely enter new markets faster and utilize the available resources more effectively, i.e., they display higher productivity and provide higher remuneration. At the same time, HT industries expand into global markets at a quicker pace. This classification was drafted for the OECD in

M. A. Weresa (ed.), *Innovation, Human Capital and Trade Competitiveness*, Innovation, Technology, and Knowledge Management, DOI 10.1007/978-3-319-02072-3,

Table A1 Classification of manufacturing industries by technology level (ISIC Revision 2). (Source: Hatzichronoglou (1997))

Sectors		ISIC 2
HT		
1.	Aerospace	3845
2.	Computers, office, machinery	3825
3.	Electronics–communications	3832
4.	Pharmaceuticals	3522
MHT		
5.	Scientific instruments	385
6.	Motor vehicles	3843
7.	Electrical machinery	383-3832
8.	Chemicals	351+352+3522
9.	Other transport equipment	3842+3844+3849
10.	Nonelectrical machinery	382-3825
MLT		
11.	Rubber and plastic products	355+356
12.	Shipbuilding	3841
13.	Other manufacturing	39
14.	Nonferrous materials	372
15.	Nonmetallic mineral products	36
16.	Fabricated metal products	381
17.	Petroleum refining	351+354
18.	Ferrous metals	371
LT		
19.	Paper printing	34
20.	Textile and clothing	32
21.	Food, beverages, and tobacco	31
22.	Wood and furniture	33

the mid-1990s and is used by OECD experts to perform competitiveness analyses (Hatzichronoglou 1997).

Classification Based on Utilization Level of Input Factors

The classification based on the utilization level of input factors was developed by experts of the Austrian Institute of Economic Research—WIFO—at the end of the 1990s, among others, to enable the performance of EU competitiveness analyses (Peneder 1999). This classification differentiates between five industry groups depending on the share of various types of expenditures in the total manufacturing costs (Table A2):

- Mainstream-manufacturing (MM).
- Labor-intensive (LI).
- Capital-intensive industries (CI).
- Technology-driven industries (TDI).
- Marketing-driven industries (MDI).

Table A2 Classification of manufacturing industries according to relative abundance of capital or labor, and endogenously created firm-specific advantages resulting from intangible investments in marketing or innovation. (Source: Peneder (1999))

MM industries	
1730	Finishing of textiles
1770	Knitted and crocheted articles
1750	Other textiles
1760	Knitted and crocheted fabrics
2120	Articles of paper and paperboard
2430	Paints, coatings, printing ink
2510	Rubber products
2520	Plastic products
2610	Glass and glass products
2660	Articles of concrete, plaster, and cement
2680	Other nonmetallic mineral products
2720	Tubes
2870	Other fabricated metal products
2910	Machinery for production, use of mech. power
2920	Other general purpose machinery
2930	Agricultural and forestry machinery
2950	Other special purpose machinery
2960	Weapons and ammunition
2970	Domestic appliances n. e. c.
3110	Electric motors, generators, and transformers
3130	Isolated wire and cable
3140	Accumulators, primary cells, and primary batteries
3150	Lighting equipment and electric lamps
3540	Motorcycles and bicycles
3550	Other transport equipment n. e. c.
LI industries	
1720	Textile weaving
1740	Made-up textile articles
1810	Leather clothes
1820	Other wearing apparel and accessories
1830	Dressing and dyeing of fur; articles of fur
2010	Sawmilling, planing, and impregnation of wood
2020	Panels and boards of wood
2030	Builders' carpentry and joinery
2040	Wooden containers
2050	Other products of wood and articles of cork
2620	Ceramic goods
2640	Bricks, tiles, and construction products
2670	Cutting, shaping, and finishing of stone
2810	Structural metal products
2830	Steam generators

Table A2 (continued)

2840	Forging, pressing, stamping, and roll forming of metal
2750	Casting of metals
2850	Treatment and coating of metals
2940	Machine tools
3160	Electrical equipment n. e. c.
3420	Bodies for motor vehicles, trailers
3510	Ships and boats
3520	Railway locomotives and rolling stock
3610	Furniture
3620	Jewelry and related articles
CI industries	
1710	Textile fibers
2110	Pulp, paper, and paperboard
2310	Coke oven products
2320	Refined petroleum products
2410	Basic chemicals
2470	Man-made fibers
2630	Ceramic tiles and flags
2650	Cement, lime, and plaster
2710	Basic iron and steel, ferroalloys (ECSC)
2730	Other first processing of iron and steel
2740	Basic precious and nonferrous metals
3430	Parts and accessories for motor vehicles
MDI	
1510	Meat products
1520	Fish and fish products
1530	Fruits and vegetables
1540	Vegetable and animal oils and fats
1550	Dairy products; ice cream
1560	Grain mill products and starches
1570	Prepared animal feeds
1580	Other food products
1590	Beverages
1600	Tobacco products
1910	Tanning and dressing of leather
1920	Luggage, handbags, saddlery, and harness
1930	Footwear
2210	Publishing
2220	Printing
2230	Reproduction of recorded media
2450	Detergents, cleaning and polishing, perfumes
2820	Tanks, reservoirs, central heating radiators, and boilers
2860	Cutlery, tools, and general hardware

Table A2 (continued)

3350	Watches and clocks
3630	Musical instruments
3640	Sports goods
3650	Games and toys
3660	Miscellaneous manufacturing n. e. c.
TDI	
2420	Pesticides, other agrochemical products
2440	Pharmaceuticals
2460	Other chemical products
3000	Office machinery and computers
3120	Electricity distribution and control apparatus
3210	Electronic valves and tubes, other electronic comp.
3220	TV, and radio transmitters, apparatus for line telephony
3230	TV, radio and recording apparatus
3310	Medical equipment
3320	Instruments for measuring, checking, testing, navigating
3330	Industrial process control equipment
3340	Optical instruments and photographic equipment
3410	Motor vehicles
3530	Aircraft and spacecraft

Classification Based on Similarities in Skills of Labor

The classification based on labor force skills, similarly as the previous taxonomy, was developed by experts of the Austrian Institute of Economic Research—WIFO—at the end of the 1990s, among others, to enable the performance of EU competitiveness analyses (Peneder 1999). It divides processing industries into four groups (Table A3):

- High skills (HS).
- Medium/white-collar skills (MW).
- Medium/blue-collar skills (MB).
- Low skills (LS).[1]

[1] Peneder (1999).

Table A3 Manufacturing industries grouped according to similarities in labor skills. (Source: Peneder (1999))

HS industries	
2910	Machinery for production, use of mech. power
2920	Other general purpose machinery
2930	Agricultural and forestry machinery
2950	Other special purpose machinery
2960	Weapons and ammunition
2940	Machine tools
3510	Ships and boats
2440	Pharmaceuticals
3000	Office machinery and computers
3530	Aircraft and spacecraft
LS industries	
1730	Finishing of textiles
1770	Knitted and crocheted articles
1750	Other textiles
1760	Knitted and crocheted fabrics
2510	Rubber products
2520	Plastic products
2610	Glass and glass products
2660	Articles of concrete, plaster, and cement
2680	Other nonmetallic mineral products
2720	Tubes
1720	Textile weaving
1740	Made-up textile articles
1810	Leather clothes
1820	Other wearing apparel and accessories
1830	Dressing and dyeing of fur; articles of fur
2620	Ceramic goods
2640	Bricks, tiles, and construction products
2670	Cutting, shaping, and finishing of stone
2750	Casting of metals
3620	Jewelry and related articles
1710	Textile fibers
2630	Ceramic tiles and flags
2650	Cement, lime, and plaster
2710	Basic iron and steel, ferroalloys (ECSC)
2730	Other first processing of iron and steel
2740	Basic precious and nonferrous metals
1510	Meat products
1520	Fish and fish products
1530	Fruits and vegetables
1540	Vegetable and animal oils and fats
1550	Dairy products; ice cream
1560	Grain mill products and starches
1570	Prepared animal feeds

Table A3 (continued)

1580	Other food products
1590	Beverages
1600	Tobacco products
1910	Tanning and dressing of leather
1920	Luggage, handbags, saddlery, and harness
1930	Footwear
3630	Musical instruments
3640	Sports goods
3650	Games and toys
3660	Miscellaneous manufacturing n. e. c.
MB industries	
2870	Other fabricated metal products
3540	Motorcycles and bicycles
3550	Other transport equipment n. e. c.
2010	Sawmilling, planing, and impregnation of wood
2020	Panels and boards of wood
2030	Builders' carpentry and joinery
2040	Wooden containers
2050	Other products of wood and articles of cork
2810	Structural metal products
2830	Steam generators
2840	Forging, pressing, stamping, and roll forming of metal
2850	Treatment and coating of metals
3420	Bodies for motor vehicles, trailers
3520	Railway locomotives and rolling stock
3610	Furniture
3430	Parts and accessories for motor vehicles
2820	Tanks, reservoirs, central heating radiators, and boilers
2860	Cutlery, tools, and general hardware
3410	Motor vehicles
MW industries	
2120	Articles of paper and paperboard
2430	Paints, coatings, printing ink
2970	Domestic appliances n. e. c.
3110	Electric motors, generators, and transformers
3130	Isolated wire and cable
3140	Accumulators, primary cells, and primary batteries
3150	Lighting equipment and electric lamps
3160	Electrical equipment n. e. c.
2110	Pulp, paper, and paperboard
2310	Coke oven products
2320	Refined petroleum products
2410	Basic chemicals
2470	Man-made fibers
2210	Publishing

Table A3 (continued)

2220	Printing
2230	Reproduction of recorded media
2450	Detergents, cleaning and polishing, perfumes
3350	Watches and clocks
2420	Pesticides, other agrochemical products
2460	Other chemical products
3120	Electricity distribution and control apparatus
3210	Electronic valves and tubes, other electronic comp.
3220	TV, and radio transmitters, apparatus for line telephony
3230	TV, radio, and recording apparatus
3310	Medical equipment
3320	Instruments for measuring, checking, testing, and navigating
3330	Industrial process control equipment
3340	Optical instruments and photographic equipment

MIX
Papier aus verantwortungsvollen Quellen
Paper from responsible sources
FSC® C105338
FSC
www.fsc.org

If you have any concerns about our products,
you can contact us on
ProductSafety@springernature.com

In case Publisher is established outside the EU,
the EU authorized representative is:
Springer Nature Customer Service Center GmbH
Europaplatz 3, 69115 Heidelberg, Germany

Printed by Libri Plureos GmbH
in Hamburg, Germany